Lecture Notes in Mathematics

Edited by A. Dold and B. Eckmann

1192

Equadiff 6

Proceedings of the International Conference on Differential Equations and their Applications held in Brno, Czechoslovakia, Aug. 26–30, 1985

Edited by J. Vosmanský and M. Zlámal

Springer-Verlag
Berlin Heidelberg New York Tokyo

Editors

Jaromír Vosmanský
J. E. Purkyně University, Department of Mathematics
Janáčkovo nám 2a, 662 95 Brno, Czechoslovakia

Miloš Zlámal
Technical University, Computing Centre
Obránců míru 21, 602 00 Brno, Czechoslovakia

Published in co-edition with *Equadiff 6*, J. E. Purkyně University, Department of Mathematics, Brno, Czechoslovakia.

Sole distribution rights outside the East European Socialist Countries, China, Cuba, Mongolia, Northern Korea, USSR, and Vietnam:
Springer-Verlag Berlin, Heidelberg, New York, Tokyo

Mathematics Subject Classification (1980): 34-02, 35-02, 65-02, 73-02, 76-02, 80-02

ISBN 3-540-16469-3 Springer-Verlag Berlin Heidelberg New York Tokyo
ISBN 0-387-16469-3 Springer-Verlag New York Heidelberg Berlin Tokyo

Printed in Czechoslovakia

Printing: Tisk, Brno
Binding: Beltz Offsetdruck, Hemsbach/Bergstr.
2146/3140-543210

PREFACE

Following the tradition of the previous Conference EQUADIFF 1–5, held periodically in Prague (1962, 1977), Bratislava (1966, 1981) and Brno (1972), The 6th Czechoslovak Conference on Differential Equations and Their Applications EQUADIFF 6 was held in Brno from August 26 to August 30, 1985. The Conference was organized by the University of J. E. Purkyně in Brno with support of the International Mathematical Union in cooperation with the Technical University in Brno, the Mathematical Institut of the Czechoslovak Academy of Sciences, Society of Czechoslovak Mathematicians and Physicists, sponsored by the Faculty of Mathematics and Physics of the Charles University in Prague, the Faculty of Mathematics and Physics of the Comenius University in Bratislava, the Czech Technical University in Prague, the Faculty of Science of the Palacký University in Olomouc, the Faculty of Science of the University of P. J. Šafarik in Košice, the School of Mechanical and Electrotechnical Engineering in Plzeň and the School of Transport and Communications in Žilina.

EQUADIFF 6 was prepared by the Organizing Committee president by M. Zlámal, chairman, and J. Vosmanský, executive secretary, with the help of the local organizing staff.

The topic of this meeting were differential equations in the broad sense including numerical methods of their solutions and applications. The main goal was to stimulate cooperation among various branches in differential equations.

The Conference was attended by 473 participants (207 from Czechoslovakia, 266 from abroad) and 62 accompanying persons from 31 countries. 36 participants from abroad were granted the financial support.

92 invited mathematicians from abroad took part in the Conference and together with Czechoslovak scientists delivered plenary lectures and other invited lectures and communications in sections. The participants had the opportunity to deliver their papers as communications, at the poster session or in the form of the enlarged abstracts (without oral presentation).

The scientific program comprised 10 plenary lectures and 64 main lectures in the following sections:

1. Ordinary Differential Equations (20)
2. Partial Differential Equations (16)
3. Numerical Methods (14)
4. Applications (14)

In addition 251 papers were presented

a) as communications in 9 simultaneous subsections (136)
b) at the poster session (46)
c) in the form of enlarged abstracts (70)

Besides the scientific program the participants and the accompanying persons could enjoy a rich social program.

Two slightly different parallel editions of this volume are published. The Springer-Verlag edition contains 9 plenary lectures and 48 main lectures in sections representing the substantial part of lectures presented at the Conference. The EQUADIFF 6 edition for the participants of the Conference and for the socialist countries contains also Supplement consisting of 7 additional contributions. These contributions are not fully compatible with the conditions for the Lecture Notes publication and their revised version could not be arranged.

Editors

CONTENTS

B. *Partial differential equations*

C. *Numerical methods*

SUPPLEMENT (in the EQUADIFF 6 edition only)

LIST OF FURTHER MAIN LECTURES PRESENTED AT THE CONFERENCE

PLISS V.: Stable and unstable manifolds of hyperbolic systems (plenary lecture)
ATKINSON F. V.: Critical cases of certain ground-state problems for nonlinear wave equations
BOBROWSKI D.: Boundary-value problems for random differential equations
EVERITT W. N.: On linear ordinary quasi-differential equations
HEDBERG L. I.: Sobolev spaces and nonlinear potential theory
LAZAROV R.: Superconvergence of the gradient for triangular finite elements
MARKOWICH P.: The semiconductor device equations
MASLENNIKOVA V. N.: Boundary value problems for second order elliptic equations in domains having non-compact and non-smooth boundaries
PŮŽA B.: Ob odnom metode analiza razreshimosti kraevykh zadach dlja obyknovennykh differentsialjnykh uravnenii
SELL G.: Lyapunov exponents and oscillatory behavior equations with negative feedback
SUSSMANN H. J.: A theory of envelopes and high order optimality condition for bang-bang controls

LIST OF PAPERS PRESENTED AT THE CONFERENCE

I. PAPERS PRESENTED AS COMMUNICATIONS IN SECTIONS

A. *Ordinary differential equations*

ANGELOV V.: A coincidence theorem in uniform spaces and applications
ANGELOVA D.: Asymptotic and oscillation properties on functional - differential equations
ANDRES J.: Higher kind periodic orbits
AUGUSTYNOWICZ A.: On the existence of continuous solutions of operator equations in Banach spaces
BERKOVIČ L. M.: A constructive approach in the theory of differential equations: Factorization and transformations
BIHARI I.: A second order nonlinear differential inequality
BRESQUAR A. M.: Asymptotic solutions for the oscillatory differential equation
BIANCHINI R. M., CONTI R.: Local and global controllability
ČADEK M.: Pointwise transformations of linear differential equations
DŁOTKO T.: Initial functions as controls
DOŠLÁ Z.: Differential equations and higher monotonicity
DOŠLÝ O.: Transformations of linear differential systems
ELBERT A.: Eigenvalue estimations for the halflinear second order differential equations
FENYŐ I.: On the interrodifferential equation $x(t)+\lambda\int^{\infty} J_n(2\sqrt{tz})\,(t/z)^{n/2}\,x^{(k)}(z)dz = F(t)$
FISHER A.: Almost periodic solutions of systems of linear and quasilinear differential equations with almost periodic coefficients and with time lag
FOFANA M. S.: The stability of a special differential equation
FOLTÝŇSKÁ I.: An oscillation of solutions of nonlinear integro-differential equations system
GARAY B. M.: Parallelizability in Banach spaces: Examples and counterexamples
GRAEF J. R., SPIKES P. W., ZHANG B. G.: Sufficient conditions for the oscillatory solutions of a delay differential equation to converge to zero
GREGUŠ M.: Nontrivial solutions of a nonlinear boundary value problem
HABETS P.: On periodic solutions of nonlinear second order differential equations
HADDOCK J.: Phase spaces for functional differential equations
HALICKÁ M.: Existence of regular synthesis for two classes of optimal control problems
HATVANI L.: A generalization of the invariance principle to nonautonomous differential systems
JAROŠ J.: Oscillation criteria for forced functional differential inequalities
KARTÁK K.: Generalized absolutely continuous solutions of ODE
KHEKIMOVA M.: Periodicheskie i kraevye zadachy dlya singulyarno vozmushchennykh sistem s impulsnym vozdeystvem

B. *Partial differential equations*

BIROLI M.: Wiener obstacles for Δ^2
BOJARSKI B.: Microlocal analysis of linear transmission problems
DŁOTKO T.: Geometric description of quasilinear parabolic equations
DRÁBEK P.: Destabilizing effect of certain unilateral conditions for the system of reaction-diffusion type
DZIUK G.: A simple climate modell
FILO J.: On a nonlinear diffusion equation with nonlinear boundary conditions: Method of lines
FILA M.: Connecting orbits in certain reaction diffusion equations
HEGEDÜS J.: Zadachi sopryazheniya dlya nekotorykh ellipticheskikh i giperbolicheskikh uravnenii
HUEBER H.: Dirichlets problem for some hypoelliptic differential operators
KAMONT Z.: Weak solutions of first order partial differential equations with a retarded argument
KAWOHL B.: Starshaped rearrangement and applications
KOLOMÝ J.: On accretive operators
LEWIS R. T.: The eigenvalues of elliptic differential operators
LORENZI A.: An inverse problem for a quasilinear parabolic equation in divergence form
MUSTONEN V.: Topological degree of mappings of monotone type and applications
NARAZAKI T.: Global classical solutions of semilinear evolution equation
NAUMANN J.: Liouville property and regularity for parabolic systems
NETUKA I.: The best harmonic approximation
ÔTANI M.: Existence and non-existence of non-trivial solutions of some nonlinear degenerate elliptic equations
PULTAR M.: Numerical methods of solution of hyperbolic equations
ROTHER W.: Generalized Thomas-Fermi-von Weizsäcker equations
SALVI R.: The equations of viscous incompressible non-homogenous fluids: On the existence and regularity
SHOPOLOV N.: The first boundary problem of a parabolic equation with arguments reversing their roles
SOKOLOWSKI J.: Differential stability of solutions to constrained optimization problems for p.d.e.
SPECK F.-O.: Boundary value problems for elliptic convolution type equations
SZULKIN A.: Minimax principles for lower semicontinuous functions and applications to elliptic boundary value problems
ŠVEC A.: Spectrum of spheres
TERSIAN S.: Characterizations of the range of Neumann problem for semilinear elliptic equations
TIBA D.: Control of nonlinear hyperbolic equations
TURO J.: A boundary value problem for quasilinear hyperbolic systems of differential-functional equations
VERHULST F.: The Galerkin-averaging method for a nonlinear Klein-Gordon equation

C. *Numerical methods*

AMIRALIEV G.: Towards the numerical solution of the system of Boussinesq equation

D. *Applications*

II. PAPERS PRESENTED AT THE POSTER SESSION

The papers of the following authors were also presented at the Poster Session:
BOGDANOV R., BOUZNASKI E., VASSILEVSKI P.

III. PAPERS PRESENTED IN THE FORM OF ENLARGED ABSTRACTS

LIST OF PARTICIPANTS

Hatvani L.
Hegedüs J.
Karolyi K.
Krisztin T.
Moson P.
Reti P.
Stephan G.
Terjeki J.

IRAN
Mamourian A.
Mehri B.

IRAQ
Al-Faiz M.

ITALY
Biroli M.
Bresquar A. M.
Brezzi F. 285
Conti R.
Gatteschi A.
Gatteschi L.
Giaquinta C. G. 215
Invernizzi C. G.
Invernizzi C. S.
Laforgia A.
Magnaghi-Delfino P.
Marchi V.
Nkhasama M.
Omari P.
Paparoni F.
Salvi S.
Torelli A.
Valli A. 259
Zanolin F.

JAPAN
Kusano T. (465)
Narazaki T.
Otani M.

JUGOSLAVIA
Ćurgus B.
Jovanovic B.
Vrdoljak B.

NETHERLANDS
Axelsson O. 275
Verhulst F.

POLAND
Augustynowicz A.
Bartuzel G.
Bobrowski D.
Bojarski B.
Borzymovski A.
Choczewski B.
Desperat T.
Dlotko Tad.
Dlotko Tom.
Foltýnská I.
Goncerzewicz J.
Gorowski J.
Grysa K.
Hacia L.
Hyb W.
Jankowski J.
Jedryka T. M.
Kamont Z.
Kisielewicz M.
Kubiaczik I.
Litewska M.
Matkowski J.
Mikolajski J.
Morchalo J.
Moszner Z.
Muszynski B.
Myjak Z.
Nadzieja T.
Okrasinski W.
Olech Cz.
Popenda J.
Reginska T.
Rzepecki B.
Skierczynski B.
Sokolowski J.
Sosulski W.
Stankiewicz R.
Szmanda B.
Tabisz K.
Tabor J.
Turo J.
Wakulicz A.
Werbowski J.
Wyrwinska A.

RUMANIA
Aniculaesei G.
Ionescu I.
Lungu N.
Morozan T.
Muresan M.
Potra T.
Rasvan V.
Sofonea M.
Stavre R.
Tiba D.
Varsan C.
Vernescu B.
Vornicescu N.

SINGAPORE
Agarwal R. P. 267
Chow Y. M.

SPAIN
Vega C.

SWEDEN
Daho K.
Hedberg L. I.
Karlsson T.
Szulkin A.
Thomeé V. 339

SWITZERLAND
Desloux J. 295
Schwarz H. R.

USA
Bebernes J. 193
Burton T. 115
Friedman A. 17
Graef J. R.
Haddock M.
Kreith K. 141
Lewis R.
Sussmann H.
Sell G. R.
Trench W. F. 181

USSR
Amiraltyev G.
Berkovic L.
Bogdanov R.
Chanturia T. A. (431)
Cherkas L.
Chusainov D. Ja.
Gorochowik S.
Gudovic N. N.
Koshelev A. I. (459)
Kurzhanski A. B.
Lamzyuk V. D.

PLENARY LECTURES

MATHEMATICAL AND NUMERICAL STUDY OF NONLINEAR PROBLEMS IN FLUID MECHANICS

M. FEISTAUER
Faculty of Mathematics and Physics, Charles University
Malostranské nám. 25, 118 00 Prague 1, Czechoslovakia

INTRODUCTION

The study of flow problems in their generality is very difficult since real flows are three-dimensional, nonstationary, viscous with large Reynolds numbers, rotational, turbulent, sometimes also more-fase and in regions with a complicated geometry. Therefore, we use simplified, usually two-dimensional and non-viscous models. (The effects of viscosity are taken into account additionally on the basis of the boundary layer theory.)

Here we give a surway of results obtained in the study of boundary value problems describing two-dimensional, non-viscous, stationary or quasistationary incompressible or subsonic compressible flows with the use of a stream function.

1. STREAM FUNCTION FORMULATION OF THE PROBLEM

On the basis of a detailed theoretical and numerical analysis of various types of flow fields (plane or axially symmetric channel flow, flow past an isolated profile, cascade flow etc.) a unified conception for the stream function-finite element solution of flow problems was worked out.

We start from the following assumptions:

1) The domain $\Omega \subset R_2$ filled by the fluid is bounded with a piecewise smooth, Lipschitz-continuous boundary $\partial\Omega$. (Usually Ω has the form of a curved channel with inserted profiles.)

2) $\partial\Omega = \Gamma_D \cup \Gamma_N \cup \left(\bigcup_{j=1}^{K_I} \Gamma_I^j\right) \cup \left(\bigcup_{j=1}^{K_T} \Gamma_T^j\right) \cup \Gamma_P^- \cup \Gamma_P^+$, where Γ_I^j, Γ_T^j

are arcs or simple closed curves, Γ_P^+, Γ_P^- are piecewise linear arcs, Γ_P is obtained by translating Γ_P^- in a given direction by a given distance. This translation is represented by a one-to-one mapping $Z_P: \Gamma_P^- \xrightarrow{\text{onto}} \Gamma_P^+$. Γ_D and Γ_N are formed by finite numbers of arcs. Of course, all these arcs and simple closed curves are mutually disjoint, except neighbouring arcs that have common initial or terminal points. We assume that $\Gamma_D \neq \emptyset$.

3) The differential equation has the form

$$\sum_{i=1}^{2} (b(x,u,(\nabla u)^2)u_{x_i})_{x_i} = f(x,u,(\nabla u)^2) \quad \text{in} \quad \Omega. \tag{1.1}$$

4) We admit the following boundary conditions:

$$u|\Gamma_D = u_D \quad \text{(Dirichlet condition)}, \tag{1.2}$$

$$b(.,u,(\nabla u)^2)\frac{\partial u}{\partial n}|\Gamma_N = -\varphi_N \quad \text{(Neumann condition)}, \tag{1.3}$$

$$u(Z_P(x)) = u(x) + Q, \tag{1.4,a}$$

$$(b(.,u,(\nabla u)^2)\frac{\partial u}{\partial n})(Z_P(x)) = -(b(.,u,(\nabla u)^2)\frac{\partial u}{\partial n})(x) \tag{1.4,b}$$

$$\forall\, x \in \Gamma_P^- \quad \text{(periodicity conditions)},$$

$$u|\Gamma_I^j = u_I^j + q_I^j, \quad q_I^j = \text{const}, \tag{1.5,a}$$

$$\int_{\Gamma_I^j} b(.,u,(\nabla u)^2)\frac{\partial u}{\partial n}dS = -\gamma_I^j \quad \text{(velocity circulation conditions)} \tag{1.5,b}$$

$$j = 1,\dots,K_I,$$

$$u|\Gamma_T^j = u_T^j + q_T^j, \quad q_T^j = \text{const}, \tag{1.6,a}$$

$$\frac{\partial u}{\partial n}(z_j) = 0, \quad z_j \in \Gamma_T^j, \quad \text{(Kutta-Joukowski trailing conditions)} \tag{1.6,b}$$

$$j = 1,\dots,K_T.$$

u_D, u_N, u_I^j, u_T^j are given functions, Q, γ_I^j - given constants, $z_j \in \Gamma_T^j$ are given trailing stagnation points. u is an unknown function and q_I^j, q_T^j are unknown constants.

The contact of some boundary conditions is prohibited e.g. (1.2) and (1.5,a-b). It is also necessary to consider the consistency of some types of these conditions as e.g. (1.2) and (1.4). For concrete examples see [7 - 10] .

2. THE PROBLEM WITHOUT TRAILING CONDITIONS

Since the problem (1.1) - (1.5) without trailing conditions has better properties from the mathematical point of view than the general problem (1.1) - (1.6), we shall treat these problems separately.

2.1. Variational formulation of the problem (1.1) - (1.5). We shall seek a weak solution in the well-known Sobolev space $H^1(\Omega) = W_2^1(\Omega)$. We define the set

$$\mathcal{V} = \{v \in C^\infty(\bar{\Omega});\ v|\Gamma_D = 0,\ v(Z_P(x)) = v(x),\ x \in \Gamma_P^-,\ v|\Gamma_I^j = \text{const}\} \tag{2.1}$$

and the space

$$V = \{v \in H^1(\Omega);\ v|\Gamma_D = 0,\ v(Z_P(x)) = v(x),\ x \in \Gamma_P^-,\ v|\Gamma_I^j = \text{const (in the sense of traces on } \partial\Omega)\}. \tag{2.2}$$

The validity of the following assertion is important:

The set $\mathcal{V}$ is dense in V, i.e.

$$\bar{\mathcal{V}}^{H^1(\Omega)} = V. \tag{2.3}$$

It is not easy to prove this. For a cascade flow problem see [10].

Further, let $u^* \in H^1(\Omega)$ satisfy

$$\text{a) } u^*|\Gamma_D = u_D, \quad \text{b) } u^*|\Gamma_I^j = u_I^j, \tag{2.4}$$
$$\text{c) } u^*(Z_P(x)) = u^*(x) + Q,\ x \in \Gamma_P^-.$$

Very often the existence of this u^* follows from the fact that u_D and u_I^j are indefinite integrals of functions from $L_2(\Gamma_D)$ and $L_2(\Gamma_I^j)$, respectively (cf. [20]).

Under the above notation the problem (1.1) - (1.5) is (formally) equivalent to the following variational formulation: Find u such that

$$\text{a) } u \in H^1(\Omega), \quad \text{b) } u - u^* \in V, \tag{2.5}$$
$$\text{c) } a(u,v) = m(v) \quad \forall v \in V,$$

where

$$a(u,v) = \int_\Omega (b(.,u,(\nabla u)^2)\nabla u.\nabla v + f(.,u,(\nabla u)^2)v)dx \tag{2.6,a}$$
$$\forall u,v \in H^1(\Omega),$$

$$m(v) = -\sum_{j=1}^{K_I} \gamma_I^j v|\Gamma_I^j - \int_{\Gamma_N} \phi_N v dS, \quad v \in V. \tag{2.6,b}$$

2.2. Finite element discretization. Let Ω be approximated by a polygonal domain Ω_h and let $\mathcal{T}_h$ be a triangulation of Ω_h with usual properties. We denote by $\sigma_h = \{P_1,\dots,P_N\}$ the set of all vertices of $\mathcal{T}_h$. Let the common points of Γ_D, Γ_N etc. and also the points of $\partial\Omega$, where the condition of smoothness of $\partial\Omega$ is not satisfied, belong to σ_h. Moreover, let $\sigma_h \cap \partial\Omega_h \subset \partial\Omega$, and $P_j \in \Gamma_P^- \cap \sigma_h \Leftrightarrow Z_P(P_j) \in \Gamma_P^+ \cap \sigma_h$. Hence, the sets Γ_D, Γ_N etc. are approximated by arcs or curves Γ_{Dh}, Γ_{Nh} etc. $\subset \partial\Omega_h$ in a natural way.

An approximate solution is sought in the space of linear conforming triangular elements

$$W_h = \{ v_h \in C(\bar{\Omega}_h);\ v_h|T \text{ is affine } \forall T \in \mathcal{T}_h \} \qquad (2.7)$$

The <u>discrete problem</u> is written down quite analogously as the continuous problem (2.5,a-c): Find u_h such that

$$\text{a) } u_h \in W_h, \qquad \text{b) } u_h - u_h^* \in V_h \qquad (2.8)$$

$$\text{c) } a_h(u_h, v_h) = m_h(v_h) \qquad \forall v_h \in V_h,$$

where

$$V_h = \{ v_h \in W_h;\ v_h|\Gamma_{Dh} = 0,\ v_h(Z_P(P_j)) = v_h(P_j),\ P_j \in \sigma_h \cap \Gamma_P^-,$$

$$v_h|\Gamma_{Ih}^j = \text{const} \}. \qquad (2.9)$$

The function $u_h^* \in W_h$ has the properties

$$\text{a) } u_h(P_i) = u_D(P_i),\ P_i \in \sigma_h \cap \Gamma_D,$$

$$\text{b) } u_h(P_i) = u_I^j(P_i),\ P_i \in \sigma_h \cap \Gamma_I^j, \qquad (2.10)$$

$$\text{c) } u_h(Z_P(P_i)) = u_h(P_i) + Q,\ P_i \in \sigma_h \cap \Gamma_P^- \quad \text{and}$$

plays the same role as u^* in the continuous problem. Further,

$$a_h(u_h, v_h) = \int_{\Omega_h} (b(.,u_h,(\nabla u_h)^2)\nabla u_h.\nabla v_h + f(.,u_h,(\nabla u_h)^2)v_h)dx \qquad (2.11,a)$$

$$m_h(v_h) = -\sum_{j=1}^{K_I} \gamma_I^j v_h|\Gamma_{Ih}^j - \int_{\Gamma_{Nh}} \varphi_h v_h dS. \qquad (2.11,b)$$

Usually, the integrals in (2.11,a-b) are evaluated by convenient <u>numerical quadratures</u>. Then, instead of u_h, a_h and m_h we have u_h^{int}, a_h^{int} and m_h^{int} in (2.8,a-c).

The problem (2.8,a-c) leads to a <u>system of algebraic equations</u>

$$A(\bar{u})\bar{u} = F(\bar{u}). \qquad (2.12)$$

Here $\bar{u} = (u_1,\dots,u_n)^T$ is a vector with components defining the approximate solution, $A(\bar{u})$ is an $n\times n$ $(n < N)$ symmetric positive definite matrix for all $\bar{u} \in R_n$ and $F: R_n \to R_n$.

Now let us introduce the <u>properties of the functions b and f:</u>

1) b and f depend on $x \in \bar{\Omega}$, $u \in R_1$, $\eta \geq 0$ ($\eta := (\nabla u)^2$).

2) b, f and their derivatives $\partial b/\partial x_i$, $\partial b/\partial u$, $\partial b/\partial \eta$, $\partial f/\partial x_i$ etc. are continuous and bounded.

3) $b \geq b_1 > 0$, $b_1 = \text{const}$, $\partial b/\partial \eta \geq 0$.

4) $\frac{\partial b}{\partial \eta}(x,u,s^2)s^2$, $\left|\frac{\partial b}{\partial u}(x,u,s^2)s\right| \leq \text{const}$ $\forall x \in \bar{\Omega}$, $\forall u$, $s \in R_1$.

5) $b(Z_P(x), u+Q, \eta) = b(x,u,\eta)$ $\forall x \in \Gamma_P^-$, $u \in R_1$, $\eta \geq 0$.

f satisfies the second inequality in 4) and the assumption 5).

<u>2.3. The solvability of the problem (2.5,a-c)</u> is a consequence

of the monotone and pseudomonotone operator theory ([19, 22]). If the flow is irrotational ($b = b(x, \eta)$, $f = 0$), then the form $a(u,v)$ satisfies the condition of strong monotony and the solution is unique. These results for various types of flows are contained in [1, 4, 5, 10, 15].

2.4. The study of the discrete problem. Its solvability easily follows from the Brower's fixed point theorem and the properties of b and f (cf. [19, 13]). Much more complex is the question on the convergence of the finite element method, since by Strang ([23]) we have commited three variational crimes (approximation of Ω by a polygonal domain; $W_h \not\subset H^1(\Omega)$, $V_h \not\subset V$; numerical integration), the problem is nonlinear and boundary conditions are nonhomogeneous and nonstandard.

We shall consider numerical quadratures of precision d=1 with nonnegative coefficients. Let φ_N and $\partial\Omega$ be piecewise of the class C^2.

Let us consider a regular system of triangulations $\{T_h\}_{h\in(0,h_o)}$ of Ω_h ($h_o > 0$ is sufficiently small) and study the behaviour of u_h, if $h \to 0+$.

By $\|.\|_{1,\Omega_h}$ we denote the usual norm in $H^1(\Omega_h)$ and put

$$|v|_{1,\Omega_h} = \left(\int_{\Omega_h} (\nabla v)^2 dx\right)^{1/2} \tag{2.13}$$

It is important that

$$\|v_h\|_{1,\Omega_h} \leq c\,|v_h|_{1,\Omega_h} \quad \forall v_h \in V_h \quad \forall h \in (0,h_o) \tag{2.14}$$

with a constant $c > 0$ independent of v_h and h (see [13] or [24]).

By [21], the solution u of the continuous problem and the function u^* posses the Calderon extensions from Ω to a domain $\tilde{\Omega}$ such that $\Omega, \Omega_h \subset \tilde{\Omega}$ $\forall h \in (0,h_o)$ and $u, u^* \in H^1(\tilde{\Omega})$.

Further, let us assume that

$$\|u^* - u_h^*\|_{1,\Omega_h} \to 0, \text{ if } h \to 0+. \tag{2.15}$$

In some cases (cf. e.g. [10]) $u^* \in W_2^{1+\varepsilon}(\tilde{\Omega})$, $\varepsilon > 0$, and we can put $u_h^* = r_h u^*$ (= the Lagrange interpolation of u^*). Then, since

$$\|u^* - r_h u^*\|_{1,\Omega_h} \leq c h^{\varepsilon} \|u^*\|_{W_2^{1+\varepsilon}(\tilde{\Omega})} \tag{2.16}$$

(with c independent of u^* and h) we have (2.15).

First let us consider an irrotational flow. The study of the convergence $u_h \to u$, if $h \to 0+$, is based on the following results.

2.4.1. Theorem. There exist α, $K > 0$ such that

$$a_h(u_1,u_1-u_2) - a_h(u_2,u_1-u_2) > \alpha\,|u_1-u_2|^2_{1,\Omega_h}, \tag{2.17}$$

$$|a_h(u_1,v) - a_h(u_2,v)| \leq K\,\|u_1-u_2\|_{1,\Omega_h}\,\|v\|_{1,\Omega_h} \tag{2.18}$$

$$\forall u_1,u_2,v \in H^1(\Omega_h), \quad \forall h \in (0,h_o)$$

and

$$a_h^{int}(u_1,u_1-u_2) - a_h^{int}(u_2,u_1-u_2) \geq \alpha |u_1-u_2|^2_{1,\Omega_h} \qquad (2.17^*)$$

$$|a_h^{int}(u_1,v) - a_h^{int}(u_2,v)| \leq K \|u_1-u_2\|_{1,\Omega_h} \|v\|_{1,\Omega_h} \qquad (2.18^*)$$

$$\forall u_1,u_2,v \in W_h , \quad \forall h \in (0,h_o) .$$

Proof follows easily from the properties of the functions b and f, the Mean Value Theorem and [3, Theorem 4.1.5] .

Now let us introduce abstract error estimates.

<u>2.4.2. Theorem</u>. There exist constants A_1, A_2, A_3 independent of h such that

$$\|u-u_h\|_{1,\Omega_h} \leq A_1 \inf_{w_h \in u_h^*+V_h} \|u-w_h\|_{1,\Omega_h} + \qquad (2.19)$$

$$+ A_2 \sup_{v_h \in V_h} (|a_h(u,v_h)-m_h(v_h)|/ \|v_h\|_{1,\Omega_h})$$

and

$$\|u_h-u_h^{int}\|_{1,\Omega_h} \leq \qquad (2.20)$$

$$A_3 \sup_{v_h \in V_h} (|a_h(u_h,v_h)-a_h^{int}(u_h,v_h)|+|m_h(v_h)-m_h^{int}(v_h)|)/\|v_h\|_{1,\Omega_h} .$$

Proof is a consequence of Theorem 2.4.1.

<u>2.4.3. Theorem</u>. Let $u,u^* \in H^2(\tilde{\Omega})$. Then $\|u-u_h\|_{1,\Omega_h} = O(h)$.

Proof. We apply the technique common in linear problems (cf. [3]) based on estimates (2.19) and (2.16) with $\varepsilon = 1$ and a similar estimate for u. This, the use of Green's theorem and the fact that $\text{meas}((\Omega-\Omega_h)\cup(\Omega_h-\Omega)) \leq ch^2$ give the result. (Another approach avoiding the use of Green's theorem is used in [18] .)

<u>2.4.4. Theorem</u>. Let $u \in H^1(\tilde{\Omega})$ and let (2.15) be satisfied. Then

$$\lim_{h\to 0+} \|u-u_h\|_{1,\Omega_h} = 0.$$

Proof. From (2.3) and (2.15) we get

$$\lim_{h\to 0+} \inf_{w_h \in u_h^*+V_h} \|u-w_h\|_{1,\Omega_h} = 0.$$

The convergence of the second term in (2.19) to zero is proved by introducing convenient modifications $\hat{v}_h \in V$ of $v_h \in V_h$. Hence,

$$a(u,\hat{v}_h) = m(\hat{v}_h).$$

This, the estimates of $v_h-\hat{v}_h$ derived in [24] and the estimates of $a(u,\hat{v}_h)-a_h(u,v_h)$ and $m_h(v_h)-m(\hat{v}_h)$ imply the desired result.

<u>2.4.5. Theorem</u>. If we use numerical integration of precision d=1 with nonnegative coefficients and $\partial\Omega$, φ_N are piecewise of class C^2, then $\|u_h-u_h^{int}\|_{1,\Omega_h} = O(h)$.

Proof follows from the estimate (2.20), [3, Theorem 4.1.5] and the boundedness of $\{u_h\}_{h\in(0,h_o)}$.

2.4.6. Remark. The convergence of the finite element solution with numerical integration applied to a nonlinear elliptic problem was proved in [13]. A more complex analysis will be given in [18].

For general rotational flows, instead of strong monotony, we have pseudomonotony only. Then by the application of methods from the pseudomonotone operator theory ([19, 22]) we get the following result:

2.4.7. Theorem. Let Ω be a polygonal domain and let the conditions (2.3) and (2.15) be satisfied. Let the forms a and m are evaluated by means of numerical quadratures of precision d=1 with nonnegative coefficients. Hence, a and m are approximated by $a_h := a_h^{int}$ and $m_h := m_h^{int}$, respectively. Then it holds:
1) To each $h\in(0,h_o)$ at least one solution u_h of (2.8,a-c) exists.
2) There exists $c>0$ such that $\|u_h\|_{1,\Omega} \le c$ for all $h\in(0,h_o)$.
3) If $\{u_{h_n}\}_{n=1}^{\infty}$ is a subsequence of the system $\{u_h\}_{h\in(0,h_o)}$, $h_n \to 0$ and $u_{h_n} \rightharpoonup u$ weakly in $H^1(\Omega)$ for $n\to\infty$, then u is a solution of the continuous problem (2.5,a-c) and $u_{h_n} \to u$ strongly in $H^1(\Omega)$.

Proof. Let $A:H^1(\Omega) \to (H^1(\Omega))^*$ be the operator defined by the relation

$$\langle A(u),v\rangle = a(u,v), \quad u, v \in H^1(\Omega) . \tag{2.21}$$

From the properties of b and f it follows:
a) A is Lipschitz-continuous and bounded.
b) A satisfies the generalized property (S), i.e. it holds:
$z_n, z\in V$, $z_n \rightharpoonup z$ weakly, $u_n^* \to u^*$ strongly, $\langle A(u_n^*+z_n)-A(u^*+z), z_n-z\rangle \to 0 \Rightarrow u_n^*=u_n+z_n \to u=u^*+z$ strongly.

The proof of the assertion a) follows from the properties of the functions b and f. Let us show that also b) is valid. We assume that $z_n, z\in V$, $z_n \rightharpoonup z$ weakly, $u_n \to u$ strongly, $u=u^*+z$, $u_n=u_n^*+z_n$ and

$$J_n = \langle A(u_n) - A(u), z_n - z\rangle \to 0.$$

If we put

$$I_n = a(u_n,u_n-u) - a(u,u_n-u),$$

then

$$J_n = I_n + a(u,u_n^*-u^*) - a(u_n^*,u_n-u^*).$$

Since $u_n^* \to u^*$ we find out that

$$a(u,u_n^*-u^*) - a(u_n,u_n^*-u^*) \to 0.$$

Hence, $I_n \to 0$.

From the definition of the form a it follows that

$$I_n = \int_\Omega \{ \sum_{i=1}^{2} (b(.,u_n,(\nabla u_n)^2) \frac{\partial u_n}{\partial x_i} - b(.,u,(\nabla u)^2)\frac{\partial u}{\partial x_i}) \frac{\partial (u_n-u)}{\partial x_i} +$$
$$+ (f(.,u_n,(\nabla u_n)^2) - f(.,u,(\nabla u)^2))(u_n-u)\} dx .$$

As $u_n \rightharpoonup u$ weakly in $H^1(\Omega)$, $u_n \to u$ strongly in $L_2(\Omega)$. The properties of f and b imply the relations

$$I_n \geq \alpha |u_n-u|^2_{1,\Omega} + c_n, \qquad \alpha > 0,$$
$$c_n = \int_\Omega \{(b(.,u_n,(\nabla u_n)^2) - b(.,u,(\nabla u)^2))\nabla u \cdot \nabla(u_n - u) +$$
$$+ (f(.,u_n,(\nabla u_n)^2) - f(.,u,(\nabla u)^2))(u_n - u)\} dx,$$
$$c_n = (b(.,(\nabla u_n)^2) - b(.,u,(\nabla u)^2))\nabla u \cdot \nabla(u_n - u) +$$
$$+ (f(.,u_n,(\nabla u_n)^2) - f(.,u,(\nabla u)^2))(u_n - u) dx,$$

(The sequence $\{\|u_n-u\|_{1,\Omega}\}_{n=1}^{\infty}$ is bounded.) From this and equivalence of the norms $\|.\|_{1,\Omega}$ and $|.|_{1,\Omega}$ in the space V we already conclude that $u_n \to u$ strongly.

Since $\{u_h\}_{h\in(0,h_o)}$ is a bounded set and A is a bounded operator, we can assume that we have a sequence $u_n := u_{h_n}$ such that

$$h_n \to 0, \; u_n \rightharpoonup u \text{ weakly in } H^1(\Omega) \quad , \tag{2.22}$$
$$A(u_n) \rightharpoonup X \text{ weakly in } (H^1(\Omega))^* .$$

In view of (2.15), it is evident that $u = u^* + z$, $z \in V$ and $z_n = z_{n_h} \rightharpoonup z$ weakly.

Similarly as in [13] or [18] we derive the estimates

$$|a(u_h,v_h) - a_h(u_h,v_h)| \leq ch\|v_h\|_{1,\Omega} \quad \forall v_h \in V_h \quad \forall h \in (0,h_o) \tag{2.23}$$

$$|m(v_h) - m_h(v_h)| \leq ch\|v_h\|_{1,\Omega} \quad \forall v_h \in V_h \quad \forall h \in (0,h_o) \tag{2.24}$$

with c independent of v_h and h.

Let $v \in V$. By (2.16) $v_n := r_{h_n} v \to v$ in $H^1(\Omega)$, $v_n \in V_{h_n}$. We have

$$\langle A(u_n),v_n \rangle = m(v_n) + (a(u_n,v_n) - a_{h_n}(u_n,v_n)) +$$
$$+ (m_{h_n}(v_n) - m(v_n)).$$

From this, (2.22)-(2.24) and (2.3) we derive the relation

$$\langle X,v \rangle = m(v) \qquad \forall v \in V. \tag{2.25}$$

Further, by (2.21)-(2.25),

$$\langle A(u_n) - A(u), z_n - z \rangle \to 0.$$

Now, if we use the generalized property (S) of the operator A, we find out that $z_n \to z$ and thus, $u_n \to u$ (strongly). As the operator A is continuous, $A(u) = \lim_{n\to\infty} A(u_n) = X$. By (2.25),

$$\langle A(u), v \rangle = m(v) \qquad \forall v \in V,$$

which we wanted to prove.

2.4.8. Remark. Instead of Lipschitz-continuity of the operator A it is sufficient to use its demicontinuity: "$u_n \to u$ strongly $\Rightarrow$ $\Rightarrow$ $A(u_n) \to A(u)$ weakly. The proof of the convergence of the approximate solution obtained without numerical integration is similar (and of course more simple). The case of the problem in a nonpolygonal domain Ω remains open.

3. ON THE GENERAL PROBLEM (1.1) - (1.6)

In practice the complete problem (1.1) - (1.6) is very important, but its mathematical study is unfortunately much more difficult because of the discrete trailing conditions (1.6,b). Therefore, the results are not so complete as in the case of the problem (1.1) - (1.5) and we present here only a brief surway.

3.1. The solvability of the continuous problem has to be studied in classes of classical solutions. The main tool for proving the solvability are appriori estimates of solutions to linear and nonlinear elliptic equations and the strong maximum principle. The study was successful for incompressible irrotational and rotational flows ([6, 8]) and for irrotational compressible flows ([9]). The solvability of the general rotational compressible flow problem remains open.

3.2. Finite element discretization. Let us consider a triangulation $\mathcal{T}_h$ of the domain Ω_h with the properties from 2.2. Moreover, we assume that to each trailing point $z_j \in \Gamma_T^j$ there exists a triangle $T_j \in \mathcal{T}_h$ with vertices $\tilde{P}_j = z_j$ and $P_j^* \in \Omega_h$ such that the side $S_j = \tilde{P}_j P_j^*$ is normal to Γ_T^j. Then, if we discretize the condition (1.6,b) by its finite-difference analogue and consider (1.6,a), we derive the conditions (for simplicity we assume that $u_T^j = 0$)

$$u_h(P_k) = q_T^j = u_h(P_j^*) \qquad \forall P_k \in \sigma_h \cap \Gamma_T^j \tag{3.1}$$

Now the discrete problem to (1.1) - (1.6) is written down in the following way: Find u_h such that

$$\text{a) } u_h \in W_h, \qquad \text{b) } u_h - u_h^* \in \tilde{V}_h, \tag{3.2}$$

$$\text{c) } a_h(u_h, v_h) = m_h(v_h) \qquad \forall v_h \in V_h.$$

Here,

$$V_h = \{ v_h \in W_h;\ v_h|_{\Gamma_{Dh}} = 0,\ v_h(Z_P(P_i)) = v_h(P_i), \tag{3.3}$$
$$P_i \in \sigma_h \cap \Gamma_P^-,\ v_h|_{\Gamma_{Ih}^j} = \text{const},\ v_h|_{\Gamma_{Th}^j} = 0\} ,$$

$$\tilde{V}_h = \{v_h \in W_h;\ v_h|_{\Gamma_{Dh}} = 0,\ v_h(Z_P(P_i)) = v_h(P_i), \tag{3.4}$$
$$P_i \in \sigma_h \cap \Gamma_P^-,\ v_h|_{\Gamma_{Ih}^j} = \text{const},\ v_h|_{\Gamma_{Th}^j \cup S_j} = \text{const}\} ,$$

a) $u_h^* \in W_h,\ u_h^*(P_i) = u_D(P_i),\ P_i \in \sigma_h \cap \Gamma_D$, (3.5)

b) $u_h^*(P_i) = u_I^j(P_i),\ P_i \in \sigma_h \cap \Gamma_I^j$,

c) $u_h^*(Z_P(P_i)) = u_h^*(P_i) + Q,\ P_i \in \sigma_h \cap \Gamma_P^-$,

d) $u_h^*|_{\Gamma_{Th}^j \cup S_j} = 0$.

a_h and m_h are again defined by (2.11,a-b).

The problem (3.2,a-c) is equivalent to a system (2.12). Since $V_h \neq \tilde{V}_h$, the matrix $A(\bar{u})$ is not more symmetric. However, if <u>all angles of all $T \in \mathcal{T}_h$ are less then or equal to 90°</u>, then $A(\bar{u})$ is an irreducibly diagonally dominant matrix and the system (2.12) has a solution. Under the same assumption, with the use of the discrete maximum principle, we can prove the <u>convergence of the method</u>: if $u \in C^2(\bar{\Omega})$ and the problem is linear, then $\|u-u_h\|_{L_\infty(\Omega_h)} \leq ch$ for all $h \in (0,h_o)$. For details see [14].

4. ITERATIVE SOLUTION OF THE DISCRETE PROBLEM

It is convenient to distinguish several cases:

<u>4.1. Irrotational incompressible flow</u> ($b = b(x)$, $f = 0$): The system (2.12) is linear and we use the SOR method.

<u>4.2. Irrotational compressible flow</u> ($b = b(x,\eta)$, $f = 0$): Among the methods we have tested the following iterative process occurs as an effective one:

a) $\bar{u}^o \in R_n$ (a convenient initial approximation) (4.1)

b) $B\bar{u}^{k+1} = B\bar{u}^k - \omega(A(\bar{u}^k)\bar{u}^k - F(\bar{u}^k)),\ k \geq 0,\ \omega > 0.$

The speed of the convergence depends on the choice of ω (its estimate can be obtained on the basis of the behaviour of the function b) and of a preconditioning positive definite matrix B.

<u>4.3. Rotational incompressible flow</u> ($b = b(x)$, $f = f(x,u)$): Similarly as in [7] we can apply a Newton relaxation method. If the vorticity is too strong, it is better to proceed as in the following

case.

4.4. Rotational compressible flow: As a sufficiently robust the method of least squares and conjugate gradients by Glowinski et al. appears (see [2]). The details will be the subject matter of an intended paper.

5. EXAMPLES

As a simple test problem we introduce a flow through a plane channel. On the inlet (left side of the boundary - see Fig. 1) and outlet (right side of the boudary) we consider the Neumann condition $\partial u/\partial n = 0$. On the lower wall we put $u = 0$ and on the upper wall $u = 25$. We consider a rotational flow described by the equation

$$\Delta u = -200 \text{ arctg } u.$$

The uniqueness of this boundary value problem is not sure.

This problem was successfully solved by the method of least squares and conjugate gradients starting from the solution of the corresponding linear irrotational flow ($\Delta u = 0$). In Fig. 1 we see the triangulation used. The iterative process was stopped after 6 conjugate gradient iterations, when the resulting value of the cost functional was 10^{-5}. For one-dimensional minimization the golden-section method was applied. In Fig. 2 the calculated velocity field is plotted. It is interesting with backward flows caused by a strong vorticity.

The second example represents an industrial application of the presented theory and numerical methods - a result of a cascade flow calculation (cf. [10 - 14, 16, 17]). In Fig. 3 we show velocity vectors plotted in the domain representing one period of a cascade of profiles.

For other examples see [11, 12, 17].

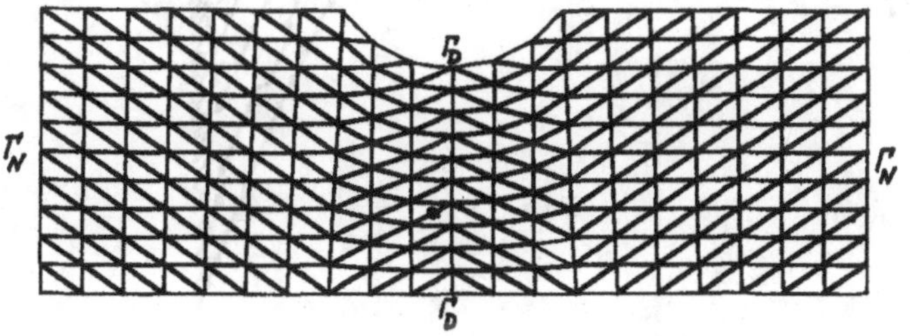

Fig. 1

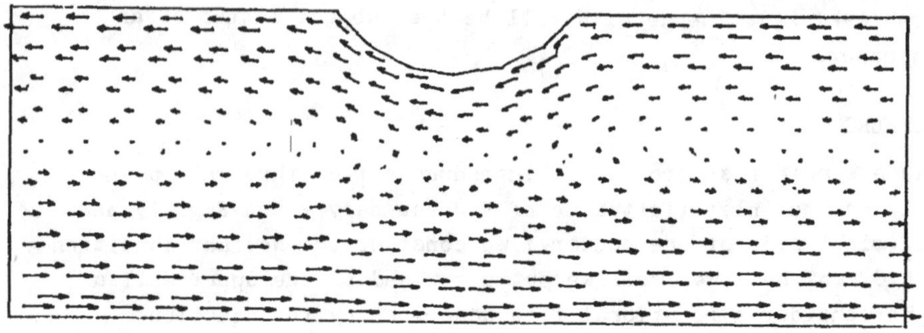

Fig. 2

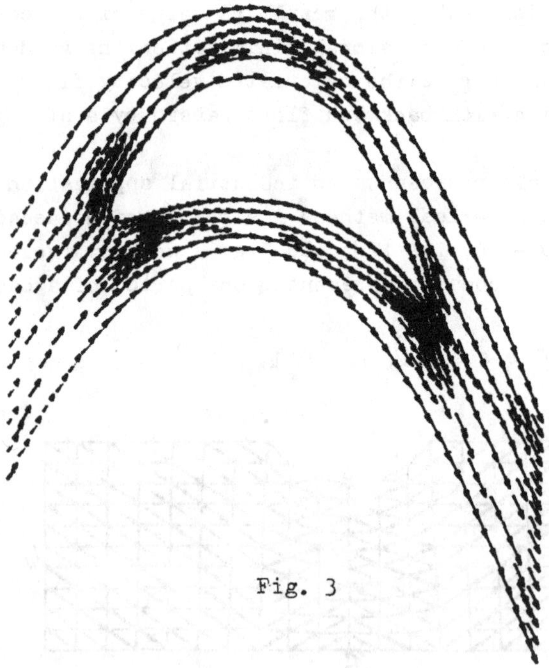

Fig. 3

REFERENCES

[1] J.Benda, M.Feistauer: Rotational subsonic flow of an ideal compressible fluid in axially symmetric channels. Acta Polytechnica, 7(IV,3), 1978, 95-105 (in Czech).

[2] M.O.Bristeau, R.Glowinski, J.Periaux, P.Perrier, O.Pironneau, G. Poirier: Application of optimal control and finite element methods to the calculation of transonic flows and incompressible viscous flows. Rapport de Recherche no. 294 (avril 1978), LABORIA IRIA.

[3] Ph.G.Ciarlet: The Finite Element Method for Elliptic Problems. North-Holland, Amsterdam-New York-Oxford, 1978.

[4] M.Feistauer: On two-dimensional and three-dimensional axially symmetric flows of an ideal incompressible fluid. Apl.mat. 22 (1977), 199-214.

[5] M.Feistauer: Mathematical study of three-dimensional axially symmetric stream fields of an ideal fluid. Habilitation Thesis, Faculty of Math.and Physics, Prague, 1979 (in Czech).

[6] M.Feistauer: Solution of elliptic problem with not fully specified Dirichlet boundary value conditions and its application in hydrodynamics. Apl.mat. 24(1979), 67-74.

[7] M. Feistauer: Numerical solution of non-viscous axially symmetric channel flows. In: Methoden und Verfahren der mathematischen Physik, Band 24, 65-78, P.Lang-Verlag, Frankfurt am Main-Bern, 1982.

[8] M.Feistauer: Mathematical study of rotational incompressible non-viscous flows through multiply connected domains. Apl.mat. 26 (1981) 345-364.

[9] M.Feistauer: Subsonic irrotational flows in multiply connected domains. Math.Meth.in the Appl.Sci. 4(1982), 230-242.

[10] M.Feistauer: On irrotational flows through cascades of profiles in a layer of variable thickness. Apl.mat. 29(1984), 423-458.

[11] M.Feistauer: Finite element solution of non-viscous flows in cascades of blades. ZAMM 65(1985),4, T191-T194.

[12] M.Feistauer: Mathematical and numerical study of flows through cascades of profiles. In: Proc. of "International Conference on Numerical Methods and Applications" held in Sophia, August 27-September 2, 1984 (to appear).

[13] M.Feistauer: On the finite element approximation of a cascade flow problem. Numer.Math. (to appear).

[14] M.Feistauer: Finite element solution of flow problems with trailing conditions (to appear).

[15] M.Feistauer, J.Římánek: Solution of subsonic axially symmetric stream fields. Apl.mat. 20(1975), 266-279.

[16] M.Feistauer, J.Felcman, Z.Vlášek: Calculation of irrotational flows through cascades of blades in a layer of variable thickness. Research report, ŠKODA Plzeň, 1983 (in Czech).

[17] M.Feistauer, J.Felcman, Z.Vlášek: Finite element solution of flows through cascades of profiles in a layer of variable thickness. Apl.mat. (to appear).

[18] M.Feistauer, A.Ženíšek: Finite element solution of nonlinear elliptic problems (submitted to Numer. Math.)

[19] J.L.Lions: Quelques Methodes de Résolution des Problémes aux Limites non Linéaires. Dunod, Paris, 1969.

[20] J.Nečas: Über Grenzwerte von Funktionen, welche ein endliches Dirichletsches Integral haben. Apl.mat. 5(1960), 202-209.

[21] J.Nečas: Les Méthodes Directes en Théorie des Équations Elliptiques. Academia, Prague, 1967.

[22] J.Nečas: Introduction to the Theory of Nonlinear Elliptic Equations.Teubner-Texte zur Mathematik, Band 52, Leipzig, 1983.

[23] G.Strang: Variational crimes in the finite element method. In: The Mathematical Foundations of the Finite Element Method with Applications to Partial Differential Equations (A.K.Aziz, Ed.), Academic Press, New York, 1972, 689-710.

[24] A.Ženíšek: How to avoid the use of Green's theorem in the Ciarlets's and Raviart's theory of variational crimes (to appear).

FREE BOUNDARY PROBLEMS IN FLUID DYNAMICS

A. FRIEDMAN
Northwestern University
Evanston, Illinois 60201, U.S.A.

The velocity potential of a 2-dimensional ideal incompressible and irrotational fluid satisfies $\Delta\phi = 0$; further, Bernoulli's law $|\nabla\phi|^2 + 2p = \text{const.}$ yields $|\nabla\phi| = \text{const.}$ on the (free) boundary of the fluid in contact with air. Since $\nabla\phi$ is tangential to the free boundary, the stream function u (i.e., the harmonic conjugate of ϕ) satisfies:

$$\begin{aligned} &\Delta u = 0 \quad \text{in the fluid} \\ &u = c,\ \frac{\partial u}{\partial \nu} = \lambda \quad \text{on the free boundary} \end{aligned} \tag{1}$$

where c,λ are constants. If we take gravity into account, then λ is replaced by $\sqrt{a + gy}$ $(a > 0,\ g > 0)$ where the gravitational force is in the upward direction.

In addition to (1) we must impose boundary conditions

$$u = u_0 \quad \text{or} \quad \frac{\partial u}{\partial \nu} = u_1 \tag{2}$$

on the fixed boundary

as well as a condition at infinity. For example

(i) for a symmetric jet flow from a nozzle ℓ we have:

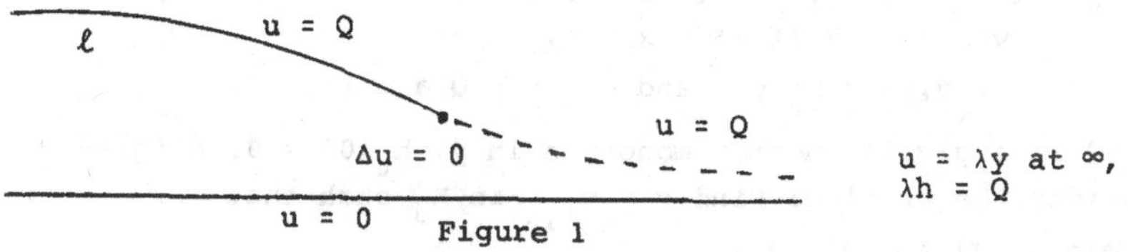

Figure 1

where h is the asymprotic height of the free boundary as $x \to \infty$;

(ii) for a symmetric cavitational flow with nose ℓ we have:

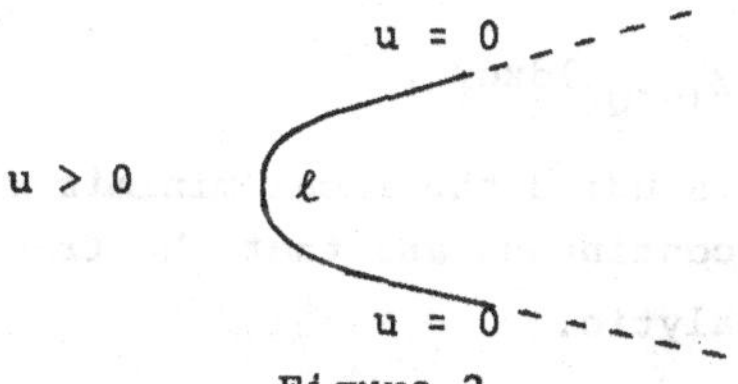

Figure 2

$u = y(1 + o(1))$. $\nabla u = \vec{e}(1 + o(1))$ where $\vec{e} = (0,1)$ and $o(1) \to 0$ if $x^2 + y^2 \to \infty$.

Problems such as (i), (ii) have been solved by several methods over the last 100 years. The general procedure has been to use conformal mappings or the hodograph transformation in order to reduce problems such as (i), (ii) to nonlinear integral equations (of a rather complicated type) and then apply the Leary-Schauder fixed point theorem; for details see [14][23][24] and the references in [12],[22]. Another approach based on a variational principle was developed in [19],[20].

In the last few years Alt, Caffarelli and Friedman have developed a new variational approach to establish existence of solutions for free boundary problems of general ideal fluids [2-4,8,9]. This work has also been extended to two fluids (flowing side-by-side)[2-4,8,9]. We shall explain the essence of the method in the simplest case (i)(Figure 1, above).

Consider the functional

$$J(v) = \int_{\Omega_\mu} |\nabla v - \lambda\vec{e}\chi_{\{v<Q\}}\chi_{E_\mu}|^2 dxdy, \quad v \in K_\mu$$

where

$$\ell : y = g(x), \ -\infty < x < 0, \quad g \text{ monotone } (b = g(0),\ A = (0,b)),$$

$$E_\mu = \{(x,y);\ -\mu < x < \infty,\ 0 < y < b\},\ \Omega_* = \{(x,y);\ 0 < y < g(x),\ -\infty < x \le 0\},$$

$$R^2_+ = \{(x,y);\ x > 0,\ y > 0\},$$

$$\Omega = \Omega_* \cup R^2_+,\ \Omega_\mu = \Omega \cap \{x > -\mu\},$$

$$K_\mu = \{v \in H^1(R^2),\ v = Q \text{ if } y \ge g(x),\ x < 0 \text{ or } y \ge b,\ x > 0;$$
$$v(x,0) = 0 \text{ if } -\infty < x < \infty,$$
$$v(-\mu,y) = h_\mu(y), \text{ and } 0 \le v \le Q \text{ a.e.}\};$$

here $h_\mu(y)$ is a given function monotone in y, $h_\mu(0) = 0$, $h_\mu(g(-\mu)) = Q$.

Consider the problem: Find $v = v_{\lambda,\mu}$ in K_μ such that

$$\min_{v\in K_\mu} J(v) = J(u_{\lambda,\mu}).$$

It is easily seen that this problem has a solution. The solution is harmonic in $\Omega_*\setminus E_\mu$ and is a local minimizer in E_μ of

$$\tilde{J}(v) \equiv \int(|\nabla v|^2 + \lambda^2\chi_{\{v<Q\}})dxdy.$$

Alt and Caffarelli [1] studied the local minimizer $\tilde{v}$ of $\tilde{J}$ and proved that $\tilde{v}$ is Lipschitz continuous and that the free boundary $\partial\{\tilde{v} < Q\} \cap E_\mu$ is locally analytic.

LEMMA 1. The minimizer is unique.

Indeed, suppose u_1, u_2 are two minimizers and introduce $u_1^\varepsilon(x,y) = u_1(x - \varepsilon, y)$ and

$$v_1 = u_1^\varepsilon \wedge u_2, \quad v_2 = u_1^\varepsilon \vee u_2 .$$

Denote by J^ε the functional $J = J_{\lambda,\mu}$ corresponding to the translation $x \to x + \varepsilon$ of Ω_μ, K_μ. One verifies that

$$J^\varepsilon(u_1^\varepsilon) + J(u_2) = J^\varepsilon(v_1) + J(v_2) ,$$

which implies that $J(u_2) = J(v_2)$, i.e., $u_1^\varepsilon \vee u_2$ is a minimizer. Consequently $u_1^\varepsilon \vee u_2$ is smooth, which implies that either $u_1^\varepsilon \geq u_2$ or $u_1^\varepsilon \leq u_2$ everywhere. Since $u_1^\varepsilon < u_2$ near the boundary, we deduce that $u_1^\varepsilon < u_2$ throughout the domain. Taking $\varepsilon \to 0$ we get $u_1 \leq u_2$, and similarly $u_2 \leq u_1$.

Taking $u_1 = u_2$ in the above argument we get:

$$\frac{\partial}{\partial x} u_{\lambda,\mu} \geq 0 .$$

Thus the analytic free boundary $\Gamma = \Gamma_{\lambda,\mu}$ has the form

$$\Gamma_{\lambda,\mu} : x = f_{\lambda,\mu}(y) .$$

Next we have:

LEMMA 2. $f_{\lambda,\mu}(y)$ is continuous and finite if and only if $h < y < b$, where $h = Q/\lambda$.

LEMMA 3. $\lambda \to f_{\lambda,\mu}(b)$ is continuous ($f_{\lambda,\mu}(b) = f_{\lambda,\mu}(b + 0)$).

LEMMA 4. If λ is sufficiently small then $f_{\lambda,\mu}(b) < 0$; if $\lambda < Q/b$ and $|\lambda - Q/b|$ is small enough, then $f_{\lambda,\mu}(b) > 0$.

From Lemmas 3, 4 we deduce that there is a value $\lambda = \lambda(\mu)$ such that $f_{\lambda,\mu}(b) = 0$, i.e., there is a "continuous fit" at A. Fro this value of λ, $(u_{\lambda,\mu}, \Gamma_{\lambda,\mu})$ "almost" solves the jet problem. In order to complete the construction of a solution we let $\mu \to \infty$, $\lambda(\mu) \to \lambda$ and denote the limiting $u_{\lambda,\mu}$, $\Gamma_{\lambda,\mu}$ by u, Γ.

LEMMA 5. Continuous fit implies smooth fit.

More precisely the curve $\ell \cup \Gamma$ is not only continuous at the point of detachment A, but it is also C^1 at A, and ∇u is uniformly C^1 in $\{u < Q\}$-neighborhood of A.

THEOREM. There exists a unique classical solution of the symmetric jet problem (i).

Existence was already outlined above; uniqueness is proved by a comparison argument [21].

The above procedure has been extended to three-dimensional axially symmetric jets [2], 2-dimensional asymmetric flows [3], to flows in a gravity field [4], to rotational flows [16] and to compressible fluids [8][9]; some cavity problems are treated in [13][18].

Two-fluid problems are treated in [5-7]. Here u^+ and u^- are harmonic and

$$\frac{\partial u^+}{\partial \nu} - \frac{\partial u^-}{\partial \nu} = \lambda \quad \text{on the free boundary.} \tag{3}$$

In a two-fluid flow in a porous media of a rectangular dam, the free boundary condition can be reduced to

$$\frac{\partial u^+}{\partial \nu} - \frac{\partial u^-}{\partial \nu} = \cos(x,\nu) \tag{4}$$

which is similar to (3); in (3) λ is not a priori given, whereas in (4) a degeneracy occurs at points where $\cos(x,\nu) = 0$. Problem (4) is studied in [10] where existence of a solution is proved having a C^1 free boundary.

Lemma 5 has been extended in [11] to quasilinear elliptic operators and to more general boundary conditions $\partial u/\partial \nu = f$. The assertion is that either $\Gamma \cup \ell$ is C^2 at A or it is only in $C^{3/2}$ and the curvature of Γ goes to $\pm\infty$ as $x \downarrow 0$.

Other physical flow problems lead to free boundary conditions as above. We mention the problem of freezing in a channel because of heat sink at the origin [25]. Thus

$$\Delta u = -M\delta \quad \text{in} \quad \{u > 0\}$$

where δ = Dirac measure, $-u$ is the temperature, and

$$u > 0 \quad \text{in the ice,}$$

$$\frac{\partial u}{\partial \nu} = \lambda \text{ on the free boundary;}$$

λ and M are given positive constants. Assuming that the channel Ω is symmetric with respect to teh y-axis it was recently proved by Friedman and Stojanovic [17] that the problem has a unique solution with free boundary concave to the ice. This implies that if $\partial\Omega$ consists of p curves ℓ_i convex to Ω then the free boundary consists of at most p arcs ("fingers") concave to Ω, each connecting an adjacent pair ℓ_i, ℓ_{i+1}.

References

[1] H.W. Alt and A.Caffarelli, *Existence and regularity for a minimum problem with free boundary*. J. Reine Angew. Math. 105 (1981), 105-144.

[2] H.W. Alt, L.A. Caffarelli and A. Friedman, *Axially symmetric jet flows*, Arch. Rat. Mech. Anal. 81 (1983), 97-149.

[3] H.W.Alt, L.A.Caffarelli and A. Friedman, *Asymmetric jet flows*, Comm. Pure Appl. Math. 35 (1982), 29 - 68.

[4] H.W.Alt, L.A.Caffarelli and A. Friedman, *Jet flows with gravity*, J. Reine Angew. Math. 331 (1982), 58-103,

[5] H.W.Alt, L.A. Caffarelli and A. Friedman, *Variational problems with two phases and their free boundaries*, Trans. Amer. Math. Soc., 282 (1984), 431-461.

[6] H.W.Alt, L.A.Caffarelli and A. Friedman, *Jets with two fluids I: one free boundary*, Indiana Univ. Math. J., 33 (1984), 213-247.

[7] H.W.Alt, L.A.Caffarelli and A. Friedman, *Jets with two fluids II: two free boundaries*, Indiana Univ. Math. J., 33 (1984), 367-391.

[8] H.W.Alt, L.A.Caffarelli and A. Friedman, *A free boundary problem for quasi-linear elliptic equations*, Ann. Sc. Norm. Sup. Pisa, 11 (1984), 1-44.

[9] H.W.Alt, L.A.Caffarelli and A. Friedman, *Compressible flows of jets and cavities*, J. Diff. Eqs., 56 (1985), 82-141.

[10] H.W.Alt, L.A.Caffarelli and A. Friedman, *The dam problem with two fluids*, Comm. Pure Appl. Math., 37 (1984), 601-646.

[11] H.W.Alt, L.A.Caffarelli and A. Friedman, *Abrupt and smooth separation of free boundaries in flow problems*, Scu. Norm. Sup. Pisa, to appear.

[12] G.D. Birkhoff and E.H. Zarantonello, *Jets, wakes and cavities*, Academic Press, New York, 1957.

[13] L.A. Caffarelli and A. Friedman, *Axially symmetric in finite cavities*, Indiana Univ. Math. J. 30 (1982), 135-160.

[14] R. Finn, *Some theorems on discontinuity of plane fluid motion*, J.D'Analyse Math. 4 (1954/6), 246-291.

[15] A. Friedman, *Variational principles and free boundary problems*, John Wiley, New York, 1982.

[16] A. Friedman, *Axially symmetric cavities in rotational flows*, Comm. in P.D.E., 8 (1983), 949-997.

[17] A. Friedman and S. Stojanovic, *A free boundary problem associated with icing in a chanel*, to appear.

[18] A. Friedman and T.I. Vogel, *Cavitational flow in a chanel with oscillatory wall*, Nonlinear Analysis, 7 (1983), 1157-1192.

[19] P.R. Garabedian, H. Lewy and M. Schiffer, *Axially symmetric cavitational flow*, Ann. of Math. 56 (1952), 560-602.

[20] P.R.Garabedian and D.C. Spencer, *Extremal methods in cavitational flow*, J. Rat. Mech. Anal. 1 (1952), 350-409.

[21] D. Gilbarg, *Uniqueness of axially symmetric flows with free boundaries*, J. Ra. Mech. Anal. 1 (1952), 309-320.

[22] D. Gilbarg, *Jets and cavities*, in Handbuch der Physik, vol. 9, Springer-Verlag New York, 1960.

[23] J. Leray, *Les Problèmes de représentation conforme de Helmholtz I, II*, Comm. Math. Helv. 8 (1935), 149-180; 250-263.

[24] J. Serrin, *Existence theorems for some hydrodynamical free boundary porblems*, J. Rat. Mech. Anal. 1 (1952), 1-48.

[25] L. Ting, D.S. Ahluwalia and M.J. Miksis, *Solutions of a class of mixed free boundary problems*, SIAM J. Appl. Math., 43 (1983), 759-775.

METHOD OF ROTHE IN EVOLUTION EQUATIONS

J. KAČUR
Institute of Applied Mathematics, Comenius University
Mlynská dolina, 842 15 Bratislava, Czechoslovakia

The aim of this paper is to present Rothe's method (also called method of lines , or the method of semidiscretization) as an efficient theoretical tool for solving a broad scale of evolution problems . Using time discretization , evolution problems are approximated by corresponding elliptic problems by means of which an approximate solution for the original evolution problem is constructed . By a relatively simple technique convergence of the approximate solution to the solution of the original evolution problem is proved . Thus , unlike some abstract methods for the analysis of existence and uniqueness problems for evolution equations , Rothe's method has a strong numerical aspect . At the same time it gives a first good insight into the structure of the solution of the investigated evolution problems .

Rothe's method introduced by E.Rothe in 1930 has been used and developed by many authors , e.g. O.A. Ladyženskaja ; T.D. Ventcel' ; A.M. Il-jin , A.S. Kalašnikov , O.A. Olejnik ; Š.J. Ibragimov ; P.P. Mosolov ; K. Rektorys[10] in linear and quasilinear parabolic problems . Nonlinear and abstract parabolic problems has been studied by J. Kačur[2]-[6]; J. Nečas[9] ; A.G. Kartsatos , E.M. Parrott[7],[8] etc. Linear and quasilinear hyperbolic equations has been considered by J. Jerome ; E. Martensen ; M. Pultar ; J. Kačur , etc. A modification of Rothe's method (discretization in x- variable) has been used by V.N. Faddeeva ; W. Walter ; C. Corduneanu , etc. Time and space discretization for solving evolution problems has been employed by many authors ,e.g. R. Glowinski , J.L. Lions , R. Trémoliéres[1] ; M. Zlámal[11] ; A. Ženíšek[12]; H.W. Alt , S. Luckhaus etc. , using similar technique to that of Rothe's method . For the more complete references we refer the reader to [6] .

Efficiency of Rothe's method we demonstrate in solving :

I. A nonlinear parabolic problems ;
II. Variational inequalities ;
III. Higher order equations .

I. A nonlinear parabolic problems.

Let V be a reflexive B-space with its dual V^* and let H be a Hilbert space. Let $||.||$, $|.|$ be the norms in V, H, respectively. We assume that $V \cap H$ is a dense set in V and H. By $\langle f,v\rangle$ we denote the duality for $f \in V^*$ and $v \in V$. Scalar product in H we denote by $(.,.)$. Let S_t be the interval $[-q,t]$ for $t \in [0,T] \equiv I$, $q \geq 0$.

An operator $F : L_\infty(S_T,H) \to L_\infty(S_T,H)$ is a Volterra operator iff $u(s) = v(s)$ a.e. in S_t implies $F(u)(t) = F(v)(t)$ for any $t \in S_T$. We assume $A : V \to V^*$ to be a coercive maximal monotone operator. Consider the equation

$$(1.1)\qquad \frac{du(t)}{dt} + Au(t) = f(t,F(u)(t)) \quad \text{a.e. } t \in I\ ,\ u = \phi \ \text{ in } S_o$$

where $\phi : S_o \to H$ is a given Lipschitz continuous function ($\phi \in Lip(S_o \to H)$) and $f \in Lip(I \times H \to H)$. Coerciveness of A we assume in the form

$$(1.2)\qquad \langle Au,u\rangle \geq ||u||p(||u||) - C_1|u|^2 - C_2 \qquad \forall\, u \in V$$

where $p : R_+ \to R_+$ and $p(s) \to \infty$ for $s \to \infty$. Lipschitz continuity of f is expressed in the form

$$(1.3)\qquad |f(t,v) - f(t',v')| \leq C(|t-t'| + |t-t'||v| + |v-v'|)\ \forall\, t,t' \in I,$$

$\forall\, v,v' \in H$. We assume that F maps $Lip(S_T \to H)$ into $Lip(S_T \to H)$ and

$$(1.4)\qquad ||F(u) - F(v)||_{C(S_T,H)} \leq C||u-v||_{C(S_T,H)}$$

$$(1.5)\qquad |F(u)(t) - F(u)(t')| \leq |t-t'|L(||u||_{C(S_T,H)})(1 + ||\frac{du}{dt}||_{L_\infty(S_t,H)})$$

$\forall\ t,t' \in S_T$, $t' < t$ and $u \in Lip(S_T \to H)$ where $L:R_+ \to R_+$ is cont.f.

Solving (1.1) we apply Rothe's method in the following way: Let n be a positive integer, $h = Tn^{-1}$, $t_i = ih$. Successively for $i=1,\dots,n$ we look for the solution $u_i \equiv u_{i,n} \in V \cap H$ of the elliptic equation

$$(1.6)\qquad \left(\frac{u-u_{i-1}}{h},v\right) + \langle Au,v\rangle = (f(t_i,F(\tilde{u}_{i-1})(t_i)),v) \qquad \forall\, v \in V \cap H$$

where $u_o = \phi(0)$ and $\tilde{u}_{i-1} \in Lip(S_T \to H)$ is defined by

$$\tilde{u}_{i-1} \equiv \tilde{u}_{i-1,n} = \begin{cases} \phi & \text{on } S_o \\ \phi(0) & \text{on } [0,h] \\ u_{j-1} + (t-t_{j-1})h^{-1}(u_j-u_{j-1}), & t_j \leq t \leq t_{j+1} \end{cases}$$

$$\begin{cases} & \text{for } j=1,\dots,i-1 \\ u_{i-1} & \text{on } [t_i, T] \end{cases}.$$

The existence of $u_i \in V \cap H$ is assured by the following argument. The element $u_{i-1}h^{-1} + f(t_i, F(\tilde{u}_{i-1})(t_i))$ is in H and the operator $A_h : V \cap H \to V^* + H$ defined by $[A_h u, v] = \frac{1}{h}(u,v) + \langle Au, v\rangle$ is a coercive maximal monotone . Hence theory of monotone operators guarantee the existence of u_i . Uniqueness of u_i is a consequence of strict monotonicity of A_h . By means of $u_i \equiv u_{i,n}$ we construct Rothe´s function $u_n(t)$ and the corresponding step function $\bar{u}_n(t)$

(1.7) $u_n(t) = u_{i-1} + (t-t_{i-1})h^{-1}(u_i - u_{i-1}),\ t_{i-1} \le t \le t_i$, $i=1,\dots,n$

(1.8) $\bar{u}_n(t) = u_i$ for $t_{i-1} < t \le t_i$, $i=1,\dots,n$, $\bar{u}_n(0) = u_o$.

Then (1.6) can be rewritten in the form

(1.9) $\left(\frac{du_n(t)}{dt}, v\right) + \langle A\bar{u}_n(t), v\rangle = (f_n(t, F(\tilde{u}_{n-1})(t)), v) \quad \forall\, v \in V \cap H$

where $f_n(t,v) = f(t_i, v)$ for $t_{i-1} < t \le t_i$, $i=1,\dots,n$. First , we prove a priori estimates for $\{u_n\}$ (see Lemmas 1,2,3) and then we take the limit as $n \to \infty$ in (1.9) . We obtain

Theorem 1. Let $A : V \to V^*$ be maximal monotone and let $A\phi(0) \in H$. If (1.2) - (1.5) are satisfied then there exists the unique solution u of (1.1) in the following sense : $u \in L_\infty(I,V)$, $u \in Lip(I \to H)$, $\frac{du}{dt} \in L_\infty(I,H)$ and $Au \in L_\infty(I,H)$. Moreover , the estimate

$$||u_n - u||^2_{C(I,H)} \le \frac{C}{n}$$

takes place where $\{u_n\}$ is from (1.7).

A priori estimates we obtain in the following way.

Lemma 1. The estimate $|u_i| \le C$ takes place for all n , $i=1,\dots,n$.

Proof. We put $u = u_i$, $v = hu_i$ into (1.6) . We sum it up for $i=1,\dots,j$. Using (1.2) -(1.4) we estimate

$$|u_j|^2 \le C_1 + C_2 \sum_{i=1}^{j} \max_{1 \le k \le i} |u_k|^2 h$$

and hence

$$\max_{1\le k\le j} |u_k|^2 \le C_1 + C_2 \sum_{i=1}^{j} \max_{1\le k\le i} |u_k|^2 h \quad .$$

Thus Gronwall's Lemma implies the required result.

Lemma 2. The estimates

(1.10) $$\left|\frac{u_i-u_{i-1}}{h}\right| \le C \, , \quad \|u_i\| \le C$$

hold for all n , $i=1,\dots,n$.

Proof. We subtract (1.6) for $u=u_j$, $v=\delta_h u_j \equiv \frac{u_j-u_{j-1}}{h}$ from (1.6) for $u=u_{j-1}$, $v=\delta_h u_j$. Owing to the monotonicity of A and (1.3) -(1.5) we obtain

(1.11) $$|\delta_h u_j| \le |\delta_h u_{j-1}| + C(h + \max_{1\le k\le j} |\delta_h u_k| h)$$

because of Lemma 1 . From (1.6) for $i=1$, $u=u_1$, $v=\delta_h u_1$ we conclude

$$|\delta_h u_1| \le C$$

since $u_o=\phi(0)$ and $Au_o \in H$. Thus successively from (1.11) we obtain

$$|\delta_h u_j| \le C_1 + C_2 \sum_{i=1}^{j} \max_{1\le k\le i} |\delta_h u_k| \, h$$

which similarly as above implies $|\delta_h u_i| \le C$. Using this estimate and (1.10) in (1.6) for $u=u_i$, $v=u_i$ we obtain $\|u_i\| \le C$ because of (1.2) .

As a consequence of (1.10) and (1.6) for $u=u_i$ we have

$$|\langle Au_i,v\rangle| \le C|v| \qquad \forall\, n \, , \; i=1,\dots,n \, .$$

The previous a priori estimates can be rewritten in the form

(1.12) $$\left|\frac{du_n(t)}{dt}\right| \le C \, , \; \|u_n(t)\|_{V\cap H} \le C \, , \; |\langle A\bar{u}_n(t),v\rangle| \le C|v|$$

(1.13) $$|u_n(t) - u_n(t')| \le C|t - t'| , \quad |u_n(t) - \bar{u}_n(t)| \le \frac{C}{n} \quad .$$

Lemma 3. There exists an $u \in L_\infty(I,V)$, $u \in Lip(I \to H)$ with $\frac{du}{dt} \in L_\infty(I,H)$ such that

$$\|u_n-u\|^2_{C(I,H)} \le \frac{C}{n} \, , \quad \frac{du_n}{dt} \rightharpoonup \frac{du}{dt} \text{ in } L_2(I,H) \text{ and } A\bar{u}_n(t) \rightharpoonup Au(t) \text{ in } V^*$$

(also in H) $\forall\, t \in I$.

Proof. Subtract (1.9) for $n=r$ from (1.9) for $n=s$ where $v = \bar{u}_r(t) - \bar{u}_s(t)$. Using (1.12) and (1.13) we estimate

$$(1.14)\qquad \frac{d}{dt}|u_r-u_s|^2 \le C\Big(\frac{1}{r} + \frac{1}{s} + ||\tilde{u}_{r-1}-u_r||^2_{C(S_t,H)} + ||\tilde{u}_{s-1}-u_s||^2_{C(S_t,H)} + ||u_r-u_s||^2_{C(S_t,H)}\Big).$$

Integrating (1.14) over $(0,t)$ and taking into account the estimate

$$||\tilde{u}_{r-1} - u_r||^2_{C(S_t,H)} \le \frac{C}{r^2} \sup_{\tau\in[0,t]} \Big|\frac{du_r(\tau)}{d\tau}\Big|^2 \le \frac{C}{r^2}$$

we conclude $u_n \to u$ in $C(I,H)$ and the estimate $||u_n-u||^2_{C(I,H)} \le \frac{C}{n}$.
Hence and from (1.12), (1.13) we obtain
$\bar{u}_n(t) \rightharpoonup u(t)$, $||A\bar{u}_n(t)||_* \le C$, ($|A\bar{u}_n(t)| \le C$) and

$$\langle A\bar{u}_n(t), \bar{u}_n(t) - u(t)\rangle \to 0 \quad \forall\, t\in I .$$

Hence maximal monotonicity of A implies $A\bar{u}_n(t) \rightharpoonup Au(t)$ in V^* (moreover in H).

Proof of Theorem 1. We integrate (1.9) over (τ_1,τ_2) and take the limit as $n \to \infty$. Owing to Lemma 3 we conclude that u is a solution of (1.1) since $\tau_1,\tau_2 \in I$ are arbitrary. Uniqueness follows from (1.1) by standard arguments.

Remark 1. Theorem 1 holds true also when $A : V \to V^*$ is nonstationary under the following assumptions:

$A(t) : V \to V^*$ is maximal monotone $\forall\, t\in I$;

$A(t)u = \nabla\phi(t,u)$, i.e. $A(t)$ are potential (ϕ: $I\times V \to R$)

$\langle A(t)u,u\rangle \ge ||u||p(||u||)$, $p(t) \to \infty$ for $t \to \infty$

$$||\frac{d}{dt} A(t)u||_* + ||\frac{d^2}{dt^2} A(t)u||_* \le C_1 + C_2p(||u||) .$$

For the proof it suffices to combine the techniques used in the proof of Theorem 1 with those used in [3].

Remark 2. A modification of Theorem 1 with m-accretive operators $A(t) : D \subset V \to V^*$ $(t\in I)$ satisfying

$$||A(t)v - A(t')v|| \le |t-t'|L(||u||)(1 + ||A(t)v||)$$

and with the right hand side $f = G(t,u_t)$ (at fixed t the operator G transforms the values of $u_t(s) = u(t + s)$, s $[-q,0]$ into H) satisfying Lipschitz like condition has been obtained by A.G. Kartsatos and M.E. Parrott in [7].

II. Variational inequalities.

Let Φ be a proper ($\Phi: V \to (-\infty,\infty]$, $\Phi \not\equiv \infty$), convex and lower semi-continuous (l.s.c.) function on V . We assume $A : V \to V^*$ to be a bounded maximal monotone operator . Consider the variational inequality

$$(2.1) \qquad (\frac{du(t)}{dt}, v-u(t)) + \langle Au(t), v-u(t)\rangle + \Phi(v) - \Phi(u(t)) \geq$$
$$(f(t,F(u)(t)), v-u(t)) \quad \forall\, v \in V \cap H \text{ , a.e. } t \in I \text{ , } u = \phi \text{ on } S_o$$

where ϕ and F are the same as in Section I. We use the approximation scheme

$$(2.2) \qquad (\delta_h u_i, v-u_i) + \langle Au_i, v-u_i\rangle + \Phi(v) - \Phi(u_i) \geq (f(t_i, F(\tilde{u}_{i-1})(t_i)), v-u_i) \text{ , } \forall\, v \in V \cap H$$

where $\tilde{u}_{i-1}$ is the same as in Section I . It is elliptic variational inequality with respect to u_i provided $u_1,\dots,u_{i-1}$ are known . Coerciveness of A is assumed in the form : There exists $v_o \in V$ with $\Phi(v_o) < \infty$ such that

$$(2.3) \qquad (\langle Au, u-v_o\rangle + \Phi(u)).||u||^{-1} \to \infty \quad \text{for} \quad ||u|| \to \infty \text{ .}$$

Then , existence of $u_i \in V \cap H$ satisfying (2.2) is guaranteed by the well-known results from elliptic variational inequalities . Here , instead of $Au_o \in H$ we assume : There exists $z_o \in H$ such that

$$(2.4) \qquad (z_o, v-u_o) + \langle Au_o, v-u_o\rangle + \Phi(v) - \Phi(u_o) \geq (f(0, F(\tilde{u}_o)(0)), v-u_o)$$
$$\forall\, v \in V \cap H \quad \text{where} \quad u_o = \phi(0)$$

Since we have less possibilities in the test function v than in Section I , we assume

$$(2.5) \qquad \begin{cases} \text{either} \quad \Phi(0) < \infty \quad \text{, or} \\ |F(u)(t) - F(u)(t')| \leq C|t-t'| \; ||\frac{du}{dt}||_{L_\infty(S_t,H)} \end{cases}$$

for $u \in Lip(S_T \to H)$. Let us put $i=j$, $v=u_{j-1}$ into (2.2) and then $i=j-1$, $v=u_j$. Adding these inequalities the values with Φ are eliminated and we are in the same situation as in the case of equations . Thus , we obtain the same a priori estimates (except of $|\langle Au_i, v\rangle| \leq C|v|$) as in Section I . Since $\bar{u}_n(t) \rightharpoonup u(t)$ we have $\Phi(u(t)) \leq \liminf \Phi(\bar{u}_n(t))$ (Φ is also weakly l.s.c. on V). From this information and from

$$(2.6) \qquad \langle A\bar{u}_n(t), \bar{u}_n(t)-v\rangle \leq \Phi(v) - \Phi(\bar{u}_n(t)) + (\frac{du_n(t)}{dt}, v-\bar{u}_n(t)) -$$
$$(f(t, F(\tilde{u}_{n-1})(t)), v-\bar{u}_n(t))$$

for $v=u(t)$ we conclude $\limsup \langle A\bar{u}_n(t),\bar{u}_n(t)-u(t)\rangle \le 0$. Since A is a pseudomonotone operator we obtain

$$\langle Au(t),u(t)- v \rangle \le \liminf \langle A\bar{u}_n(t),\bar{u}_n(t) - v\rangle \quad \forall v\in V .$$

Then integrating (2.6) and taking $\liminf_{n\to\infty}$ in (2.6) we obtain the solution of (2.1) by the same arguments as in Section I .

Theorem 2. Let $A : V \to V^*$ be a bounded maximal monotone operator . If (1.3) , (1.4), (2.3)-(2.5) are satisfied then there exists the unique solution of (2.1) with the same properties as in Theorem 1 .

Remark 3. Theorem 2 holds true also in the case of the operator A being nonstationary under the assumptions of Remark 1 .

Similarly , the following types of evolution variational inequalities can be solved

a) $$\left(\frac{d^2u(t)}{dt^2},v-\frac{du(t)}{dt}\right) + b\left(t;\frac{du(t)}{dt},v-\frac{du(t)}{dt}\right) + a\left(t;u(t),v-\frac{du(t)}{dt}\right) +$$
$$\Phi(v) - \Phi\left(\frac{du(t)}{dt}\right) \ge \left(f(t),v-\frac{du(t)}{dt}\right) \qquad u(0) = U_o , \ \frac{du(0)}{dt} = U_1 ;$$

b) $$\left(\frac{du(t)}{dt},v-\frac{du(t)}{dt}\right) + a\left(t;u(t),v-\frac{du(t)}{dt}\right) + \Phi(v) - \Phi\left(\frac{du(t)}{dt}\right) \ge$$
$$\left(f(t),v-\frac{du(t)}{dt}\right) \qquad u(0) = U_o ;$$

c) $$a\left(t;u(t),v-\frac{du(t)}{dt}\right) + \Phi(v) -\Phi\left(\frac{du(t)}{dt}\right) \ge \left(f(t),v-\frac{du(t)}{dt}\right) ,u(0)=U_o$$
$\forall \ v\in V\cap H$, a.e. $t\in I$.

Here V , H are Hilbert spaces and $b(t;u,v)$, $a(t;u,v)$ are continuous bilinear forms in $u,v\in V$ $(t\in I)$. We use the approximation scheme

a_1) $$\frac{1}{h}(\delta_h u_i-\delta_h u_{i-1},v-\delta_h u_i) + b(t_i;\delta_h u_i,v-\delta_h u_i) + a(t_i;u_i,v-\delta_h u_i) +$$
$$\Phi(v) - \Phi(\delta_h u_i) \ge (f(t_i),v-\delta_h u_i)$$

b_1) $$(\delta_h u_i,v-\delta_h u_i) + a(t_i;u_i,v-\delta_h u_i) + \Phi(v) - \Phi(\delta_h u_i) \ge (f(t_i),v-\delta_h u_i)$$

c_1) $$a(t_i;u_i,v-\delta_h u_i) + \Phi(v) - \Phi(\delta_h u_i) \ge (f(t_i),v-\delta_h u_i)$$

$\forall \ v\in V\cap H$, $i=1,\dots,n$. If we express $u_i = u_o + \delta_h u_i h + \sum_{j=1}^{i-1} \delta_h u_i h$ and $a(t_i;u_i,v) = ha(t_i;\delta_h u_i,v) + \sum_{j=1}^{i-1} ha(t_i;\delta_h u_j,v) + a(t_i;u_o,v)$ in a_1), b_1), c_1) then we obtain the elliptic variational inequalities with respect to $\delta_h u_i$. We shall assume

(2.7) $a(t;u,v) = a(t;v,u)$;

(2.8) $a(t;u,u) + \alpha|u|^2 \geq C||u||^2$ $(\alpha \geq 0)$;

(2.9) $b(t;u,u) \geq -C|u|^2$;

(2.10) $\left|\frac{d^p}{dt^p} a(t;u,v)\right| \leq C||u|| \, ||v||$ (p=1,2 in the cases a),b) p=1 in the case c)) ;

(2.11) $\left|\frac{d}{dt} b(t;u,v)\right| \leq C||u|| \, ||v||$

We assume that there exist $s_o \in H$, $z_o \in H$ such that

(2.12)$_a$ $(s_o, v-U_1) + b(0;U_1,v-U_1) + a(0;U_o,v-U_1) + \Phi(v) - \Phi(U_1) \geq (f(0),v-U_1)$;

(2.12)$_b$ $(z_o, v-z_o) + a(0;v-z_o) + \Phi(v) - \Phi(z_o) \geq (f(0),v-z_o)$;

(2.12)$_c$ $a(0;U_o,v-z_o) + \Phi(v) - \Phi(z_o) \geq (f(0),v-U_o)$

$\forall\, v \in V \cap H$. By the same way as above (see also [5]) we obtain

Theorem 3. If (2.7) - (2.12) are satisfied and if $f, \frac{df}{dt}, \frac{d^2f}{dt^2} \in L_2(I,V^*)$ (or $f, \frac{df}{dt} \in L_2(I,H)$ in a), b)) then there exists the unique solution of a), b) , c) , respectively , with the following properties :

$$u \in C(I,V) \;,\; \frac{du}{dt} \in L_\infty(I,V) \;,\; ||u_n - u||^2_{C(I,V)} \leq \frac{C}{n}$$

where $\{u_n\}$ is the corresponding sequence of Rothe's function . Moreover , in the cases a) , b) we have

$$\frac{du}{dt} \in C(I,H) \;,\; \frac{d^2u}{dt^2} \in L_\infty(I,H) \;,\; \left|\frac{du_n(t)}{dt} - \frac{du(t)}{dt}\right|^2 \leq \frac{C}{n} \qquad \forall\, t \in I \;.$$

Remark 4. In the fact in the cases a) , b) a perturbed symmetry of $a(t;u,v)$ can be assumed . Let $a_o(t;u,v)$ be continuous bilinear form in $u,v \in V$ $(t \in I)$ satisfying $|a_o(t;u,v)|$ $C||u|| \, |v|$. It suffices to assume $a(t;u,v) + a_o(t;u,v)$ is symmetric .

Remark 5. In the cases of variational inequalities a) , b) a more general problem (corresponding to problem (2.1)) with a right hand side $f(t,F(u)(t))$ can be considered . If (1.3),(1.4),(2.5) are satisfied then Theorem 3 holds true .

Remark 5. Using time and space discretization the variational inequalities a) , b) have been solved in [1]. A special case of (2.1) (A is asymptotically linear , $\Phi \equiv \Phi_K$ - indicatrix of the closed convex set K in V) have been solved in [12] .

III. Higher oder evolution equations.

In this section we apply Rothe's method to the equations of the form

$$(3.1)\quad G(t)\,\frac{d^m w(t)}{dt^m} + \sum_{k=0}^{m-1} A_k(t)\,\frac{d^k w(t)}{dt^k} = g(t,w,\dots,\frac{d^{m-1}w}{dt^{m-1}})$$

$$\frac{d^k w(0)}{dt^k} = W_k\ ,\ k=0,\dots,m-1\ ,\quad \text{where}\ A_k(t)\in\mathcal{L}(V,V^*),\ G(t)\in\mathcal{L}(H,H)$$

$(\in\mathcal{L}(V,V^*))$, $g\in \mathrm{Lip}(I\times[V]^m\to H)$ and V, H being Hilbert spaces with $V\cap H$ dense in V and H. The equations of type (3.1) include the governing equations of quasistatic and dynamic problems of viscoelastic plates and shallow shells (see [13]). We assume that either i) $A_{m-1}(t)$ is V-elliptic, or ii) $A_{m-2}(t)$ is V-elliptic. Operator $G(t)$ is supposed to be symmetric and H-elliptic. Using transformation

$$u = \frac{d^{m-1}w}{dt^{m-1}}\ \text{in the case i), or}\ u = \frac{d^{m-2}w}{dt^{m-2}}\ (m\ge 2)\ \text{in the case ii)}$$

the equation (3.1) can be reduced to the form

$$E)_i\quad G(t)\,\frac{du(t)}{dt} + A(t)\,u(t) = f(t,u,F(u)(t))\quad \text{or}$$

$$E)_{ii}\quad G(t)\,\frac{d^2u(t)}{dt^2} + B(t)\,\frac{du(t)}{dt} + A(t)u(t) = f(t,u,\frac{du}{dt},F(u)(t))$$

where

$$(3.2)\quad F(u)(t) = (\int_0^t u\,ds,\dots,\int_0^t (t-s)^p\,u(s)\,ds)\quad (p=m-2\ \text{in i)},\ p=m-3\ \text{in ii)})$$

The problem $E)_i$ has been considered in Section I. Now, we formulate Problem 3.1 which includes the problem $E)_{ii}$.

Let V, V_1; H, H_1 be Hilbert spaces and let $\langle u,v\rangle_V$, $\langle x,y\rangle_H$ be the continuous pairings between $u\in V_1$, $v\in V$ and $x\in H_1$, $y\in H$, respectively. Let $a(t;u,v)$, $b(t;u,v)$ be the same as in Section I and let $G(t;u,v)$ be a continuous bilinear form for $u,v\in H$. Consider the operators $f_V\in \mathrm{Lip}(I\times V\to V_1)$, $f_H\in \mathrm{Lip}(I\times V\times H\to H_1)$ and Volterra type operators F_V: $\mathrm{Lip}(S_T\to V)\to \mathrm{Lip}(S_T\to V)$, F_H: $\mathrm{Lip}(S_T\to H)\to \mathrm{Lip}(S_T\to H)$.

<u>Problem 3.1.</u> To find $u\in C(I,V\cap H)$ with $\frac{du}{dt}\in L_\infty(I,V\cap H)$, $\frac{du}{dt}\in C(I,H)$ $\frac{d^2u}{dt^2}\in L_\infty(I,H)$ such that

$$(3.3)\quad G(t;\frac{d^2u(t)}{dt^2},v) + b(t;\frac{du(t)}{dt},v) + a(t;u(t),v) =$$

$$\langle f_V(t,F(u)(t)),v\rangle_V + \langle f_H(t,F_V(u)(t),F_H(\tfrac{du}{dt})(t)),v\rangle_H$$

holds for all $v\in V\cap H$ and $u=\phi$, $\frac{du}{dt}=\psi$ on S_0 where $\phi\in \mathrm{Lip}(S_0\to V)$, $\psi\in \mathrm{Lip}(S_0\to H)$ are given functions.

To solve Problem 3.1 we use the approximation scheme

$$(3.4)\quad \frac{1}{h}\, G(t_i;\delta_h u_i-\delta_h u_{i-1},v) + b(t_i;\delta_h u_i,v) + a(t_i;u_i,v) =$$
$$\langle f_V(t_i,F(\tilde u_{i-1})(t_i)),v\rangle_V + \langle f_H(t_i,F_V(\tilde u_{i-1})(t_i),F_H(\delta_h\tilde u_{i-1})(t_i)),v\rangle_H$$

$\forall\, v\in V\cap H$ where $\tilde u_{i-1}$ is the same as in Section I and $\delta_h\tilde u_{i-1}$ is constructed by means of ψ, $\delta_h u_1,\dots,\delta_h u_{i-1}$ by the same way as $\tilde u_{i-1}$. Similarly as above (3.4) can be transformed to the elliptic equation with respect to $\delta_h u_i$ provided $\delta_h u_1,\dots,\delta_h u_{i-1}$ are known .

The solution of Problem 3.1 and the convergence of our approximation scheme we obtain under the following assumptions

(3.5) $G(t;u,v) = G(t;v,u)$;

(3.6) $C_1|u|^2 \le G(t;u,u) \le C_2|u|^2$;

(3.7) $|\frac{d}{dt}\, G(t;u,v)| \le C\,|u|\,|v|$;

(3.8) $\|F_R(u) - F_R(v)\|_{C(S_T,R)} \le C\|u - v\|_{C(S_T,R)}$ for $R = V$, H ;

(3.9) $\|F_R(u)(t)-F_R(u)(t')\|_R \le |t-t'|\,L(\|u\|_{C(S_T,R)})(1+\|\frac{du}{dt}\|_{L_\infty(S_t,R)})$

where $L: R_+\to R_+$ is continuous , $t,t'\in I$, $t'<t$ and $F(u)$ is from (3.2). Analogously to (2.12) we assume : There exists $s_o\in H$ such that

$$(3.10)\quad G(0;s_o,v) + b(0;\psi(0),v) + a(0;\phi(0),v) = \langle f_V(0,0),v\rangle_V +$$
$$\langle f_H(0,F_V(\tilde\phi)(0),F_H(\tilde\psi)(0)),v\rangle_H \quad \forall\, v\in V\cap H .$$

Theorem 4. Suppose $f_V\in \mathrm{Lip}(I\times V\to V_1)$, $f_H(I\times V\times H\to H_1)$ (see (1.3)) and $\psi(0)\in V$. If (3.5) -(3.10) are satisfied then there exists the unique solution of Problem 3.1 . Moreover, the estimates

$$(3.11)\quad \|u_n - u\|^2_{C(I,V)} \le \frac{C}{n}\ , \quad \|\frac{du_n}{dt} - \frac{du}{dt}\|^2_{L_\infty(I,H)} \le \frac{C}{n}$$

hold where $\{u_n\}$ is the corresponding sequence of Rothe's functions .

By a similar technique used in Sections I and II, successively we obtain a priori estimates

$$|u_i| \le C\ ,\ |\delta_h u_i| \le C$$

and then

$$|\delta_h^2 u_i| \le C\ ,\ \|\delta_h u_i\| \le C\ ,\ \|u_i\| \le C\ .$$

Similarly as in Lemma 3 a priori estimates (3.11) can be proved .Then taking the limit as $n\to\infty$ in approximation scheme (3.4) we conclude Theorem 4 .

Example . Problem 3.1 can be interpreted in the following way . We put $V = \mathring{W}_2^2(\Omega)$, $V_1 = W_2^{-2}$, $H = L_2(\Omega) = H_1$ where $\Omega \subset R^N$. Consider

$$a(t;u,v) = \sum_{|i|\ |j|\le 2} \int_\Omega a_{ij}(x,t)\, D^i u\, D^j v\, dx \quad \text{for } u,v \in \mathring{W}_2^2(\Omega) \;;$$

$$b(t;u,v) = \sum_{|i|\ |j|\le 2} \int_\Omega b_{ij}(x,t)\, D^i u\, D^j v\, dx \quad (\text{or } b(t;u,v) = \int_\Omega uv\, dx);$$

$$\langle f_V(t,F(u)(t)),v\rangle_V = \int_\Omega \Delta v \int_0^t (t-s)^p\, \Delta u(s)\, ds\, dx \quad (p \ge 1)\;;$$

$$\langle f_H(t,F_V(u)(t),F_H(\tfrac{du}{dt})(t)),v\rangle_H = \begin{cases} \int_\Omega v\, \Delta u(\omega(t))\, dx & \omega \in \mathrm{Lip}(S_T \to S_T),\ \omega(t) \le t\;; \\ \int_\Omega v \int_{-q}^{\omega(t)} K(s,t)\, \Delta u(s)\, ds\, dx\;; \\ \int_\Omega v \int_{-q}^{\omega(t)} K(s,t)\, \frac{du(s)}{ds}\, ds\, dx\;. \end{cases}$$

Bilinear form $G(t;u,v)$ can be interpreted in the following way .

1) $H = L_2(\Omega) = H_1$, $G(t;u,v) = \int_\Omega uv\, dx$.
Then the first term in (3.1) is of the form $\frac{d^m w}{dt^m}$;

2) $H = L_{2,\alpha}(\Omega) = \{u\ ;\int_\Omega \alpha u^2\, dx < \infty\}$, $H_1 = L_{2,\frac{1}{\alpha}}(\Omega)$, $\langle u,v\rangle_H = \int_\Omega uv\, dx$ where $\alpha(x) > 0$, $\alpha \in L_1(\Omega)$. We consider $C_1\alpha(x) \le g(x,t) \le C_2\alpha(x)$. Then $G(t;u,v) = \int_\Omega g(x,t)\, uv\, dx$ $(u,v \in L_{2,\alpha}(\Omega))$ generates a degenerate first term in (3.1) in the form $g(x,t)\,.\,\frac{d^m w}{dt^m}$;

3) $H = V = \mathring{W}_2^2$, $H_1 = V_1 = W_2^{-2}$. Then $G(t;u,v)$ generates the first term in (3.1) in the form $G(t)\,\frac{d^m w}{dt^m}$ where $G(t) \in \mathcal{L}(V,V^*)$ is a symmetric, V-elliptic operator .

References

[1] R. Glowinski, J.L. Lions, R. Tremoliéres : Analyse numérique des des inéquations variationnelles . Dunod, Paris 1976 .

[2] J. Kačur : The Rothe method and nonlinear parabolic equations of arbitrary order . Theory of nonlinear operators - Summer school-Neuendorf 1972 . Akademie-Verlag . Berlin 1974, 125 - 131 .

[3] — : Application of Rothe's method to nonlinear evolution equations . Mat. Časopis Sloven. Akad. Vied 25, 1975 , 63 - 81 .

[4] — : Method of Rothe and nonlinear parabolic boundary value problems of arbitrary order . Czech. Math. J., 28 103 , 1978

[5] — : On an approximate solution of variational inequalities . Math. Nachr. , 123 , 1985 , 63 - 82 .

[6] — : Method of Rothe in Evolution Equations . TEUBNER-TEXTE zur Mathematik , Leipzig , to appear .

[7] A.G. Kartsatos , M.E. Parrott : A method of lines for a nonlinear abstract functional differential equations . Trans . Am. Mth. Soc. V-286 , N-1 , 1984 , 73 - 91 .

[8] — : Functional evolution equations involving time dependent maximal monotone operators in Banach spaces . Nonlinear analysis Theory , Methods and Applications . Vol.8 , 1984 , 817-833 .

[9] J. Nečas : Applications of Rothe's method to abstract parabolic equations . Czech. Math. J. 24 , 1974 , 496-500 .

[10] K. Rektorys : On application of direct variational methods to the solution of parabolic boundary value problems of arbitrary order in the space variables . Czech. Math. J. ,21, 1971 , 318-339 .

[11] M. Zlámal : Finite element solution of quasistationary nonlinear magnetic fields . RAIRO , Anal. Num., V-16, 1982, 161-191 .

[12] A. Ženíšek : Approximation of parabolic variational inequalities . Aplikace matematiky, to appear .

[13] J. Brilla : New functional spaces and linear nonstationary problems of mathematical physics . EQUADIFF 5 - Proceedings of the conference held in Bratislava, 1981 . TEUBNER-TEXTE zur Mathematik , Band 47 , Leipzig , 1982 .

BOUNDARY VALUE PROBLEMS IN WEIGHTED SPACES

A. KUFNER
Mathematical Institute, Czechoslovak Academy of Sciences
115 67 Prague 1, Czechoslovakia

1. Introduction

Let us consider a linear differential operator of order $2k$ of the form

$$(Lu)(x) = \sum_{|\alpha|, |\beta| \leq k} (-1)^{|\alpha|} D^\alpha \left(a_{\alpha\beta}(x) D^\beta u(x)\right) , \quad x \in \Omega , \tag{1.1}$$

together with the associated bilinear form

$$a(u,v) = \sum_{|\alpha|, |\beta| \leq k} \int_\Omega a_{\alpha\beta}(x) D^\beta u(x) D^\alpha v(x) \, dx . \tag{1.2}$$

Here $a_{\alpha\beta}$ are given (real) functions defined on the domain $\Omega \in \mathbb{R}^N$.

The usual procedure for solving a boundary value problem for the operator L proceeds in the following fundamental steps:

(i) Choose an appropriate Banach space V such that the form $a(u,v)$ is *defined* and *continuous* on $V \times V$ and *elliptic* on V , i.e., that there exist constants $c_1 > 0$ and $c_0 > 0$ such that the following conditions are fulfilled:

$$|a(u,v)| \leq c_1 ||u||_V ||v||_V \quad \text{for every} \quad u, v \in V \tag{1.3}$$

(continuity of $a(u,v)$), and

$$a(u,u) \geq c_0 ||u||_V^2 \quad \text{for every} \quad u \in V \tag{1.4}$$

(ellipticity of $a(u,v)$).

(ii) Use the *Lax-Milgram Lemma* in order to obtain assertions about the existence and uniqueness of a (*weak*) solution in the space V .

If the coefficients $a_{\alpha\beta}$ of the operator L are *bounded*, i.e., if

$$a_{\alpha\beta} \in L^\infty(\Omega) \quad \text{for} \quad |\alpha|, |\beta| \leq k , \tag{1.5}$$

and if the operator L is (for simplicity) *uniformly elliptic*, i.e., if there exists a constant $c_0 > 0$ such that

$$\sum_{|\alpha|,|\beta|\leq k} a_{\alpha\beta}(x)\xi_\alpha\xi_\beta \geq c_0|\xi|^2 \tag{1.6}$$

for every $\xi \in \mathbb{R}^M$ ($\xi = \{\xi_\alpha, |\alpha| \leq k\}$), then the first step mentioned above can be realized if we choose for the space V the *Sobolev space* $W^{k,2}(\Omega)$ or one of its subspaces selected according to the type of the boundary conditions.

If one or both of the conditions (1.5) and (1.6) are violated, i.e., if operators with *singular coefficients* appear - condition (1.5) is not fulfilled - or if the operator becomes *degenerate* - the quadratic form on the left-hand side in (1.6) is only positive semidefinite - then the Sobolev spaces $W^{k,2}(\Omega)$ cannot be used in general. In these cases, appropriate *weighted Sobolev spaces* can be constructed which replace the classical spaces $W^{k,2}(\Omega)$. The *weight functions* appearing in these new spaces *are determined by the coefficients of the operator*, and the method of the proof of the corresponding existence and uniqueness theorem for weak solutions is the same as in the case of classical Sobolev spaces, the main tool being the Lax-Milgram Lemma.

On the other hand, there appear boundary value problems in which the operator L satisfies conditions (1.5) and (1.6) but the right-hand side in the equation $Lu = f$ or the right-hand sides in the boundary conditions $B_i u = g_i$, $i = 1,\dots,k$ (B_i being boundary operators) behave in such a way that the classical Sobolev spaces cannot be used: the function f is not an element of the dual space $(W^{k,2}(\Omega))^*$ or some of the functions g_i are not traces of functions from $W^{k,2}(\Omega)$ on the boundary $\partial\Omega$ of Ω . Also in such cases, weighted spaces can be sometimes used for obtaining assertions about existence and uniqueness of weak solutions. The bilinear form $a(u,v)$ is considered to be defined on a *weighted space* V or on a *product of two weighted spaces* $V_1 \times V_2$, and it is necessary to show for *which* such spaces conditions (1.3), (1.4) or their certain modifications are fulfilled.

In what follows, we give a survey of results obtained in these two directions of application of weighted Sobolev spaces to the solution of boundary value problems.

1.1. Definition of the weighted space. Let $1 < p < \infty$ and let Ω be a domain in $\mathbb{R}^N$. For $k \in \mathbb{N}$ and for N-dimensional multiindices α such that $|\alpha| \leq k$ let $\sigma_\alpha = \sigma_\alpha(x)$ be *weight functions*, i.e. measurable and a.e. in Ω positive functions, and let us denote $S = \{\sigma_\alpha, |\alpha| \leq k\}$. The weighted Sobolev space

$$W^{k,p}(\Omega;S) \tag{1.7}$$

is defined as the set of all functions $u = u(x)$, $x \in \Omega$, such that

$$||u||^p_{k,p,S} = \sum_{|\alpha| \leq k} \int_\Omega |D^\alpha u(x)|^p \sigma_\alpha(x) \, dx < \infty , \qquad (1.8)$$

the derivatives $D^\alpha u$ being considered in the sense of distributions. Further, let

$$W_0^{k,p}(\Omega;S) \qquad (1.9)$$

be the closure (if it is meaningful) of the set $C_0^\infty(\Omega)$ with respect to the norm (1.8).

<u>1.2. Theorem.</u> *Let us suppose that*

$$\sigma_\alpha^{-1/(p-1)} \in L^1_{loc}(\Omega) \quad \text{for} \quad |\alpha| \leq k . \qquad (1.10)$$

Then the linear set $W^{k,p}(\Omega;S)$ *is a* B a n a c h *space with respect to the norm* $||\cdot||_{k,p,S}$ *defined by* (1.8). - *If, moreover,*

$$\sigma_\alpha \in L^1_{loc}(\Omega) \quad \text{for} \quad |\alpha| \leq k , \qquad (1.11)$$

then the linear set $W_0^{k,p}(\Omega;S)$ *is a* B a n a c h *space with respect to the same norm.*

<u>1.3. Remark.</u> Conditions (1.10), (1.11) are rather restrictive. In the paper A. KUFNER, B. OPIC [2] it is shown how to modify the definition of the weighted spaces if (1.10) and/or (1.11) are not fulfilled. - The most frequent type of weight functions σ_α are the so called *power type weights*

$$\sigma_\alpha(x) = [\text{dist}(x,M)]^\varepsilon$$

with $M \subset \bar{\Omega}$ and $\varepsilon = \varepsilon(\alpha)$ real numbers. If M is a subset of the boundary $\partial\Omega$ of Ω , then conditions (1.10) and (1.11) are obviously satisfied.

<u>2. Operators with singular or degenerating coefficients</u>

Let us consider the operator L from (1.1) and let us suppose that its coefficients fulfil the following conditions:

$$a_{\alpha\alpha} > 0 \ \text{a.e. in} \ \Omega ; \quad a_{\alpha\alpha}, \ a^{-1}_{\alpha\alpha} \in L^1_{loc}(\Omega) , \quad |\alpha| \leq k ; \qquad (2.1)$$

there exist constants $c_1 > 0$, $c_0 > 0$ such that

$$|a_{\alpha\beta}(x)| \leq c_1\sqrt{a_{\alpha\alpha}(x)a_{\beta\beta}(x)} \quad \text{a.e. in } \Omega \tag{2.2}$$

for $|\alpha|, |\beta| \leq k$, $\alpha \neq \beta$;

$$\sum_{|\alpha|,|\beta|\leq k} a_{\alpha\beta}(x)\xi_\alpha\xi_\beta \geq c_0 \sum_{|\alpha|\leq k} a_{\alpha\alpha}(x)\xi_\alpha^2 \quad \text{a.e. in } \Omega \tag{2.3}$$

for all $\xi \in \mathbb{R}^M$.

Conditions (2.1) indicate that the weighted spaces $W^{k,2}(\Omega;S)$ and $W_0^{k,2}(\Omega;S)$ with $\sigma_\alpha = a_{\alpha\alpha}$, $|\alpha| \leq k$, i.e. with

$$S = \{a_{\alpha\alpha} = a_{\alpha\alpha}(x) \ , \ |\alpha| \leq k\} \tag{2.4}$$

are Banach spaces. From the following theorem we see that these weighted spaces are just the right tool for solving boundary value problems.

<u>2.1. Theorem.</u> *Let the operator* L *from* (1.1) *fulfil conditions* (2.1) - (2.3). *Let* S *be given by* (2.4); *let* $f \in (W_0^{k,2}(\Omega;S))^*$ *and* $u_0 \in W^{k,2}(\Omega;S)$. *Then there exists one and only one weak solution* $u \in W^{k,2}(\Omega;S)$ *of the Dirichlet problem for the equation* $Lu = f$, *i.e., such a function* u` *that*

$$u - u_0 \in W_0^{k,2}(\Omega) \tag{2.5}$$

and

$$a(u,v) = \langle f,v \rangle \quad \textit{for every} \quad v \in W_0^{k,2}(\Omega) \ . \tag{2.6}$$

Moreover, there is a constant $c > 0$ *such that*

$$||u||_{k,2,S} \leq c(||f||_* + ||u_0||_{k,2,S}) \ . \tag{2.7}$$

Idea of the p r o o f : Condition (2.2) implies the continuity of the bilinear form $a(u,v)$ from (1.2) on $V \times V$ with $V = W^{k,2}(\Omega;S)$ and conditions (2.3) imply its ellipticity while (2.1) guarantees that the space V and its subspace $W_0^{k,2}(\Omega;S)$ are well defined. A standard application of the Lax-Milgram Lemma then yields the existence and uniqueness of a weak solution $u \in W^{k,2}(\Omega;S)$ as well as the estimate (2.7) which expresses the continuous dependence of the solution on the data of the boundary value problem.

<u>2.2. Remarks.</u> (i) In the sequel, we shall give two examples of boundary value problems which *go beyond the frame of conditions* (2.1) - (2.3), but for which again existence and uniqueness of a weak solution can be proved. These examples indicate that conditions (2.1) - (2.3) can be substantially weakened and that the adequate weighted space can be constructed in a much more sophisticated way. A detailed description of

the (rather complicated) construction of these spaces can be found in A. KUFNER B. OPIC [1], [3].

(ii) Although Theorem 2.1 is a simplified version of an application of weighted Sobolev spaces to the solution of boundary value problems, some of its conditions can be weakened: E. g. condition (2.3) follows from (2.2) if the constant c_1 in (2.2) is sufficiently small, i. e. if $c_1 < 1/(M-1)$ where M is the number of multiindices α such that $|\alpha| \leq k$.

(iii) The restriction to the Dirichlet problem in Theorem 2.1 is not substantial, either; other boundary value problems can be handled in the same manner.

2.3. Example. Let us consider the differential operator of order two, i.e. $k = 1$:

$$(Lu)(x) = -\sum_{i=1}^{N} \frac{\partial}{\partial x_i}\left(a_i(x)\frac{\partial u}{\partial x_i}\right) + a_0(x)u$$

where $a_i > 0$ for $i = 1,\dots,N$ but $a_0 < 0$, $a_0 = -\lambda b_0$ with $b_0 > 0$, $\lambda \geq 0$. We suppose that $a_i, a_i^{-1} \in L^1_{loc}(\Omega)$ for $i = 0,1,\dots,N$. Here, one of the conditions in (2.1) is not fulfilled, but if we take $S = \{b_0, a_1, \dots, a_N\}$, then $W^{1,2}(\Omega;S)$ and $W_0^{1,2}(\Omega;S)$ are the adequate spaces which can be used for deriving existence and uniqueness theorems *provided the following inequality holds for all* $u \in C_0^\infty(\Omega)$:

$$\int_\Omega |u(x)|^2 b_0(x)\,dx \leq c\sum_{i=1}^{N}\int_\Omega \left|\frac{\partial u}{\partial x_i}\right|^2 a_i(x)\,dx \tag{2.8}$$

with a constant c independent of u, and *provided the constant* λ *in* $a_0 = -\lambda b_0$ *is sufficiently small*, namely $\lambda < 1/c$.

2.4. Example. Let us consider the plane domain $\Omega = (0,\infty)\times(0,\infty)$ (i.e. $N = 2$) and the *fourth* order operator

$$(Lu)(x) = \frac{\partial^2}{\partial x_1 \partial x_2}\left(x_1^{\delta_1} x_2^{\delta_2}\frac{\partial^2 u}{\partial x_1 \partial x_2}\right) -$$

$$- \frac{\partial}{\partial x_1}\left(x_1^{\gamma_1} x_2^{\gamma_2}\frac{\partial u}{\partial x_1}\right) - \frac{\partial}{\partial x_2}\left(x_1^{\beta_1} x_2^{\beta_2}\frac{\partial u}{\partial x_2}\right).$$

Here we have two possibilities:

(i) We can prove existence and uniqueness of a weak solution of the Dirichlet problem in the *anisotropic* space $W^{E,2}(\Omega;S)$ normed by

$$||u||^2 = \int_\Omega |u|^2 x_1^{\delta_1-2} x_2^{\delta_2-2}\,dx + \int_\Omega \left|\frac{\partial u}{\partial x_1}\right|^2 x_1^{\gamma_1} x_2^{\gamma_2}\,dx$$
$$+ \int_\Omega \left|\frac{\partial u}{\partial x_2}\right|^2 x_1^{\beta_1} x_2^{\beta_2}\,dx + \int_\Omega \left|\frac{\partial^2 u}{\partial x_1 \partial x_2}\right|^2 x_1^{\delta_1} x_2^{\delta_2}\,dx \tag{2.9}$$

provided $\delta_1 \neq 1$, $\delta_2 \neq 1$ (these last conditions are caused by the fact that the ellipticity constant c_0 equals $16(\delta_1 - 1)^{-2}(\delta_2 - 1)^2 + 1$).

(ii) We can prove existence and uniqueness in another anisotropic space $W^{\tilde{E},2}(\Omega;\tilde{S})$ normed by the expression obtained by omitting the second and third integrals in (2.9) provided $\gamma_1 = \delta_1$, $\gamma_2 = \delta_2 - 2$, $\beta_1 = \delta_1 - 2$, $\beta_2 = \delta_2$.

2.5. Remarks. (i) Example 2.4 shows that the structure of the operators as well as of the weighted spaces can be more general than that mentioned in formula (1.1) and in Definition 1.1; in particular, *anisotropic* operators and spaces can be treated by our method.

(ii) In Example 2.3, the estimate (2.8) played an important role. Estimates of such a type, which can be viewed as continuous imbeddings of $W^{1,2}(\Omega;S)$ into the weighted L^2-space $L^2(\Omega;b_0)$, are very useful tools both in the theory and in applications of weighted Sobolev spaces.

2.6. Nonlinear operators. Let us consider the *nonlinear* operator

$$(Lu)(x) = \sum_{|\alpha| \leq k} (-1)^{|\alpha|} D^\alpha a_\alpha(x;\delta_k u(x)), \quad x \in \Omega, \tag{2.10}$$

where $\delta_k u = \{D^\beta u, |\beta| \leq k\}$. Using the *theory of monotone operators*, results concerning existence of weak solutions of boundary value problems for the equation $Lu = f$ in the weighted space $W^{k,p}(\Omega;S)$ with $1 < p < \infty$ and $S = \{\sigma_\alpha, |\alpha| \leq k\}$ can be derived provided the "coefficients" $a_\alpha(x;\xi)$ of the operator L satisfy the following three conditions:

(i) the *weighted growth condition*

$$|a_\alpha(x;\xi)| \leq \sigma_\alpha^{1/p}(x)\left[g_\alpha(x) + c_\alpha \sum_{|\beta| \leq k} |\xi_\beta|^{p-1} \sigma_\beta^{1/p}(x)\right], \tag{2.11}$$

$|\alpha| \leq k$, for a.e. $x \in \Omega$ and all $\xi \in \mathbb{R}^M$ with given constants $c_\alpha \geq 0$ and functions $g_\alpha \in L^q(\Omega)$, $q = p/(p - 1)$;

(ii) the usual *monotonicity condition*

$$\sum_{|\alpha|\leq k} [a_\alpha(x;\xi) - a_\alpha(x;\eta)][\xi_\alpha - \eta_\alpha] \geq 0 \tag{2.12}$$

for a.e. $x \in \Omega$ and all $\xi, \eta \in \mathbb{R}^M$;

(iii) *a weighted coercivity condition*

$$\sum_{|\alpha|\leq k} a_\alpha(x;\xi)\xi_\alpha \geq c_0 \sum_{|\alpha|\leq k} |\xi_\alpha|^p \sigma_\alpha(x) \tag{2.13}$$

for a.e. $x \in \Omega$ and all $\xi \in \mathbb{R}^M$ with a given constant $c_0 > 0$.

The following assertion is a nonlinear analogue of Theorem 2.1.

<u>**2.7. Theorem.**</u> *Let* $S = \{\sigma_\alpha, |\alpha| \leq k\}$ *be a given family of weight functions and* $W^{k,p}(\Omega;S)$ *the corresponding weighted space with* $p > 1$. *Let the operator* L *from* (2.10) *fulfil conditions* (2.11) - (2.13). *Let* $f \in (W_0^{k,p}(\Omega;S))^*$ *and* $u_0 \in W^{k,p}(\Omega;S)$ *be given. Then there exists at least one weak solution* $u \in W^{k,p}(\Omega;S)$ *of the Dirichlet problem for the equation* $Lu = f$, *i.e., such a function* u *that*

$$u - u_0 \in W_0^{k,p}(\Omega;S)$$

and

$$\sum_{|\alpha|\leq k} \int_\Omega a_\alpha(x;\delta_k u(x)) D^\alpha v(x)\, dx = \langle f,v\rangle \quad \textit{for every} \quad v \in W_0^{k,p}(\Omega) .$$

If the inequality in (2.12) *is strict, then the solution* u *is uniquely determined.*

Idea of the proof: Let us consider the form

$$a(u,v) = \sum_{|\alpha|\leq k} \int_\Omega a_\alpha(x;\delta_k u(x)) D^\alpha v(x)\, dx .$$

The operator $\hat{T}$, defined by the formula $a(u,v) = \langle\hat{T}u,v\rangle$, is, in view of condition (2.11), a bounded operator from $W^{k,p}(\Omega;S)$ into its dual. To find a solution u of the Dirichlet problem means to find a function $\hat{u} \in X = W_0^{k,p}(\Omega;S)$ such that $a(\hat{u} + u_0, v) = \langle f,v\rangle$ for every $v \in X$, i.e., that $\hat{T}(\hat{u} + u_0) = f$. If we denote by T the operator from X to X^* defined by $T\hat{u} = \hat{T}(\hat{u} + u_0)$, then our problem reduces to the solution of the equation $Tu = f$ in X with a given $f \in X^*$. Conditions (2.11) - (2.13) guarantee that the operator T is bounded, demicontinuous, monotone and coercive, and so, the existence of at least one solution of the Dirichlet problem follows by applying the method of monotone operators. Uniqueness follows by contradiction if we assume that the inequality in (2.12) is strict.

2.8 Example. As a typical example of a nonlinear operator closely connected with the weighted Sobolev space $W^{k,p}(\Omega;S)$, $S = \{\sigma_\alpha = \sigma_\alpha(x); |\alpha| \le k\}$, we can consider the operator

$$(Lu)(x) = \sum_{|\alpha|\le k} (-1)^{|\alpha|} D^\alpha\{|D^\alpha u(x)|^{p-1} \operatorname{sgn} D^\alpha u(x)\, \sigma_\alpha(x)\} .$$

3. Elliptic operators with "bad" right-hand sides

Let us now suppose that the operator L from (1.1) satisfies condition (1.4) with a space $V \subset W^{k,2}(\Omega)$ (V is chosen in accordance with the type of the boundary conditions: for instance, $V = W_0^{k,2}(\Omega)$ for the Dirichlet problem and $V = W^{k,2}(\Omega)$ for the Neumann problem).

Further, let $S = \{\sigma_\alpha, |\alpha| \le k\}$ be a family of weight functions and let us denote by $1/S$ the family $\{1/\sigma_\alpha, |\alpha| \le k\}$. The functions $1/\sigma_\alpha$ are weight functions as well and consequently, we can consider the pair of weighted Sobolev spaces

$$H_1 = W^{k,2}(\Omega;S) \quad \text{and} \quad H_2 = W^{k,2}(\Omega;1/S) . \tag{3.1}$$

Rewriting the bilinear form $a(u,v)$ from (1.2) in the form

$$a(u,v) = \sum_{|\alpha|,|\beta|\le k} \int_\Omega a_{\alpha\beta}(x) D^\beta u(x)\sqrt{\sigma_\alpha(x)}\; D^\alpha v(x)\sqrt{\frac{1}{\sigma_\alpha(x)}}\, dx \tag{3.2}$$

we conclude immediately from Hölder's inequality that $a(u,v)$ *is a continuous bilinear form on* $H_1 \times H_2$.

This last property replaces the continuity condition (1.3). If we now replace the ellipticity condition (1.4) by the *pair of conditions*

$$\sup_{\|u_i\|_{H_i}\le 1} |a(u_1,u_2)| \ge c_j\|u_j\|_{H_j} , \quad i, j = 1,2 , \quad i \ne j , \tag{3.3}$$

– we say in this case that $a(u,v)$ *is* (H_1,H_2)*-elliptic* – we can again derive assertions about existence and uniqueness of weak solutions of the equation $Lu = f$ in H_1 (or its subspaces selected according to the type of the boundary conditions), $f \in H_2^*$. The main tool is a *modified version of the Lax-Milgram Lemma* due to J. NEČAS [1],[2], who also proposed the method roughly described above.

3.1. Problem. For what weight functions $\sigma_\alpha, |\alpha| \le k$, the conditions (3.3) are satisfied? More precisely: For what weight functions σ_α does the ellipticity condition (1.4) with V a subspace of the *classical* Sobolev space $W^{k,2}(\Omega)$ imply the *weighted* (H_1,H_2)-ellipticity?

3.2. Power type weights. Let us consider weight functions of the form

$$\sigma_\alpha(x) = [\mathrm{dist}(x,M)]^\varepsilon \quad \text{for all} \quad |\alpha| \leq k \tag{3.4}$$

where M is an m-dimensional manifold, $M \subset \partial\Omega$, and ε is a real number. The corresponding weighted space $W^{k,2}(\Omega;S)$ will be denoted by $W^{k,2}(\Omega;(\mathrm{dist})^\varepsilon)$, so that we have $W^{k,2}(\Omega;(\mathrm{dist})^{-\varepsilon})$ for $W^{k,2}(\Omega;1/S)$.

In the case of the *Dirichlet problem* the solution of Problem 3.1 is given by the following statement.

3.3. Theorem. *There exists an interval* J *containing* 0 *such that for* $\varepsilon \in J$ *the* (H_1,H_2)*-ellipticity conditions* (3.3) *are satisfied with* $H_1 = W_0^{k,2}(\Omega;(\mathrm{dist})^\varepsilon)$, $H_2 = W_0^{k,2}(\Omega;(\mathrm{dist})^{-\varepsilon})$.

The p r o o f is based on the imbedding

$$W_0^{1,2}(\Omega;(\mathrm{dist})^\varepsilon) \subsetneq L^2(\Omega;(\mathrm{dist})^{\varepsilon-2}) \tag{3.5}$$

which holds for $\varepsilon \neq 2+m-N$ ($m = \dim M$). Using the ellipticity condition (1.4) and repeatedly the imbedding (3.5) (also for higher derivatives and for both ε and $-\varepsilon$) we obtain the lower estimates (3.3) with constants c_j which depend on the coefficients of the operator L (i.e. on the L^∞-norm of $a_{\alpha\beta}$, on the ellipticity constant c_0), on geometrical properties of Ω (especially on $m = \dim M$ and on the smoothness of M) and on the norm of the imbedding operator from (3.5) (i.e. mainly on the value of the parameter ε). The interval J is determined by the requirement of the positivity of the constants c_j. A detailed derivation can be found in A. KUFNER [1] who extended to arbitrary $M \subset \partial\Omega$ the ideas developed by J. NEČAS [1] for the case $M = \partial\Omega$.

3.4. The size of J. Theorem 3.3 states the *existence of an interval* J; consequently, the existence and uniqueness of a weak solution of the Dirichlet problem for the equation $Lu = f$ in $W^{k,2}(\Omega;(\mathrm{dist})^\varepsilon)$ is guaranteed provided $\varepsilon \in J$. For applications it is necessary to know the exact *size* of the interval J of admissible powers ε in the weight function (3.4). It depends on L, Ω and M, but the estimates derived in the proof of Theorem 3.3 are very rough, and therefore, it is necessary to evaluate the interval J in every particular case separately. For example, for the operator $L = -\Delta$ it can be shown that Theorem 3.3 holds for $|\varepsilon| < 1$ so that we have $J = (-1,1)$ but in the case of $M = \{x_0\}$ with $x_0 \quad \partial\Omega$ (i.e., the case of the weight function $|x - x_0|^{\varepsilon}$) where Ω is a plane domain with the *outer cone property* at the point x_0 (the cone being characterized by the angle ω) we have a better estimate $|\varepsilon| < 2\pi/(2\pi - \omega)$. In this connection, let us mention the recent result of J. VOLDŘICH [1] who has shown that for any given $\varepsilon \neq 0$, $|\varepsilon|$ arbitrary small, a second order elliptic differential operator L can be constructed (depending on ε) such that *the Dirichlet problem for* L *has no solution in the space* $W^{1,2}(\Omega;(\mathrm{dist})^{\varepsilon})$.

In the case of other than power type weights, certain results concerning weight functions of the type

$$\sigma_{\alpha}(x) = s(\mathrm{dist}(x,M)) ,$$

$s = s(t)$ a positive function on $(0,\infty)$, have been derived by B. OPIC. These results as well as other examples concerning the Dirichlet problem can be found in the book A. KUFNER [1].

3.5. Other boundary value problems. For non-Dirichlet problems, Problem 3.1 has been investigated only for power type weights, and results similar to Theorem 3.3 have been established. The fundamental difference as compared to the Dirichlet problem consists in the fact that *serious restrictive conditions appear.* E. g., in the case of the Neumann problem, where one has to work with the space $W^{k,2}(\Omega;S)$ instead of $W_0^{k,2}(\Omega;S)$, the following analogue of Theorem 3.3 holds.

3.6. Theorem. *Let for* $\Omega \subset \mathbb{R}^N$ *and* $M \subset \partial\Omega$ *with* $m = \dim M$ *the following condition hold:*

$$N - m \geq 2k + 1 . \tag{3.6}$$

Then there exists an interval J *containing* 0 *such that for* $\varepsilon \in J$ *the* (H_1,H_2)*-ellipticity conditions* (3.3) *are satisfied with* $H_1 = W^{k,2}(\Omega;(\mathrm{dist})^{\varepsilon})$, $H_2 = W^{k,2}(\Omega;(\mathrm{dist})^{-\varepsilon})$ (and consequently, existence and uniqueness of a weak solution $u \in H_1$ of the Neumann problem for an elliptic equation of order $2k$ is guaranteed).

The p r o o f uses the same ideas as the proof of Theorem 3.3, but it is based on the imbedding

$$W^{1,2}(\Omega;(\mathrm{dist})^{\varepsilon}) \hookrightarrow L^2(\Omega;(\mathrm{dist})^{\varepsilon-2})$$

which, in contrary to the imbedding (3.5), holds only for $\varepsilon > 2 + m -$

N . This difference leads to the unpleasant restriction (3.6).

3.7. Remark. For *second order* equations, i.e. for $k = 1$, condition (3.6) has the form

$$N - m \geq 3$$

and excludes many important and interesting special cases of M as points (vertices - $m = 0$) or lines (edges - $m = 1$) on the boundaries of domains Ω of dimension $N = 2$ or $N = 3$, respectively. Nonetheless, for some *special domains* (cubes) and *special operators* $(- \Delta)$, J. VOLDŘICH derived results analogous to Theorem 3.3, even in the case if (3.6) is violated. For details see A. KUFNER, J. VOLDŘICH [1].

3.8. Another approach. The method described above is a little more complicated than the usual method mentioned in Introduction: It needs a *pair* of Banach spaces and *two* "ellipticity" conditions (3.3) instead of one simpler condition (1.4) and involves the Lax-Milgram-Nečas Lemma mentioned in the beginning of this section. In the paper of A. KUFNER, J. RÁKOSNÍK [1], another method is proposed which uses only *one* (weighted) space and requires the classical version of the Lax-Milgram Lemma.

Let us describe the method for the *Dirichlet problem*. We introduce a *new* bilinear form b by the formula

$$b(u,v) = a(u,\sigma v) \tag{3.7}$$

where σ is a (sufficiently smooth) weight function, and consider the weighted space $W^{k,2}(\Omega;S)$ with the family $S = \{\sigma_\alpha(x) = \sigma(x)$ for all $|\alpha| \leq k\}$ as well as the corresponding space $W_0^{k,2}(\Omega;S)$. For a given functional $f \in (W_0^{k,2}(\Omega;S))^*$ and a given function $u_0 \in W^{k,2}(\Omega;S)$, we say that the function $u \in W^{k,2}(\Omega;S)$ is a *σ-weak solution of the Dirichlet problem for the operator* L if

$$u - u_0 \in W_0^{k,2}(\Omega;S)$$

and

$$b(u,v) = \langle f,v \rangle \quad \text{for every} \quad v \in W_0^{k,2}(\Omega;S)$$

(provided $b(u,v)$ is meaningful for $u, v \in W^{k,2}(\Omega;S)$).

Let us further consider a weight function σ which satisfies the following conditions: There exist a weight σ_0 and constants c_1 , c_2 such that

$$\int_\Omega |u(x)|^2 \sigma_0(x)\, dx \leq c_1 \sum_{i=1}^N \int_\Omega \left|\frac{\partial u}{\partial x_i}\right|^2 \sigma(x)\, dx \tag{3.8}$$

for every $u \in C_0^\infty(\Omega)$ and

$$|\nabla\sigma(x)|^2/\sigma(x) \leq c_2\sigma_0(x) \quad \text{a.e. in } \Omega . \tag{3.9}$$

Then it can be shown that the form $b(u,v)$ is bounded on $W^{1,2}(\Omega;S) \times W^{1,2}(\Omega;S)$. Further, it can be shown that if the constant c_1c_2 is sufficiently small then the form $b(u,v)$ satisfies the ellipticity condition $b(u,u) \geq c_0\|u\|^2_{1,2;S}$, so that the existence and uniqueness of a σ-weak solution follows by a standard application of the Lax-Milgram Lemma.

3.9. Remarks. (i) The last result was derived for $k = 1$, i.e. for the *second order* operators only. For $k > 1$, we have to consider weights σ which fulfil conditions (3.8), (3.9) repeatedly (i.e. for σ there must exist the corresponding σ_0, for σ_0 the corresponding $(\sigma_0)_0$ etc. k-times).

(ii) The pair of conditions (3.8), (3.9) on σ can be replaced by the single condition

$$|\nabla\sigma(x)| \leq c_2\sigma(x) \quad \text{a.e. in } \Omega \tag{3.10}$$

(in the case $k = 1$). For such weights we again deduce that $b(u,v)$ is continuous and, moreover, for $c_2 > 0$ sufficiently small also elliptic, so that existence and uniqueness of a σ-weak solution in $W^{1,2}(\Omega;S)$ follows in a standard way.

3.10. Examples. (i) For $\sigma(x) = [\text{dist}(x,M)]^\varepsilon$, condition (3.9) is satisfied with $\sigma_0(x) = [\text{dist}(x,M)]^{\varepsilon-2}$ and $c_2 = \varepsilon^2$ and condition (3.8) is satisfied with $c_1 \approx |\varepsilon - 1|^{-\frac{1}{2}}$ if $\varepsilon \neq 1$ and with $c_1 \approx |\varepsilon + N - m - 2|^{-\frac{1}{2}}$ if $\varepsilon \neq m + 2 - N$ ($m = \dim M$, $N = \dim \Omega$). Consequently, we obtain an assertion about the existence and uniqueness of a $(\text{dist})^\varepsilon$-weak solution u of the Dirichlet problem in the space $W^{1,2}(\Omega;(\text{dist})^\varepsilon)$ *provided* $|\varepsilon|$ *is sufficiently small*, i.e. $\varepsilon \in J^*$ where J^* is an interval containing the origin. Thus, we have obtained a result similar to Theorem 3.3, and a comparison of the interval J from Theorem 3.3 with the interval J^* shows that (at least in some special cases) $J^* \supset J$ so that our second approach improves the set of admissible powers.

(ii) The weight $\sigma(x) = \exp(\varepsilon\, \text{dist}(x,M))$ satisfies condition (3.10) with the constant $c_2 = |\varepsilon|$. Weights of such a type are suitable for *unbounded* domains Ω and the existence and uniqueness of a σ-weak solution is guaranteed for $|\varepsilon|$ small.

3.11. Other boundary value problems can be dealt with in the same manner, and similar difficulties arise as in the first approach. E. g., if we consider the Neumann problem in the space $W^{1,2}(\Omega;S)$ with $\sigma(x) = [\mathrm{dist}(x,M)]^{\varepsilon}$, we obtain a result about the existence and uniqueness for $\varepsilon \in J^*$ which is the same interval as in the case of the Dirichlet problem, but under the restrictive condition $N - m \geq 3$. Further, one can show that for $N - m = 1$ our method cannot be used while for $N - m = 2$ we find that admissible values are positive ε's from J^*. - On the other hand, the mixed boundary value problem admits existence for $\varepsilon \in J^*$ without restriction on the dimension of M.

3.12. Remark. Since the form $a(u,v)$ was derived from the operator L by using Green's formula for the integral $\int_\Omega Lu\, v\, dx$, $v \in C_0^\infty(\Omega)$, the form $b(u,v) = a(u,\sigma v)$ can be derived in the same way from the integral $\int_\Omega Lu(\sigma v)\, dx = \int_\Omega \sigma Lu\, v\, dx$. Consequently, we can treat our σ-weak solution as the solution of a boundary value problem for the operator $\sigma L u$. Since $\sigma(x) > 0$ a.e. in Ω, the difference between a weak and a σ-weak solution is more or less formal.

References

KUFNER, A.:
[1] *Weighted Sobolev spaces.* J. Wiley & Sons, Chichester-New York -Brisbane-Toronto-Singapore 1985

KUFNER, A.; OPIC, B.:
[1] *The Dirichlet problem and weighted spaces I.* Časopis Pěst. Mat. 108(1983), 381-408
[2] *How to define reasonably weighted Sobolev spaces.* Comment.Math. Univ. Carolinae 25(3)(1984), 537-554
[3] *The Dirichlet problem and weighted spaces II.* To appear in Časopis Pěst. Mat.

KUFNER, A.; VOLDŘICH, J.:
[1] *The Neumann problem in weighted Sobolev spaces.* Math. Rep. Roy. Soc. Canada

KUFNER, A.; RÁKOSNÍK, J.:
[1] *Linear elliptic boundary value problems and weighted Sobolev spaces: A modified approach.* Math. Slovaca 34(1984), No.2 185-197

NEČAS, J.:
[1] *Sur une méthode pour résoudre les équations aux dérivées partielles du type elliptique, voisine de la variationnelle.* Ann. Scuola Norm. Sup. Pisa 16(1962), 305-326

[2] *Les méthodes directes en théorie des équations elliptiques.* Academia, Prague & Masson et Cie, Paris 1967

VOLDŘICH, J.:
[1] *A remark on the solvability of boundary value problems in weighted spaces.* To appear in Comment Math. Univ. Carolinae

CRITICAL POINT THEORY AND NONLINEAR DIFFERENTIAL EQUATIONS

J. MAWHIN
Institut Mathématique, Université de Louvain
B-1348 Louvain-la-Neuve, Belgium

1. INTRODUCTION

The variational approach to boundary value problems for differential equations consists in writing the problem, whenever it is possible, as an abstract equation of the form

$$(1) \qquad \Phi(u) = 0$$

where $\Phi : E \to E^*$ is of the form $\Phi = \varphi'$, with φ' the Gâteaux derivative of a real function φ defined on a Banach space E. In this way the search of solutions for (1) is equivalent to the determination of the critical points of Φ, i.e. of the zeros of φ'. Such a viewpoint can be traced at least to Fermat, with his minimal type principle to find the law of refraction for the light.

Since Fermat also, we know that the points at which φ achieves its extremums are critical points of φ. Thus, any way which succeeds in proving, directly, that φ has a maximum or a minimum provides a way of proving the existence of a solution of (1). This is the so-called direct method of the calculus of variations which goes back to Gauss, Kelvin, Dirichlet, Hilbert, Tonelli and others. More recent work deals with proving the existence of critical points at which φ does not achieve an extremum (saddle points). This paper surveys some of the recent work in this direction. A systematic exposition of many aspects of the variational approach to boundary-value problems for ordinary differential equations will be given in [11].

For definiteness, we shall consider a system of ordinary differential equations of the form

$$(2_\alpha) \qquad u'' + \alpha u = \nabla F(x,u) \qquad (\nabla = D_u)$$

on a compact interval $I = [a.b]$, submitted to homogeneous boundary conditions, say, of Dirichlet, Neumann or periodic type. For simplicity, we assume here that F and ∇F are continuous on $I \times \mathbf{R}^N$. We could as well consider elliptic partial differential equations. It is well known that the spectrum of $-d^2/dt^2$ submitted on I to the above boundary conditions has the form

$$(0\leq)\ \lambda_1 < \lambda_2 < \ldots.$$

Moreover, (2) is the Euler-Lagrange equation associated to the functional

$$\varphi : H \to \mathbf{R},\ u \mapsto Q_\alpha(u) + \int_I F(.,u(.))$$

where

$$Q_\alpha(u) = \int_I (1/2)(|u'|^2 - \alpha|u|^2),$$

and $H = H_0^1(I;\mathbf{R}^N)$, $H^1(I;\mathbf{R}^N)$ or $H_\#^1(I,\mathbf{R}^N) = \{u \in H^1(I,\mathbf{R}^N) : u(a) = u(b)\}$ with their usual norm denoted by $\|.\|$. Solving (2) with one of the above boundary condition is thus equivalent to finding a *critical point* of φ on H, i.e. a point $u \in H$ such that

$$(3) \qquad \varphi'(u) = 0.$$

If $c = \varphi(u)$ with u a critical point, c is called a *critical value* for φ.

The simplest situation for (3) to hold is when φ has a global minimum (which requires of course φ to be bounded from below).

Since Hammerstein [6] in 1930 (in the Dirichlet case) we know that φ will have a global minimum whenever

$$(4) \qquad \alpha < \lambda_1$$

and

$$(5) \qquad F(x,u) \geq -(\beta/2)|u|^2 - \gamma(x)$$

for some $\beta < \lambda_1 - \alpha$, $\gamma \in L^1(I)$ and all $(x,u) \in I \times \mathbf{R}^N$. In fact, φ is coercive ($\varphi(u) \to +\infty$ for $\|u\| \to \infty$) because, by (4) and (5), φ is bounded from below by a coercive quadratic form. Moreover φ is weakly lower semi-continuous so that φ has a global minimum by a classical result. We shall discuss now situations where (4) and (5) do not hold.

2. THE CASE OF $\alpha = \lambda_1$ AND $\int_I F$ COERCIVE ON THE KERNEL

The situation is already more complicated when $\alpha = \lambda_1$ (*resonance at the lowest eigenvalue*) and condition (5) is no more sufficient for the existence of a critical points as shown by a linear equation violating the Fredholm alternative condition. To motivate the introduction of a new-sufficient condition, let us first consider the case where ∇F is bounded.

a) The case where ∇F is bounded

Writing $u(x) = \bar{u}(x) + \tilde{u}(x)$ with $\bar{u} \in \bar{H}_1$ the eigenspace of λ_1 and $\tilde{u} \in \tilde{H}_1 = H_1^{\perp}$, we have

$$\varphi(u) = \Omega_{\lambda_1}(\tilde{u}) + \int_I [F(.,\bar{u}(.)) + F(.,u(.)) - F(.,\bar{u}(.))] \geq \Omega_{\lambda_1}(\tilde{u}) + \int_I F(.,\bar{u}(.)) - M\|\tilde{u}\|_{L^2} \geq c_1\|\tilde{u}\|^2 - c_2\|\tilde{u}\| + \int_I F(.,\bar{u}(.)),$$

where M is an upper bound for $|\nabla F|$ on $I \times \mathbb{R}^N$, and we shall recover coercivity for φ if we assume that

$$(6) \qquad \int_I F(.,v(.)) \to +\infty \quad \text{as } \|v\| \to \infty \text{ in } \bar{H}_1$$

(*coercivity of the averaged* F *on the kernel*). Such a condition was first introduced by Ahmad, Lazer and Paul [1] and it generalizes the classical Landesman-Lazer conditions. As φ is again w.l.s.c., the existence of a minimum is insured.

b) The case where F is convex

The boundedness of ∇F can be replaced by the convexity of $F(x,.)$ for each $x \in I$. In this case, if (6) also holds, there exists $\bar{u}_0 \in \bar{H}_1$ such that

$$(7) \qquad \int_I \nabla F(.,\bar{u}_0(.))\bar{v} = 0 \quad \text{for all } \bar{v} \in \bar{H}_1.$$

Moreover, by convexity and using (7) we have

$$\varphi(u) \geq Q_{\lambda_1}(\tilde{u}) + \int_I [F(.,\bar{u}_0(.)) + (\nabla F(.,\bar{u}_0(.)),u - \bar{u}_0)]$$

$$(8) \qquad = Q_{\lambda_1}(\tilde{u}) + \int_I F(.,\bar{u}_0(.)) + \int_I(\nabla F(.,\bar{u}_0(.)),\tilde{u})$$

$$\geq c_1\|\tilde{u}\|^2 - c_2\|\tilde{u}\| - c_3$$

Thus each minimizing sequence (u_k') for φ has $(\tilde{u}_k)$ bounded in the norms $\|.\|$ and $\|.\|_{L^\infty}$. On the other hand, by convexity again

$$F(x,\bar{u}_k/2) \le (1/2)F(x,u_k) + (1/2)F(x,-\tilde{u}_k)$$

and hence,

$$\varphi(u_k) \ge 2\int_I F(.,\bar{u}_k/2) - \int_I F(.,-\tilde{u}_k) \ge$$

$$\ge 2\int_I F(.,\bar{u}_k/2) - c_4,$$

which, by (6), implies that $(\bar{u}_k)$ is bounded and φ has a minimum.

Let us remark that when $F(x,.)$ is strictly convex for each $x \in I$ and $\alpha = \lambda_1$, it can be shown that (6) is necessary and sufficient for the existence of a solution [10].

c) <u>The case where $-\int_I F$ is coercive on the kernel</u>

Let us assume now that

(9) $$\int_I F(.,N(.)) \to -\infty \quad \text{as} \quad \|v\| \to \infty \quad \text{in } \bar{H}_1.$$

As this situation only holds in trivial situations when $F(x,.)$ is convex, let us assume again that ∇F is bounded. By (9), we have

$$\varphi(v) = \int_I F(.,v(.)) \to -\infty \quad \text{as} \quad \|v\| \to \infty$$

in $\bar{H}_1$, so that φ is no more bounded from below and has no global minimum. On the other hand, on $\tilde{H}_1$,

$$\varphi(w) = Q_{\lambda_1}(w) + \int_I [F(.,0) + (F(.,w(.)) - F(.,0))] \ge$$

$$\ge c_1\|w\|^2 - c_2\|w\| - c_3$$

and hence $\varphi|_{\tilde{H}_1}$ is bounded from below (even coercive). Consequently, there exists $R > 0$ such that

$$\sup_{\bar{H}_1 \cap \partial B(R)} \varphi < \inf_{\tilde{H}_1} \varphi$$

This suggests the use of the following *saddle type* or *minimax theorem* of Rabinowitz [15], introduced to give a variational proof of the Ahmad-Lazer-Paul results [1].

LEMMA 1. *Let* E *be a Banach space and* $\varphi \in C^1(B,\mathbb{R})$. *Assume that there exists a decomposition* $E = E_1 \oplus E_2$ *with* $\dim E_1 < \infty$ *and* $R > 0$ *such that*

$$\sup_{E_1 \cap \partial B(R)} \varphi < \inf_{E_2} \varphi ,$$

Let

$$\Sigma = \{\sigma \in C(E,E) \mid \sigma(u) = u \ \text{ on } \ \partial B(R)\}$$

and

$$c = \inf_{\sigma\in\Sigma} \ \max_{s\in B(R)\cap E_1} \varphi(\sigma(s)) \qquad (\geq \inf_{E_2} \varphi) \tag{10}$$

Assume that if there is a (u_k) *such that* $\varphi(u_k) \to c$ *and* $\varphi'(u_k) \to 0$, *then* c *is a critical value. (Palais-Smale type condition* PS^* *at* c). *Then* φ *has a critical point with critical value* c.

This theorem can be proved by deformation techniques [12] or Ekeland's variational lemma [4].

In the above case with $E = H$, $E_1 = \bar{H}_1$, $E_2 = \tilde{H}_1$, the PS^*-condition holds for each c and φ has a critical point.

The above results are summarized in the following.

THEOREM 1. *Assume that*

$$\int_I F(.,v(.)) \to +\infty \quad \text{as } \|v\| \to \infty \ \text{ in } \bar{H}_1$$

(the eigenspace of λ_1*) and that either* ∇F *is bounded or* F *is convex in* u. *Then* (2_{λ_1}) *with the suitable boundary conditions has at least a solution which minimizes* φ. *Assume that*

$$\int_I F(.,v(.)) \to -\infty \quad \text{as } \|v\| \to \infty \ \text{ in } \bar{H}_1$$

and that ∇F *is bounded. Then* (2_{λ_1}) *with the suitable boundary conditions has at least a solution* u *with* $\varphi(u) = c$ *given by (10) with* $E_1 = \bar{H}_1$ *the eigenspace of* λ_1.

3. THE CASE OF $\alpha = \lambda_1$ AND F PERIODIC

An interesting situation in which (6) does not hold occurs when

$$F(x,u + T_i e_i) = F(x,u) \qquad (1 \le i \le N)$$

for all $x \in I$, $u \in \mathbb{R}^N$ and some $T_i > 0$. $(1 \le i \le N)$.
This implies that F and ∇F are bounded on $I \times \mathbb{R}^N$. Therefore

$$\begin{aligned} \varphi(u) &= Q_{\lambda_1}(\tilde{u}) + \int_I F(.,u(.)) \\ &\ge c_1 \|\tilde{u}\|^2 - c_2 , \end{aligned} \tag{11}$$

so φ is bounded from below and any minimizing sequence (u_k) is such that $(\tilde{u}_k)$ is bounded in the norms $\|.\|$ and $\|.\|_{L^\infty}$.

a) The case of Neumann or periodic boundary conditions

Then, $\lambda_1 = 0$ and $\overline{H}_1 \approx \mathbb{R}^N$ is the space of constant mappings from $[a,b]$ into $\mathbb{R}^N$. Moreover,

$$\varphi(u + T_i e_i) = \varphi(u) \qquad (1 \le i \le N) \tag{12}$$

for all $u \in H$, so that any minimizing sequence can be supposed, without loss of generality, such that

$$|\bar{u}_k| \le \left(\sum_{i=1}^{N} T_i^2\right)^{1/2}.$$

Thus φ has a bounded minimizing sequence and hence a minimum. This result is due to Willem [18] and (independently and in special cases) Hamel [5] and Dancer [3]. The existence of a second solution was proved by Mawhin-Willem [8,9] using the mountain pass lemma, a variant of Lemma 1. Their approach was extended to systems of the form

$$\frac{d}{dt}\frac{\partial L}{\partial \dot{u}}(u,\dot{u}) - \frac{\partial L}{\partial u}(u,\dot{u}) = 0$$

by Capozzi, Fortunato and Salvatore [2]. See also Pucci-Serrin [13, 14] for abstract critical point theorems motivated by this situation.

b) The case of Dirichlet boundary conditions

The Dirichlet case strongly differs from the other ones because

$\lambda_1 = \frac{\pi^2}{(b-a)^2} > 0$ and $\tilde{H}_1 = \text{span}(\sin \frac{\pi x}{b-a})$ which imply that we loose the periodicity property (12) of φ. The problem has been studied by Ward [17] for N = 1 and

(13) $$F(x,u) = G(u + E(x))$$

where G is continuous and T-periodic and $E : I \to \mathbf{R}$ is continuous. Indeed, Ward considered explicitely the problem

$$v'' + \lambda_1 v = g(v) + e(t)$$

$$v(a) = v(b) = 0$$

when $g(v + T) = g(v)$, $\int_0^T g = 0$ and $e \in \tilde{H}_1$, which reduces to the above case by a trivial change of variables.

A possible way of approach, slightly different from Ward's one, makes use of the following lemma which can be proved by a deformation technique or Ekeland variational lemma.

LEMMA 2. *Let E be a Banach space and $\varphi \in C^1(E,\mathbf{R})$ be bounded from below and satisfy PS* at $c = \inf \varphi$. Then φ has a minimum.*

Using an extension of the Riemann-Lebesgue lemma, one can prove that φ associated to F given in (13)satisfies the PS*-condition at each $b \neq 0$ and that $\varphi|_{\tilde{H}_1}$ satisfies PS* at each $b \in \mathbf{R}$. Thus the existence of a critical point is insured by Lemma 2 except when

$$0 = \inf_H \varphi < \inf_{\tilde{H}_1} \varphi$$

The above mentioned Riemann-Lebesgue type lemma also implies that, on $\tilde{H}_1$, $\varphi(v) \to 0$ as $\|v\| \to \infty$. Thus, there exists some $R > 0$ such that

$$\max_{\bar{H}_1 \cap \partial B(R)} \varphi < \inf_{\tilde{H}_1} \varphi$$

and then c given by the Rabinowitz lemma is greater or equal to $\inf \varphi$ and hence nonzero. Consequently, this c is a critical value for φ.

The above results can be summarized in the following.

THEOREM 2. *Assume that*

$$F(x,u+T_ie_i) = F(x,u) \qquad (1 \le i \le N)$$

with N = 1 *and* F *of the form (13) in the Dirichlet case. Then* (2_{λ_1}) *with the suitable boundary condition has at least one solution.*

4. THE CASE OF $\lambda_{i-1} < \alpha \le \lambda$ $(i \ge 2)$

In this case, φ is neither bounded from below nor from above, as $Q_\alpha(v) \to -\infty$ on $\bar{H}_{i-1}$ = span of eigenfunctions of $\lambda_1,\dots,\lambda_{i-1}$ and $Q_\alpha(v) \to +\infty$ on $\tilde{H}_{i+1}$ = span of eigenfunctions of $\lambda_{i+1},\dots$.

a) The case where ∇F is bounded

Then one can use the Rabinowitz Lemma in a way similar to the case where $\alpha = \lambda_1$ and $\int_I F(.,v(.)) \to -\infty$ as $\|v\| \to \infty$ if the extra condition

$$(14) \qquad \int_I F(.,v(.))dx \to +\infty \text{ or } -\infty \text{ as } \|v\| \to \infty \text{ in the eigenspace of } \lambda_i$$

holds when $\alpha = \lambda_i$. One choose in this case $E_1 = \bar{H}_i$, $E_2 = \tilde{H}_i$ or $E_1 = \bar{H}_{i+1}$, $E_2 = \tilde{H}_{i+1}$ according to the sign of ∞ in (14). Under these conditions (2_α) has at least one solution. This is essentially a result of Ahmad-Lazer-Paul [1] and Rabinowitz [15].

b) The case where F is convex

Then, sharper results can be obtained without boundedness assumption of ∇F through the use of the Clarke-Ekeland dual least action principle which reduce the study of the critical points of φ to that of an associate dual function ψ involving the (possibly generalized) inverse of $\frac{d^2}{dt^2} + \lambda_i I$ and the Legendre-Fenchel transform of F(x,.). Under reasonable conditions on F, ψ is bounded from below and, in this way the existence of a solution is in particular insured when

$$\limsup_{|u|\to\infty} \frac{F(x,u)}{|u|^2} \le \beta < \frac{\lambda_{i+1} - \lambda_i}{2} \quad \text{(unif. in } x \in I)$$

and (if $\lambda_i = \alpha$),

$$\int_I F(x,f(x))dx \to +\infty \text{ as } \|v\| \to \infty \text{ in the eigenspace of } \lambda_i.$$

See [10] for general results in this direction.

c) The case where F is periodic and $\alpha = \lambda_i$

Results are known only when N = 1 and F has the form (13). The proof, due to Lupo and Solimini [16,7] is such more delicate because the PS* is not satisfied at c = 0. This requires, in addition to the classical Rabinowitz saddle point theorem, other saddle point theorems of the same type and some topological arguments (together with the Riemann-Lebesgue-type lemma mentioned above).

The above results can be summarized in the following

THEOREM 3. *Assume that* $\lambda_{i-1} < \alpha \le \lambda_i$ $(i \ge 2)$ *and that one of the following conditions holds:*

i) ∇F *is bounded and, whenever* $\alpha = \lambda_i$,

$$\int_I F(.,v(.)) \to +\infty \text{ or } -\infty \text{ as } \|v\| \to \infty \text{ in the eigenspace of } \lambda_i$$

ii) $F(x,.)$ *is convex,* $\limsup_{|u| \to \infty} \frac{F(x,u)}{|u|^2} \le \beta < \frac{\lambda_{i+1} - \lambda_i}{2}$ *(unif. in* $x \in I$*),*

and, whenever $\alpha = \lambda_i$,

$$\int_I F(.,v(.)) \to +\infty \text{ as } \|v\| \to \infty \text{ in the eigenspace of } \lambda_i \tag{14}$$

iii) $\alpha = \lambda_i$, N = 1 *and* F *has the form (13) with* G *T-periodic.*

Then the problem (2_α) *with any of the boundary conditions has at least one solution.*

One can show that (14) is necessary and sufficient when F(x,.) is strictly convex [10].

References

[1] S.AHMAD, A.C.LAZER and J.L.PAUL, Elementary critical point theory and perturbations of elliptic boundary value problems at resonance, Indiana Univ. Math. J. 25 (1976) 933-944.

[2] A.CAPOZZI, D.FORTUNATO and A.SALVATORE, Periodic solutions of Lagrangian systems with bounded potential, to appear.

[3] E.N.DANCER, On the use of asymptotic in nonlinear boundary value problems, Ann. Mat. Pura Appl. (4) 131 (1982) 167-185.

[4] I.EKELAND, Nonconvex minimization problems, Bull. Amer. Math. Soc. (NS) 1 (1979) 443-474.

[5] G.HAMEL, Über erzwungene Schwingungen bei endlichen Amplituden, Math. Ann. 86 (1922) 1-13.

[6] A.HAMMERSTEIN, Nichtlineare Integralgleichungen nebst Anwendungen, Acta Math., 54 (1930) 117-176.

[7] D.LUPO and S.SOLIMINI, A note on a resonance problem, Proc. Royal Soc. Edinburgh, Ser. A, to appear.

[8] J.MAWHIN and M.WILLEM, Multiple solutions of the periodic boundary value problem for some forced pendulum-type equations, J. Diff. Equations 52 (1984) 264-287.

[9] J.MAWHIN and M.WILLEM, Variational methods and boundary value problems for vector second order differential equations and applications to the pendulum equation, in "Nonlinear Analysis and Optimization", Lect. Notes in Math. No 1107, Springer, Berlin, 1984, 181-192.

[10] J.MAWHIN and M.WILLEM, Critical points of convex perturbations of some indefinite quadratic forms and semi-linear boundary value problems at resonance, Ann. Inst. H. Poincaré, Analyse non-linéaire, to appear.

[11] J.MAWHIN and M.WILLEM, "Critical Point Theory and Hamiltonian Systems", in preparation.

[12] P.S.PALAIS, Critical point theory and the minimax principle, in Proc. Symp. Pure Math. vol. 15, Amer. Math. Soc., Providence, 1970, 185-212.

[13] P.PUCCI and J.SERRIN, Extensions of the mountain pass theorem, J. Funct. Anal. 59 (1984) 185-210.

[14] P.PUCCI and J.SERRIN, A mountain pass theorem, J. Differential Equations, 57 (1985).

[15] P.RABINOWITZ, Some minimax theorems and applications to nonlinear partial differential equations, in "Nonlinear Analysis, a volume dedicated to E.H.Rothe", Academic Press, New York, 1978, 161-178.

[16] S.SOLIMINI, On the solvability of some elliptic partial differential equations with the linear part at resonance, to appear.

[17] J.R.WARD, A boundary value problem with a periodic nonlinearity, J. Nonlinear Analysis, to appear.

[18] M.WILLEM, Oscillations forcées de systèmes hamiltoniens, Publ. Sémin. Analyse non linéaire Univ. Besançon, 1981.

ORDINARY LINEAR DIFFERENTIAL EQUATIONS – A SURVEY OF THE GLOBAL THEORY

F. NEUMAN
Mathematical Institute of the Czechoslovak Academy of Sciences, branch Brno
Mendlovo nám. 1, 603 00 Brno, Czechoslovakia

I. History

Investigations of linear differential equations from the point of their transformations, canonical forms and invariants started in the last century. In 1834 E.E. Kummer [6] studied transformations of the second order equations in the form involving a change of the independent variable and multiplication of the dependent variable. Till the end of the last century several mathematicians dealt also with higher order equations. Let us mention at least E. Laguerre, A.R. Forsyth, F.Brioschi, G.H.Halphen from many others. Perhaps the most known result from this period is the so called Laguerre-Forsyth canonical form of linear differential equations characterized by the vanishing of the coefficients of the (n - 1)st and (n - 2)nd derivatives.

However as late as in 1892 P. Stäckel (and one year later independently S.Lie) proved that the form of transformation considered by Kummer (as well as all his successors) is the most general pointwise transformation that converts solutions of any linear homogeneous differential equation of the order greater than one into solutions of an equation of the same kind. In fact, only this result justified backwards the whole previous investigations.

Already in 1910 G.D. Birkhoff [1] pointed out that the investigations, considered in the real domain, were of local character. He presented an example of the third order linear differential equation that cannot be transformed into any equation of the Laguerre-Forsyth canonical form on its whole interval of definition.

The local nature of methods and results is not suitable for dealing with problems of global character, as boundedness, periodicity, asymptotic or oscillatory behavior and other properties of solutions that necessarily involve investigations on the whole intervals of definition.

Only to demonstrate that even in the middle of this century there were just isolated results of a global character and no systematic theory, let me mention G. Sansone's example of the third order linear differential equation with all oscillatory solutions. This result occured as late as in 1948 in spite of the fact that the question about the existence or nonexistence of such an equation is as old as the problem of factorization of linear differential operators.

It is now some 35 years ago that O. Borůvka started the systematic study of global properties of the second order linear differential equations. He deeply developed this theory and summarized his original methods and results in his monograph [3] that appeared in 1967 in Berlin and in an extension version in 1971 in London.

For linear differential equations of the second and higher orders there have occurred results of a global character in papers of several mathematicians. Let me mention at least N.V.Azbelev, J.H.Barrett, E. Barvínek, L.M.Berkovič, T.A.Burton, Z.B.Caljuk, T.A.Chanturija, W.A. Coppel, W.N.Everitt, M.Greguš, H.Guggenheimer, G.B.Gustafson, M.Hanan, Z.Hustý, I.P.Kiguradze, V.A. Kondratjev, M.K. Kwong, M.Laitoch, A.C. Lazer, A.Ju.Levin, W.T.Patula, M.Ráb, G.Sansone, S.Staněk, J.Suchomel, C.A.Swanson, V.Šeda, M.Švec, M.Zlámal from several others. However, there was still no unified and systematic theory of global properties of linear differential equations of an arbitrary order enabling us to fortell what can and what cannot happen in global behavior of solutions.

In the last 15 years we discovered enough general approach and methods, we introduced new useful notions and derived results giving answers to substantial questions and solving basic problems in the area of global properties of linear differential equations of an arbitrary order. O.Borůvka's methods and results for the second order equations were at the beginning of our approach to equations of arbitrary orders and they still play an important role in the whole theory. We cannot see the possibility how to handle the general situation without having had his results at our disposal.

Algebraic, topological, analytical and geometrical tools together with methods of the theory of dynamical systems and functional equations make it possible to deal with problems concerning global properties of solutions by contrast to the previous local investigations or isolated results. Theory of categories, Brandt and Ehresmann groupoids, Cartan's moving-frame-of-reference method among other differential geometry methods, and functional equations are some of the means used in our approach.

The theory in question includes also effective methods for solving several special problems, e.g. concerning the global equivalence of two given equations, or from the area of questions on distribution of zeros of solutions, disconjugacy, oscillatory behavior, etc.

II. Global Transformations

For $n \geq 2$, let $P_n(y,x;I)$ denote a linear homogeneous ordinary linear differential equation

$$y^{(n)} + p_{n-1}(x)y^{(n-1)} +\dots+ p_o(x) = 0 ,$$

where $p_i \in C^o(I)$, $i = 0,1,\dots,n-1$, are real continuous functions defined on an open interval I of reals. Similarly, $Q_n(z,t;J)$ denotes

$$z^{(n)} + q_{n-1}(t)z^{(n-1)} +\dots+ q_o(t) = 0 , \quad q_i \in C^o(J) ,$$

$i = 0,1,\dots,n-1$, $J \subset \mathbf{R}$ being an open interval.

We say that $P_n(y,x;I)$ is *globally transformable* into $Q_n(z,t;J)$ if there exist

a function $f \in C^n(J)$, $f(t) \neq 0$ on J, and
a C^n-diffeomorphism h of J into I,

such that

$$z(t) = f(t) . y(h(t)), \quad t \in J$$

is a solution of $Q_n(z,t;J)$ whenever y is a solution of $P_n(y,x;I)$.

This definition complies with the most general form of a pointwise transformation derived by Stäckel. The bijectivity of h guarantees the transformation of solutions on their whole intervals of definition, i.e. the globality of the transformation. Let me remark also, that recently M.Čadek derived Stäckel's result without any differentiability assumption, [4].

It appears to be convenient to write the global transformation in the following form. Let $y = (y_1,\dots,y_n)^T$ be the vector column function whose coordinates y_i are linearly independent solutions of the equation $P_n(y,x;I)$ for $i = 1,\dots,n$. Let us call the y a fundamental solution of $P_n(y,x;I)$. Similarly, let z denote a fundamental solution of the equation $Q_n(z,t;J)$. Then there exists a nonsingular n by n constant matrix C such that

(α) $\quad z(t) = C.f(t).y(h(t)), \quad t \in J$.

The global transformation expressed explicitely by this formula will be denoted by $\alpha = \langle Cf,h \rangle_y$, and we shall write

$$P_n(y,x;I)\alpha = Q_n(z,t;J) ,$$

or shortly

$P\alpha = Q$.

The relation of global transformability is an *equivalence relation*. Hence the set $\mathcal{A}$ of all linear homogeneous differential equations of all orders greater than and equal to two, is decomposed into the classes of globally equivalent equations.

Let $\mathcal{B}$ be one of the classes of the equivalence. For each three equations P, Q and T of the class $\mathcal{B}$ there exist global tranformations α and β such that

$P\alpha = Q$ and $Q\beta = T$.

If we define a composition $\alpha\beta$ of the tranformations α and β by

$(P\alpha)\beta = P(\alpha\beta) = T$,

we introduce a certain algebraic structure into each class $\mathcal{B}$ of globally equivalent equations. This algebraic structure considered on the whole set $\mathcal{A}$ is a special *category*, called the *Ehresmann groupoid*. Linear differential equations are objects and global transformations are morphisms of the category. The same algebraic structure restricted to any class $\mathcal{B}$ of globally equivalent equations is a special Ehresmann groupoid, called the *Brandt groupoid*.

The basic (and in fact, the only) structural notion of a Brandt groupoid is the so called *stationary group* of any of its objects. In our case of differential equations, the stationary group G(P) of an equation P is formed by all global transformations that transform the equation P into itself, i.e.

$G(P) = \{\alpha;\ P\alpha = P\}$.

It can be shown that the stationary groups of any two equations P and Q from the same equivalent class $\mathcal{B}$ are conjugate:

if $P\alpha = Q$ then $G(P) = \alpha G(Q)\alpha^{-1}$.

Having a special (*canonical*) object (*equation*) $S_{\mathcal{B}}$ in the class $\mathcal{B}$ of equivalent equations, all global transformations transforming P into Q are described by the formula

$\gamma^{-1}G(S_{\mathcal{B}})\delta$, where $P = S_{\mathcal{B}}\gamma$ and $Q = S_{\mathcal{B}}\delta$.

We could observe that in each area of mathematics where a structure of an Ehresmann groupoid occurs as it is also in our case, the following basic problems have to be solved in order to describe the structure of sets of objects and transformations in this area, and in this manner, to form a foundation of the corresponding theory:

1. Find sufficient and/or necessary conditions (if even effective, the better) under which two given objects, two given equations are

equivalent, i.e. *criterion of global equivalence.*

2. Characterize all possible *stationary groups* according to the classes of equivalence.

3. Find, construct *canonical objects*, equations in *each* class of equivalent equations.

In what follows we shall answer the mentioned questions for linear differential equations of arbitrary orders.

First, let us introduce also a *geometrical representation* of our global transformations very useful in the sequel when different geometrical approaches are applied.

Again, let an equation P be represented by its (arbitrary, but fixed) fundamental solution y, considered now as a curve in n-dimensional vector space V_n, the independent variable x ranging through the interval I and being the parameter of the curve. Due to the form (α) of a global transformation,

the change $x = h(t)$ is only a reparametrization,

the factor $f(t)$ selects only another curve but on the same cone K formed by straight lines going through the origin $0 \in V_n$ and all points of the original curve y,

the matrix C performs a centroaffine mapping.

We may conclude that each fundamental solution, or curve z of any equation Q globally equivalent to the equation P is a section of a cone in n-dimensional vector space obtained as a centroaffine image of a fixed cone determined by a fixed curve y.

Now, let us come to answer the above mentioned basic questions.

III. Global Equivalence

A sufficient and necessary condition for global equivalence of the *second order* linear differential equations was found by O. Borůvka [3] in the sixties. First, some definitions:

The maximal number of zeros of nontrivial solutions of an equation of the second order P_2 gives the *type* of the equation: either *finite*, an integer m, or *infinite*. Moreover, the equation P_2 being of finite type m is called of *general kind*, if it admits two linearly independent solutions with m - 1 zeros, everything considered on the whole interval of definition. Otherwise, P_2 being of finite type m is called of *special kind*. If the equation P_2 is of infinite type then its kind is either *one-side oscillatory* or *both-side oscillatory*.

Now Borůvka's criterion reads as follows:

Two second order linear differential equations are globally equivalent if and only if they are of the same type and at the same time of the same kind.

Our criterion of global equivalence of equations of higher orders needs the following notion. Let

(p) $\quad u'' + p(x)u = 0$

be an equation of the second order whose coefficient p belongs to the class $C^{n-2}(I)$, and let u_1 and u_2 denote two of its independent solutions. Define n functions

$$y_1 := u_1^{n-1},\ y_2 := u_1^{n-2}.u_2,\ldots,\ y_n := u_2^{n-1} .$$

These functions are of the class $C^n(I)$ and they are linearly independent. Hence they can be considered as solutions of the uniquely determined n-th order linear differential equation, called the *iterative equation* iterated from the equation (p). We denote the iterative equation by $p^{[n]}(y,x;I)$, or simply by $p^{[n]}$. The differential expression of the iterative equation normalized by the unit leading coefficient will be denoted as $|p^{[n]}|$. It can be shown (e.g. [5]) that

$$|p^{[n]}| = y^{(n)} + \binom{n+1}{3}p(x)y^{(n-2)} + 2\binom{n+1}{4}p'(x)y^{(n-3)}+\ldots .$$

In order to find whether two given linear differential equations of the n-th order, $P_n(y,x;I)$ and $Q_n(z,t;J)$ with sufficiently smooth coefficients are globally equivalent, we rewrite them in the form

$$P_n(y,x;I) = |p^{[n]}| + r_{n-3}(x)y^{(n-3)} + r_{n-4}(x)y^{(n-4)}+\ldots = 0$$

and

$$Q_n(z,t;J) = |q^{[n]}| + s_{n-3}(t)z^{(n-3)} + s_{n-4}(t)z^{(n-4)}+\ldots = 0 ,$$

where the first three coefficients of P_n and Q_n coincide with the coefficients of the iterative expressions $|p^{[n]}|$ and $|q^{[n]}|$, respectively. *If the equation P_n is globally transformable into the equation Q_n by means of a global transformation with the change $x = h(t)$, then*

A. *the second order equation $u'' + p(x)u = 0$ on I is globally transformable into $v'' + q(t)v = 0$ on J with the same change $x = h(t)$ of the independent variable,*

B. *the following relations are satisfied*

$r_{n-3}(h(t))h'^3(t) = s_{n-3}(t)$ *on J*

$r_{n-4}(h(t))h'^4(t) = s_{n-4}(t)$ *on J where* $s_{n-3}(t) = 0$,

$r_{n-5}(h(t))h'^5(t) = s_{n-5}(t)$ *on J where* $s_{n-3}(t) = s_{n-4}(t) = 0$,

etc.

Due to condition B the criterion is *in general effective*, that means, that it is expressible in terms of quadratures of coefficients of given equations. Let us recall that for the second order equations the criterion is not effective in this sense, since it requires the number of zeros of solutions.

IV. Stationary Groups

Stationary groups for the second order equations, called groups of *dispersions*, were studied and completely described by O. Borůvka [3] in the sixties. Some results on stationary groups of linear differential equations of an arbitrary order were obtained in 1977 mainly by using the theory of functional equations [11].

In 1979 J. Posluszny and L.A. Rubel [15] characterized (up to conjugacy) those transformations, called *motions*, of a linear differential equation into itself that consist in a change of the independent variable only.

Finally, in 1984 on the basis of our criterion of global equivalence a *complete characterization* of all possible stationary groups was derived [14]. Here is the list of the groups up to conjugacy of linear differential equations of all orders considered with respect to global transformations in the most general form, i.e., involving changes both the independent and the dependent variables:

1. The functions $h : \mathbf{R} \to \mathbf{R}$, $h(x) = \operatorname{Arctan} \frac{a \tan x + b}{c \tan x + d}$, $|ad - bc| = 1$

2. $h : \mathbf{R}_+ \to \mathbf{R}_+$, $h(x) = \operatorname{Arctan} \frac{a \tan x}{c \tan x + 1/a}$, $a \neq 0$

3m. For each positive integer m, $h : (0, m\pi) \to (0, m\pi)$,
$h(x) = \operatorname{Arctan} \frac{a \tan x}{c \tan x \pm 1/a}$, $a \neq 0$

4m. For each positive integer m, $h : (0, m\pi - \pi/2) \to (0, m\pi - \pi/2)$,
$h(x) = \operatorname{Arctan}(k \tan x)$ and $h(x) = \operatorname{Arctan}(k \cot x)$, $k > 0$

5. The functions $h : \mathbf{R} \to \mathbf{R}$, $h(x) = x + c$ and $h(x) = -x + c$, $c \in \mathbf{R}$

6. The increasing functions from 5

7. The functions $h : \mathbf{R} \to \mathbf{R}$, $h(x) = x + k$ and $h(x) = -x + k$, $k \in \mathbf{Z}$

8. The increasing functions from 7

9. $id_{\mathbf{R}}$ and $-id_{\mathbf{R}}$

10. Only $id_{\mathbf{R}}$.

These groups range from the maximal one, a three-parameter group in case 1, through an infinite cyclic group in case 8, to the trivial group in case 10 consisting from the identity only. Let me point out that the maximal group has already occured as the fundamental group in

Borůvka's investigations of the second order equations.

For each case of the stationary groups we can characterize the corresponding equations and each of the cases listed here actually occurs. E.g., the case 1 takes place exactly when the equation is an iterative equation of an arbitrary order iterated from a both-side oscillatory second order equation.

Let us note that if we consider global transformations with only *increasing* changes of the independent variable then, up to conjugacy, there are 5 *possible cases* of stationary groups *with respect to the number of parameters* as announced in 1982 [12].

V. Canonical Forms

The next important notion is the notion of canonical forms of linear differential equations. Such forms were studied from the early beginning of investigation of the equations in the middle of the last century.

We have mentioned that already in 1910 G.D. Birkhoff pointed out that the so called *Laguerre-Forsyth canonical form is not global*. It can be shown [13] that also the other canonical form that has occurred in the literature, the so called *Halphen canonical form is not global either*.

For constructions of *global canonical forms* we may proceed in two ways, either we use a certain geometrical approach, or we may apply the criterion of global equivalence.

First let us explain shortly our geometrical approach. We have seen that fundamental solutions z, considered as curves in an n-dimensional vector space, corresponding to all equations globally equivalent to one equation with a fundamental solution y, a curve y, are obtained as sections of a cone determined by the curve y. To find a canonical, that means, a special equation in the class of equivalent equations, we need a special section of the cone. By applying Cartan's moving-frame-of-reference method we come unfortunately again to the Halphen forms that are not global. However, if we consider the euclidean n-dimensional space and take the central projection of our curves and then their length parametrization, we obtain special sections of the cone, special curves. Fortunately, this can be done without any restrictions on the whole intervals of definition. Then by using differential geometrical methods the explicit forms of the special, canonical equations corresponding to the special curves are obtained.

These global canonical forms are

$n = 2: \qquad y'' + y = 0 \quad \text{on (different)} \quad I \subset \mathbf{R},$

$n = 3$: $\quad y''' - (p'(x)/p(x))y'' + (1 + p^2(x))y' - (p'(x)/p(x))y = 0$ on $I \subset \mathbf{R}$,

(one) arbitrary function $p \in C^1(I)$, $p(x) \neq 0$ on I,

etc.

For $n = 2$ the canonical equations coincide with the canonical forms studied by O.Borůvka.

There is also another procedure producing global canonical forms. This procedure is analytical and the construction is based on our criterion of global equivalence. Among many different global canonical forms obtained by this approach [13] the following equations

$$y^{(n)} + 0.y^{(n-1)} + 1.y^{(n-2)} + p_{n-3}(x)y^{(n-3)} + \ldots + p_0(x)y = 0, I \subset \mathbf{R},$$

are global canonical forms. for equations with sufficiently smooth coefficients. They are characterized by their first three coefficients

1, 0, 1 .

Comparing with the local Laguerre-Forsyth canonical forms having the corresponding sequence

1, 0, 0 ,

we may conclude that if Laguerre and Forsyth had taken 1 as the coefficient of the (n-2)nd derivative instead of their zero they would have got global forms instead of their local.

VI. Invariants

Invariants of linear differential equations with respect to transformations have been derived from the middle of the last century either directly, or mainly on the basis of the Halphen canonical forms. These invariants are local.

A *global invariant* of the *second order* linear differential equations is in fact their *type*: finite (a positive integer) or infinite, and their *kind*, as introduced and derived by O.Borůvka in the sixties.

Due to the criterion of global equivalence we have now also *global invariants* for equations of an *arbitrary order*. Indeed, the *type and kind of the equation* (p): $u'' + p(x)u = 0$ *on* I *is a global invariant of the n-th order equations* P_n *rewritten in the form*

$$P_n(y,x;I) = |p^{[n]}(y,x;I)| + r_{n-3}(x)y^{(n-3)} + \ldots = 0 .$$

Another interesting invariants have occurred recently. It is a bit misleading fact that each second order equation with only continuous coefficients can be globally transformed into an equation with even analytic coefficients, e.g., into $y'' + 1.y = 0$ on some $I \subset \mathbf{R}$. For higher order equations the degree of the smoothness of their coefficients is in some respect an invariant property. From many results of this kind let me introduce at least the following simplest one:

If the coefficients of the equation $P_n(y,x;I)$ satisfy

$$p_{n-1} \in C^{n-2}(I),\ p_{n-2} \in C^{n-3}(I),\dots,\ p_j \in C^{j-1}(I) \text{ for some } j \le n-1 ,$$

then the *coefficients of any globally equivalent equation to the* $P_n(y,x;I)$ *have the same order of differentiability.*

VII. Equations with Solutions of Prescribed Properties

The main idea how to construct linear differential equations with solutions of some prescribed properties is based on the following "*coordinate approach*".

Having global canonical forms (the globality is essential), each linear differential equation P of an arbitrary order can be "coordinated" by a couple $\{S,\alpha\}$ consisting of its global canonical form S and of the global transformation α converting S into P, i.e., $P = S\alpha$.

If we succeed to reformulate a given property of solutions of P equivalently into properties of S and α, we may construct all required equations. Also problems concerning relations among certain properties are then converted into (sometimes simple, or even already solved) problems from the theory of functions.

By using this approach there were constructed linear differential equations that have important applications in differential and integral geometries. E.g., it was possible *to generalize Blaschke's and Santaló's isoperimetric theorems*, [8].

Connections between boundedness of solutions and their L^2-properties were easily explained by the above method [7].

Relations between *distributions of zeros* and *asymptotic behavior* of the solutions were also deeply studied by means of the coordinate approach.

There is also another way, a geometrical one, how to see what happens with zeros of solutions and how to construct equations with prescribed distribution of zeros of their solutions.

VIII. Zeros of Solutions

This geometrical approach is based on the representation of a fundamental solution y of an equation $P_n(y,x;I)$ as a curve in n-dimensional vector or even euclidean space V_n mentioned in the previous sections.

Let the curve v be the central projection of the curve y onto the unit sphere S_{n-1} in the space V_n without a change of parameter x. Each solution y of $P_n(y,x;I)$ can be written as a scalar product c . y where c is a nonzero constant vector in V_n. Let H(y) denote the hyperplane

$$H(y) := \{d \in V_n ;\ c \cdot d = 0\}$$

going through the origin and corresponding to the vector c. Evidently

$$0 = y(x_0) = c \cdot y(x_0) = c \cdot v(x_0)|y(x_0)| \Leftrightarrow c \cdot v(x_0) = 0$$

since $|y(x_0)| \neq 0$. Thus we have shown that

to each solution y *of the equation* P_n *there corresponds a hyperplane* H(y) *in* V_n *going through the origin such that*

zeros of the solution y *occur as parameters of intersections of the particular hyperplane* H(y) *with the curve* v, *and vice versa.*

Multiplicities of zeros occur as orders of contacts, [9].

Let us recall that all this happens on the unit sphere, a compact space, where strong topological tools are at our disposal.

Several open problems were solved and many complicated constuctions were easily explained by using this approach, [10]. As a simple demonstration of the method let us present Sansone's result by constructing a *third order linear differential equation with all oscillatory solutions.*

For this purpose it is sufficient to have an enough smooth (of the class C^3) curve u on the unit aphere S_2 in 3-dimensional space without points of inflexion (that means, that Wronskian of u is nonvanishing) such that each plane going through the origin intersects u for infinitely many values of parameter. The picture of a closed "prolonged cycloid" infinitely many times surrounding the equator as its parameter ranges from $-\infty$ to $+\infty$ may serve as an example of a curve with the required property.

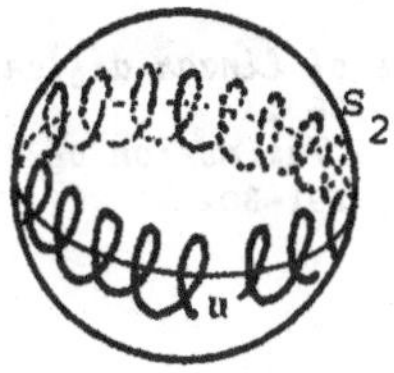

IX. Applications

To the end of my survey let me mention some fruitful applications of the presented theory.

The above methods were succesfully applied to *systems of linear differential equations*. E.g., construction of certain second order systems with only periodic solutions, [10], plays an important role in geometry of manifolds whose all geodesics are closed [2].

By using the above approach there were solved some problems concerning *linear and nonlinear differential equations and systems with one or several delays*. There are useful applications in generalized differential equations and *linear differential expressions with quasi-derivatives* as well. Last but not least, there are many fruitful connections with the *theory of functional equations*.

References

[1] Birkhoff, G.D.: *On the solutions of ordinary linear homogeneous differential equations of the third order*, Annals of Math. 12 (1910/11), 103-124.

[2] Besse, A.L.: *Manifolds All of Whose Geodesics are Closed*, Ergenisse, Vol. 93, Springer, Berlin & New York, 1978.

[3] Borůvka, O.: *Linear differentialtransformationen 2. Ordnung*, VEB Berlin 1967; *Linear Differential TRansformations of the Second Order*, The English Univ. Press, London 1971.

[4] Čadek, M.: *A form of general pointwise transformations of linear differential equations*, Czechoslovak Math. J. (in print).

[5] Hustý, Z.: *Die Iteration homogener linear Differentialgleichungen*, Publ. Fac. Sci. Univ. J.E. Purkyně (Brno) 449 (1964), 23-56.

[6] Kummer, E.: *De generali quadam aequatione differentiali tertii ordinis*, Progr. Evang. Königl. & Stadtgymnasiums Liegnitz 1834.

[7] Neuman, F.: *Relation between the distribution of the zeros of the solutions of a 2nd order linear differential equation and the boundedness of these solutions*, Acta Math. Acad. Sci. Hungar. 19 (1968), 1-6.

[8] Neuman, F.: *Linear differential equations of the second order and their applications*, Rend. Math. 4 (1971), 559-617.

[9] Neuman, F.: *Geometrical approach to linear differential equations of the n-th order*, Rend. Mat. 5 (1972), 579-602.

[10] Neuman, F.: *On two problems about oscillation of linear differential equations of the third order*, J. Diff. Equations 15 (1974), 589-596.

[11] Neuman, F.: *On solutions of the vector functional equation* $y(\xi(x)) = f(x).A.y(x)$, Aequationes Math. 16 (1977), 245-257.

[12] Neuman, F.: *A survey of global properties of linear differential equations of the n-th order*, in: Lecture Notes in Math. 964, 543-563.

[13] Neuman, F.: *Global canonical forms of linear differential equations*, Math. Slovaca 33, (1983), 389-394.

[14] Neuman, F.: *Stationary groups of linear differential equations*, Czechoslovak Math. J. 34 (109) (1984), 645-663.

[15] Posluszny, B. and Rubel, L.A.: *The motion of an ordinary differential equation*, J. Diff. Equations 34 (1979), 291-302.

Details will appear in

Neuman, F.: *Ordinary Linear Differential Equations*, Academia Publishing House, Prague & North Oxford Academic Publishers Ltd., Oxford.

NUMERICAL AND THEORETICAL TREATING OF EVOLUTION PROBLEMS BY THE METHOD OF DISCRETIZATION IN TIME

K. REKTORYS
Technical University Prague
Thákurova 7, 166 29 Prague 6, Czechoslovakia

More than fifty years ago, E. Rothe suggested an approximate method of solution of parabolic problems. He divided the interval $I = [0,T]$ for the variable t into p subintervals I_j of the length $h = T/p$ and at each point $t_j = jh$, $j = 1,\dots,p$, he approximated the function $u(x,t_j)$ by the function $z_j(x)$ and the derivative $\partial u/\partial t$ by the difference quotient $[z_j(x) - z_{j-1}(x)]/h$. Starting with z_0 given by $z_0(x) = u(x,0) = u_0(x)$, he found, successively for $j = 1,\dots,p$, the approximations $z_j(x)$ as solutions of the so arisen ordinary boundary value problems. The problem, solved originally by E. Rothe, was a very simple one. However, his method turned out to be a very useful tool for solution of substantially more complicated evolution problems (at first linear and quasilinear parabolic problems of the second order in n dimensions, later parabolic problems of arbitrary order, nonlinear problems, hyperbolic problems, the Stephan problem, integrodifferential problems, mixed parabolic-hyperbolic problems, etc.). The development of the Rothe method, called also the method of discretization in time, or the horizontal method of lines, is connected with such names as O. A. Ladyženskaja, T. D. Ventcel, A. M. Iljin, A. S. Kalašnikov, O. A. Olejnik, J. I. Ibragimov, P. S. Mosolov, O. A. Liskovec, R. D. Richtmayer, N. N. Janěnko, M. Zlámal, J. Nečas, J. Kačur, A. G. Kartsatos, M. E. Parrot, W. Ziegler, J. W. Jerome, E. Martensen and his school, U. v. Welck, J. Naumann, C. Corduneanu, etc. Theoretical as well as numerical questions have been examined (existence and convergence theorems, regularity questions, numerical aspects, etc.). Many of the obtained results were obtained as well by other methods - method of compactness, theory of semigroups, method of monotone operators, Fourier transform, etc. (A. Friedman, M. Krasnoselskij, P. E. Sobolevskij, F. E. Browder, J. L. Lions, E. Magenes, H. Brézis, V. Barbu, D. Pascali, M. G. Crandal, W. v. Wahl, etc.). As concerns numerical methods, related to the Rothe method, the methods of space - or time-space discretization were applied (V. N. Fadějeva, J. Douglas, T. Dupont, M. Zlámal, R. Glowinski, J. L. Lions,

R. Tremolière, P. A. Raviart, W. Walter, K. Gröger, etc.). Each of the mentioned methods, including the Rothe method, has its preferences and its drawbacks. However, the Rothe method has its significance both as a numerical method and theoretical tool. Existence theorems are proved in a constructive way. Thus no other methods are needed to give preliminary information on existence, or regularity of the solution as required in many other numerical methods when questions on convergence, or order of convergence, etc. are to be answered. The Rothe method is a stable method. To the solution of elliptic problems generated by this method, current methods, especially the variational ones, can be applied. As concerns theoretical results, they are obtained in a relatively simple way, as usual. Moreover, the Rothe method, being a very natural one, makes it possible to get a particularly good insight into the structure of the solutions. Often a brief inspection of the corresponding elliptic problems gives an information what can be expected as concerns properties of the solution. This is why I prefer it.

In 1971, a slightly different technics than that applied currently in this method appeared in my work [2], making it possible to treat corresponding elliptic problems in a particularly simple way. This technics was followed by other authors (in our country J. Nečas, J. Kačur) and became a base for work of my seminar at the Technical University in Prague. Results obtained in this seminar were summarized in my monograph [1] in 1982. I would like to present some of them here, pointing out the very simple way in which they have been obtained.

1. Existence and convergence theorem. Let us start with a relatively simple parabolic problem

$$\frac{\partial u}{\partial t} + Au = f \quad \text{in} \quad G \times (0,T) \; , \tag{1}$$

$$u(x,0) = 0 \; , \tag{2}$$

$$B_i u = 0 \quad \text{on} \quad \Gamma \times (0,T) \; , \quad i = 1,\dots,\mu \; , \tag{3}$$

$$C_i u = 0 \quad \text{on} \quad \Gamma \times (0,T) \; , \quad i = 1,\dots,k-\mu \; . \tag{4}$$

Here, G is a bounded region in E_N with a Lipschitz boundary Γ ,

$$A = \sum_{|i|,|j|\le k} (-1)^{|i|} D^i\big(a_{ij}(x)D^j\big) \tag{5}$$

with a_{ij} bounded and measurable in G , $f \in L_2(G)$; (3), or (4) are (linear) boundary conditions, stable (thus containing derivatives of orders $\le k - 1$), or unstable with respect to the operator A , respectively. Denote

$$V = \{v;\ v \in \overset{o}{W}{}_2^{(k)}(G),\quad B_i v = 0 \ \text{ on } \Gamma \ \text{ in the sense of traces},\ i = 1,\dots,\mu\}\ , \tag{6}$$

let $((\cdot,\cdot))$ be the bilinear form, corresponding to the operator A and to the boundary conditions (3), (4), familiar from the theory of variatiinal methods. (Roughly speaking, $((v,u))$ is obtained of (v,Au), applying to every integral $\int_G v\, D^i(a_{ij}\, D^j u)\, dx$ i-times the Green theorem in the usual way, see e.g. [1], or [3]. For example, if $A = -\Delta$ and $u = 0$ on Γ is prescribed, then

$$V = \overset{o}{W}{}_2^{(1)}(G) \quad \text{and} \quad ((v,u)) = \sum_{i=1}^{N} \int_G \frac{\partial v}{\partial x_i} \frac{\partial u}{\partial x_i}\, dx\ .)$$

Let the form $((\cdot,\cdot))$ be bounded in $V \times V$ and V-elliptic, i.e. let two positive constants K and α (independent of v and u) exist such that the inequalities

$$|((v,u))| \le K\|v\|_V \|u\|_V\ , \tag{7}$$

$$((v,v)) \ge \alpha \|v\|_V^2 \tag{8}$$

hold for all $v, u \in V$. Let to the solution of the problem (1) - (4) the Rothe method be applied. Denote

$$Z_i(x) = \frac{z_i(x) - z_{i-1}(x)}{h}\ , \quad i = 1,\dots,p\ . \tag{9}$$

(Thus $Z_i(x)$ "corresponds" to the derivative $\partial u/\partial t$ for $t = t_i$.) In the weak formulation, we have to find successively for $j = 1,\dots,p$, the functions $z_j \in V$, satisfying,

$$((v,z_j)) + (v,Z_j) = (v,f) \quad \forall\ v \in V\ , \tag{10}$$

with $z_0(x) = u(x,0) = 0$. Under the assumptions (7), (8), each of these problems has exactly one solution $z_j \in V$. <u>Apriori estimates</u>: Put $v = Z_1 = (z_1 - z_0)h = z_1/h$ into (10) written for $j = 1$. We obtain

$$h((Z_1,Z_1)) + (Z_1,Z_1) = (Z_1,f)\ . \tag{11}$$

Because of (8) and $|(Z_1,f)| \le \|Z_1\|\ \|f\|$, (11) yields

$$\|Z_1\|^2 \le \|Z_1\|\ \|f\| \Longrightarrow \|Z_1\| \le \|f\|\ . \tag{12}$$

Subtracting (10), written for $j - 1$, from (10) gives

$$h((v,Z_j)) + (v,\ Z_j - Z_{j-1}) = 0\ .$$

Putting $v = Z_j$, we obtain, in a similar way as before,

$$\|Z_j\| \le \|Z_{j-1}\|\ ,\quad j = 2,\dots,p\ ,$$

what gives, together with (12)

$$||Z_j|| \leq ||f|| = c_1 \ . \tag{13}$$

Let us refine our division, considering the divisions d_n with the steps $h_n = h_1/2^{n-1}$, $n = 1,2,\dots$, $h_1 = h$. Denote the corresponding functions

$$z_j^n \ , \quad Z_j^n = \frac{z_j^n - z_{j-1}^n}{h_n} \ .$$

The estimate (13) having been obtained independently of the lengh of the step h , it remains valid as well for the division d_n ,

$$||Z_j^n|| \leq c_1 \ . \tag{14}$$

Because $z_j^n = (z_j^n - z_{j-1}^n) + \dots + (z_1^n - z_0^n)$, it follows

$$||z_j^n|| \leq jh_n\big(||Z_j|| + \dots + ||Z_1||\big) \leq Tc_1 = c_2 \ . \tag{15}$$

Putting then $v = Z_j^n$ into (10) written for the functions z_j^n and Z_j^n and using (8), we get

$$||z_j^n||_V \leq c_3 \ . \tag{16}$$

(14), (15) and (16) are the basic needed a priori estimates. They have actually been obtained in a very, very simple way. What follows, is a standard procedure, now. Let

$$u_n(t) = z_{j-1}^n + (z_j^n - z_{j-1}^n)\,\frac{t - t_{j-1}^n}{h} \qquad \text{for } t_{j-1}^n \leq t \leq t_j^n \ , \quad j = 1,\dots,p\cdot 2^{n-1} \tag{17}$$

($n = 1,2,\dots$) (the so-called Rothe functions), or

$$U_n(t) = \begin{cases} Z_1^n & \text{for } t = 0 \ , \\ Z_j^n & \text{for } t_{j-1}^n < t \leq t_j^n \ , \quad j = 1,\dots,p\cdot 2^{n-1} \end{cases} \tag{18}$$

($n = 1,2,\dots$) be abstract funcions, considered as functions from $I = [0,T]$ into V , or $L_2(G)$, respectively. In consequence of their form and of (16) and (14), they are uniformly bounded (with respect to n). in $L_2(I,V)$, or $L_2(I,L_2(G))$, respectively (even in $C(I,V)$, or $L_\infty(I,L_2(G))$). The space $L_2(I,V)$, and $L_2(I,L_2(G))$ being Hilbert spaces, a subsequence $\{u_{n_k}\}$, or $\{U_{n_k}\}$ can be found such that

$$u_{n_k} \rightharpoonup u \ \text{ in } L_2(I,V) \ , \quad U_{n_k} \rightharpoonup U \ \text{ in } L_2(I,L_2(G)) \ . \tag{19}$$

Now, (17), (18) imply

$$\int_0^t U_n(\tau)d\tau = u_n(t) \ \forall n ,$$

yielding easily

$$\int_0^t U(\tau)d\tau = u(t) ,$$

and thus $U = u'$ in $L_2(I,L_2(G))$. Consequently, $u \in AC(I,L_2(G))$ and $u(0) = 0$ in $C(I,L_2(G))$. So the function u satisfies

$$u \in L_2(I,V) \cap AC(I,L_2(G)), \tag{20}$$

$$u' = U \in L_2(I,L_2(G)), \tag{21}$$

$$u(0) = 0 \text{ in } C(I,L_2(G)). \tag{22}$$

Moreover, on base of integral identities (10) and of (19), one comes to the integral identities

$$\int_0^T ((v,u))\, dt + \int_0^T (v,u')\, dt = \int_0^T (v,f)\, dt. \tag{23}$$

A function with the properties (20) - (23) is called the <u>weak solution</u> of the problem (1) - (4).

<u>Uniqueness</u>: Let u_1, u_2 be two functions satisfying (20) - (23). Then their difference $u = u_2 - u_1$ has the properties (20) - (22) and satisfies

$$\int_0^T ((v,u))\, dt + \int_0^T (v,u')\, dt = 0 \quad \forall\, v \in L_2(I,V).$$

Let $a \in I$ be arbitrary. Choose

$$v(t) = \begin{cases} u(t) & \text{for } 0 \leq t \leq a, \\ 0 & \text{for } a < t \leq T. \end{cases}$$

We have

$$\int_0^T (v,u')\, dt = \int_0^a (u,u')\, dt = \frac{1}{2}||u(a)||^2 - \frac{1}{2}||u(0)||^2 = \frac{1}{2}||u(a)||^2.$$

In consequence of (8) we thus obtain $||u(a)||^2 = 0$; the point a having been chosen arbitrarily, $u = 0$ in I.

Uniqueness implies in the familiar way that $u_n \to u$ (not only $u_{n_k} \to u$) in $L_2(I,V)$. Moreover, for every $t \in I$ the sequence $\{u_n(x)\}$ is bounded in V and thus compact in $L_2(G)$. The functions $u_n(t)$ being uniformly

bounded in I, in the metric of the space $L_2(G)$, and equicontinuous on base of (13), the Ascoli theorem can be applied, implying (strong) uniform convergence, in I, of $\{u_n\}$ to u.
Summarising, we thus have:

Theorem 1. Let (7), (8) be satisfied, let $f \in L_2(G)$. Then there exists exactly one weak solution of the problem (1) - (4) and

$$u_n \rightharpoonup u \text{ in } L_2(I,V), \quad u_n \rightrightarrows u \text{ in } C(I,L_2(G)). \tag{24}$$

Remark 1. By a more detailed treatment it can be proved that even $u_n \rightrightarrows u$ in $C(I,V)$, $u' \in L_\infty(I,L_2(G))$. We shall not go into details here. See [1].

2. Error Estimates. As can be excepted, to get an efficient error estimate, some supplementary assumptions are needed: Let the assumptions of Theorem 1 be fulfilled. Let, moreover,

$$f \in V, \quad Af \in L_2(G), \quad ((v,f)) = (v,Af) \;\forall\; v \in V. \tag{25}$$

Then

$$\|u(x,t_j) - z_j(x)\| \leq \frac{Mjh^2}{2}, \quad j = 1,\dots,p, \tag{26}$$

where $M = \|Af\|$. If, moreover, the coefficient C of positive definiteness can be easily found, for which thus

$$((v,v)) \geq C^2 \|v\|^2 \;\forall\; v \in V$$

holds, then the following (slightly better) estimate can be used:

$$\|u(x,t_j) - z_j(x)\| \leq \frac{Mh}{2C^2}(1 - e^{-C^2 jh}), \quad j = 1,\dots,p. \tag{27}$$

Proof of (26) (the proof of (27) is similar): Let us investigate the division d_2 (for the notation see the text following (13)) and denote

$$z^2_{2i} - z_i = q^1_i, \quad i = 0,1,\dots,p$$

(with $z^1_i = z_i$). The functions z^2_{2i} satisfy the integral identities

$$((v,z^2_{2i})) + \frac{1}{h/2}(v,z^2_{2i} - z^2_{2i-1}) = (v,f) \;\forall\; v \in V.$$

Subtracting the integral identity (10), with i written for j, we obtain

$$((v,z^2_{2i} - z_i)) + \frac{1}{h}(v,(z^2_{2i} - z_i) - (z^2_{2i-2} - z_{i-1})) =$$

$$= -\frac{1}{h}(v,z^2_{2i} - 2z^2_{2i-1} + z^2_{2i-2}) \;\forall\; v \in V,$$

or, denoting

$$s_i^2 = \frac{z_i^2 - 2z_{i-1}^2 + z_{i-2}^2}{(h/2)^2} \quad (s_i^n = \frac{z_i^n - 2z_{i-1}^n + z_{i-2}^n}{h_n^2}, \text{ in general}),$$

$$((v,q_i^1)) + \frac{1}{h}(v,q_i^1 - q_{i-1}^1) = -\frac{h}{4}(v,s_{2i}^2) \quad \forall\, v \in V,$$

with $q_0^1 = z_0^2 = z_0 = 0$. Putting $v = q_1^1$ for $i = 1$ and taking (8) into account, we get

$$\|q_1^1\| \leq \frac{h^2}{4} \|s_2^2\| \leq \frac{h^2 M}{4},$$

because under the assumptions (25)

$$\|s_i^n\| \leq M \text{ for all } n \text{ and } i.$$

Similarly, for $i = 2$ we obtain

$$\|q_2^1\| \leq \|q_1^1\| + \frac{h^2 M}{4} \leq \frac{2h^2 M}{4},$$

and, finally,

$$\|q_i^1\| \leq \frac{ih^2 M}{4}, \qquad i = 0,1,\ldots,p.$$

In general, we have, for $q_i^n = z_{2i}^{n+1} - z_i^n$,

$$\|q_i^n\| \leq \frac{i(h/2^{n-1})^2 M}{4}, \quad i = 0,1,\ldots,2^{n-1}p.$$

Now,

$$\|u_n(x,t_j) - z_j(x)\| = \|z^n_{2^{n-1}j} - z_j^1\| \leq$$

$$\leq \|q^{n-1}_{2^{n-2}j} + q^{n-2}_{2^{n-3}j} + \ldots + q_{2j}^2 + q_j^1\| \leq$$

$$\leq \frac{h^2 M}{4}(j + \frac{2j}{2^2} + \ldots + \frac{2^{n-1}j}{(2^{n-1})^2}) < \frac{jh^2 M}{2}.$$

Coming to the limit for $n \to \infty$ (what is allowed because of Theorem 1), we obtain (26).

The estimate (26) is very sharp (and the more is the estimate (27)), as can be seen from the following simple example:

$$\frac{\partial u}{\partial t} - \frac{\partial^2 u}{\partial x^2} = \sin x \text{ in } (0,\pi) \times (0,1), \tag{28}$$

$$u(x,0) = 0, \tag{29}$$

$$u(0,t) = 0, \quad u(\pi,t) = 0. \tag{30}$$

The assumptions (25) are easily established. Further, we have

$$M = ||Af|| = ||-(\sin x)''|| = ||\sin x|| = \sqrt{(\pi/2)} .$$

Choosing, for, example, $h = 0.01$ and $j = 20$, or $j = 40$, the Rothe method yields

$$z_{20} = 0.1805 \sin x ,$$
$$z_{40} = 0.3284 \sin x ,$$

respectively. The exact solution is known (this was the reason why such a simple example has been chosen): $u = (1 - e^{-t}) \sin x$. Thus

$$u(x, 0.20) = 0.1813 \sin x ,$$

$$u(x, 0.40 = 0.3297 \sin x .$$

The actual errors then are

$$\begin{aligned} ||u(x, 0.20) - z_{20}|| &= 0.0008||\sin x|| = 0.00101 , \\ ||u(x, 0.40) - z_{40}|| &= 0.0013||\sin x|| = 0.00163 , \end{aligned} \tag{31}$$

while (26) gives

$$\begin{aligned} ||u(x, 0.20) - z_{20}|| &\leq \frac{20 \cdot 0.01^2}{2} \sqrt{\tfrac{\pi}{2}} = 0.00125 , \\ ||u(x, 0.40) - z_{40}|| &\leq \frac{40 \cdot 0.01^2}{2} \sqrt{\tfrac{\pi}{2}} = 0.00251 . \end{aligned} \tag{32}$$

Finding $C = 1$ (see [1], p. 90) and using (27), we get the estimates

$$\begin{aligned} ||u(x, 0.20) - z_{20}|| &\leq \frac{0.01}{2} (1 - e^{-0.2}) \sqrt{\tfrac{\pi}{2}} = 0.00113 , \\ ||u(x, 0.40) - z_{40}|| &\leq \frac{0.01}{2} (1 - e^{-0.4}) \sqrt{\tfrac{\pi}{2}} = 0.00206 \end{aligned} \tag{33}$$

which are still better then the estimates (32). The example demonstrates very well the sharpness of the estimates (26), (27) and the fact that they cannot be substantially improved.

3. Nonhomogeneous initial and boundary conditions. Let us first investigate the problem (1) - (4) with homogeneous equation and nonhomogeneous initial condition $u(x,0) = u_0 \in L_2(G)$, i.e. the problem

$$\frac{\partial u}{\partial t} + Au = 0 \quad \text{in} \quad G \times (0,T) , \tag{34}$$

$$u(x,0) = u_0(x) \tag{35}$$

$$B_i u = 0 \quad \text{on} \quad \Gamma \times (0,T) , \quad i = 1,\dots,\mu , \tag{36}$$

$$G_i u = 0 \quad \text{on} \quad \Gamma \times (0,T) , \quad i = 1,\dots,k-\mu . \tag{37}$$

The corresponding integral identities are

$$((v,z_j)) + (v,Z_j) = 0 \quad \forall v \in V , \quad j = 1,\dots,p , \tag{38}$$

with $Z_j = (z_j - z_{j-1})/h$, $z_0 = u_0$. (39)

Similarly as in the case of the problem (1) - (4), we come to the inequality

$$||Z_j|| \leq ||Z_{j-1}|| .$$

However, if only $u_0 \in L_2(G)$ is assumed, it is not possible to put $v = Z_1$ into the first of the integral identities (38) to obtain a simple estimate for $||Z_1||$ as in (11), (12), because we have not $Z_1 \in V$ here, in general. Thus an existence theorem is derived, first for "sufficiently smooth" $u_0 = s \in V$, more precisely for u_0 from the set M of such functions $s \in V$ for which a unique $g \in L_2(G)$ exists satisfying

$$((v,s)) = (v,g) \quad \forall\ v \in V.$$

Putting then

$$z_j = r_j + s$$

into the integral identities

$$((v,z_j)) + \frac{1}{h}(v,z_j - z_{j-1}) = 0 \quad \forall\ v \in V,$$

with $z_0 = s$, we come to the identities

$$((v,r_j)) + \frac{1}{h}(v,r_j - r_{j-1}) = -(v,g) \quad \forall\ v \in V,$$

with $r_0 = 0$, corresponding to the problem (1) - (4) in which u is replaced by r and f by -g. Having obtained its weak solution r(t), the weak solution of (34) - (37) with $u_0 = s \in M$ is defined by $u(t) = r(t) + s$. Moreover, if we put z_j for v into the original integral identities, we obtain, subsequently,

$$||z_1|| \leq ||s||, \quad ||z_2|| \leq ||z_1|| \leq ||s||, \text{ etc.}$$

The function u(t) being the limit, in $C(I,L_2(G))$, of the corresponding Rothe sequence, it follows

$$||u(t)|| \leq ||s|| \quad \text{for all } t \in I.$$

Now, the form ((v,u)), being V-elliptic, the set M is dense in V, thus as well in $L_2(G)$. Let $u_0 \in L_2(G)$ and let $s_i \in M$, i = 1,2,..., be (an arbitrary) sequence converging to u_0 in $L_2(G)$. Then the sequence of corresponding weak solutions $u^{(i)}(t)$ is a Cauchy sequence in $C(I,L_2(G))$, because

$$||u^{(j)}(t) - u^{(k)}(t)|| \leq ||s_i - s_k|| \quad \forall\ t \in I.$$

Its limit is then called the very weak solution of the problem (34) - (37). Obviously, this very weak solution is uniquely determined by the initial function $u_0 \in L_2(G)$.
About convergence, in $C(I,L_2(G))$, of the corresponding Rothe sequence to this very weak solution as well about nonhomogeneous boundary conditions see [1].

4. The Ritz-Rothe method. Let us investigate the problem (34) - (37) with $f \in L_2(G)$ on the right-hand side of (34) instead of zero. The solution $u(t)$ of this problem is the sum of solutions of the problems (1) - (4) and (34) - (37). (The problems are linear.) The corresponding integral identities when applying the Rothe method are:

$$((v,z_j)) + \frac{1}{h}(v, z_j - z_{j-1}) = (v,f) \quad \forall v \in V, \quad j = 1,\dots,p, \qquad (40)$$

with $z_0 = u_0$. Let us solve each of these problems approximately - to be concrete, by the Ritz method (or by a method with similar properties) So let $v_1,\dots,v_n$ be the first n terms of a base in V and let z_1^* be the Ritz approximation of the function z_1. Put z_1^* instead of z_1 into the second of the identities (40),

$$((v,\tilde{z}_2)) + \frac{1}{h}(v, \tilde{z}_2 - z_1^*) = (v,f) \qquad (41)$$

and let z_2^* be the Ritz approximation of the function $\tilde{z}_2$, etc. We thus can construct the function

$$u_1^*(t) = z_{j-1}^* + \frac{z_j^* - z_{j-1}^*}{h}(t - t_{j-1}) \quad \text{for} \quad t_{j-1} \leq t \leq t_j, \quad j = 1,\dots,p \qquad (42)$$

which is an analogue of the Rothe function $u_1(t)$. (41) announces that using the Ritz method, the errors become cumulated with increasing j. Fortunately, according to a very simple law: Subtract (41) from the second of the identities (40). We obtain

$$((v, z_2 - \tilde{z}_2)) + \frac{1}{h}(v, (z_2 - \tilde{z}_2) - (z_1 - z_1^*)) = 0 \quad \forall v \in V.$$

Putting $v = z_2 - \tilde{z}_2$, we get

$$||z_2 - \tilde{z}_2|| \leq ||z_1 - z_1^*||.$$

Etc. Using this result, convergence of this "Ritz-Rothe" method is easily proved: Let $\varepsilon > 0$ be given. According to Theorem 1, such a (fine) division of the interval I into p subintervals can be found - let us preserve the notation h for the length of these subintervals - that

$$||u(t) - u_1(t)|| < \frac{\varepsilon}{2} \quad \forall\ t \in I$$

(where $u_1(t)$ is the corresponding Rothe function). Denote $\eta = \varepsilon/2p$. Let the number of terms in the Ritz approximation be sufficiently large so that

$$||z_1 - z_1^*|| < \eta$$

be fulfilled. Then - as just shown -

$$||z_2 - \tilde{z}_2|| < \eta .$$

Let the Ritz approximation z_2^* of $\tilde{z}_2$ be such that

$$||\tilde{z}_2 - z_2^*|| < \eta$$

again. Thus

$$||z_2 - z_2^*|| < 2\eta .$$

In a similar way we come to the estimates

$$||z_j - z_j^*|| < j\eta = \frac{\varepsilon}{2}\ , \quad j = 3,\ldots,p.$$

Because of the form of the Rothe functions $u_1(t)$ and $u_1^*(t)$ (they are piecewise linear in t), we have

$$||u(t) - u_1^*(t)|| < \varepsilon \quad \forall\ t \in I.$$

5. Regularity of the solution. a) Regularity with respect to t, smoothing effect. In [1], Chap. 12 and 13, regularity properties of the weak, or very weak solutions with respect to t are examined. We shall not go into details and show the very simple idea of these investigations on the example of the problem (34) - (37). Let the form $((.,.))$ satisfy (7) and (8) (boundedness and V-ellipticity) and let it be, moreover, symmetric in V, i.e. let

$$((v,u)) = ((u,v)) \quad \forall\ v,\ u \in V \tag{43}$$

be fulfilled. Thus $((.,.))$ has the properties of a scalar product. Let h be sufficiently small (in order that the points t^0, $2t^0$, etc., investigated below, lie in the interval $[0,T]$) and choose an arbitrary $t^0 \in (0,T)$ such that $t^0 = jh$ (j being a positive integer). Take the first of the integral identities (38) and put $v = z_1$. We obtain

$$h((z_1,z_1)) + (z_1,\ z_1 - u_0) = 0 .$$

Writing $(z_1,\ z_1 - u_0)$ in the form $\frac{1}{2}\left(||z_1||^2 + ||z_1 - u_0||^2 - ||u_0||^2\right)$, we get

$$h((z_1,z_1)) + \frac{1}{2}||z_1||^2 \leq \frac{1}{2}||u_0||^2 .$$

Similarly,

$$h((z_2,z_2)) + \frac{1}{2} ||z_2||^2 \leqq \frac{1}{2} ||z_1||^2 ,$$
$$\cdots\cdots\cdots$$
$$h((z_j,z_j)) + \frac{1}{2} ||z_j||^2 \leqq \frac{1}{2} ||z_{j-1}||^2 .$$

Making the sum, we obtain

$$h \sum_{i=1}^{j} ((z_i,z_i)) \leqq \frac{1}{2} ||u_0||^2 . \tag{44}$$

Putting, in the second of the identities (38), $v = z_2 - z_1$, we get, similarly,

$$((z_2 - z_1, z_2)) + \frac{1}{h} (z_2 - z_1, z_2 - z_1) = 0 ,$$

$$\frac{1}{2} [((z_2,z_2)) + ((z_2 - z_1, z_2 - z_1)) - ((z_1,z_1))] \leqq 0 ,$$

$$((z_2,z_2)) \leqq ((z_1,z_1)) .$$

Going on in this way, we obtain

$$((z_j,z_j)) \leqq ((z_{j-1},z_{j-1})) \leqq \ldots \leqq ((z_2,z_2)) \leqq ((z_1,z_1)) . \tag{45}$$

Thus replacing in (44) all the summands by $((z_j,z_j))$ and taking into account that $jh = t^0$, we have

$$((z_j,z_j)) \leqq \frac{1}{2t^0} ||u_0||^2 , \tag{46}$$

and, because of the V-ellipticity of the form $((.,.))$ (see (8)),

$$||z_j||_V \leqq \frac{1}{\sqrt{(2\alpha t^0)}} ||u_0|| . \tag{47}$$

In consequence of (45), this result holds for all larger indices, too, and also for all divisions d_n with $n \geqq 1$,

$$||z_i^n||_V \leqq \frac{1}{\sqrt{(2\alpha t^0)}} ||u_0|| \quad \forall\, t_i^n \geqq t^0 . \tag{48}$$

Using this result, interchanging the role of z_i and Z_i and assuming $2t^0 \in (0,T)$, we get, similarly,

$$||z_i^n|| \leqq \frac{1}{2t^0} ||u_0|| \quad \forall\, t_i^n \geqq 2t^0 . \tag{49}$$

A simple consideration leads then to the conclusion that for the restrictions $\tilde{u}(t)$ and $\tilde{u}'(t)$ of the functions $u(t)$ and $u'(t)$ on the interval $[2t^0,T]$, (48), (49) imply

$$\tilde{u} \in L_2([2t^0,T],V) , \quad \tilde{u}' \in L_2([2t^0,T],L_2(G)) .$$

Going on in the same way, we prove similarly (assuming $4t^0 \in (0,T)$)

$$\tilde{u}' \in L_2([4t^0,T],V) , \quad \tilde{u}'' \in L_2([4t^0,T],L_2(G)) ,$$

etc. Let $\eta \in (0,T)$ be arbitrary, $q > 0$ an arbitrary integer. Chosing $t^0 \leq \eta/(2q + 4)$, we come, in this way, to the result that

$$\tilde{u}^{(i+1)} \in L_2([\eta,T],V) \ , \quad \tilde{u}^{(i+2)} \in L_2([\eta,T],L_2(G)) \ , \quad i = 0,\ldots,q \ ,$$

what implies, among others,

$$\tilde{u}^{(i)} \in AC([\eta,T],V) \ , \quad \tilde{u}^{(i+1)} \in AC([\eta,T],L_2(G)) \ , \quad i = 0,\ldots,q \ .$$

The numbers η and q having been chosen arbitrarily, we have come, in this very simple way, to the following

Theorem 2. Let (7), (8), (43) be fulfilled, $u_0 \in L_2(G)$. Then the very weak solution $u(t)$ of the problem (34) - (37), considered as an abstract function $[0,T] \to V$, has on the interval $(0,T]$ continuous derivatives of all orders.

Let us remark that applying the just shown idea in a properly modified way, J. Kačur obtained rather strong regularity results for the equation $\partial u/\partial t + A(t)u = f(t)$. See [4].

b) Regularity with respect to x. While the basic method how to examine regularity with respect to t has been shown in [1], the idea how to obtain regularity results with respect to x belongs to J. Kačur. Regularity results known for elliptic problems are utilized. For details see [4].

6. Other parabolic problems. Using the same technics as above, linear parabolic equations of the form $\partial u/\partial t + A(t)u = f(t)$, nonlinear equations, integrodifferential equations as well as some nontraditional problems (problem with an integral condition, for example) can be examined. For details see [1].

7. Hyperbolic problems. Also hyperbolic problems can be treated in the same way. Under the assumption of boundedness in $V \times V$, V-ellipticity and V-symmetry of the form $((.,.))$, an existence and convergence theorem has been derived, in [1], and convergence of the "Ritz-Rothe method" proved. Moreover, for $f \in V$, the following error estimate has been found, in a similar way as in the case (26):

$$||u(x,t_j) - z_j(x)|| \leq Mh^3 \, j(j + 1) \quad \text{with} \quad M = \sqrt{\tfrac{3}{2}((f,f))}. \tag{50}$$

For some regularity results see [5]. For generalization to the case of quasilinear hyperbolic equations see [4].

References

[1] REKTORYS, K.: *The Method of Discretization in Time and Partial Differential Equations.* Dordrecht-Boston-London, D. Reidel 1982.

[2] REKTORYS, K.: *On Application of Direct Variational Methods to the Solution of Parabolic Boundary Value Problems of Arbitrary Order.* Czech. Math. J. 21 (1971), 318-339.

[3] REKTORYS, K.: *Variational Methods in Mathematics, Science and Engineering,* 2nd Ed. Dordrecht-Boston-London, D. Reidel 1979.

[4] KAČUR, J.: *Method of Rothe in Evolution Equations.* Leipzig, Teubner. To appear.

[5] PULTAR, M.: *Solution of Abstract Hyperbolic Equations by Rothe Method.* Aplikace matematiky 29, (1984), 23-39.

ALGORITHMS FOR THE INCLUSION OF SOLUTIONS OF ORDINARY INITIAL VALUE PROBLEMS

H. J. STETTER
Technical University Vienna
A-1040 Wien, Austria

Introduction

Customary numerical algorithms do not produce bounds for the true solution of the specified problem but an approximate solution. Information about the remaining error is obtained from a *secondary problem:*

Given the original problem and an approximate solution, find an approximation to its error.

It is obvious that this does not eliminate the uncertainty about the quality of the approximate solution. This is tolerable because most problems are only approximations of real-life situations. Nevertheless, there arise situations where rather concise information about the error of an approximate solution must be obtained. In the following, we will analyze the structure of this task for initial value problems for systems of first order ordinary differential equations.

The Problem

We formulate our task in analogy to the secondary problem above. The original problem is ($y(t) \in \mathbb{R}^s$)

$$y'(t) = f(t,y(t)), \quad y(0) = y_0 , \quad t \in [0,T] , \tag{1.1}$$

with sufficient regularity in a sufficiently large neighborhood of the unique solution trajectory $(t,y(t))$.

Inclusion problem: Given an approximate solution $\tilde{y} : [0,T] \to \mathbb{R}^s$ of (1.1)

Find $\tilde{E} : [0,T] \to \mathbb{P}\mathbb{R}^s$ such that

$$e(t) := \tilde{y}(t) - y(t) \in \tilde{E}(t) , \quad t \in [0,T] . \tag{1.2}$$

Here $\mathbb{P}$ denotes the power set; normally we have to restrict the range of

$\tilde{E}$ to an easily representable subset of $\mathbb{P}\mathbb{R}^s$ like the set $\mathbb{I}\mathbb{R}^s$ of all intervals in $\mathbb{R}^s$. While the computation of norm bounds for e is a special case of (1.2), we will primarily be interested in componentwise lower and upper bounds which may well be of equal sign. Often we will be satisfied with producing values of $\tilde{E}$ at a sequence of arguments $t_o, t_1, \ldots, t_n \in [0,T]$.

It is clear that the inclusion (1.2) of e implies an inclusion

$$y(t) \in \tilde{y}(t) - \tilde{E}(t) \tag{1.3}$$

for the true solution y of (1.1). The algorithms which we will consider are also immediately applicable to the case of *strips* of true solutions ($y(t) \in \mathbb{P}\mathbb{R}^s$) as they appear for a set-valued initial condition $y(0) \in Y_o \in \mathbb{P}\mathbb{R}^s$ in (1.1).

Naturally, the inclusion problem becomes the more delicate the tighter an inclusion we request. It is clear, however, that we cannot generate an inclusion of a prespecified maximal width in a one-pass step-by-step procedure for a general initial value problem.

The generation of numerical solutions for the inclusion problem (1.2) has been studied by many scientists and a good number of algorithms have been proposed. One of the early investigations is by N.J. Lehmann [4]; it is remarkable that he has already suggested the use of symbol manipulation systems in this connection.

For lack of space, we cannot systematically list and comment the various contributions. A very extensive bibliography on the subject is to be found, e.g., in Nickel [7]. Our own bibiliography contains only some typical examples of specific approaches.

Rather than sketching the historic development, we will attempt to display a common conceptual framework for most of the algorithms which have been proposed. This should help in their understanding and evaluation and stimulate the further analysis and development of the area.

Local Analysis

Except in trivial situations, a numerical algorithm for (1.2) cannot cover the interval [0,T] at once. Hence we consider at first *one step* in a forward stepping algorithm: We have arrived at $t_{\nu-1}$ and obtained a set $\tilde{E}_{\nu-1}$ such that $e(t_{\nu-1}) \in \tilde{E}_{\nu-1}$. In the construction of a corresponding set $\tilde{E}_\nu$ at $t_\nu = t_{\nu-1}+h_\nu$, we have to regard *all* solutions $y(t;t_{\nu-1},y_{\nu-1})$ of (1.1) which pass through an admissible value $y_{\nu-1}$ at $t_{\nu-1}$.

Local problem: Find $\tilde{E}_\nu$ such that (see Fig.1)

$$\tilde{y}(t_\nu) - y(t_\nu;t_{\nu-1},y_{\nu-1}) \in \tilde{E}_\nu \quad \text{for all } y_{\nu-1} \in \tilde{y}(t_{\nu-1}) - \tilde{E}_{\nu-1} \,. \tag{2.1}$$

Obviously, the use of the *local exact inclusion*

$$E_\nu := \{\tilde{y}(t_\nu)-y(t_\nu;t_{\nu-1},y_{\nu-1}) \;:\; y_{\nu-1} \in \tilde{y}(t_{\nu-1})-E_{\nu-1}\} \tag{2.2}$$

for $\tilde{E}_\nu$ would keep the inclusion optimally tight. By (2.1), we have $E_\nu \subset \tilde{E}_\nu$ and we can use the *interior difference*

$$D_\nu := \tilde{E}_\nu \mathbin{\underset{\cdot}{-}} E_\nu \quad \in \mathbb{IP}\mathbb{R}^s \tag{2.3}$$

to represent the excess of $\tilde{E}_\nu$ over E_ν.

(For two sets in a linear space, with $A \subset B$, the interior difference $B \mathbin{\underset{\cdot}{-}} A$ is the unique set C which satisfies $A + C = B$. Obviously, $0 \in B \mathbin{\underset{\cdot}{-}} A$. The norm of $B \mathbin{\underset{\cdot}{-}} A$ is the Hausdorff distance of A and B.)

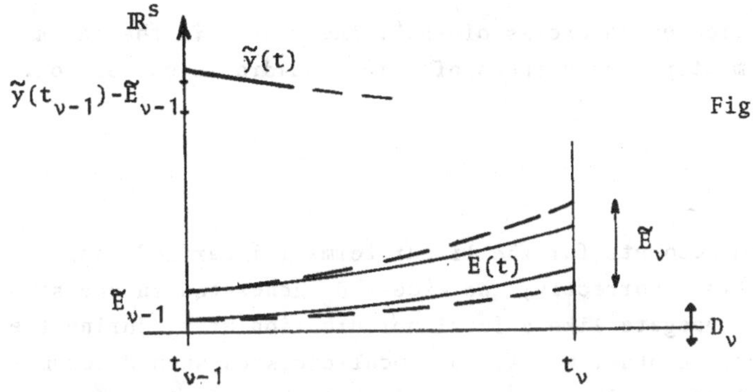

Fig. 1

It appears that the *local excess* D_ν of (2.3) is the natural analogon to the *local error* in a stepwise algorithm for (1.1). Its size, expressed e.g. by

$$\|D_\nu\| := \max_{d \in D_\nu} \|d\|$$

may be used as a quantitative measure of the (local) accuracy of an inclusion algorithm. For $s > 1$ and $\tilde{E}_{\nu-1}, \tilde{E}_\nu \in \mathbb{I}\mathbb{R}^s$, D_ν will generally not be an interval because $E_\nu \notin \mathbb{I}\mathbb{R}^s$.

Typically, the computation of $\tilde{E}_\nu$ will be based, at best, on

- correct representation of a few derivatives w.r.t. t
- correct representation of *linear* terms in the deviation e between $\tilde{y}$ and y (first order perturbation analysis)
- strict bounding of remainder terms, nonlinearities, etc.

Round-off error effects will be caught by the use of directed rounding. We assume that their influence is negligible compared to the leading local excess terms.

A *local excess analysis* for an algorithm of this kind leads to

$$D_\nu = O(h_\nu^{p+1}) + O(h_\nu \varepsilon_{\nu-1}^2) + \text{higher order terms} \tag{2.5}$$

where $\varepsilon_{\nu-1} := \operatorname{diam} \tilde{E}_{\nu-1}$. The appearance of the second term seems unavoidable, even if quadratic terms in e are evaluated:

Take $y'=y^2$ so that $y'(y_0+e) = y_0^2 + 2y_0 e + e^2$. Assume $e \in [-\varepsilon,\varepsilon] =: E$ and compute bounds for $y'(y_0+e)$ by interval evaluation of $y_0^2 + 2y_0 E + E^2$. With $E^2 = [0,\varepsilon^2]$, one obtains

$$hy'(y_0+e) \in h \cdot \left[y_0^2 - 2y_0\varepsilon,\ y_0^2 + 2y_0\varepsilon + \varepsilon^2\right]$$

whose lower bound creates an excess of $-h\varepsilon^2$. The reason is the *dependence* between the multiple occurences of E in a quadratic expression.

Stability

If our algorithm accounts for the linear terms (linearized about $\tilde{y}(t)$) in the deviations correctly, the excess D_ν generated in the step towards t_ν should propagate like a local perturbation at t_ν during the further integration. In other words, our local excesses should accumulate like the local errors in a one-step algorithm for (1.1), at least for sufficiently small steps h_ν and a sufficiently narrow inclusion strip.

However, computationally there arises the necessity to *represent* inclusions in terms of a *semiorder* of the $\mathbb{R}^s$ *based on components*, e.g. by componentwise intervals. Not always such a semiorder is preserved by the differential system (1.1): The initial value problem (1.1) is called *quasimonotone* w.r.t. a semiorder $\geq$ in $\mathbb{R}^s$ if

$$w'(t) \geq f(t,w(t)) , \quad t \in [0,T] , \qquad \text{implies} \quad w(t) \geq y(t),\ t \in [0,t] .$$
$$w(0) \geq y_0$$

(Criteria for quasimonotony and related theorems may be found in Walter [9].)

If (1.1) is *not* quasimonotone w.r.t. componentwise semiorder, the following happens (see Fig.2):

Consider an inclusion interval $\tilde{E}_{\nu-1} = \tilde{e}_{\nu-1} + [-1,+1] \cdot d_{\nu-1}$, with $d_{\nu-1} \geq 0$. The variational equation of (1.1) (near the solution trajectory) maps $\tilde{E}_{\nu-1}$ into $E_\nu = G\,\tilde{E}_{\nu-1} + \ldots$, but E_ν is not represented or included by $e_\nu + [-1,+1] \cdot Gd_{\nu-1}$.

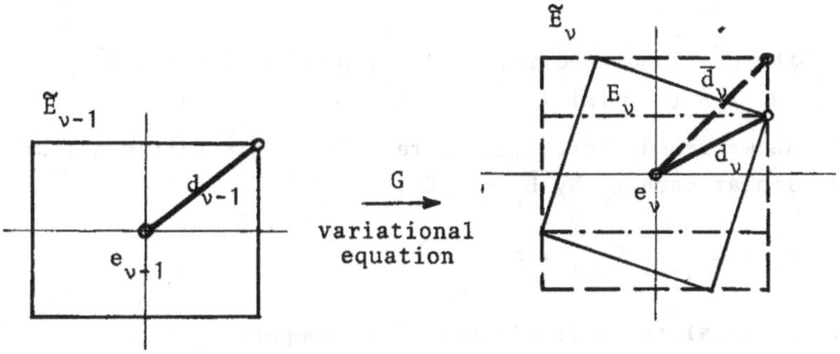

Fig. 2

The smallest interval including E_ν is rather specified by

$$\bar{d}_\nu = |G| \cdot d_{\nu-1} . \tag{3.1}$$

Since $G \approx I + h_\nu J_{\nu-1}$, (3.1) simulates the use of the modified Jacobian

$$J^+_{\nu-1} := \begin{bmatrix} \diagdown & & |R| \\ & D & \\ |R| & & \diagdown \end{bmatrix} \tag{3.2}$$

i.e. only diagonal elements are not replaced by their modulus. For a non-quasimonotone problem we have $J^+ \neq J$ which implies

$$\rho(|G|) > \rho(G) ; \tag{3.3}$$

hence the excess of an inclusion is amplified *more violently* than small perturbations.

This "wrapping" effect (see e.g. Jackson [3]) equally appears in a direct use of the variational equation: To include a solution of

$$e'(t) = J(t)\, e(t) + \dots ,$$

the $\genfrac{}{}{0pt}{}{\text{lower}}{\text{upper}}$ bound of e must be used with negative elements of J in the computation of the $\genfrac{}{}{0pt}{}{\text{upper}}{\text{lower}}$ bound of e. This corresponds to the use of the 2s×2s-matrix

$$\begin{pmatrix} D+R^+ & R^- \\ R^- & D+R^+ \end{pmatrix}$$

which has the combined spectrum of $J = (D + R^+ + R^-)$ and $J^+ = (D + R^+ - R^-) = (D + |R|)$.

A well-known remedy (cf. e.g. Moore [6]) is the following: Represent the inclusion at each t_μ by $\tilde{E}_\mu = A_\mu \hat{E}_\mu$, $A_\mu \in \mathbb{R}^{s\times s}$. If

$$E_\nu = G_{\nu,\nu-1}\, \tilde{E}_{\nu-1} + C_\nu$$

is the exact local inclusion (see (2.2)), compute

$$A_\nu := G_{\nu,\nu-1}\, A_{\nu-1}\, , \qquad (3.4)$$

$$\hat{E}_\nu := \hat{E}_{\nu-1} + W[A_\nu^{-1} C_\nu]\, , \qquad (3.5)$$

where W is the *wrapping* operation which maps a bounded set in $\mathbb{R}^s$ into the smallest enclosing interval. Now the correct propagation is maintained by (3.4), and each inclusion *increment* C_ν is only wrapped twice (instead of n-ν times), by (3.5) and in the "output" formation

$$\tilde{E}_\nu = W[A_\nu \hat{E}_\nu]\, . \qquad (3.6)$$

(3.6) is needed for the evaluation of various terms in the step proceeding from t_ν. Hereby, C_ν may be distorted by a factor $\|A_\nu\|\,\|A_\nu^{-1}\| = \text{cond}(A_\nu)$, see (3.5) and (3.6).

Hence, in the use of (3.4)/(3.5) the growth of $\text{cond}(A_\nu)$ has to be monitored and the representation must be restarted if necessary:

$$\tilde{E}_\nu = A_\nu\, \hat{E}_\nu =: I\, \hat{E}_\nu\, .$$

Other tricks to counteract the effect of (3.2) - see e.g. Conradt [1, section 6] - also suffer if the condition of A_ν of (3.4) grows with ν.

None of the presently suggested algorithms is suitable for *stiff problems* (1.1) because polynomial approximations in t are used.

The Accumulated Excess

At some t_ν, we compare the computed inclusion $\tilde{E}_\nu$ to the true error $e(t_\nu)$ (cf. (1.2)) or the true inclusion set $E(t_\nu)$ in case of a set initial condition Y_0 for (1.1) and define

$$\text{global (=accumulated) excess } X_\nu := \tilde{E}_\nu \ominus E(t_\nu) \ . \qquad (4.1)$$

The computational continuation of the inclusions from $t_{\nu-1}$ to t_ν

- propagates the excess $X_{\nu-1}$ present in $\tilde{E}_{\nu-1}$,
- generates the additional excess $D_\nu =: h_\nu \bar{D}_\nu$, cf. (2.3),
- may introduce further excess by wrapping.

Let us assume that (1.1) is quasimonotone w.r.t. componentwise semiorder so that we may disregard the wrapping effects and that all deviations are sufficiently small so that a perturbation approach is justified. Then we have (cf. "Local Analysis" and "Stability")

$$X_\nu = X_{\nu-1} + h_\nu J_\nu X_{\nu-1} + h_\nu \bar{D}_\nu \ . \qquad (4.2)$$

For sufficiently small h_ν, $X_\nu \approx X(t_\nu)$ where

$$X'(t) = J(t)\ X(t) + \bar{D}(t) \ , \qquad X(0) = \{0\} \ , \qquad (4.3)$$

if we assume that we start the inclusion correctly. From (2.5) we have, with appropriate Λ and Γ,

$$\bar{D}(t) = \Lambda(t)\ h^p + \Gamma(t)\ (\text{diam}\ \tilde{E}(t))^2 + \text{ small terms} \ . \qquad (4.4)$$

Case 1: Point initial condition for (1.1)

Here $\tilde{E}(t) = e(t) + X(t)$ so that $\text{diam}\ \tilde{E}(t) = \text{diam}\ X(t)$. This leads, with (4.4) and (4.3), to

$$X(t;h) = \overline{X}(t)\, h^p + O(h^{p+1}) \tag{4.5}$$

as long as the "small terms" do not become dominant.

This means that the *tightness* of our inclusions (of the error as well as of the true solution) depends on the order p of the algorithm and on the stepsize used, as in usual in o.d.e. algorithms.

Case 2: Interval initial condition Y_o for (1.1)

$$\begin{aligned} \operatorname{diam} \tilde{E}(0) &= \operatorname{diam} E(0) = \operatorname{diam} Y_o =: \varepsilon_o > 0 , \\ \operatorname{diam} \tilde{E}(t) &\geq \operatorname{diam} E(t) = \operatorname{diam} Y(t) = r(t)\varepsilon_o + O(\varepsilon_o^2) . \end{aligned} \tag{4.6}$$

Substitution of (4.4)/(4.6) into (4.3) now yields

$$X(t;h,\varepsilon_o) \geq \overline{X}_1(t)h^p + \overline{X}_2(t)\varepsilon_o^2 + \text{higher order terms} . \tag{4.7}$$

This means that a reduction of h cannot improve the inclusion beyond the second term. However, this *unavoidable* excess from the quadratic terms in the deviation is $O(\varepsilon_o^2)$ while the error (and solution) tube diameter is $O(\varepsilon_o)$, see (4.6).

The behavior (4.5) and (4.7) is well displayed in numerical computation; results of some experiments are shown in Table 1.

Methods

We can only sketch the two fundamental approaches and must refer to the literature for more detail:

1) Defect Correction: Let $d(t) := \tilde{y}'(t) - f(t,\tilde{y}(t))$ denote the *defect* of $\tilde{y}$. Then we have (cf. (1.2) and (2.2))

$$e(t_\nu) = e(t_{\nu-1}) + \int_{t_{\nu-1}}^{t_\nu} [f(\tau,\tilde{y}(\tau)) - f(\tau,\tilde{y}(\tau)-e(\tau))]d\tau + \int_{t_{\nu-1}}^{t_\nu} d(\tau)\, d\tau ,$$

$$\begin{aligned} E_\nu \subseteq \tilde{E}_{\nu-1} &+ \int_{t_{\nu-1}}^{t_\nu} J^+(\tau)E(\tau)\, d\tau + \int_{t_{\nu-1}}^{t_\nu} \text{incl. } \{d(\tau)\}\, d\tau \\ &+ \int_{t_{\nu-1}}^{t_\nu} \text{incl. } \{\text{2nd deriv. terms w.r.t. } e\}\, d\tau \end{aligned} \tag{5.1}$$

The solution of the integral inequality (5.1) may be bounded by approximating the resolvent kernel and bounding the remainder. In the last term of (5.1), an a priori estimate for e in $[t_{\nu-1},t_\nu]$ must be used.

This approach was initiated by Schröder (e.g. [8]); an elaborate algorithm has been described by Marcowitz [5] and Conradt [1].

2) <u>Local Expansion</u>: Let y be a truncated Taylor-expansion about $t_{\nu-1}$; denote $y^{(i)}(\bar{t};\bar{t},\bar{y}) =: f_i(\bar{t},\bar{y})$. Then

$$e(t_\nu) = e(t_{\nu-1}) + \sum_{i=1}^{p-1} \frac{h_\nu^i}{i!} [f_i(t_{\nu-1},y_{\nu-1}) - f_i(t_{\nu-1},\tilde{y}_{\nu-1}-e_{\nu-1})] - \frac{h_\nu^p}{p!} f_p(\tau,\bar{y}(\tau))$$

$$E_\nu \subset \tilde{E}_{\nu-1} + \sum_{i=1}^{p-1} \frac{h_\nu^i}{i!} f_i'(t_{\nu-1},\tilde{y}_{\nu-1})\tilde{E}_{\nu-1}$$

$$+ \sum_{i=1}^{p-1} \frac{h_\nu^i}{i!} [f_i'(t_{\nu-1},\tilde{y}_{\nu-1}-\tilde{E}_{\nu-1}) - f_i'(t_{\nu-1},\tilde{y}_{\nu-1})]\,\tilde{E}_{\nu-1} - \frac{h_\nu^p}{p!} f_p([t_{\nu-1},t_\nu],\bar{Y}_\nu) \qquad (5.2)$$

where $\bar{Y}_\nu$ is an a priori estimate for y in $[t_{\nu-1},t_\nu]$. The approach was initiated by Moore (e.g.[6]); a detailed analysis of an algorithm based upon (5.2) has been presented by Eijgenraam [2].

Obviously, an efficient *implementation* of an inclusion algorithm for (1.2) must rely on a powerful *Computer Algebra system* for the automatic generation of procedures for derivatives and bounds of various kinds, and it must also use an *Interval Arithmetic system* which automatically handles intervals properly (with correct rounding). As both kinds of programming tools are becoming more widely available in standardized forms, the design of transportable and easily usable software for the inclusion problem (1.2) should now become feasible.

References

[1] J. CONRADT, *Ein Intervallverfahren zur Einschließung des Fehlers einer Näherungslösung...*, Freiburger Intervall-Berichte 80/1, 1980.

[2] P. EIJGENRAAM, *The solution of initial value problems using interval arithmetic*, Math. Centre Tracts 144, 1981.

[3] L.W. JACKSON, *Interval arithmetic error-bounding algorithms*, SINUM 12(1975) 223-238.

[4] N.J. LEHMANN, *Fehlerschranken für Näherungslösungen bei Differentialgleichungen*, Numer. Math. 10(1967) 261-288.

[5] U. MARCOWITZ, *Fehlerschätzung bei Anfangswertaufgaben von gew. Diffgln...*, Numer. Math. 24 (1975) 249-275.

[6] R.E. MOORE, *Interval Analysis*, Prentice Hall Inc., 1966.

[7] K. NICKEL, *Using interval methods for the numerical solution of ODEs*, MRC Tech. Summary Rep. #2590, 1982.

[8] J. SCHRÖDER, *Fehlerabschätzung mit Rechenanlagen bei gew. Diffgln. 1. Ordn.*, Numer. Math. 3 (1961) 39-61.

[9] W. WALTER, *Differential - und Integralungleichungen*, Springer-Tracts in Nat. Phil. vol. 2, 1964.

h	$\varepsilon_o = 0$	$\varepsilon_o = 2^{-6}$		$\varepsilon_o = 2^{-4}$		$\varepsilon_o = 2^{-2}$	
2^{-1}	.25 (-2)	.27 (-2)	17.3	.38 (-2)	6.11		
	27.7	*26.9*		*18.2*			
2^{-2}	.91 (-4)	.10 (-3)	.64	.21 (-3)	.34	.29 (-2)	1.16
	18.7	*11.8*		*3.23*		*2.27*	
2^{-3}	.48 (-5)	.85 (-5)	.054	.65 (-4)	.10	.13 (-2)	.51
	17.4	*2.79*		*1.39*		*1.39*	
2^{-4}	.28 (-6)	.31 (-5)	.020	.47 (-4)	.075	.93 (-3)	.37
	16.7	*1.23*		*1.13*		*1.16*	
2^{-5}	.16 (-7)	.25 (-5)	.016	.41 (-4)	.066	.80 (-3)	.32
	16.4	*1.06*		*1.06*		*1.07*	
2^{-6}	.10 (-8)	.23 (-5)	.015	.39 (-4)	.062	.74 (-3)	.29
	16.2	*1.03*		*1.03*		*1.04*	
2^{-7}	.63 (-9)	.23 (-5)	.015	.38 (-4)	.060	.72 (-3)	.28
	diam $E(\varepsilon_o)$ =	.157 (-3)	*4.0*	.626 (-3)	*4.0*	.253 (-2)	

Table 1. Excess X as a function of h and ε_o.

The problem was $y' = -y^2$, $Y_o = [1 - \frac{\varepsilon_o}{2}, 1 + \frac{\varepsilon_o}{2}]$, $t \in [0,9]$.

The algorithm used was an implementation of (5.2), with p = 4.

The main figures display diam X at t = 9, cf. (4.1). The *italic* figures are quotients of their two neighbors. The right-hand figures in the $\varepsilon_o > 0$ columns are the quotients diam $X(h,\varepsilon_o)$ / diam $E(\varepsilon_o)$.

RECENT DEVELOPMENTS IN THE THEORY OF FUNCTION SPACES

H. TRIEBEL
Sektion Mathematik, Universität Jena
DDR-6900 Jena, Universitäts Hochhaus

1. Introduction

The word "function spaces" covers nowadays rather different branches and techniques. In our context function spaces means spaces of functions and distributions defined on the real euclidean n-space R_n which are isotropic, non-homogeneous and unweighted. More precisely, this survey deals with the spaces $B^s_{p,q}$ and $F^s_{p,q}$ on R_n which cover Hölder-Zygmund spaces, Sobolev-Slobodeckij spaces, Besov-Lipschitz spaces, Bessel-potential spaces and spaces of Hardy type. First we try to describe how the different approaches are interrelated, inclusively few historical remarks. Secondly, we outline some very recent developments which, by the opinion of the author, not only unify and simplify the theory of function spaces under consideration considerably, but which also may serve a starting point for further studies.

2. How to Measure Smoothness?

Let R_n be the real euclidean n-space. The classical devises to measure smoothness are derivatives and differences. If one wishes to express smoothness not only locally but globally, in our case on R_n, then function spaces, e.g. of L_p-type, seem to be an appropriate tool. We use standard notations for the derivatives D^α and the differences Δ^m_h,

$$D^\alpha = \frac{\partial^{|\alpha|}}{\partial x_1^{\alpha_1} \cdots \partial x_n^{\alpha_n}} \quad \text{if } x = (x_1,\dots,x_n) \in R_n,\ \alpha = (\alpha_1,\dots,\alpha_n), |\alpha| = \sum_{j=1}^n \alpha_j$$

and

$$\Delta^1_h f(x) = f(x+h) - f(x),\quad \Delta^m_h = \Delta^{m-1}_h \Delta^1_h$$

if $x \in R^n$, $h \in R_n$, and $m = 2,3,\dots$ Furthermore,

$$\|f|L_p\| = \left(\int_{R_n} |f(x)|^p dx\right)^{1/p},\quad 0 < p \le \infty,$$

with the usual modification if $p = \infty$. Recall that S and S' stand for the Schwartz space of all complex-valued infinitely differentiable rapidly decreasing functions on R_n and the space of all complex-

-valued tempered distributions on R_n, respectively. Of course, the spaces L_p with $0 < p \le \infty$ have the usual meaning (complex-valued functions).

Definition 1. (i) (Hölder-Zygmund spaces). Let s be a positive number and let m be an integer with $0 < s < m$. Then

$$C^s = \{f \mid f \in L_\infty, \|f|C^s\|_m = \|f|L_\infty\| + \sup_{\substack{x\in R_n \\ 0\neq h\in R_n}} |h|^{-s}|\Delta_h^m f(x)| < \infty\}. \qquad (1)$$

(ii) (Sobolev spaces). Let $1 < p < \infty$ and let m be a natural number. Then

$$W_p^m = \{f \mid f \in L_p, \|f|W_p^m\| = \sum_{|\alpha|\le m} \|D^\alpha f|L_p\| < \infty\}. \qquad (2)$$

Remark 1. Let $0 < s < 1$. Then

$$\|f|C^s\|_1 = \sup_{x\in R_n} |f(x)| + \sup_{x\neq y} \frac{|f(x) - f(y)|}{|x-y|^2} \qquad (3)$$

are the familiar norms in the Hölder spaces C^s. If s is a positive fractional number, i.e. $0 < s = [s] + \{s\}$ with $[s]$ integer and $0 < \{s\} < 1$ then (3) can be extended by

$$\sum_{0\le|\alpha|\le[s]} \|D^\alpha f|L_\infty\| + \sum_{|\alpha|=[s]} \|D^\alpha f|C^{\{s\}}\|_1 .$$

The corresponding spaces are the well-known Hölder spaces (on R_n) as they had been used since the twenties. It had been discovered by A. Zygmund [29] in 1945 that it is much more effective to use higher differences than derivatives combined with first differences. Definition 1(i) must be understood in this sense. In particular if s is given then all the admissible norms $\|f|C^s\|_m$ are equivalent to each other. The spaces W_p^m have been introduced by S.L. Sobolev [16] in 1936. The derivatives involved must be understood in the sense of distributions.

In the fifties several attemps hade been made to extend the spaces from Definitio 1, to fill the gaps between L_p, W_p^1, W_p^2,... and to replace the sup-norm in (1) by other norms. On the basis of quite different motivations S.M. Nikol'skij introduced in the early fifties the spaces $\Lambda^s_{p,\infty}$ with $s > 0$, $1 < p < \infty$ (we always prefer the notations used below which are different from the original ones) and L.N.Slobodeckij, N.Aronszajn and E.Gagliardo defined the spaces $\Lambda^s_{p,p}$ with $s > 0$, $1 < p < \infty$. The next major step came around 1960. Let F and F^{-1} be the Fourier transform and its inverse on S', respectively. Let

$$I_s f = F^{-1}[(1 + |\xi|^2)^{\frac{s}{2}} Ff], \quad f \in S', -\infty < s < \infty . \qquad (4)$$

Definition 2. (i) (Besov-Lipschitz spaces). Let $s > 0$, $1 < p < \infty$

and $1 \le q \le \infty$. Let m be an integer with $m > s$. Then

$$\Lambda^s_{p,q} = \{f \mid f \in L_p, \|f|\Lambda^s_{p,q}\|_m = \|f|L_p\| + \left(\int_{R_n} |h|^{-sq} \|\Delta^m_h f(.)|L_p\|^q \frac{dh}{|h|^n}\right)^{\frac{1}{q}} < \infty\}$$

(usual modification if $q = \infty$).

(ii) (Bessel-potential spaces). Let $-\infty < s < \infty$ and $1 < p < \infty$. Then

$$H^s_p = \{f \mid f \in S', \|f|H^s_p\| = \|I_s f|L_p\| < \infty\} . \tag{6}$$

Remark 2. The Besov-Lipschitz spaces $\Lambda^s_{p,q}$ have been introduced by O.V. Besov [2,3] (following the way paved by S.M.Nikol'skij). They proved to be one of the most successful scales of function spaces. The two sup-norms in (1) (with respect to $x \in R_n$ and $h \in R_n$) are splitted in (5) in an L_p-norm and an L_q-norm. In some sense these spaces are the appropriate extensions of the spaces C^s in the way described above and they fill the gaps between the Sobolev spaces in a reasonable way, although the Sobolev spaces are not special cases of the spaces $\Lambda^s_{p,q}$ if $p \neq 2$. As in the case of the spaces C^s all the admissible norms $\|f|\Lambda^s_{p,q}\|_m$ (with different m's) are pairwise equivalent. The spaces H^s_p have been introduced bv A.P.Calderón [5] and N.Aronszajn, K.T. Smith [1]. First we remark that

$$H^s_p = W^s_p \quad \text{if } s = 0,1,2,\dots \text{ and } 1 < p < \infty .$$

In other words, also the spaces H^s_p fill the gaps between the Sobolev spaces and extend these spaces to negative values of s. But more important, successful method, the Fourier-analytic approach, or the spectral approach, which we discuss in the next section.

3. The Fourier-Analytical Approach

We return to (4) and (6). Let Δ be the Laplacian on R_n and let E be the identity. Recall that

$$(E - \Delta)f = F^{-1}[(1 + |\xi|^2)Ff], \quad f \in S' .$$

More general, the fractoonal powers of $E - \Delta$ are given by

$$(E - \Delta)^{\frac{s}{2}} f = F^{-1}[(1 + |\xi|^2)^{\frac{s}{2}} Ff], \quad f \in S', \quad -\infty < s < \infty .$$

In other words, $f \in H^s_p$ if and only if $(E - \Delta)^{s/2} f \in L_p$. This gives a better feeling what is going on in (6). In particular, smoothness is measured in the Fourier image by the weight-function $g(\xi) = (1 + |\xi|^2)^{s/2}$, and the growth of this weight-function at infinity represents the degree of smoothness. Let $h(\xi)$ be another positive smooth weight-function, not necessarily of the above polynomial type. In order to provide a better understanding of the Fourier-analytical method we dare a bold speculation: If $h_1(\xi)$ and $h_2(\xi)$ are two weight-functions with the same behaviour at infinity then they generate the same smoothness class in the above sense. It comes out that something of this type is correct (via Fourier multiplier theorems), but we shall not try to make this vague assertion more precise. But on the basis of this speculation we try to replace the above weight-function $g(\xi) = (1 + |\xi|^2)^{s/2}$ by more handsome weight-functions which offer a greater flexibility. If $|\xi| \sim 2^j$ with $j = 0,1,2,\ldots$ then $g(\xi) \sim 2^{js}$. Hence one can try to replace $g(\xi)$ by a step function $\tilde{g}(\xi)$ with $\tilde{g}(\xi) \sim 2^{js}$ if $|\xi| \sim 2^j$. This replacement is a little bit too crude, but a smooth version of this idea is just what we want. We give a precise formulation. Let $\varphi(\xi) \in S$ with

$$\operatorname{supp} \varphi \subset \{\xi \mid \tfrac{1}{2} \le |\xi| \le 2\}$$

and

$$\sum_{j=-\infty}^{\infty} \varphi(2^{-j}\xi) = 1 \quad \text{if} \quad \xi \neq 0 .$$

Functions with these properties exist. Let $\varphi_j(\xi) = \varphi(2^{-j}\xi)$ if $j = 1,2,3,\ldots$ and $\varphi_0(\xi) = 1 - \sum_{j=1}^{\infty} \varphi_j(\xi)$. Then $\varphi_0(\xi)$ has also a compact support. The desired substitute of $(1 + |\xi|^2)^{\frac{s}{2}}$ is now given by $\sum_{j=0}^{\infty} 2^{js}\varphi_j(\xi)$. We introduce the pseudodifferential operators

$$\varphi_j(D)f(x) = F^{-1}[\varphi_j(\xi)Ff](x),\ x \in R_n,\ j = 0,1,2,\ldots,f \in S'. \tag{7}$$

This makes sense because by the Paley-Wiener-Schwartz theorem $\varphi_j(D)f(x)$ is an analytic function in R_n for any $f \in S'$. Furthermore, by a theorem of Paley-Littlewood type we have

$$\|f|H_p^s\| \sim \|(\sum_{j=0}^{\infty} |2^{js}\varphi_j(D)f(.)|^2)^{1/2}|L_p\|,\ -\infty < s < \infty,\ 1 < p<\infty, \tag{8}$$

(in the sense of equivalent norms). This is the substitute we are looking for. Now we can ask questions.Does it make sense to replace the l_2-norm in (8) by an l_q-norm (or quasi-norm), $0 < q \le \infty$? Is it reasonable to interchange the roles of L_p and l_2 (or more general l_q) in (8)?

Definition 3. (i) Let $-\infty < s < \infty$, $0 < p \le \infty$ and $0 < q \le \infty$. Then

$$B^s_{p,q} = \{f \mid f \in S', \|f|B^s_{p,q}\|_\varphi = (\sum_{j=0}^{\infty} 2^{jsq}\|\varphi_j(D)f(.)|L_p\|^q)^{1/q} < \infty\}, \tag{9}$$

(usual modification if $q = \infty$).

(ii) Let $-\infty < s < \infty$, $0 < p < \infty$ and $0 < q \le \infty$. Then

$$F^s_{p,q} = \{f \mid f \in S', \|f|F^s_{p,q}\|_\varphi = \|(\sum_{j=0}^{\infty} 2^{jsq}|\varphi_j(D)f(.)|^q)^{1/q}|L_p\| < \infty\} \tag{10}$$

(usual modification if $q = \infty$).

Remark 3. For all admissible values s,p,q the spaces $B^s_{p,q}$ and $F^s_{p,q}$ are quasi-Banach spaces (Banach spaces if $p \geq 1$ and $q \geq 1$), and they are independent of the chosen function φ (in the sense of equivalent quasi-norms). Maybe this fact is not so astonishing if p and q are restricted by $1 < p < \infty$ and $1 < q < \infty$, because in those cases the Fourier multiplier theory for L_p with $1 < p < \infty$ and its vector-valued counterparts can be taken as hints that something of this type may be valid. But it was a big surprise, also for the creators of this theory, that these definitions make sense even if $0 < p \leq 1$ (and $0 < q \leq 1$). The only exception is $p = \infty$ in the case of the spaces $F^s_{p,q}$ (but even in this case one can do something after appropriate modifications). The above definition of the spaces $B^s_{p,q}$ is due to J.Peetre [11,12]. The spaces $F^s_{p,q}$ have been introduced by the author [19], P.I.Lizorkin [10] and J.Peetre [13]. Fro the greater part of the theory of these spaces a restriction to $p \geq 1$, $q \geq 1$ would be artifical. But from a technical point of view such a restriction often simplifies the proofs because one has the elaborated technique of Banach space theory at hand (and one avoids a lot of pitfalls which are so abundant if $p<1$). Systematic treatments of the theory of the spaces $B^s_{p,q}$ and $F^s_{p,q}$ have been given in [14] (mostly restricted to $B^s_{p,q}$ with $1 \leq p \leq \infty$) and [23] (with [21,22] as forerunners, cf. also [20]). Again one can ask questions. What is the use of these spaces? What is the connection of these spaces and those ones introduced in Section 2? As far as the latter question is concerned one has the following answer.

Theorem 1. (i) Let $s > 0$. Then

$$C^s = B^s_{\infty,\infty} . \tag{11}$$

(ii) Let $1 < p < \infty$ and $-\infty < s < \infty$. Then

$$H^s_p = F^s_{p,q} , \tag{12}$$

(in particular, $W^m_p = F^s_{p,2}$ if $m = 0,1,2,\ldots$ and $1 < p < \infty$).

(iii) Let $s > 0$, $1 < p < \infty$ and $1 \leq q \leq \infty$. Then

$$\Lambda^s_{p,q} = B^s_{p,q} . \tag{13}$$

(iv) Let $0 < p < \infty$. Then $F^0_{p,2}$ is a (non-homogeneous) space of Hardy type.

Remark 4. Proofs may be found in [23], cf. also Sections 6 and 7.

4. Points Left Open

The Fourier-analytical approach proved to be very useful in connection with applications to linear and non-linear partial differential equations, cf. [20,23] as far as linear equations are concerned. In the recently developed method of para-multiplications by J.M.Bony and Y.Meyer (in order to obtain local and microlocal smoothness assertions for non-linear partial differential equations) characterizations of type (11) play a crucial role. An extension of these methods to the full scales $B^s_{p,q}$ and $F^s_{p,q}$ has been given by T.Runst [15] (there one can also find the necessary references to the papers by Bony, Meyer).

There is no claim that this paper gives a systematic description of the history of those function spaces which are treated here. We omitted few important developments. But we wish to mention at least few key-words and some milestone-papers. Interpolation theory plays a crucial role in the theory of function spaces since the sixties. The outstanding papers are those ones of J.-L.Lions, J.Peetre [9] and A.P. Calderón [6]. A systematic approach to the theory of function spaces from the standpoint of interpolation theory has been given in [20]. Another important approach to the theory of function spaces is the real variable method in the theory of Hardy spaces and the elaboration of the technique of maximal functions. The milestone-paper in this field is C.Fefferman, E.M.Stein [7].

5. Harmonic and Thermic Extensions

The interest in Hardy spaces has its origin in complex function theory: traces of holomorphic functions in the unit disc or the upper half-plane on the respective boundaries. A generalization of this idea yields a characterization of functions and distributions of the spaces $B^s_{p,q}$ and $F^s_{p,q}$ on R_n as traces of harmonic functions or temperaturs in $R^+_{n+1} = \{(x,t) | x \in R_n, t > 0\}$ on the hyperplane $t = 0$, which is identified with R_n. We reformulate this problem as follows. Let $\Delta = \sum_{j=1}^{n} \frac{\partial^2}{\partial x_j^2}$ be the Laplacian in R_n and let $f \in B^s_{p,q}$ or $f \in F^s_{p,q}$. What can be said (in the sense of characterizing properties) about the solutions $u(x,t)$ and $v(x,t)$ of the problems

$$\left(\frac{\partial^2 u}{\partial t^2} + \Delta u\right)(x,t) = 0 \text{ if } (x,t) \in R^+_{n+1};\ u(x,0) = f(x) \text{ if } x \in R_n \quad (14)$$

(harmonic extension) and

$$\left(\frac{\partial v}{\partial t} - \Delta v\right)(x,t) = 0 \text{ if } (x,t) \in R^+_{n+1};\ v(x,0) = f(x) \text{ if } x \in R_n \quad (15)$$

(thermic extension)? At least in a formal way the solutions $u(x,t)$ and

$v(x,t)$ are known,

$$u(x,t) = P(t)f(x) = c \int_{R_n} \frac{t}{(|x-y|^2 + t^2)^{\frac{n+1}{2}}} f(y)dy, \quad x \in R_n,\ t>0 \qquad (16)$$

(Cauchy-Poisson semigroup) and

$$v(x,t) = W(t)f(x) = ct^{-\frac{n}{2}} \int_{R_n} e^{-\frac{|x-y|^2}{4t}} f(y)dy, \quad x \in R_n,\ t > 0 \qquad (17)$$

(Gauss-Weierstrass semigroup). If $f \in S'$ is given, then (17) makes sense. Furthermore, (16) must be understood in the following theorem via limiting procedures. If a is a real number we put $a_+ = \max(0,a)$.

Theorem 2. Let $\varphi_0 \in S$ with $\varphi_0(0) \neq 0$.

(i) Let $-\infty < s < \infty$, $0 < p \leq \infty$, and $0 < q \leq \infty$. Let k and m be non-negative integers with $k > n(\frac{1}{p} - 1)_+ + \max(s, n(\frac{1}{p} - 1)_+)$ and $2m > s$. Then

$$\|\varphi_0(D)f|L_p\| + \left(\int_0^1 t^{(k-s)q} \left\|\frac{\partial^k P(t)f}{\partial t^k}|L_p\right\|^q \frac{dt}{t}\right)^{\frac{1}{q}} \qquad (18)$$

and

$$\|\varphi_0(D)f|L_p\| + \left(\int_0^1 t^{(m-\frac{s}{2})q} \left\|\frac{\partial^m W(t)f}{\partial t^m}|L_p\right\|^q \frac{dt}{t}\right)^{\frac{1}{q}} \qquad (19)$$

(modification if $q = \infty$) are equivalent quasi-norms in $B^s_{p,q}$. If $s > n(\frac{1}{p} - 1)_+$ then $\|\varphi_0(D)f|L_p\|$ in (18),(19) can be replaced by $\|f|L_p\|$.

(ii) Let $-\infty < s < \infty$, $0 < p < \infty$ and $0 < q \leq \infty$. Let k and m be non-negative integers with $k > \frac{n}{\min(p,q)} + \max(s, n(\frac{1}{p} - 1)_+)$ and $2m > s$. Then

$$\|\varphi_0(D)f|L_p\| + \left\|\left(\int_0^1 t^{(k-s)q} \left|\frac{\partial^k P(t)f}{\partial t^k}(\cdot)\right|^q \frac{dt}{t}\right)^{\frac{1}{q}}|L_p\right\| \qquad (20)$$

and

$$\|\varphi_0(D)f|L_p\| + \left\|\left(\int_0^1 t^{(m-\frac{s}{2})q} \left|\frac{\partial^m W(t)f}{\partial t^m}(\cdot)\right|^q \frac{dt}{t}\right)^{\frac{1}{q}}|L_p\right\| \qquad (21)$$

(modification if $q = \infty$) are equivalent quasi-norms in $F^s_{p,q}$. If $s > n(\frac{1}{p} - 1)_+$ then $\|\varphi_0(D)f|L_p\|$ in (20), (21) can be replaced by $\|f|L_p\|$.

Remark 5. Characterizations of the above type have a long history. As far as the classical Besov-Lipschitz spaces $\Lambda^s_{p,q}$ and the Bessel-potential spaces H^s_p are concerned the first comprehensive treatment in the sense of the above theorem has been given by M.H.Taibleson [18], cf. also T.M.Flett [8]. In this context we mention also the books by P.L.Butzler, H.Berens [4] and E.M. Stein [17] where one can find many informations about characterizations of the above type (for the classical space) and the semigroups from (16) and (17), cf. also [20,

2.5.2, 2.5.3]. More recent results (characterizations of the spaces $B^s_{p,q}$ and $F^s_{p,q}$ in the sense of the above theorem) have been obtained by G.A.Kaljabin, B.-H.Qui and the author. The above formulation has been taken over from [25] (cf. also [23, 2.12.2] where we also gave references to the papers by B.-H.Qui and G.A.Kaljabin).

6. Unified Approach

Up to this moment we said nothing how to understand that the apparently rather different approaches via derivatives, differences, Fourier-analytical decompositions, harmonic and thermic extensions, always yield the same spaces $B^s_{p,q}$ and $F^s_{p,q}$. In [23] we proved equivalence assertions of the above type mostly by rather specific arguments, cf. also [14,22]. But recently it became clear that there exists a unified approach which covers all these methods, at least in principle, and which sheds some light on the just-mentioned problem. We follow [25] where [24] may be considered as a first step in this direction. The basic idea is to extend the admissible functions φ and φ_j in (7) and (9), (10), such that corresponding (quasi-)norms in the sense of (9), (10) cover automatically characterizations of type (18), (19) and (5). We recall that

$$\varphi(tD)f(x) = F^{-1}[\varphi(t.)Ff](x) = ct^k \frac{\partial^k P(t)}{\partial t^k} f(x) \text{ if } \varphi(\xi) = |\xi|^k e^{-|\xi|} \tag{22}$$

and

$$\varphi(\sqrt{t}\, D)f(x) = ct^m \frac{\partial^m W(t)f(x)}{\partial t^m} \text{ if } \varphi(\xi) = |\xi|^{2m} e^{-|\xi|^2} \tag{23}$$

Furthermore we remark that the discrete quasi-norms in (9) and (10) have always continuous counterparts, i.e.

$$\|\varphi_0(D)f|L_p\| + \left(\int_0^1 t^{-sq}\|\varphi(td)f(.)|L_p\|^q \frac{dt}{t}\right)^{\frac{1}{q}} \tag{24}$$

is the continuous substitute of the quasi-norm in (9) and

$$\|\varphi_0(D)f|L_p\| + \left\|\left(\int_0^1 t^{-sq}|\varphi(tD)f(.)|^q \frac{dt}{t}\right)^{\frac{1}{q}}\Big|L_p\right\| \tag{25}$$

is the continuous substitute of the quasi-norm in (10). This replacement of "discrete" quasi-norms by "continuous" ones is a technical matter and has nothing to do with the extension of the class of admissible φ's which we have in mind. If one puts (22),(23) in (24),(25) then one obtains (18)-(21). Of course one has to clarify under what conditions for the parameters involved this procedure is correct. However before giving some details we ask how to incorporate derivatives and differences in this Fourier-analytical concept. We have

$$\varphi(D)f(x) = cD^{\alpha}\Delta_h^m f(x) \text{ if } \phi(\xi) = \xi^{\alpha}(e^{i\xi h}-1)^m, \tag{26}$$

with $\alpha = (\alpha_1,\ldots,\alpha_n)$, m natural number, and $\xi h = \sum_{j=1}^{n} \xi_j h_j$, $\xi^{\alpha} = \xi_1^{\alpha_1}\ldots$ $\ldots\xi_n^{\alpha_n}$. The three functions φ in (29),(23),(26) have in common that they tend to tero if $|\xi| \to 0$ (even if $\alpha = 0$ in (26)). In addition the functions φ from (22),(23) have the same property if $|\xi| \to \infty$. If one compares these functions φ with the function φ from Section 3 used in Definition 3 then it seems to be at least plausible that one can substitute φ in (9),(10) by the functions φ from (22),(23) if k and m are chosen sufficiently large. As for the function φ form (26) this question is more delicate. First one has no decay if ξ tends to infinity and secondly one has not only to handle an isolated function φ but a family of functions parametrized by $h \in R_n$ (and, maybe, by α). We return to these questions later on and formulate a result which covers in principle all cases of interest.

Let $h(x) \in S$ and $H(x) \in S$ with supp $h \subset \{y \mid |y| \le 2\}$, supp $H \subset \{y \mid \frac{1}{4} \le |y| \le 4\}$, $h(x) = 1$ if $|x| \le 1$, and $H(x) = 1$ if $\frac{1}{2} \le |x| \le 2$.

Theorem 3. Let $0 < p \le \infty$, $0 < q \le \infty$ and $-\infty < s < \infty$. Let s_0 and s_1 be two real numbers with

$$s_0 + n(\tfrac{1}{p}-1)_+ < s < s_1 \quad \text{and} \quad s_1 > n(\tfrac{1}{p}-1)_+ . \tag{27}$$

Let $\varphi_0(\xi)$ and $\varphi(\xi)$ be two infinitely differentiable complex-valued functions on R_n and $R_n - \{0\}$, respectively, which satisfy the Tauberian conditions

$$|\varphi_0(\xi)| > 0 \text{ if } |\xi| \le 2 \text{ and } |\varphi(\xi)| > 0 \text{ if } \tfrac{1}{2} \le |\xi| \le 2. \tag{28}$$

let $\tilde{p} = \min(1,p)$ and

$$\int_{R_n} |(F^{-1} \frac{\varphi(z)h(z)}{|z|^{s_1}})(y)|^{\tilde{p}} dy < \infty , \tag{29}$$

$$\sup_{m=1,2,\ldots} 2^{-ms_0\tilde{p}} \int_{R_n} |(F^{-1}\varphi(2^m.)H(.))(y)|^{\tilde{p}} dy < \infty , \tag{30}$$

and (30) with φ_0 instead of φ. Then

$$\|\varphi_0(D)f|L_p\| + \left(\int_0^1 t^{-sq}\|\varphi(tD)f(.)|L_p\|^q \frac{dt}{t}\right)^{\frac{1}{q}} \tag{31}$$

(modification if $q = \infty$) is an equivalent quasi-norm in $B^s_{p,q}$.

Remark 6. This formulation coincides essentially with Theorem 3 in [25]. Of course, $\varphi(tD)f = F^{-1}[\varphi(t.)Ff](x)$ and (31) coincides with (24). This theorem has a direct counterpart for the spaces $F^s_{p,q}$. Furthermore there are some modifications (both for $B^s_{p,q}$ and $F^s_{p,q}$) where not only a single function φ but families of these functions are involved, cf. the considerations in front of the above theorem. Maybe the crucial conditions (29) and (30) look somewhat complicated and seem to be hard to check. But this is not the case, in particular for functions of type (26) the formulations (29),(30) are well adapted. Furthermore, if one uses

$$\|F^{-1}\lambda|L_v\| \le c\|\lambda|H_2^\delta\|, \quad 0 < v \le 1, \quad \delta > n(\tfrac{1}{v} - \tfrac{1}{2}), \tag{32}$$

then one can replace (29),(30) by more handsome-looking conditions, where only Bessel-potential spaces H_2^δ (or even Sobolev spaces W_2^δ) are involved.

Remark 7. Theorem 2 follows from Theorem 3 and its $F^s_{p,q}$-counterpart. One has to use the functions φ from (22),(23).

7. Characterizations via Differences

In principle one can put φ from (26) in Theorem 3 and its $F^s_{p,q}$-counterpart. One can calculate under what conditions for the pamaterers (29),(30) are satisfied. However as we pointed out in front of Theorem 3 one has to modify Theorem 3, because one needs now theorems with families of functions φ instead of a single function φ. This can be done, details may be found in [25]. We formulate a result what can be obtained on this way.

Theorem 4. (i) Let $0 < p \le \infty$, $0 < q \le \infty$ and $n(\frac{1}{p} - 1)_+ < s < m$, where m is a natural number. Then

$$\|f|L_p\| + \Big(\int_{|h|\le 1} |h|^{-sq}\|\Delta_h^m f|L_p\|^q \frac{dh}{|h|^n}\Big)^{\frac{1}{q}} \tag{33}$$

(modification if $q = \infty$) is an equivalent quasi-norm in $B^s_{p,q}$.

(ii) Let $0 < p < \infty$, $0 < q \le \infty$ and $\frac{n}{\min(p,q)} < s < m$, where m is a natural number. Then

$$\|f|L_p\| + \Big\|\Big(\int_{|h|\le 1} |h|^{-sq}|(\Delta_h^m f)(\cdot)|^q \frac{dh}{|h|^n}\Big)^{\frac{1}{q}}|L_p\Big\| \tag{34}$$

(modification if $q = \infty$) is an equivalent quasi-norm in $F^s_{p,q}$.

Remark 8. We refer for details to [25] where we proved many other theorems of this type via Fourier-analytical approach from Section 6 and few additional considerations. However the theorem itself is not new, it may be found in [23, 2.5.10, 2.5.12]. But the proof in [23] is more complicated and not so clearly based on Fourier-analytical results in the sense of Theorem 3. On the basis of Theorem 4 one has now also a better understanding of (11) and (13). We prefered in the above theorem a formulation via differences only. But one can replace some differences by derivatives, as it is also suggested by (26).

8. The Local Approcah

The original Fourier-analytical approach as described in Section 3 does not reflect the local nature of the spaces $B^s_{p,q}$ and $F^s_{p,q}$. If $x \in R_n$ is given then one needs a knowledge of f on the whole R_n in order to calculate $\varphi_j(D)f(x)$ in (7). This stands in sharp contrast to the derivatives $D^\alpha f(x)$ and the differences $\Delta_h^m f(x)$ with $|h| \le 1$ as they have been used above. However the extended Fourier-analytical method as described in Section 6 gives the possibility to combine the advantages of the original Fourier-analytical approach and of a strictly local procedure. We give a description. Let $k_0 \in S$, and $k \in S$ with

$$\operatorname{supp} k_0 \subset \{y \mid |y| \le 1\}, \quad \operatorname{supp} k \subset \{y \mid |y| \le 1\},$$
$$(Fk_0)(0) \neq 0 \quad \text{and} \quad (Fk)(0) \neq 0.$$

Let $k_N = \left(\sum_{j=1}^{n} \frac{\partial^2}{\partial x_j^2}\right)^N k$, where N is a natural number. We introduce the means

$$K(k_N,t)f(x) = \int_{R_n} k_N(y)f(x+ty)dy, \; x \in R_n, \; t > 0, \tag{35}$$

where now N = 0,1,2,... This makes sense for any $f \in S'$.

Theorem 5. (i) Let $-\infty < s < \infty$, $0 < p \le \infty$ and $0 < q \le \infty$. Let $0 < \varepsilon < \infty$, $0 < r < \infty$ and $2N > \max\,(s, n(\frac{1}{q} - 1)_+)$. Then

$$\|K(k_0,\varepsilon)f|L_p\| + \left(\int_0^r t^{-sq}\|K(k_N,t)f|L_p\|^q \frac{dt}{t}\right)^{\frac{1}{q}} \tag{36}$$

(modification if $q = \infty$) is an equivalent quasi-norm in $B^s_{p,q}$.

(ii) Let $-\infty < s < \infty$, $0 < p < \infty$ and $0 < q \le \infty$. Let $0 < \varepsilon < \infty$, $0 < r < \infty$ and $2N > \max\,(s, n(\frac{1}{p} - 1)_+)$. Then

$$\|K(k_0,\varepsilon)f|L_p\| + \left\|\left(\int_0^r t^{-sq}|K(k_N,t)f(\cdot)|^q \frac{dt}{t}\right)^{\frac{1}{q}}\Big|L_p\right\| \tag{37}$$

(modification if $q = \infty$) is an equivalent quasi-norm in $F^s_{p,q}$.

Remark 9. It comes out that the above theorem can be obtained from Theorem 3 and its $F^s_{p,q}$-counterpart. On the other hand it is clear that (35) describes a local procedure.

Remark 10. With the help of Theorem 5 one can simplify and unify several proofs in [23], cf. e.g. [26]. But it is also an appropriate tool to handle psudodifferential operators, cf. [28], and to introduce spaces of $B^s_{p,q}$ and $F^s_{p,q}$ type on complete Riemannian manifolds (which are not necessarily compact), cf. [27].

References

[1] Aronsazaj,N., Smith,K.T., *Theory of Bessel potentials*, I. Ann. Inst. Fourier (Grenoble) 11 (1961), 385-476.
[2] Besov,O.V., *On a family of function spaces. Embeddings and extensions*, (Russian) Dokl. Akad. Nauk SSSR 126 (1959), 1163-1165.
[3] Besov,O.V., *On a family of function spaces in connections with embeddings and extensions*, (Russian) Trudy Mat. Inst. Steklov 60 (1961), 42-81.
[4] Butzer,P.P., Berens,H., *Semi-Groups of Operators and Approximation*, Springer; Berlin, Heidelberg, New York, 1967.
[5] Calderòn,A.P., *Lebesgue spaces of functions and distributions*, "*Part. Diff. Eq.*", Proc. Symp. Math. 4, AMS (1961), 33-49.
[6] Calderòn,A.P., *Intermediate spaces and interpolation, the complex method*, Studia Math. 24 (1964), 113-190.
[7] Fefferman,C., Stein,E.M., *H^p spaces of several variables*, Acta Math. 129 (1972), 137-193.
[8] Flett,T.M., *Temperatures, Bessel potentials and Lipschitz spaces*, Proc. London Math. Soc. 32 (1971), 385-451.
[9] Lions,J.-L., Peetre,J., *Sur une class d' espaces d' interpolation*, Inst. Hautes Etudes Sci. Publ. Math. 19 (1964), 5-68.
[10] Lizorkin,P.I., *Properties of functions of the spaces $\Lambda^r_{p,\theta}$*, (Russian) Trudy Mat. Inst. Steklov 131 (1974), 158-181.
[11] Peetre,J., *Sur les espaces de Besov*, C.R. Acad. Sci. Paris, Sér. A-B 264 (1967), 281-283.
[12] Peetre,J., *Remarques sur les espaces de Besov, Le cas $0 < p < 1$*, C.R. Acad. Sci. Paris, Sér. A-B 277 (1973), 947-950.
[13] Peetre,J., *On spaces of Triebel-Lizorkin type*, Ark. Mat. 13 (1975),123-130.
[14] Peetre,J., *New Thoughts on Besov Spaces*, Duke Univ. Math. Series, Durham, 1976.
[15] Runst,T., *Para-differential operators in spaces of Triebel-Lizorkin and Besov type*, Z. Analysis Anwendungen.
[16] Sobolev,S.L., *Méthode nouvelle à resoudre le problème de Cauchy pour les équations linéaires hyperboliques normales*, Mat. Sb. 1 (1936),39-72.
[17] Stein,E.M., *Singular Integrals and Differentiability Properties of Functions*, Princeton Univ. Press, Princeton, 1970.
[18] Taibleson,M.H., *On the theory of Lipschitz spaces of distributions on euclidean n-space, I,II*, J. Math. Mechanics 13 (1964), 407-479; (1965), 821-839.
[19] Triebel,H., *Spaces of distributions of Besov type on euclidean n-space, Duality, Interpolation*, Ark. Mat. 11 (1973), 13-64.
[20] Triebel,H., *Interpolation Theory, Function Spaces, Differential Operators*, North-Holland,,Amsterdam, New York, Oxford, 1978.
[21] Triebel,H., *Fourier Analysis and Function Spaces*, Teubner, Leipzig, 1977.
[22] Triebel,H., *Spaces of Besov-Hardy-Sobolev Type*, Teubner, Leipzig, 1978.
[23] Triebel,H., *Theory of Function Spaces*, Birkhäuser, Boston 1983, and Geest & Porting, Leipzig, 1983.
[24] Triebel,H., *Characterizations of Besov-Hardy-Sobolev spaces via harmonic functions, temperatures, and related means*, J. Approximation Theory 35 (1982), 275-297.
[25] Triebel,H., *Characterizations of Besov-Hardy-Sobolev spaces*, a unified approach.
[26] Triebel,H., *Diffeomorphism properties and pointwise multipliers for spaces of Besov-Hardy-Sobolev type*.
[27] Triebel,H., *Spaces of Besov-Hardy-Sobolev type on complete Riemannian manifolds*.
[28] Triebel,H., *Pseudo-differential operators in $F^s_{p,q}$-spaces*.
[29] Zygmund,A., *Smooth functions*, Duke Math. J. 12 (1945), 47-76.

LECTURES PRESENTED IN SECTIONS

Section A

ORDINARY DIFFERENTIAL EQUATIONS

ON PROPERTIES OF OSCILLATORY SOLUTIONS OF NON-LINEAR DIFFERENTIAL EQUATIONS OF THE n–TH ORDER

M. BARTUŠEK
Department of Applied Mathematics, Fac. of Science, University of J. E. Purkyně
Janáčkovo nám. 2a, 662 95 Brno, Czechoslovakia

Consider the differential equation

$$(1) \qquad y^{(n)} = f(t,y,\dots,y^{(n-1)}), \qquad n \geq 2$$

where $f : D \to R$ is continuous, $D = R_+ \times R^n$, $R_+ = [0,\infty)$, $R = (-\infty,\infty)$, there exists a number $\alpha \in \{0,1\}$ such that

$$(2) \qquad (-1)^\alpha f(t,x_1,\dots,x_n)x_1 \geq 0 \quad \text{in } D.$$

Definition. The solution of (1) defined on R_+ is called proper if y is not trivial in any neighbourhood of ∞. The solution of (1) defined on $[0,b)$ is called non-continuable if either $b = \infty$ or $b < \infty$ and $\sum_{i=0}^{n-1} |y^{(i)}(t)| = \infty$.

The solution y of (1) defined on $[0,b)$, $b \leq \infty$ is called oscillatory if there exists a sequence of its zeros tending to b and y is not trivial in any left neighbourhood of b.

Denote the set of all oscillatory solutions of (1), defined on $[0,b)$ by $O_{[0,b)}$. Let $O_{[0,\infty)} = O$ and $N = \{1,2,\dots\}$.

Many papers (see e.g. [6]) are devoted to the study of conditions under which oscillatory solutions exist. But the problem of behaviour of such solutions for $n > 2$ is not solved in a suitable way. We touch some problems concerning the behaviour of oscillatory solutions.

I. Definition. The point $c \in [0,b)$ is called H-point of y if there exist sequences $\{t_k\}_1^\infty$, $\{\bar{t}_k\}_1^\infty$ of numbers of $[0,b)$ such that $(t_k - c)(\bar{t}_k - c) > 0$, $y(t_k) = 0$, $y(\bar{t}_k) \neq 0$, $k \in N$.

In [4] some properties of zeros of $y \in O_{[0,\infty)}$ were studied for the linear case of (1). Especially, it was shown, that every zero of $y^{(i)}$, $i = 0,1,\dots,n-1$ is simple in some neighbourhood of $+\infty$. This result is generalized in [1] for the equation (1) if the interval $(0,b)$ does not contain H-points. Moreover, the following statement was proved:

Theorem 1. *Let either $n = 2n_0$, $n_0 + \alpha$ be* odd, *or n be* odd. *and let $y \in O_{[0,b)}$. Then there exist at most two H-points in the interval $[0,b)$.*

If there exist two ones $c_1 < c_2$, then $y(t) \equiv 0$ on $[c_1, c_2]$.

If $n = 2n_0$, $n_0 + \alpha$ is even the statement of the theorem 1 is not valid as it is shown by the following

Theorem 2. *Let $n = 2$, $\alpha = 1$. There exist continuous functions $f : D \to R$ with the property (2), $y \in O_{[0,\infty)}$ and a sequence $\{\tau_k\}_1^\infty$ of numbers such that $\tau_k \in R_+$, $\lim_{k\to\infty} \tau_k = \infty$ and τ_k is the H-point of y.*

Proof. In [5] it is shown that there exist continuous function $a : R_+ \to (-\infty, 0)$ and numbers $b \in R_+$, $\lambda \in (0,1)$ such that the differential equation $y'' = a(t)|y(t)|^\lambda \operatorname{sgn} y(t)$ has an oscillatory solution on $[0,b)$ and $y(t) \equiv 0$ on $[b,\infty)$.

Let $\tau \in [0,b)$ be an arbitrary zero of y' and denote $h = b - \tau$. Define $\bar{a} : R_+ \to (-\infty,0)$ and $\bar{y} : R_+ \to R_+$ in the following way: $\bar{a}, \bar{y}$ are periodic on $[\tau,\infty)$ with the period $2h$,

$$\bar{a}(t) = a(t),\ \bar{y}(t) = y(t) \quad \text{for } t \in [0,b)$$

$$\bar{a}(t) = a(2b - t),\ \bar{y}(t) = y(2b - t) \quad \text{for } t \in [b, b + h].$$

From this, according to $y'(\tau) = 0$ we get that $\bar{a} \in C^0(R_+)$, $\bar{y} \in C^1(R_+)$. By use of substitutions $t \to x$, $x = 2(b + ih) - t$, $t \in [b + (i - 1)h, b + ih]$, $i = 0,1,2,\dots$ can be proved that $\bar{y}$ is a solution of $y'' = \bar{a}(t)|y(t)|^\lambda \operatorname{sgn} y(t)$. As b is H-point of y and $\bar{y}$, too, we can put $\tau_k = b + 2kh$, $k \in N$. The theorem is proved.

II. Let n_0 be the entire part of $\frac{n}{2}$. Put for $y \in C^{n_0}(R_+)$, $m \in N$ $J_m(t;y) = {}_0\!\int^t {}_0\!\int^{\tau_m} \dots {}_0\!\int^{\tau_2} y(\tau_1)d\tau_1 \dots d\tau_m$, $J_0(t;y) = y(t)$, $t \in R_+$

$$Z(t;y) = \sum_{i=0}^{n-n_0-1} (-1)^{\alpha+i} \binom{n-i}{n} \frac{n}{2(n-i)} J_{2i}(t;[y^{(i)}]^2). \tag{3}$$

The following Lemma was proved in [1]:

Lemma. *Let y be a solution of (1) defined on R_+ and let either $n = 2n_0$, $n_0 + \alpha$ be odd or n be odd. Then*

$$Z^{(n-1)}(t;y) = \sum_{i=0}^{n_0-1} (-1)^{\alpha+i} y^{(n-i-1)} y^{(i)}(t) + \frac{1}{2}(-1)^{n_0+\alpha}(n - 2n_0)[y^{(n_0)}(t)]^2,$$

$$Z^{(n)}(t;y) = (-1)^\alpha y^{(n)}(t)\,y(t) + (-1)^{n_0+\alpha}[y^{(n_0)}(t)]^2(n - 2n_0 - 1) \ge 0, \quad t \in R_+.$$

In the present part we shall study the asymptotic behaviour of proper oscillatory solutions of (1) under the assumptions

(4) $\quad n = 2n_0 + 1,\ n_0 \in N$.

Definition. Let $y \in O$ and $\lim_{t\to\infty} z^{(n-1)}(t;y) = c$. Then $y \in O^1$ ($y \in O^2$) if $c = \infty$ ($c < \infty$).

It is shown in [1] that for $y \in O^1$ $\limsup_{t\to\infty} |y(t)| = \infty$ holds. The behaviour of $y \in O^2$ is different.

Theorem 3. *Let (4) be valid and let continuous functions* g, $g_1 : R_+ \to R_+$ *exist such that* $g(x) > 0$ *in some neighbourhood of* $x = 0$, $\liminf_{x\to\infty} g(x) > 0$ *and*

(5) $\quad g(|x_1|) \le |f(t,x_1,\ldots,x_n)| \le g_1(|x_1|)$ *in* D

holds. Let $y \in O^2$. *Then* $c = 0$, $\lim_{t\to\infty} y^{(i)}(t) = 0$, $i = 0,1,\ldots,n-2$ *and* $y^{(n-1)}$ *is bounded.*

Proof. Let $M \in (0,\infty)$ be a number such that $M_1 = \min_{M\le x<\infty} g(x) > 0$.

Let $D_1 = \{t : t \in R_+,\ |y(t)| \le M\}$, $D_2 = R_+ - D_1$, $y_i(t) = y(t)$ for $t \in D_i$, $y_i(t) = 0$ for $t \in R_+ - D_i$, $i = 1,2$. It is clear (by use of (5)) that $y_1 \in L^\infty(R_+)$, $y_1^{(n)} \in L^\infty(R_+)$. According to Lemma and (5)

(6) $$\infty > z^{(n-1)}(\infty;y) - z^{(n-1)}(0;y) = \int_0^\infty (-1)^\alpha y^{(n)}(t)y(t)dt \ge$$

$$\ge \int_0^\infty g(|y(t)|)|y(t)|dt \ge \int_0^\infty g(|y_2(t)|)|y_2(t)|dt \ge$$

$$\ge M_1 \int_0^\infty |y_2(t)|dt\ ;$$

$$\int_0^\infty |y_2^{(n)}(t)|dt \le \frac{1}{M}\int_0^\infty |y_2^{(n)}(t)y_2(t)|dt \le \frac{1}{M}\int_0^\infty |y^{(n)}(t)y(t)|dt <$$

$$< \infty .$$

Thus $y_2 \in L^1(R_+)$, $y_2^{(n)} \in L^1(R_+)$ and according to [3, V, §4] and (5)

(7) $\quad |y^{(i)}(t)| \le K < \infty,\ t \in R_+,\ i = 0,1,2,\ldots,n-1.$

Let $\{t_k\}_1^\infty$, $\{\tau_k\}_1^\infty$ be sequences, such that $0 \le t_k < \tau_k < t_{k+1}$, $\lim_{k\to\infty} t_k = \infty$, $y(t_k) = 0$, $y'(\tau_k) = 0$, $y(t) \neq 0$ on (t_k,τ_k), $k \in N$. Then, by use of (6) and (7)

$$\infty > \int_0^\infty g(|y(t)|)|y(t)|dt \ge \frac{1}{K}\sum_{k=1}^\infty \int_{t_k}^{\tau_k} g(|y(t)|)|y(t)||y'(t)|dt \le$$

$$\leq \frac{1}{K} \sum_{k=1}^{\infty} \int_0^{|y(\tau_k)|} g(s)\,ds \,.$$

Thus $\lim_{t\to\infty} y(t) = 0$ and according to Kolmogorov-Horny Theorem ([4], p. 167) and (7) we can conclude that

(8) $$\lim_{t\to\infty} y^{(i)}(t) = 0, \quad i = 0,1,2,\dots,n-2 \,.$$

Let $c \neq 0$. By integration and by use of Lemma we get the existence of $\bar{t} \in R_+$ such that

(9) $$|Z(t;y)| \geq \frac{|c|}{4(n-1)!} t^{n-1}, \quad t \in [\bar{t},\infty) .$$

As according to (8) $\lim y^{(n_0)}(t) = 0$, it follows from (3) that

$$|Z(t;y)| \leq A(t)t^{n-1}, \quad \lim_{t\to\infty} A(t) = 0$$

which contradicts to (9). The theorem is proved.

III. This paragraph contains some remark concerning the existence of proper oscillatory solutions of (1). The case $\alpha = 0$ was investigated in [7].

Definition. The equation (1) has Property A_0 if every proper solution of (1) is oscillatory for n even and is either oscillatory or

(10) $$\lim_{t\to\infty} y^{(i)}(t) = 0 \,,$$

$i = 0,1,\dots,n-1$ for n odd. The equation (1) has Property A_1 if every proper solution is either oscillatory or (10) holds for $i = 1,2,\dots$ $\dots,n-1$.

The following theorem gives us sufficient conditions for the existence of proper oscillatory solutions if $\alpha = 1$.

Theorem 4. *Let $\alpha = 1$ and both n, n_0 be even (n be odd). Let (1) have Property A_0 (Property A_1). Let continuous functions $h : R_+ \to R_+$, $\omega : R_+ \to (0,\infty)$ exist such that ω is non-decreasing, $\int_0^{\infty} \frac{dt}{\omega(t)} = \infty$ and*

(11) $$|f(t,x_1,\dots,x_n)| \leq h(t)\omega\left(\sum_{i=1}^{n} |x_i|\right) \quad in\ D$$

hold. Then every non-continuable solution y of (1), satisfying $Z^{(n-1)}(0;y) > 0$ is oscillatory and proper.

Proof. Let y be a non-continuable solution of (1) for which

(12) $$Z^{(n-1)}(0;y) > 0.$$

According to the assumptions of Theorem and [6, Th. 12.1] y is either proper or $\lim_{t\to\infty} y^{(i)}(t) = 0$, $i = 0,1,2,\ldots,n-1$. As by virtue of Lemma the function $Z^{(n-1)}(0;y)$ is non-decreasing, we can conclude that y is proper.

Further, in both cases, it follows from Lemma of **Kiguradze** ([5], Lemma 14.1) that in case of y be non-oscillatory $\lim_{t\to\infty} Z^{(n-1)}(t;y) = 0$ holds. Thus we get the contradiction to (12) and Lemma. The theorem is proved.

Remark 1. The conditions, under which (1) has Property A_0 or A_1 were studied by many authors, see e.g. [6].

2. For the linear case of (1) the existence of oscillatory solutions from the set 0^2 was proved in [5].

References

[1] BARTUŠEK,M., *On properties of oscillatory solutions of ordinary differential inequalities and equations*, Dif. Urav. (to appear, in Russian).

[2] BARTUŠEK,M., *On oscillatory solution of the differential equation of the n-th order*, Arch. Math. (to appear).

[3] BECKENBACH E.F., BELLMAN,R., *Inequalities*, Springer-Verlag, Berlin, 1961.

[4] ELIAS,U., *Oscillatory solutions an Extremal points for a linear differential equation*, Arch. Ration Mech. and Anal., 71, No 2, 177-198, 1979.

[5] HEIDEL,J.W., *Uniqueness, continuation, and nonoscillation for a second order nonlinear differential equation*, Pacif. J. Math., 1970, 32, No 3, 715-721.

[6] KIGURADZE,I.T., *Some singular boundary value problems for ordinary differential equations* (in Russian), Tbilisi Univ. Pres, Tbilisi 1975.

[7] KIGURADZE,I.T., *On asymptotic behaviour of solutions of nonlinear non-autonomous ordinary differential equations*, Colloq. math. soc. I. Bolgai, 30.Qualitative theory of diff. eq., Szeged, 1979, pp. 507-554.

[8] KIGURADZE I.T., *On vanishing at infinity of solutions of ordinary differential equations*, Czech. Math. J. 33 (108), 1983, 613-646.

UNIQUENESS WITHOUT CONTINUOUS DEPENDENCE

T. A. BURTON and D. P. DWIGGINS
Department of Mathematics, Southern Illinois University
Carbondale, Illinois 62901, U.S.A.

1. Introduction. In the classical theory of ordinary differential equations if solutions of a system

$$x' = h(t,x) \tag{1}$$

are uniquely determined by initial conditions, then the solutions are continuous in the initial conditions. But the situation is much different for differential equations in infinite dimensional spaces. Sufficient conditions for this to hold have been discussed in [8-16] and [19-20]. Recently, Schäffer [18] constructed a fairly abstract example of a differential equation in the Banach space ℓ^∞ of bounded sequences with the supremum norm in which solutions are unique but are not continuous in initial conditions.

We present a simple example of the same behavior and point out that the real difficulty is that there are many topologies for the initial condition space.

2. Continuity in initial conditions. Consider the system

$$x' = h(t,x) + \int_{-\infty}^{t} q(t,s,x(s))ds \tag{2}$$

in which $h\colon (-\infty,\infty) \times R^n \to R^n$, $q\colon (-\infty,\infty) \times (-\infty,\infty) \times R^n \to R^n$, with h and q continuous pointwise. To fix the function space we suppose all solutions start at $t_0 = 0$. Then, to specify a solution of (2) we require a continuous initial function $\phi\colon (-\infty,0] \to R^n$ such that

$$\Phi(t) \overset{\text{def}}{=} \int_{-\infty}^{0} q(t,s,\phi(s))ds$$

is continuous for $t \geq 0$. We may then use the Schauder fixed point theorem to show that the system

$$x' = h(t,x) + \int_{0}^{t} q(t,s,x(s))ds + \Phi(t), \quad x(0) = \phi(0) \tag{3}$$

has a solution $x(t,0,\phi)$ satisfying (3) on an interval $[0,\beta)$, for some $\beta > 0$, with $x(t,0,\phi) = \phi(t)$ on $(-\infty,0]$.

System (2) is well defined using pointwise continuity in R^n and there is an initial function set X consisting of continuous functions ϕ for which Φ is continuous for $t \geq 0$ (X may be empty). Without putting any topology at all on X the problems of existence, uniqueness, and continuation of solutions are well-defined. But to complete a classical fundamental theory for (2) we want to say that for each $\phi \in X$ if there is a unique solution $x(t,0,\phi)$ on $[0,\beta]$ and if $\{\psi_n\}$ is a sequence in X converging to ϕ then solutions $x(t,0,\psi_n)$ converge to $x(t,0,\phi)$ on $[0,\beta]$. While we are quite willing to accept any type of convergence of $x(t,0,\psi_n)$ to $x(t,0,\phi)$ on $[0,\beta]$, the meaning of ψ_n converging to ϕ must be specified.

In a given problem we frequently have a wide degree of freedom in our choice of topology for the initial condition space. Recent problems call for unbounded initial functions, plentiful compact subsets of these initial functions, and continuity of the translation map. These requirements lead us to a locally convex topological vector space (Y,ρ) with $\phi \in Y$ if $\phi\colon (-\infty,0] \to R^n$ is continuous and for $\phi,\psi \in Y$ then

$$\rho(\phi,\psi) = \sum_{k=1}^{\infty} 2^{-k}[\rho_k(\phi,\psi)/(1+\rho_k(\phi,\psi))] \tag{4}$$

where $\rho_k(\phi,\psi) = \max_{-k\le s\le 0} |\phi(s) - \psi(s)|$ and $|\cdot|$ is any norm on R^n. For motivations see [1 - 5] and [7]. Problems are also effectively treated using a Banach space with weighted norm as the same references show.

EXAMPLE 1. Consider the linear scalar equation

$$x' = x + \int_{-\infty}^{t} [x(s)/(t-s+1)^3]ds \tag{5}$$

which has the zero solution and it is unique. In fact, if $\phi\colon (-\infty,0] \to R^n$ is any continuous function for which $\Phi(t) = \int_{-\infty}^{0} [\phi(s)/(t-s+1)^3]ds$ is continuous for $t \ge 0$, then there is one and only one solution $x(t,0,\phi)$ defined on $[0,\infty)$. Note that the set X is not empty. We now show that solutions $x(t,0,\phi)$ are not continuous in (Y,ρ).

PROOF. Define a sequence $\{\phi_n\} \subset X$ by

$$\phi_n(s) = \begin{cases} 0 & \text{if } -n \le s \le 0 \\ -n(s+n) & \text{if } s \le -n. \end{cases}$$

Notice that

$$\rho(\phi_n,0) = \sum_{k=1}^{\infty} 2^{-k}\rho_k(\phi_n,0)/[1+\rho_k(\phi_n,0)]$$
$$\le \sum_{k=n}^{\infty} 2^{-k} \to 0 \quad \text{as} \quad n \to \infty$$

and so $\{\phi_n\}$ converges to the zero function in (Y,ρ). Now, for $0 \le t \le 1$

and $n \geq 2$ we have

$$\Phi_n(t) = \int_{-\infty}^{0} [\phi_n(s)/(t - s + 1)^3]ds$$

$$\geq -n \int_{-\infty}^{-n} [(s + n)/(-s + 2)^3]ds \geq 1/16.$$

Hence, we are considering the equation

$$x' = x + \int_0^t [x(s)/(t - s + 1)^3]ds + \Phi_n(t)$$

$$\geq x + (1/16)$$

so that continuity of $x(t,0,\phi)$ in ϕ fails.

Schäffer suggests that the absence of continuity in his example may be the result of his space, ℓ^∞, being neither separable nor reflexive. But our sequence $\{\phi_n\}$ is contained in a compact subset of (Y,ρ) so the subset is separable and it may be embedded in a Banach space. One can show that (Y,ρ) is not reflexive. However, since (Y,ρ) is Frechet it is barreled (cf. [17; p. 60]).

PROPOSITION 1. Let $\{\phi_n\}$ be the sequence of Example 1 in (Y,ρ). Then $\{\phi_n\}$ is contained in a compact subset of (Y,ρ).

PROOF. Define a continuous function $g: (-\infty,0] \to [0,\infty)$ by $g(s) = \sup_n \phi_n(s)$. Then g is a continuous piecewise linear function. Moreover, if $s \geq -n$, then $g(s)$ is Lipschitz with constant n. Let $\alpha: (-\infty,0] \to [0,\infty)$ be the piecewise continuous linear function defined by $\alpha(-n) = n$. Then the set

$$S = \{\phi \in Y \mid |\phi(s)| \leq g(s) \text{ on } (-\infty, 0],$$

$$|\phi(u) - \phi(v)| \leq \alpha(|u| + |v| + 1)|u - v|\}$$

is compact in (Y,ρ) (cf. [7; p. 2]) and contains $\{\phi_n\}$. This completes the proof.

To see that S can be embedded in a compact subset of a Banach space, for the function g defined in the proof of Proposition 1, define $\tilde{g}(s) = [g(s) + 1]^2$. Then define the Banach space $(Z,|\cdot|_{\tilde{g}})$ by $\phi \in Z$ if $\phi \in Y$ and if

$$|\phi|_{\tilde{g}} = \sup_{-\infty<s\leq 0} |\phi(s)|/\tilde{g}(s)$$

exists. This is a Banach space and S is compact in it.

PROPOSITION 2. The set S in the proof of Proposition 1 is contained in a reflexive subspace of Y.

PROOF. Let Q = L(S) be the linear hull of S (i.e., Q is the space formed by taking linear combinations from S.). Now Q is a subspace of the locally convex metric space Y and so Q is a locally convex metric space. Moreover, as S is closed, Q is closed and complete. Hence, Q is a Frechet space and is barreled. Q is not compact since Q is unbounded in the sense of Treves [21; pp. 136-7]. However, closed and (Treves) bounded subsets of Q satisfy boundedness and Lipschitz conditions similar to those of S, and so must be compact. Therefore, Q is a reflexive space (cf. Treves [21; p. 373]). This completes the proof.

Hence, continuity in initial conditions is not guaranteed by the separability and reflexivity of the space.

3. Fading memory. In general, (2) makes sense only when there is a fading memory. Consider the scalar equation

$$x' = A(t)x + \int_{-\infty}^{t} C(t-s)x(s)ds.$$

At the very least we wish to consider all bounded continuous $\phi \in Y$. Since we want $\Phi(t)$ to be continuous for $t \geq 0$ we need to ask that $\int_0^\infty |C(u)|du < \infty$. Then by [6] there is a continuous increasing function $r\colon [0,\infty) \to [1,\infty)$ such that $r(t) \to \infty$ as $t \to \infty$ and $\int_0^\infty |C(u)|r(u)du < \infty$. We take $g(s) = r(-s)$ so that if $|\phi(s)| \leq \gamma g(s)$ for some $\gamma > 0$, then

$$\int_{-\infty}^{0} |C(t-s)\phi(s)|ds \leq \gamma \int_t^\infty |C(u)|r(u)du$$

and this tends to 0 as $t \to \infty$. In summary, if we admit bounded ϕ then we can admit unbounded ϕ and the memory of ϕ fades in $\Phi(t)$ as $t \to \infty$. A similar result holds for nonlinear systems as may be seen in [5].

The function g is central to the study of delay equations and its role may be seen in [1 - 5] and [8]. For (2) to have meaning we expect to be able to require Φ to be continuous in t for bounded continuous ϕ. But in many problems one quickly learns that unbounded ϕ are needed; however, we show in [5] that if bounded ϕ make Φ continuous then so do certain classes of unbounded ϕ. And this gives rise to the function g which is the weight for a Banach space $(X,|\cdot|_g)$. In order to have a unified theory of existence, continuity, boundedness, stability, and periodicity we work entirely in this Banach space. The importance of that unity is illustrated in [4].

In a private communication Kaminogo informs us that he improved our Example 1 by using bounded initial functions and has obtained continuous dependence results for bounded initial functions.

As a step toward completion of a unified theory we now present a result on continuous dependence on initial conditions using unbounded initial functions. In preparation for that result we now suppose that for the equation (2) there is a continuous function $g\colon (-\infty,0] \to [1,\infty)$ which is decreasing, $g(0) = 1$, and $g(r) \to \infty$ as $r \to -\infty$. Form the Banach space $(X,|\cdot|_g)$ with $\phi \in X$ if $\phi\colon (-\infty,0] \to R^n$ is continuous,

$$|\phi|_g \overset{\text{def}}{=} \sup_{-\infty<t\leq 0} |\phi(t)|/g(t)$$

exists, and there is a nonempty subset $U \subset X$ for which the following definition holds.

DEF. A set $U \subset X$ is an existence set for (2) if $\phi \in U$ implies $\phi(t)$ is continuous for $t \geq 0$.

DEF. Let U be an existence set for (2). Then (2) has a fading memory with respect to U if for each $\phi \in U$, each $J > 0$, and each $\varepsilon > 0$ there is a $\delta > 0$, a $D > 0$, and an $M > 0$ such that if $\psi \in U$, $|\phi-\psi|_g < \delta$, and $0 \leq t \leq J$ then

(a) $\int_{-\infty}^{-D} |q(t,s,\psi(s)) - q(t,s,\phi(s))|ds < \varepsilon$ and

(b) $|\int_{-\infty}^{0} q(t,s,\psi(s))ds| \leq M.$

THEOREM. Let U be an existence set for (2) and let (2) have a fading memory with respect to U. Suppose there is a $\phi \in U$ such that $x(t,0,\phi)$ is unique on some interval $[0,t_1]$. Then $x(t,0,\phi)$ is continuous in ϕ in the following sense: If $\{\psi_n\} \subset U$ and $|\phi - \psi_n|_g \to 0$ as $n \to \infty$, then $|Q\phi - Q\psi_n|_g \to 0$ as $n \to \infty$ where $(Q\phi)(t) = x(t+t_1,0,\phi)$ for $-\infty < t \leq 0$ and $x(t,0,\psi_n)$ is any solution of (2) with initial function ψ_n.

PROOF. Let $x(t,0,\phi)$ be defined on $[0,t_1]$ and suppose it is not continuous in ϕ. Then for some $\varepsilon > 0$ and for each $\delta_k > 0$ there exists $\psi_k \in U$ and $t_k \in [0,t_1]$ with $|x(t_k,0,\phi) - x(t_k,0,\psi_k)| \geq \varepsilon$. We may assume $t_k \to S \in [0,t_1]$ by picking a subsequence if necessary. Moreover, we may assume the t_k chosen so that $\{x(t,0,\psi_k)\}$ is bounded on $[0,S]$. Thus, $\{x'(t,0,\psi_k)\}$ is bounded on $[0,S]$ and so $\{x(t,0,\psi_k)\}$ is an equicontinuous sequence with a convergent subsequence, say $\{x(t,0,\psi_k)\}$ again, with limit $\eta(t)$. We may write

$$x_k(t) = x(t,0,\psi_k) = \psi_k(0) + \int_0^t h(s,x_k(s))ds$$

$$+ \int_0^t \int_0^u q(u,s,x_k(s))ds\,du + \int_0^t \int_{-\infty}^0 q(u,s,\phi(s))ds\,du$$

$$+ \int_0^t \int_{-\infty}^{-D} [q(u,s,\psi_k(s)) - q(u,s,\phi(s))]ds\,du$$

$$+ \int_0^t \int_{-D}^0 [q(u,s,\psi_k(s)) - q(u,s,\phi(s))]ds\,du$$

for any $D > 0$. Let $\varepsilon_1 > 0$ be given and let $0 \leq t \leq S$. Then there is a $D > 0$ such that

$$\left|\int_0^t \int_{-\infty}^{-D} [q(u,s,\psi_k(s)) - q(u,s,\phi(s))]ds\,du\right| < \varepsilon_1.$$

For this $D > 0$, then $\{\psi_k(s)\}$ converges uniformly to $\phi(s)$ on $[-D,0]$. Hence, we may take the limit as $k \to \infty$ and find that $x_k(t) \to \eta(t)$ and

$$\eta(t) = \phi(0) + \int_0^t h(s,\eta(s))ds$$

$$+ \int_0^t \int_0^u q(u,s,\eta(s))ds\,du + \int_0^t \int_{-\infty}^0 q(u,s,\phi(s))ds\,du.$$

Thus, η and $x(t,0,\phi)$ satisfy the same equation. Since that equation has a unique solution, $\eta(t) = x(t,0,\phi)$. This contradicts $|x(t_k,0,\phi) - x(t_k,0,\psi_k)| \geq \varepsilon$ and completes the proof.

REFERENCES

1. Arino, O., Burton, T., and Haddock, J., Periodic solutions of functional differential equations, Royal Soc. Edinburgh, to appear.
2. Burton, T. A., Volterra Integral and Differential Equations, Academic Press, New York, 1983.
3. ________, Periodic solutions of nonlinear Volterra equations, Funkcial. Ekvac., to appear.
4. ________, Toward unification of periodic theory, in Differential Equations; Qualitative Theory (Szeged, 1984), Colloq. Math. Soc. János Bolyai, 47, North Holland, Amsterdam.
5. ________, Phase spaces and boundedness in Volterra equations, J. Integral Equations, to appear.
6. Burton, T. and Grimmer, R., Oscillation, continuation, and uniqueness of solutions of retarded differential equations, Trans Amer. Math. Soc. 179(1973), 193-209.

7. Corduneanu, C., Integral Equations and Stability of Feedback Systems, Academic Press, New York, 1973.

8. Haddock, J., A friendly space for functional differential equations with infinite delay, to appear.

9. Hale, J. K., Dynamical systems and stability, J. Math. Anal. Appl. 26(1969), 39-59.

10. Hale, J. K. and Kato, J., Phase spaces for retarded equations, Funkcial. Ekvac., 21(1978), 11-41.

11. Hino, Y., Asymptotic behavior of solutions of some functional differential equations, Tohoku Math. J. 22(1970), 98-108.

12. _______, Continuous dependence for some functional differential equations, ibid., 23(1971), 565-571.

13. _______, On stability of the solutions of some functional differential equations, Funkcial. Ekvac., 14(1970), 47-60.

14. Kappel, F. and Schappacher, W., Some considerations to the fundamental theory of infinite delay equations, J. Differential Equations, 37(1980), 141-183.

15. Naito, T., On autonomous linear functional differential equations with infinite retardations, J. Differential Equations, 21(1976), 297-315.

16. _______, Adjoint equations of autonomous linear functional diferential equations with infinite retardation, Tohoku Math. J., 28(1976), 135-143.

17. Schaefer, H. H., Topological Vector Spaces, Macmillan, New York, 1966.

18. Schäffer, J. J., Uniqueness without continuous dependence in infinite dimension, J. Differential Equations, 56(1985), 426-428.

19. Schumacher, K., Existence and continuous dependence for functional differential equations with unbounded delay, Arch. Rat. Mech. Anal., 67(1978), 315-335.

20. Seifert, G., On Caratheodory conditions for functional differential equations with infinite delays, Rocky Mt. J. Math., 12(1982), 615-619.

21. Treves, T., Topological Vector Spaces, Distributions and Kernels, Academic Press, New York, 1967.

CONNECTIONS IN SCALAR REACTION DIFFUSION EQUATIONS WITH NEUMANN BOUNDARY CONDITIONS

B. FIEDLER and P. BRUNOVSKÝ
Universität Heidelberg,
Inst. of Applied Mathematics
Im Neuenheimer Feld 294,
Heidelberg, West Germany

Inst. of Applied Mathematics
Comenius University,
Mlynská dolina, 842 15 Bratislava
Czechoslovakia

We consider the flow of a one-dimensional reaction diffusion equation

$$u_t = u_{xx} + f(u) \qquad (1)$$

on the interval $x \in (0,1)$ with Neumann boundary conditions

$$u_x(t,0) = u_x(t,1) = 0. \qquad (2)$$

Given two stationary solutions v, w of (1),(2) (i. e. solutions of

$$v'' + f(v) = 0,\ v'(0) = v'(1) = 0\ ) \qquad (3)$$

we say that v connects to w if there exists a solution u(t,x) of (1), (2) for $t \in (-\infty,\infty)$ such that

$$\lim_{t\to-\infty} u(t,.) = v, \quad \lim_{t\to\infty} u(t,.) = w. \qquad (4)$$

For ordinary differential equations trajectories connecting stationary points have been studied in the context of shock waves [3,10] and travelling waves [10]. For (1),(2) the principal motivation for studying connections is somewhat different. As argued by Hale [4] the flow on the maximal compact invariant set A displays the essential qualitative features of the flow of (1),(2). Since (1),(2) is a gradient system, under mild growth conditions on f at infinity A consists of stationary solutions and connecting trajectories. Therefore, determining all stationary solutions and their connecting trajectories, we know the essential part of the flow.

For special classes of nonlinearities the problem of identification of pairs of stationary solutions admitting connections has been studied by Conley and Smoller [2, 10] and Henry [5, 6] who solved the problem completely for f satisfying f(0) = 0 and being qualitatively cubic-like. In [1] we have given an almost complete answer to the following question concerning equation (1) with Dirichlet boundary conditions for general f:

(Q) Given a stationary solution v, which stationary solutions does it connect to?

Similarly as in [1], to distinguish the w's to which v connects we introduce a scalar characteristics of the complexity of stationary solutions. However, while in [1] this is the maximal number of sign changes (called zero number, z), in our case its role will be played by the lap number l introduce by Matano [7]. For a given function v on [0, 1] l(v) is, by definition, the minimal number of intervals I_j into which [0, 1] can be partitioned so that v is strictly monotone on each I_j and l(v) = 0 for v constant.

For v stationary we define the instability (Morse) index i(v) as the number of negative eigenvalues of the problem

$$y'' + (f'(v(x))+\lambda)\, y = 0 \tag{5}$$

$$y'(0) = y'(1) = 0. \tag{6}$$

By a Sturm-Liouville separation of zeros argument one obtains for v≢cnst

$$l(v) \le i(v) \le l(v)+1. \tag{7}$$

The stationary solution v is called hyperbolic if $\lambda = 0$ is not an eigenvalue of the problem (5),(6).

Given v hyperbolic, for $0 \le k \le l(v)$ we denote by $\tilde{v}_k$ ($\underset{\sim}{v}_k$) the stationary solution $\tilde{v}$ ($\underset{\sim}{v}$) satisfying $l(\tilde{v}) = k$ with smallest $\tilde{v}(0) >$ max Range v ($l(\underset{\sim}{v}) = k$ with largest $\underset{\sim}{v}(0) \subset$ Range v, respectively). By $\Omega(v)$ we denote the set of stationary solutions which v connects to. The following theorem is an almost complete answer to (Q):

<u>Theorem</u>. Let f be C^2 and let

$$\overline{\lim_{|s|\to\infty}}\, f(s)/s < 0 \tag{8}$$

Let v be a hyperbolic solution of (3).

(i) If v is constant or i(v) = l(v) then

$$\Omega(v) = \{\tilde{v}_k,\ \underset{\sim}{v}_k : 0 \le k < i(v)\}$$

(ii) If v(0) = max v ≠ min v and i(v) = l(v) + 1 then

$$\Omega(v) = \Omega_1 \cup \Omega_2 \cup \Omega_3,$$

where

$$\Omega_1 = \{\tilde{v}_k : 0 \le k < i(v)\},$$

$$\Omega_2 = \{\underset{\sim}{v}_k : 0 \le k < i(v) - 1\}$$

and either

$$\Omega_3 = \{\underset{\sim}{v}_k : k = i(v) - 1\}$$

or Ω_3 consists of one or several stationary solutions w with Range w ⊂ Range v and i(w) < i(v).

Note that from [7] it follows that there are no other cases possible except of (i) - (ii).

The proof of this theorem proceeds along the lines of the proof of the analogous theorem of [1] for the Dirichlet case. Therefore, the details of its outline given below can easily be completed from [1].

To establish connections we focus on the case $f(0) = 0$, $v \equiv 0$ here for simplicity. Let the zero number $z(u(t,.))$ denote the number of sign changes of $x \mapsto u(t,x)$, $0 < x < 1$ - cf. [1]. Then $z(u(t,.))$ is decreasing with t [7, 8] and we may define the dropping times

$$t_k = \inf \{t \geq 0: z(u(t_i)) \leq k\} \leq \infty$$

and

$$\tau_k = \tanh t_k \in [0, 1].$$

Note that $t_k \leq t_{k-1} \leq \ldots \leq t_0$. If $t_k < t_{k-1}$, the sign

$$\sigma_k = \operatorname{sign} u(t,0), \quad t_k < t < t_{k-1}$$

is independent of t. We collect all this information in the map $y = (y_0, \ldots, y_k, \ldots)$ where

$$y_0 = \sigma_0(1 - \tau_0)^{1/2}$$

$$y_k = \sigma_k(\tau_{k-1} - \tau_k)^{1/2}.$$

Taking $n = i(v) - 1$ and a small sphere Σ^n around v in the unstable manifold $W^u(v)$,

$$y: \Sigma^n \to S^n$$

turns out to be a continuous and essential mapping into the standard n-sphere S^n. In particular, y is surjective. Therefore, for any $0 \leq k < i(v)$, $\sigma \in \{-1, 1\}$ there exists a $u_0 \in W^u(v)$ such that $y(u_0) = \sigma e_k$ where e_k denotes the k-th unit vector. Hence the trajectory u(t) through u_0 connects v to a stationary w with

$$z(w - v) = k \text{ and } \operatorname{sign}(w(0) - v(0)) = \sigma.$$

The fact that $w = v_k$ for $\sigma = 1$ and $w = v_k$ for $\sigma = -1$ (the latter in case (i) follows from the following two lemmas:

<u>Lemma 1</u>. The stationary solution v does not connect to w if there is a $\bar{w}$ stationary with $\bar{w}(0)$ between v(0) and w(0) such that

$$z(v - \bar{w}) \leq z(w - \bar{w}).$$

<u>Lemma 2</u>. For stationary solutions v, w

$$z(v - w) = \begin{cases} l(v) \geq 1 & \text{if Range } w \subset \text{Range } v \\ 0 & \text{if Range } v \cap \text{Range } w = \emptyset \end{cases} \qquad (9)$$

Note that up to interchange of v and w all possible cases are taken care of in Lemma 2.

The proofs of these lemmas can easily be obtained by adapting those of the corresponding lemmas from [1]. The first one is based on the maximum principle, the second employs the phase plane portraits of (v, v') and (w, w'): one notes that between two successive local extrema of v there is precisely one intersection point of v and w in case Range $w \subset$ Range v.

Concluding we note that $\tilde{v}_k$ and $\underset{\sim}{v}_k$ can easily be identified from the global bifurcation diagram of the parametric equation

$$u_t = u_{xx} + a^2 f(u)$$

as given e. g. in [9]. Also, we note that by further analysis we can identify the members of Ω_3 more precisely.

Figure 1 illustrates the Theorem for a particular f. Points on one curve represent stationary solutions with the same lap number which increases from curve to curve by one from left to right starting with $l = 1$. Case (i) applies to v in the left part of Figure 1 with $l(v) = 0$, $i(v) = 3$. Case (ii) applies to v in the right part of Figure 1. In this case $l(v) = 8$, $i(v) = 9$, $v_8 \notin \Omega_3$ and all candidates for Ω_3 not excluded by (ii) are marked by "?". By further analysis we are able to show that connections do exist to those solutions marked by "!" and do not exist to those solutions marked by "x".

REFERENCES

1. P. Brunovský, B. Fiedler: Connecting orbits in scalar reaction diffusion equations, to appear
2. C. Conley, J. Smoller: Topological techniques in reaction diffusion equations. In "Biological Growth and Spread", Proc. Heidelberg a.d. 1979, Jäger, Rost, Tautu editors, Springer Lecture Notes in Biomathematics 38, 473 - 483
3. I. M. Gelfand: Some problems in the theory of quasilinear equations. Uspechi Matem. Nauk 14 (1959), English translation AMS Translation Series 2, 29 (1963), 295 - 381
4. J. Hale, L. Magalhaes, W. Oliva: An Introduction to Infinite Dimensional Dynamical Systems - Geometric Theory. Appl. Math. Sci. 47, Springer 1984
5. D. Henry: Geometric Theory of Semilinear Parabolic Equations. Lecture Notes in Mathematics 840, Springer 1981
6. D. Henry: Some infinite dimensional Morse-Smale systems defined by parabolic equations, to appear in Journal of Differential Equations
7. H. Matano: Nonincrease of the lap number of a solution for a one-dimensional semilinear parabolic equation. Publ. Fac. Sci. Univ. Kyoto Sec. 1A, 29 (1982), 401 - 441

8. K. Nickel: Gestaltaussagen über Lösungen parabolischer Differentialgleichungen. Crelle's J. für Reine und Angew. Mathematik 211 (1962), 78 - 94

9. P.Poláčik: Generic bifurcations of stationary solutions of the Neumann problem for reaction diffusion equations. Thesis, Komensky University, Bratislava 1984

10. J. Smoller: Shock Waves and Rection-Diffusion Equations. Grundlehren der Math. Wiss. 258, Springer 1982

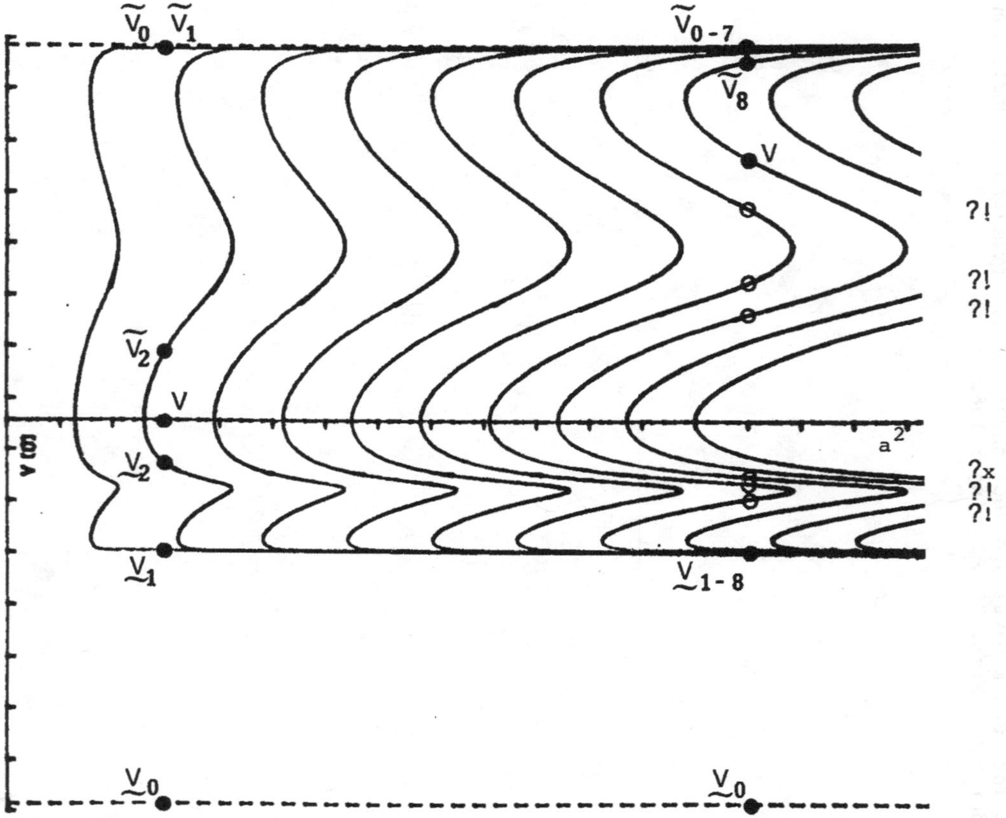

Fig. 1 Bifurcation diagram for $f(u) = (u+10.2)u((u-4)^2 + 1.75^2)(u-10)$

ON A CERTAIN BOUNDARY VALUE PROBLEM OF THE THIRD ORDER

M. GREGUŠ
Faculty of Mathematics and Physics, Comenius University
Mlynská dolina, 842 15 Bratislava, Czechoslovakia

1. The boundary value problem of the form

(1) $$y''' + [f(x) + \lambda g(x)]y' + \lambda h(x)y = 0$$

(2) $$y(-a,\lambda) = y(a,\lambda) = 0, \quad a > 0$$

(3) $$\lambda \int_{-a}^{a} [r(t)-k]\{g(t)y(t,\lambda) + \int_{-a}^{t} [h(\tau) - g'(\tau)]y(\tau,\lambda)d\tau\}dt =$$
$$= \int_{-a}^{a} [r(t)-k]\{f(t)y(t,\lambda) + \int_{-a}^{t} f(\tau)y'(\tau,\lambda)d\tau\}dt ,$$

where $f'(x)$, $g'(x)$, $h(x)$, $r''(x)$ are continuous functions on the interval $\langle -a,a\rangle$ and k is a constant, will be studied.

The boundary condition (3) is in the integral form. For the first time, such a condition was formulated in [1] and the problem (1), (2), (3) is a natural generalization of the problem discussed in [1].

It will be shown that under certain conditions on the function $r(x)$, the problem (1), (2), (3), is equivalent to the boundary problem (1), (4), where

(4) $$y(-a,\lambda) = y''(-a,\lambda) = 0, \quad y(a,\lambda) = 0 \quad a > 0 .$$

In order to solve the problem (1), (4), the theory of the third order linear differential equation [2] can be applied. Moreover some special results will be formulated.

2. Consider the problem (1), (2), (3). Let the functions f, g, h, r fulfil the conditions formulated in Section 1. Then the following theorem is true.

THEOREM 1. *The problem (1), (2), (3) is equivalent to the problem (1), (4) if the function $r = r(t)$ solves the problem*

(5) $$r'' + f(t)r = kf(t)$$

(6) $$r(-a) = k, \quad r(a) = k .$$

Proof. Integrating the differential equation (1), written in the form

$$y''' + \{[f(x) + \lambda g(x)]y\}' + \{-f'(x) + \lambda[h(x) - g'(x)]\}y = 0 ,$$

term by term from $-a$ to x, $x \le a$, and considering (2), we get

$$y'' + f(x)y + \lambda g(x)y + \int_{-a}^{x} \{-f'(\tau) + \lambda[h(\tau) - g'(\tau)]\}y(\tau,\lambda)d\tau =$$

$$= y''(-a,\lambda) .$$

Now suppose that $y''(-a,\lambda) = 0$, multiply the last equality by $r(x)-k$, where k is a constant, and integrate it from $-a$ to a. We come to the equality

$$(7) \qquad \int_{-a}^{a} [r(t)-k][y''(t,\lambda) + f(t)y(t,\lambda)]\,dt =$$

$$= \int_{-a}^{a} [r(t)-k]\{g(t)y(t,\lambda) + \int_{-a}^{t} [h(\tau)-g'(\tau)]y(\tau,\lambda)d\tau\}dt -$$

$$- \int_{-a}^{a} [r(t)-k]\{-f(t)y(t,\lambda) + \int_{-a}^{t} f(\tau)y'(\tau,\lambda)d\tau\}dt .$$

The right-hand side of (7) contains the expression which stands in condition (3). Therefore it is necessary to prove that the integral on the left-hand side of (7) is equal to zero.

Calculate this integral. Under the conditions (2), it follows that

$$\int_{-a}^{a} [r(t)-k][y''(t,\lambda) + f(t)y(t,\lambda)]\,dt = y'(a,\lambda)[r(a)-k] -$$

$$-y'(-a,\lambda)[r(-a)-k] + \int_{-a}^{a} [r''(t) + f(t)r(t)-k\,f(t)]y(t,\lambda)dt .$$

This implies that the boundary condition (3) will be fulfilled if $y''(-a,\lambda) = 0$ and if the function $r(t)$ solves the boundary problem (5), (6). Thus the theorem is proved.

3. In [3] the problem

$$(8) \qquad y''' + \{[1 + \lambda g(x)]y\}' = 0$$

$$(9) \qquad y(-a) = y(a) = 0, \qquad \int_{-a}^{a} (\cos t - \cos a)g(t)y(t)dt = 0$$

and its generalization

(10) $$y''' + [1 + \lambda g(x)]y' + \lambda h(x)y = 0$$

(11) $$y(-a) = y(a) = 0, \quad \int_{-a}^{a} (\cos t - \cos a)\{g(t)y(t) + + \int_{-a}^{x} [h(\tau)-g'(\tau)]y(\tau)d\tau\}dt = 0$$

where $a > 0$, $g'(x)$, $h(x)$ are continuous functions on $\langle -a,a\rangle$, were discussed.

REMARK 1. The problems (8), (9), and (10), (11) are special cases of the problem (1), (2), (3).

Clearly, if we suppose $f(x) = 1$, $h(x) = g'(x)$, $k = \cos a$, we get that (8), (9) is a special case of (1), (2), (3) and from Theorem 1 it follows hat $r(x) = \cos x$. Similarly, if $f(x) = 1$ and $k = \cos a$ we get that (10), (11) is a special case of (1), (2), (3) if $r(x) = \cos x$ and $k = \cos a$. But $r(x) = \cos x$ solves the problem (5), (6), where $k = \cos a$ and $f(x) = 1$.

In [3] it has been proved that under the condition $a = \pi/2$ the problems (8), (9) and (10), (11), respectively are equivalent to the problems (8), (4) and (10), (4) respectively.

Now we prove the following theorem (the formulation will be only for the equation (8), in the case of the equation (10) the equation is similar).

THEOREM 2. *Let $g(x)$ be continuous on $\langle -a,a\rangle$ and let $0 < a < \pi/2$. Then the problem (8), (13), where*

(13) $$y(-a) = y(a) = 0, \quad \int_{-a}^{a} [r(t)-1] \; g(t)y(t)dt = 0$$

is equivalent to the problem (8), (4) if

(14) $$r(x) = \int_{-a}^{a} G(x,\xi)d\xi + \varphi_1(x) + \varphi_2(x) ,$$

where $G(x,\xi)$ is the Green function of the problem

$r'' + r = 0$, $r(-a) = r(a) = 0$, $0 < a < \frac{\pi}{2}$,

$\varphi_1(x)$, and $\varphi_2(x)$, respectively, are the solutions of the problem $r'' + r = 0$, $r(-a) = 1$, $r(a) = 0$, and of the problem $r'' + r = 0$, $r(-a) = 0$, $r(a) = 1$ respectively.

The proof of Theorem 2 is similar to that of Theorem 1. Integrating (8) term by term from $-a$ to $x \le a$ and considering (2), we get $y'' + y + \lambda g(x)y = y''(-a,\lambda)$.

Let $y''(-a,\lambda) = 0$, multiply this equation by $r(x)-1$ and integrate it from $-a$ to a. We obtain

$$(15) \qquad - \int_{-a}^{a} [y''(t) + y(t)][r(t)-1]dt = \lambda \int_{-a}^{a} [r(t)-1]g(t)y(t)dt \ .$$

It is necessary to find such an $r(x)$ that the integral on the left-hand side of (15) be equal to zero.

Calculating it we get

$$\int_{-a}^{a} [y''(t) + y(t)][r(t)-1]dt = y'(a)[r(a)-1] - y'(-a)[r(-a)-1] + \\ + \int_{-a}^{a} y(t)[r''(t) + r(t)-1]dt \ .$$

From this equality it follows that $r(x)$ must solve the problem

$$r'' + r = 1$$
$$r(-a) = 1, \quad r(a) = 1 \ .$$

Thus the theorem is proved.

REFERENCES

[1] Lockschin, A., *Über die Knickung eines gekrümmten Stables*, ZAMM, 16, 1936, 49-55.

[2] Gerguš, M., *Linear Differential Equation of the Third Order* (in slovak), Veda, Bratislava 1981.

[3] Greguš, M., *On Some Application of Ordinary Differential Equations in Physics*, Proc. Third. Conf. Diff. Equ. and Appl. Rousse 1985 (to appear).

ON NONPARASITE SOLUTIONS

P. KRBEC
Aeronautical Research and Test Institute
199 05 Prague 9, Czechoslovakia

1. Introduction

We shall investigate the differential relation

(1) $\dot{x} \in F(t,x), \quad x(0) = x_0$

where $F : U \to K$, $U = \langle 0,1\rangle \times B_1$, K is the set comprising nonempty, compact subsets of some ball in R^n, B_1 is the unit ball in R^n. Jarník and Kurzweil [2] proved that if F(t,x) is convex then we can suppose F to be Scorza-Dragonian. These authors and many others (see e.g. [1], [2], [3], [10], [12]) have studied the convex case very thoroughly. The nonconvex r.h.s. has been attacked too, certain very strong results being obtained e.g. by Olech [7], Ţolstonogov [10], [11], Vrkoč [12]. It is easy to see that to obtain some reasonable existence theorem in nonconvex case it is necessary to suppose F to be continuous. It is a well known fact that the solutions of $\dot{x} \in F$ are then dense in the set of all solutions of $\dot{x} \in \text{conv } F$, see e.g. Tolstonogov [9].

It is tempting then to use the Filipov respectively Krasovskij operation to define generalized solutions of $\dot{x} \in F(t,x)$, F being possibly nonconvex. To be more specific, we can define the solution of $\dot{x} \in F(t,x)$ through the relation $\dot{x} \in G(t,x)$ where

$$G(t,x) = \bigcap_{\delta>0} \bigcap_{\mu(N)=0} \overline{\text{conv}}\, F(t,B_\delta(x) - N) \quad \text{or}$$

$$G(t,x) = \bigcap_{\delta>0} \overline{\text{conv}}\, F(t,B_\delta(x)) .$$

The main problem is that introducing even the solution of $\dot{x} = f(x)$, f discontinuous real valued function, through Filippov or even Krasovskij operation we can obtain certain meaningless solutions.

2. Example 1. (Sentis [8])

Let $f : R \to R$, $f(x) = -1$ for $x \geq 0$, $f(x) = +1$ for $x < 0$. Then $x(t) = 0$ is a (unique) Filippov solution of the Cauchy problem $\dot{x} = f(x)$, $x(0) = 0$, $t \in \langle 0,1\rangle$. This type of solution is called sliding motion and there are good reasons to consider it to be the solution.

On the other hand let $f(x) = 1$ for $x \geq 0$, $f(x) = -1$ for $x < 0$. Then the Cauchy problem $\dot{x} = f(x)$, $x(0) = 0$ has the Filippov solution $x_+(t) = t$, $x_-(t) = -t$ and $x_a(t) = 0$ for $t \in \langle 0,|a|\rangle$, $x_a(t) = \operatorname{sgn} a.(t - |a|)$ for $t \geq |a|$. All the $x_a(.)$ solutions are physically meaningless, they are called parasite solution. For the exact definition of sliding and parasite solution see [4] or Sentis [8].

3. Generalized solutions

Our aim is to define the solution of $\dot{x} \in F(t,x)$ in such a manner that all the sliding solutions are retained and all parasite are expelled. The first definition of this type was given by Sentis [8] in 1976 and it was as follows:

Definition 1. Function $y(.) : \langle 0,1\rangle \to R^n$ is a g-solution of the differential relation $\dot{x} \in F(t,x)$, $x(0) = x_0$ on $\langle 0,1\rangle$ iff there exists a sequence $\{y_n\}_{n=1}^{\infty}$ of piecewise linear functions and a sequence $\{h_n\}_{n=1}^{\infty}$ of divisions such that (denote $y_n(h_n^k)$ by x_n^k and $\nu(h_n)$ by ν_n)

i) $\lim_{n\to\infty} |h_n| = 0$,

ii) $x_n^0 = x_0$

iii) for every positive integer n and $k = 0,1,\dots,\nu_n$ there are $a_n^k \in F(h_n^k,x_n^k)$ and $\varepsilon_n^k \in R^n$ such that $x_n^{k+1} = x_n^k + a_n^k(h_n^{k+1} - h_n^k) + \varepsilon_n^k$ and $y_n(.)$ is linear on every $\langle h_n^k,h_n^{k+1}\rangle$, $k = 0,1,\dots,\nu_n$

iv) $\lim_{n\to\infty} \sum_{k=1}^{\nu_n} \|\varepsilon_n^k\| = 0$

v) $\lim_n y_n = y$ uniformly on $\langle 0,1\rangle$.

Sentis introduced this definition to cover the case (cl stands for closure)

$F(t,x) = \bigcap_{\delta>0} \bigcap_{\substack{N\subset R^{n+1}\\ \mu(N)=0}} \operatorname{cl} f(B_\delta(t,x) - N)$ and his definition works well for such right-hand sides. He proved that any classic solution of $\dot{x} \in F(t,x)$ (i.e. any absolutely continuous function $x(.)$ such that $\dot{x}(t) \in F(t,x(t))$ a.e.) is a g-solution, any g-solution of $\dot{x} \in F(t,x)$ is a classic solution of $\dot{x} \in \operatorname{conv} F(t,x)$ and there are no parasite solutions.

4. Example 2.

For $R^n = R$ set $F_1(t,x) = \{-1\}$ for $x < 0$ and every t, $F_1(t,x) =$

$= \{-1,1\}$ for $x = 0$ and every t and $F_1(t,x) = \{1\}$ for $x > 0$ and every t, $F_2(t,x) = F_1(t,x)$ for t dyadically irrational and every x. For $t = (k/2^m)$, k odd, set $F_2(t,x) = F_1(t,x)$ for $x \notin \langle -1/2^m, 1/2^m\rangle$ and $F_2(t,x) = \{-1,1\}$ for $x \in \langle -1/2^m, 1/2^m\rangle$. Then both F_1 and F_2 are u.s.c. mappings and $\mu\{t \in \langle 0,1\rangle \mid \exists_x (F_1(t,x) \neq F_2(t,x))\} = 0$.

The function $y(.)$, identically equal to zero on $\langle 0,1\rangle$ is not a g-solution of $\dot{x} \in F_1(t,x)$, $x(0) = 0$ but it is a g-solution of the relation $\dot{x} \in F_2(t,x)$, $x(0) = 0$ on $\langle 0,1\rangle$.

This example shows that even for F u.s.c. the solution does depend on values which F obtaines on a set whose projection on t-axis is of measure zero. In the sequel we shall modify the definition of the g-solution to avoid this discrepancy.

5. Regular Generalized Solutions

Let F be Scorza-Dragonian. Denote $G_M F = \{(t,x,y) \mid y \in F(t,x), t \notin M\}$ i.e. $G_M F$ is the graph of the partial mapping $F|_{(\langle 0,1\rangle - M)\times B}$. We set $G^*F = \bigcap_{\substack{\mu(M)=0 \\ M\subset\langle 0,1\rangle}} \mathrm{cl}\, G_M F$ and define a multivalued mapping F^* through its graph i.e. we set graph $F^* = G^*F$. It is possible to prove that there exists a set $M_0 \subset \langle 0,1\rangle$, $\mu(M_0) = 0$ and $G^*F = \mathrm{cl}\, G_{M_0} F$, so our definition is meaningfull. The set G^*F is closed hence F^* is u.s.c. If the mapping F is u.s.c. too then $F^* \subset F$ because graph $F^* = \mathrm{cl}\, G_{M_0} F \subset \mathrm{cl}\, GF = GF$ and $\{t \in \langle 0,1\rangle \mid \exists_x (F^*(t,x) \neq F(t,x))\} \subset M_0$ i.e. its measure is zero. We define the solution of $\dot{x} \in F(t,x)$ through the Sentis g-solution of $\dot{x} \in F^*(t,x)$; resulting type of solution being called rg-solution. It retains all the nice properties of Sentis g-solution and is independent on behaviour of F on a set of measure zero (in t). If the mapping F is supposed to be only Scorza-Dragonian we have only graph $F^* \subset \mathrm{cl}\, GF$ and $F^*(t,x) \supset F(t,x)$ for $t \notin M_0$, nonetheless the rg-solution can be defined too. There is following characterisation of rg-solution:

Theorem 1. Let F be a Scorza-Dragonian mapping. Then a function $y(.)$ is an rg-solution of $\dot{x} \in F(t,x)$ iff for every $M \subset \langle 0,1\rangle$, $\mu(M) = 0$ there are sequences $\{y_n\}_{n=1}^{\infty}$ and $\{h_n\}_{n=1}^{\infty}$ such that all conditions of Definition 1 are fulfilled and $\bigcup_{n=1} h_n \cap M = \phi$.

To prove the theorem we will use the following trivial lemma.

Lemma. Let us suppose $a \in F^*(t,x)$, $M \subset [0,1]$, $\mu(M) = 0$. Then there are sequences $\{(t_n,x_n)\}_{n=1}^{\infty}$ and $\{a_n\}_{n=1}^{\infty}$ such that $a_n \in F^*(t_n,x_n)$,

$t_n \notin M$, $\lim_{n\to\infty} (t_n, x_n, a_n) = (t, x, a)$.

Proof. From $a \in F^*(t,x)$ we obtain as a consequence of the identity $GF^* = G^*F$ and of Lemma 1 that $(t,x,a) \in GF^* = \mathrm{cl}\, G_{M_0 \cup M}F$, $\mu(M_0 \cup M) = 0$. Hence there exists a sequence $\{t_n, x_n, a_n\} \to (t,x,a)$ such that $t_n \notin M_0 \cup M$ and $a_n \in F(t_n, x_n)$. Since $F^*(\tau,\xi) = F(\tau,\xi)$ for $\tau \notin M_0$ the proof is complete.

Proof of the theorem: Since $\{t \in [0,1] \mid \exists_{x\in R^n}\ F^*(t,x) = F(t,x)\} \subset M_0$, $\mu(M_0) = 0$, the "only if" part of the theorem follows immediately. To prove the "if" part let $y(.)$ be an rg-solution and $M \subset [0,1]$, $\mu(M) = 0$. Then there is a sequence $\{y_n\} \to y$ and the sequence $\{h_n\}$ such that the conditions (i),...,(v) from Definition 1 are fulfilled with F^* instead of F. Condition (iii) written explicitly has the following form:

$$y_n(h_n^{k+1}) = y_n(h_n^k) + a_n^k(h_n^{k+1} - h_n^k) + \varepsilon_n^k, \quad a_n^k \in F^*(h_n^k, y_n(h_n^k)).$$

As a consequence of Lemma we obtain that y_n, h_n^k, a_n^k and ε_n^k can be replaced by $\bar{y}_n, \bar{h}_n^k, \bar{a}_n^k, \bar{\varepsilon}_n^k$ such that

$$(2) \qquad \bar{h}_n = \{0 = \bar{h}_n^0 < \bar{h}_n^1 < \ldots < \bar{h}_n^{\nu_n+1} = 1\} \cap M = \phi$$

for every $n = 1,2,3,\ldots, \bar{h}_n^k < h_n^{k+1}$, $(\bar{h}_n^k - h_n^k) < 1/(n\,.\,\nu_n)$, $\sum_{k=1}^{\nu_n} \|\bar{\varepsilon}_n^k\| \to 0$ as $n \to \infty$ and

$$(3) \qquad y_n(\bar{h}_n^{k+1}) = y_n(\bar{h}_n^k) + \bar{a}_n^k(\bar{h}_n^{k+1} - \bar{h}_n^k) + \bar{\varepsilon}_n^k, \quad \bar{a}_n^k \in F^*(\bar{h}_n^k, y_n(\bar{h}_n^k))$$

for $n = 1,2,\ldots$ and $k = 0,1,2,\ldots,\nu_n$.

We can proceed for example as follows. For every $n = 1,2,\ldots$ we set $\bar{h}_n^0 = h_n^0 = 0$, $y_n(\bar{h}_n^0) = x_0$, $\bar{h}_n^{\nu_n+1} = 1$, $\bar{y}_n(1) = y_n(1)$, $\bar{a}_n^0 = a_n^0$. Let us denote $1/(n\nu_n)$ by ρ. As a consequence of Lemma we can choose $\bar{h}_n^k$, $\bar{a}_n^k$ and ψ_n^k, such that (2) is fulfilled and $|\bar{h}_n^k - h_n^k| < \rho$, $\psi_n^k \in B_\rho(y_n(h_n^k))$ $\bar{a}_n^k \in F^*(\bar{h}_n^k, \psi_n^k)$, $\bar{a}_n^k \in B_\rho(a_n^k)$ holds for $k = 1,2,\ldots,\nu_n$. We set $\bar{y}_n(\bar{h}_n^k) = \psi_n^k$ and choose such $\bar{\varepsilon}_n^k$ that (3) is fulfilled. Then

$$\bar{\varepsilon}_n^k = \bar{y}_n(\bar{h}_n^{k+1}) - \bar{y}_n(\bar{h}_n^k) - \bar{a}_n^k(\bar{h}_n^{k+1} - \bar{h}_n^k)$$

and

$$\|\bar{\varepsilon}_n^k\| \le \|\bar{y}_n(\bar{h}_n^{k+1}) - y_n(h_n^{k+1})\| + \|y_n(h_n^k) - \bar{y}_n(\bar{h}_n^k)\| + \|\bar{a}_n^k - a_n^k\| \,.$$

$$.\ \|\bar{h}_n^{k+1} - \bar{h}_n^k\| + \|a_n^k\|(|\bar{h}_n^{k+1} - h_n^{k+1}| + |\bar{h}_n^k - h_n^k|) +$$

$$+ \|y_n(h_n^{k+1}) - y_n(h_n^k) - a_n^k(h_n^{k+1} - h_n^k)\| \le 3\rho + 2\rho + \|\varepsilon_n^k\| \,.$$

Hence $\lim_{n\to\infty} \Sigma\|\bar{\varepsilon}_n^k\| = 0$. Similarly we obtain $\lim \bar{y}_n = y$ uniformly on $[0,1]$

and the proof is complete.

It means that using division to construct a solution we can avoid any set of measure zero.

6. Gauge approach

To define rg-solution we need F to be Scorza-Dragonian (due to the definition of F^*) but by means of avoiding the sets of measure zero we can define the rg-solution for quite a general system. In the sequel, using gauge approach, we introduce another procedure to define solutions. Let us remind that a gauge is an arbitrary real valued positive function and a division $\Delta = \{t_\gamma\}$ is subordinated to a gauge δ (or Δ is δ-fine, $\Delta < \delta$) iff $t_{i+1} - t_i < \delta(t_i)$. We shall say that a set Ω is a gauge set iff for every positive constant c there exists a $\delta \in \Omega$ such that $\sup \delta(t) < c$ and for every $\delta_1,\ldots,\delta_n \in \Omega$ there exists a $\delta \in \Omega$ such that $\delta \le \min(\delta_1,\ldots,\delta_n)$.

There is a well known theorem about δ-fine divisions saying that for every δ there is a δ-fine division which is finite, see Kurzweil [6]. In our case this theorem doesn't hold because we operate with so called left divisions. But a similar theorem holds with a countable divisions. Let us note that using general division instead of left one we don't succeed in rejecting parasite solutions.

Let Ω be a gauge set. We shall say that y is an Ω-solution of $\dot{x} \in F(t,x)$, $x(0) = x_0$ iff all items of Definition 1 are fullfiled with δ-fine division, $\delta \in \Omega$ i.e.

$$\forall_{\varepsilon>0}\ \forall_{\delta\in\Omega}\ \exists_{\Delta<\delta}\ \exists_{\varepsilon_\Delta}\ \exists_{\xi_\Delta}\ \exists_{x_\Delta}\ (|\varepsilon_\Delta| < \varepsilon,\ |y - x_\Delta| < \varepsilon)\ .$$

The following theorem can be proved.

Theorem 2. Let F be bounded and let Ω be a gauge set. Then there exists an Ω-solution.

Proof: Let $\rho > 0$ be such that $\|y\| \le \rho$ for all $y \in F(t,x)$, $(t,x) \in [0,1] \times R^n$ and let K be the set of all $x(.) \in C(\langle 0,1\rangle)$ such that

a) $|x(t)| \le \rho$ for every $t \in [0,1]$

and

b) $|x(t_1) - x(t_2)| \le \rho|t_1 - t_2|$ for every $t_1,t_2 \in [0,1]$.

The K with the norm max is the compact metric space. Let $\delta \in \Omega$. We shall construct a set $S_\delta \subset K$. Let S_δ^J be the set of all functions fulfilling all the conditions of Definition 1 and such that (see condition iv) $\sum_i \|\varepsilon_i\| \le \sup \delta(t)$. It can be proved, by the method of transfinite sequences (see [13]), that S_δ^J is non-empty. Every function $x(.) \in S_\delta^J$ can be modified, by subtracting jumps ε_i in points t_i of division Δ, to obtain a function $y(.) \in K$. This procedure results in a set $S_\delta \subset K$. The set K is compact, hence $\bigcap_{\delta\in\Omega} \overline{S}_\delta \neq \phi$. It is easy to see that every function $x(.)$, $X \in \bigcap_{\delta\in\Omega} \overline{S}_\delta$ is an Ω-solution, which completes the proof.

Let us denote $\Omega_0 = \{\delta(.) | \delta \ge a(\delta) > 0\}$, $\Omega_r = \{\delta(.) | \delta(t) \ge a(\delta)$ a.e., $a(\delta) > 0\}$. Then it is possible to prove that Ω_0-solutions are exactly the Sentis g-solutions and Ω_r-solutions are precisely the rg-solutions. Using the results mentioned above we can say that for F u.s.c. the gauge set Ω_r is the good one to define a solution. But this is not true for F Scorza-Dragonian because Ω_r-solutions are the solutions of $\dot{x} \in F^*$, F^* being u.s.c., $F^* \supset F$ a.e. Hence we cannot expect Ω_r-solutions to be solutions of $\dot{x} \in \text{conv } F$. So a natural problem arises:
What is the smallest but sufficient gauge set for Scorza-Dragonian right-hand side?

References

[1] JARNÍK,J. *Constructing the Minimal Differential Relation with Prescribed Solutions*, Časop. pro pěst. matem. 105 (1980), 311-315.

[2] JARNÍK,J., KURZWEIL,J., *On Conditions on Right Hand Sides of Differential Relations*, Časop. pěst. matem. 102 (1977), 334-339.

[3] JARNÍK,J., KURZWEIL,J., *Sets of Solutions of Differential Relations*, Czech. Math. Journ. 106 (1981).

[4] KRBEC,P., *On Nonparasite Generalized Solutions of Differential Relations*, Časop. pěst. mat. 106 (1981).

[5] KRBEC,P., *On Nonparasite Solutions of Differential Relations*, to appear.

[6] KURZWEIL,J., *Nichtabsolut konvergente Integrale*, Teubner-Texte zur Mathematik 26, B.G. Teubner, Leipzig 1980.

[7] OLECH,C., *Existence of Solutions of non-convex orientor fields*, Difford 1974, Summer School on Ordinary Differential Equations, Brno 1979.

[8] SENTIS,R., *Equations differentielles a second membre measurable*, Boletion U.M.I. (5) 15-B (1978), 724-742.

[9] TOLSTOGONOV,A.A., *O plotnosti i graničnosti množestva rešenij differencialnogo vključenija v banachovom prostranstve*, DAN 1981, No. 2, tom. 261.

[10] TOLSTOGONOV,A.A., *K teoremam srovnenija dlja differencialnych vključenij v lokalno vypuklom prostranstve. I. Suščestvovanie rešenij*, Differencialnye uravnenija 1981, tom. XVII, No. 4, *II. Svojstva rešenij*. Differencialnye uravnenija 1981, tom. XVII, No. 6.

[11] TOLSTOGONOV,A.A., *O differencialnych vključenijach v banachovych prostranstvach i nepreryvnych selektorach*, Sibirskij matem. žurnal 1981, tom. XXII No.4.

[12] VRKOČ,I., *A new Definition and some Modifications of Filippov cone*, Lecture Notes in Math. 703, Equadiff IV Proceedings, Prague 1977, Springer Verlag, Berlin-Heidelberg-New York 1979.

[13] KRBEC,P., *Weak stability of Multivalued Differential Equations*, Czechoslovak Math. Journal 26 (101), 1976.

UNIFORM ZEROS FOR BEADED STRINGS

K. KREITH
University of California
Davis CA 95616, U.S.A.

1. Introduction. Early attempts to model the fundamental vibration of a musical string focussed on three physical properties which were believed to underlie this phenomenon: isochronism, the pendulum principle and a "simultaneous crossing of the axis" [3]. With the discovery that the small vibrations of such strings are described by hyperbolic partial differential equations, interest in these physical concepts declined. Isochronism became embodied in the "small amplitude assumption" which underlies the linearity of the resulting equation, while the pendulum principle (asserting that restoring forces are proportional to displacements from equilibrium) turned out to be incorrect for the wave equation. The notion of a simultaneous crossing of the axis has become identified with separation of variables and does not seem to have been pursued in its own right.

However given a linear hyperbolic PDE of the form

$$u_{tt} - u_{xx} + p(x,t)u = 0, \tag{1.1}$$

with $p(x,t)$ continuous and positive for $0 \leq x \leq L$ and $t \geq 0$, and given boundary conditions such as

$$u_x(0,t) = u_x(L,t) = 0, \tag{1.2}$$

the question of a simultaneous crossing of the axis is an important one. Specifically, the question arises whether it is possible to assign Cauchy data of the form

(1.3) $$u(x,0) = 0; \; u_t(x,0) = g(x)$$

for $0 \leq x \leq L$ such that the solution of (1.1)-(1.3) will satisfy

$$u(x,T) \equiv 0 \quad \text{for} \quad 0 \leq x \leq L$$

for some $T > 0$. We shall refer to such solutions as having a uniform zero at $t = T$.

Except in the case of separation of variables there seems to be no simple answer to this question. As such it is natural to consider a semi-discrete approximation to (1.1)-(1.3) corresponding to a beaded string. In this context one obtains [7] a system of ordinary differential equations of the form

(1.4) $$\frac{d^2\underline{u}}{dt^2} + G(t)\underline{u} = 0$$

subject to initial conditions of the form

(1.5) $$\underline{u}(0) = 0; \; \underline{u}'(0) = \underline{g}.$$

The problem of choosing $\underline{g}$ in (1.5) so as to satisfy $\underline{u}(T) = \underline{0}$ for some $T > 0$ is now the classical problem of establishing the existence of conjugate points of zero relative to (1.4). Such problems have been studied by M. Morse [9] and W. T. Reid [10] in the more general context of Hamiltonian systems. More recently Ahmad and Lazer [1] have also studied conjugate points under the assumption the entries of $G(t)$ satisfy appropriate positivity conditions.

In the case at hand, the matrix function $G(t)$ is a Jacobi matrix given by

(1.6)
$$\begin{aligned} g_{ii}(t) &= 2 + p_i(t) && \text{for} \quad 1 \leq i \leq n, \\ g_{i,i-1} &= g_{i-1,i} = -1 && \text{for} \quad 2 \leq i \leq n, \\ g_{ij} &= 0 && \text{for} \quad |i-j| \geq 2. \end{aligned}$$

While this matrix function has the symmetry required in [9] and [10], the variational criteria for conjugate points established therein are based on positive definiteness and provide no information regarding the sign of the solution which realizes a particular conjugate point. Also, the essential indefiniteness of $G(t)$ prevents the techniques of [1] from being brought to bear in establishing uniform zeros for solutions of (1.4). Accordingly, criteria for the existence of uniform zeros of (1.1) would seem to require the development of novel techniques for establishing the existence of conjugate points for (1.4).

2. The Oppositional Mode of Vibration. A special case of interest in connection with (1.4) and (1.6) is that where the initial data $\underline{g} = \mathrm{col}(g_1,\ldots,g_n)$ in (1.5) satisfies

$$(-1)^j g_j < 0; \quad 1 \leq j \leq n.$$

In this case the solution of (1.4) and (1.5) also satisfies $(-1)^j u_j(t) < 0$ for sufficiently small values of t and is said to be (initially) in an <u>oppositional mode of vibration</u>.

The special Jacobi form of (1.6) makes tractable the problem of establishing the existence of the conjugate point T whose corresponding solution $\underline{u}(t) = \mathrm{col}(u_1(t),\ldots,u_n(t)$ is in an oppositional mode for $0 < t < T$. Indeed, if we define $\underline{v}(t) = \mathrm{col}(v_1(t),\ldots,v_n(t))$ by

$$v_j(t) = (-1)^j u_j(t)$$

then $\underline{v}(t)$ is a solution of

$$\underline{v}'' + F(t)\underline{v} = 0$$

(2.2)

$$\underline{v}(0) = \underline{0}; \; v'(0) = \underline{f}$$

where $f_{ij} = |g_{ij}|$ and $f_i = |g_i|$ for $1 \leq i, j \leq n$. Because of the positivity properties of $F(t)$ and $\underline{f}$ one can apply the techniques of Ahmad and Lazer [1] to establish the existence of a conjugate point T for (2.2) which is realized by solution $\underline{v}(t)$ whose components are positive for $0 < t < T$.

A nonlinear version of this problem has been considered by Duffin [4] in connection with the "plucked string" (corresponding to a right focal point). Indeed given appropriate positivity conditions on $\underline{h}(t,\underline{v})$ one can also use the techniques of Krasnoselskii [6; Ch. 7.4] to establish the existence of positive solutions of boundary value problems of the form

$$\underline{v}'' + \underline{h}(t,\underline{v}) = 0 \tag{2.3}$$
$$\underline{v}(0) = \underline{v}(T) = 0,$$

leading to more general equations which allow for solutions in this oppositional mode.

While of interest, these results are of little help in establishing uniform zeros for (1.1). For as we seek to approximate (1.1) by systems such as (1.4) and let $n \to \infty$, solutions in the oppositional mode do not converge to solutions of (1.1).

For this reason one is led to the more difficult problem of establishing the existence of conjugate points for (1.4) which are realized by positive solutions.

3. Positive Solutions. In case the matrix $G(t)$ given by (1.6) is a constant Jacobi matrix, the existence of a conjugate point realized by a positive solution can be established by algebraic means. In this case (1.4) can be written as

$$G^{-1}\underline{u}'' + \underline{u} = 0 \tag{3.1}$$

where G^{-1} is <u>totally positive</u> in the sense of Gantmacher and Krein [5]. As shown in [5], it now follows that G^{-1} has n simple positive eigenvalues $\lambda_1 > \lambda_2 > \ldots > \lambda_n > 0$, where λ_1 corresponds to an eigenfunction $\underline{\xi}_1$ which may be taken to be positive. Accordingly the choice $\underline{g} = \underline{\xi}_1$ in (1.5) leads to $\underline{u}_1 = \underline{\xi}_1 \sin t/\sqrt{\lambda_1}$ and $T = \pi\sqrt{\lambda_1}$. (It also follows from [5] that the solution $\underline{u}_n = \xi_n \sin t/\sqrt{\lambda_n}$ corresponds to the oppositional mode of vibration considered in § 2).

In order to deal with non-constant $G(t)$ in (1.4) it will be necessary to give a non-algebraic argument for the existence of the above solution $\underline{u}_1(t)$. To that end we consider the case where

$$G(t) = \Gamma_0 + \Pi(t), \tag{3.2}$$

Γ_0 being a constant matrix with entries

$$\gamma_{ii} = p > 0; \quad \gamma_{i,i-1} = \gamma_{i-1,i} = -1; \quad \gamma_{ij} = 0 \quad \text{otherwise}$$

and $\Pi(t) = \mathrm{diag}(\pi_1(t),\ldots,\pi_n(t))$ playing the role of a perturbation of Γ_0. By [5] Γ_0 has positive eigenvalues $\mu_1 < \mu_2 < \ldots < \mu_n$, for which we establish the following property.

<u>3.1 Lemma.</u> For sufficiently large values of p the eigenvalues of Γ_0 satisfy

$$1 < \sqrt{\frac{\mu_i}{\mu_1}} < \frac{3}{2}\ ; \ 2 \le i \le n. \tag{3.3}$$

<u>Proof.</u> The eigenvalues of $\frac{1}{p}\Gamma_0$ satisfy $\frac{1}{p}\mu_1 < \ldots < \frac{1}{p}\mu_n$ and tend to 1 as $p \to \infty$. Therefore (3.3) follows for the eigenvalues of $p\,\Gamma_0$ and for the eigenvalues of Γ_0 as well.

In order to establish topological criteria for the existence of uniform zeros it will be useful to regard solutions of

$$\underline{u}'' + [\Gamma_0 + \Pi(t)]\underline{u} = 0; \ \underline{u}(0) = 0,\ \underline{u}'(0) = \underline{g} \tag{3.4}$$

as trajectories in $\mathbb{R}_n$ which emanate from the origin with initial velocity $\underline{g}$. We seek to show the existence of $\underline{g} > 0$ such that the corresponding trajectory exits the positive n-tant $\mathbb{R}_n^+$ through the origin. To that end we denote the normalized eigenvectors of Γ_0 (corresponding to the eigenvalues μ_i) by $\underline{\phi}_i$, requiring that $\underline{\phi}_1$ lie in $\mathbb{R}_n^+$ and, more generally, that the sum of the components of each $\underline{\phi}_i$ be nonnegative. This sign convention has the consequence that when we express any $\underline{g} \geq 0$ in the form $\underline{g} = c_1\underline{\phi}_1 + \dots + c_n\underline{\phi}_n$, then $c_i \geq 0$ for $1 \leq i \leq n$.

As in [4] we define a <u>contact point</u> of a trajectory $\underline{u}(t)$ as its first point of intersection with a coordinate plane. An <u>exit point</u> is a contact point at which the trajectory also crosses that coordinate plane. In the oppositional mode one can readily show [4] that such first contact points are also exit points, but this need not be the case for trajectories in $\mathbb{R}_n^+$. However, the following theorem shows that under the condition of Lemma 3.1 such an equivalence also exists for trajectories in $\mathbb{R}_n^+$.

<u>3.2 Theorem.</u> *If $\underline{v}(t)$ is a trajectory of*

$$\underline{v}'' + \Gamma_0\underline{v} = 0; \ \underline{v}(0) = 0, \ \underline{v}'(0) = \underline{g} > 0,$$

and if condition (3.3) is satisfied, then the point at which $\underline{v}(t)$ first intersects a coordinate plane bounding $\mathbb{R}_n^+$ is also an exit point.

<u>Proof.</u> Suppose the contact point occurs at $t = t_0$ and lies in the coordinate plane $(\underline{v},\underline{e}_j) = 0$, where $\underline{e}_j$ is a unit vector along the positive v_j-axis. Because of (3.3) and the fact that the $\underline{\phi}_1$ component has maximal amplitude among the characteristic directions, it follows that we must have

$$\frac{\pi}{2} < \sqrt{\mu_1}\, t_0 < \pi \quad \text{and} \quad \pi < \sqrt{\mu_i}\, t_0 < \frac{3\pi}{2} \tag{3.5}$$

for $2 \leq i \leq n$. Writing $\underline{v}(t_0)$ in terms of the eigenvectors of Γ_0 leads to the equation

(3.6) $$(c_1\underline{\phi}_1-c_2\underline{\phi}_2-\ldots-c_n\underline{\phi}_n,\underline{e}_j) = 0$$

for appropriate choice of positive constants $c_1,\ldots,c_n$. If now $\underline{v}(t_0)$ were not an exit point we would also have $(\underline{v}'(t_0),\underline{e}_j) = 0$ and, because of (3.5),

$$(d_1\underline{\phi}_1+d_2\underline{\phi}_2+\ldots+d_n\underline{\phi}_n,\underline{e}_j) = 0$$

for positive constants $d_1,\ldots,d_n$. This contradicts (3.6) and completes the proof.

Remarks

1. Given specific eigenvalues satisfying (3.3) the above argument remains valid under small perturbations of the trajectories. Accordingly Theorem 3.2 remains valid for (3.4) when $\Pi(t)$ is sufficiently small.
2. The fact that contact points are also exit points assures that contact points will vary continuously with initial data. This observation is crucial to the proof of Theorem 3.4 below.
3. In the case of oppositional vibrations the fact that the initial velocity vector $\underline{g}$ has $g_j = 0$ assures that the resulting trajectory exits the oppositional quadrant at $t = 0$ across $v_j = 0$. Lemma 3.3 shows that for (3.4) a very different situation exists.

3.3 Lemma (Crossover property). *If in Theorem* 3.2 *the vector* $\underline{g}$ *has* $g_j = 0$, *then the trajectory* $\underline{v}(t)$ *does not exit* $\mathbb{R}_n^+$ *across the coordinate plane* $(\underline{v},\underline{e}_j) = 0$.

Proof. The proof is similar to that of Theorem 3.2 by writing $g = c_1\phi_1 + \tilde{c}_2\psi$ where ψ lies in $\phi_1^{\perp}$. The fact that all components of ψ will satisfy $\pi < \sqrt{\mu_i}\, t_0 < \frac{3\pi}{2}$ at the time of contact precludes an exit across the plane $(\underline{v},\underline{e}_j) = 0$.

The above properties of trajectories of $\underline{v}'' + \Gamma_0 \underline{v} = 0$ lead to the existence of a conjugate point as follows. Among all initial velocity vectors g satisfying $\underline{g} > 0$ all $\|g\| = 1$ we define

$$T_j = \{\underline{g} : \underline{v}(t) \text{ exi ts } R_n^+ \text{ across } (\underline{v}, e_j) = 0\}.$$

A well known corollary to Sperner's lemma then leads to the fact that $\bigcap_{j=1}^{n} T_j \neq \phi$ and the following result.

3.4 Theorem. *Under the hypotheses of Theorem* 3.2, *and for sufficiently small perturbations* $\Pi(t)$, *the system* (3.4) *has a conjugate point of zero which is realized by a trajectory in* R_n^+.

References

1. S. Ahmad and A. Lazer, On the components of extremal solutions of second order systems, SIAM J. Math. Anal. 8(1977), 16-23.

2. P. Alexandroff and H. Hopf, Topologie, Berlin, Springer Verlag, 1935.

3. J. Cannon and S. Dostrovsky, The Evolution of Dynamics, Vibration Theory from 1687 to 1742, New York, Springer Verlag, 1981.

4. R. J. Duffin, Vibration of a beaded string analyzed topologically, Arch. Rat. Mech. and Anal. 56(1974), 287-293.

5. F. Gantmacher and M. Krein, Oscillation Matrices and Kernels and Small Vibrations of Mechanical Systems, Moscow, State Publishing House, 1950.

6. M. A. Krasnoselskii, Positive Solutions of Operator Equations, Groningen, Noordhoff, 1964.

7. K. Kreith, Picone-type theorems for semi-discrete hyperbolic equations, Proc. Amer. Math. Soc. 88(1983), 436-438.

8. K. Kreith, Stability criteria for conjugate points of indefinite second order differential systems, J. Math. Anal. and Applic., to appear.

9. M. Morse, A Generalization of the Sturm separation and comparison theorems in n-space, Math. Annalen 103(1930), 72-91.

10. W. T. Reid, Sturmian Theory for Ordinary Differential Equations. New York, Springer Verlag, 1980.

PERRON INTEGRAL, PERRON PRODUCT INTEGRAL AND ORDINARY LINEAR DIFFERENTIAL EQUATIONS

J. KURZWEIL and J. JARNÍK
Mathematical Institute, Czechoslovak Academy of Sciences
115 67 Prague 1, Czechoslovakia

1. Perron integral and Perron product integral

A finite set $\Delta = \{x_0, t_1, x_1, \ldots, t_k, x_k\}$ is called a partition of an interval $[a,b]$ if

$$a = x_0 < x_1 < \ldots < x_k = b \ , \quad x_{j-1} \leq t_j \leq x_j$$

for $j = 1,2,\ldots,k$. Let $\delta : [a,b] \to (0,\infty)$ (no continuity or measurability properties required). A partition Δ is said to be δ-fine if $[x_{j-1}, x_j] \subset (t_j - \delta(t_j),\ t_j + \delta(t_j))$.

Let $f : [a,b] \to \mathbb{R}$, put $S(f,\Delta) = \sum_{j=1}^{k} f(t_j)(x_j - x_{j-1})$. It is well known (cf. [1], [2]) that the following two conditions are equivalent:

$$\begin{array}{l} f \text{ is Perron integrable (P-integrable) over } [a,b] \ , \\ q = (P)\displaystyle\int_a^b f(t)\,dt \ ; \end{array} \tag{1.1}$$

$$\begin{array}{l} \text{for every } \varepsilon > 0 \text{ there exists such a } \delta : [a,b] \to (0,\infty) \text{ that} \\ |q - S(f,\Delta)| \leq \varepsilon \text{ for every } \delta\text{-fine partition } \Delta \text{ of } [a,b] \ . \end{array} \tag{1.2}$$

Condition (1.2) makes good sense since

$$\begin{array}{l} \text{for every } \delta : [a,b] \to (0,\infty) \text{ there exists a } \delta\text{-fine} \\ \text{partition } \Delta \text{ of } [a,b] \ . \end{array} \tag{1.3}$$

1.1. REMARK. The proof of (1.3) is easy: If (1.3) were false for a δ on $[a,b]$, it would be false either for δ on $[a, (a+b)/2]$ or for δ on $[(a+b)/2, b]$ and this procedure, if continued, leads to a contradiction.

Denote by M the ring of real or complex $n \times n$ matrices. For

$A : [a,b] \to M$ and a partition Δ of $[a,b]$ put

$$P(A,\Delta) = (I + A(t_k)(x_k - x_{k-1}))\dots(I + A(t_1)(x_1 - x_0)) ,$$

$$\tilde{P}(A,\Delta) = \exp(A(t_k)(x_k - x_{k-1}))\dots \exp(A(t_1)(x_1 - x_0)) .$$

The following result is well known (cf. [4], [5]): If A is continuous and if U is the matrix solution of

$$\dot{x} = A(t)x , \tag{1.4}$$

$U(a) = I$, then both $P(A,\Delta)$, $\tilde{P}(A,\Delta)$ converge to $U(b)$ in the following sense:

For every $\varepsilon > 0$ there exists such an $\eta > 0$ that $||U(b) - P(A,\Delta)|| \leqq \varepsilon$, $||U(b) - \tilde{P}(A,\Delta)|| \leqq \varepsilon$ for every partition Δ of $[a,b]$ satisfying $x_j - x_{j-1} < \eta$, $j = 1,\dots,k$. (1.5)

In [5] the Lebesgue product integral was introduced in a way analogous to the usual introduction of the Bochner integral and it was proved that $U(b)$ is equal to the Lebesgue product integral of $\exp(A(t)\,dt)$ over $[a,b]$ provided A is Lebesgue integrable in the usual sense. In the next definition, the limiting process from (1.2) is applied to the product $P(A,\Delta)$ - of course without any continuity or measurability condition on A .

1.2. DEFINITION. Let $Q \in M$ be regular. A is said to be *Perron product-integrable over* $[a,b]$ ($\mathcal{P}$-integrable), Q is called the *Perron product integral* ($\mathcal{P}$-integral) of A and denoted by $\mathcal{P}\int_a^b (I + A(t)\,dt)$, if for every $\varepsilon > 0$ there exists such a $\delta : [a,b] \to (0,\infty)$ that $||Q - P(A,\Delta)|| \leqq \varepsilon$ for every δ-fine partition Δ of $[a,b]$.

1.3. REMARK. The same concept of the $\mathcal{P}$-integral is obtained if $P(A,\Delta)$ is replaced by $\tilde{P}(A,\Delta)$ in Definition 1.2.

The integral $\mathcal{P}\int_a^b (I + A(t)\,dt)$ has properties analogous to those of the integral $(P)\int_a^b f(t)\,dt$. The properties of the latter are listed in Section 2, the analogous properties of the former in Section 3. In Section 4 some relations to ACG_*-functions and to the equation (1.4) are mentioned.

2. Properties of the Perron integral

If f is P-integrable over $[a,b]$ then $F(t) = (P)\int_a^t f(s)\,ds$ exists for $t \in (a,b]$. We put $F(a) = 0$. (2.1)

If f is P-integrable, then F is continuous and $\dot{F}(t) = f(t)$ a.e. Moreover, f is measurable. (2.2)

Let f be P-integrable over $[a,b]$. Then the following assertion holds: if $C \subset [a,b]$ is of measure zero and $\varepsilon > 0$, then there exists such a $\delta : C \to (0,\infty)$ that $\sum_{j=1}^{r} |F(\eta_j) - F(\xi_j)| < \varepsilon$ provided $\tau_j \in C$, $\xi_j \leq \tau_j \leq \eta_j \leq \xi_{j+1}$ and $[\xi_j,\eta_j] \subset (\tau_j - \delta(\tau_j), \tau_j + \delta(\tau_j))$ for $j = 1,\dots,r$. (2.3)

Let $F : [a,b] \to \mathbb{R}$ have derivative a.e. and satisfy the assertion from (2.3). Put $f(t) = \dot{F}(t)$ if $\dot{F}(t)$ exists, $f(t)$ arbitrary otherwise. (2.4)

Then $(P)\int_a^b f(t)\,dt$ exists and equals $F(b) - F(a)$.

If $(P)\int_a^t f(s)\,ds$ exists for $t < b$ and if $\lim_{t\to b-} (P)\int_a^t f(s)\,ds = q \in \mathbb{R}$, then $(P)\int_a^b f(s)\,ds$ exists and equals q. (2.5)

3. Properties of the Perron product integral

If A is P-integrable over $[a,b]$ then $U(t) = P\prod_a^t (I + A(s)\,ds)$ exists for $t \in (a,b]$. We put $U(a) = I$. (3.1)

If A is P-integrable, then $U(t)$ is regular at every t, U is continuous and $\dot{U}(t)U^{-1}(t) = A(t)$ a.e. Moreover, A is measurable. (3.2)

Let A be P-integrable. Then the assertion (2.3) holds with F replaced by U. (3.3)

Let $U : [a,b] \to M$ be continuous, regular at every t and differentiable a.e., and let it satisfy the modified assertion of (2.3) (cf. (3.3)). Put $A(t) = \dot{U}(t)U^{-1}(t)$ if $\dot{U}(t)$ exists, $A(t)$ arbitrary otherwise. Then $P\int_a^b (I + A(t)\,dt)$ exists and equals $U(b)U^{-1}(a)$. (3.4)

If $P\int_a^t (I + A(s)\,ds)$ exists for $t < b$ and if

$$\lim_{t \to b-} P\int_a^t (I + A(s)\,ds) = Q \in M \text{ is regular, then} \tag{3.5}$$

$P\int_a^b (I + A(s)\,ds)$ exists and equals Q.

4. ACG_*-solutions of linear ordinary differential equations

The concept of an ACG_*-function (cf. [6]) extends without complications to functions with values in finitedimensional linear spaces. A function $u : [a,b] \to \mathbb{R}^n$ $(\mathbb{C}^n)$ is called an ACG_*-solution of (1.4) if u is an ACG_*-function and satisfies (1.4) a.e. In an analogous manner the concept of a matrix ACG_*-solution of (1.4) is to be understood.

It is well known that F from (2.1) is an ACG_*-function and that every ACG_*-function is the primitive of its derivative (in the sense of (2.1), (2.2)). It follows from (2.3) and (2.4) that F is an ACG_*-function iff it satisfies (2.4). It follows from (3.1) - (3.4) that U from (3.1) is an ACG_*-function and that every ACG_*-function $U : [a,b] \to M$ can be written in the form

$$U(t)U^{-1}(a) = P\int_a^t (I + A(s)\,ds)$$

provided $U(t)$ is regular for every t. At the same time we have

$$U(t) - U(a) = (P)\int_a^t \dot{U}(s)\,ds = (P)\int_a^t A(s)\,U(s)\,ds$$

for $t \in [a,b]$, that is, U is an ACG_*-matrix solution of (1.4).

Thus we have obtained a class of LDE's the solutions of which are ACG_*-functions; these LDE's have the usual existence and uniqueness properties.

Denote by $H([a,b])$ the set of such $A : [a,b] \to M$ that $P\int_a^b (I + A(t)\, dt)$ exists. Let us find some effective conditions for $A \in H([a,b])$.

Assume that $c > 0$, $B : [-c,c] \to M$ is continuous, $I + B(t)$ is regular for $t \in [-c,c]$ and B is locally absolutely continuous on $[-c,c] \setminus \{0\}$. We have

$$\text{if } A(t) = \dot{B}(t)[I + B(t)]^{-1} \text{ a.e. then } A \in H([-c,c]) , \tag{4.1}$$

$$\text{if } k \in \{0,1,\dots\} , \ A(t) = \dot{B}(t)[I - B(t) + \dots + (-1)^k B^k(t)]$$
$$\text{a.e.}, \ \int_{-c}^{c} ||\dot{B}(t)B^{k+1}(t)||\, dt < \infty , \text{ then } A \in H([-c,c]) . \tag{4.2}$$

In the case (4.1), $I + B(t)$ is a fundamental matrix of (1.4), in the case (4.2) the substitution $x = [I + B(t)]y$ leads to the result.

Let $\alpha > 0$, $\beta > 0$, $T, S \in M$. If $\alpha < 1 + \beta$, then there exists such a continuous $B : \mathbb{R} \to M$ that $B(0) = 0$, $\dot{B}(t) = |t|^{-\alpha}[T \cos |t|^{-\beta} + S \sin |t|^{-\beta}]$ for $t \neq 0$. Let $c > 0$ be so small that $I + B(t)$ is regular for $t \in [-c,c]$. Then (4.1) may be applied. If $\alpha < 1 + \beta/2$ then (4.2) may be applied with $k = 0$. If $1 + \beta/2 \leq \alpha < 1 + 2\beta/3$ then (4.2) may be applied with $k = 1$; moreover, if $TS - ST \neq 0$ then $\int_{-1}^{t} A(s)\, ds$ is unbounded for $t \to 0-$ so that $(P)\int_{-1}^{1} A(t)\, dt$ does not exist.

5. The Saks-Henstock Lemma

In the proof of the properties (2.2) and (2.3) of the Perron integral the key part is played by the following

5.1. LEMMA (Saks, Henstock). Assume that f is P-integrable over $[a,b]$, $F(t) = (P)\int_a^t f(s)\, ds$. Let $\varepsilon > 0$ and let the gauge δ correspond to ε according to (1.2). Let

$$\begin{aligned} &\xi_j, \tau_j, \eta_j \in [a,b] , \quad \xi_j \leq \tau_j \leq \eta_j \leq \xi_{j+1} \\ &[\xi_j, \eta_j] \subset (\tau_j - \delta(\tau_j), \tau_j + \delta(\tau_j)) , \quad j = 1,2,\dots,r . \end{aligned} \tag{5.1}$$

Then

$$\sum_{j=1}^{r} |f(\tau_j)(\eta_j - \xi_j) - F(\eta_j) + F(\xi_j)| \leq 2\varepsilon .$$

For the Perron product integral, an analogous role in the proof of the properties (3.2) and (3.3) is played by

5.2. LEMMA. There exist $\varepsilon_0 > 0$ and $K > 0$ depending on n only so that the following holds:

Assume that A is P-integrable over $[a,b]$,

$$U(t) = P\int_a^t \big(I + A(s)\big)ds \ , \ U(b) = Q \ .$$

Let $0 < \varepsilon < \varepsilon_0/||Q^{-1}||$ and let the gauge δ correspond to ε according to Definition 1.2. Let (5.1) hold. Then

$$\sum_{j=1}^{r} ||I + A(\tau_j)(\eta_j - \xi_j) - U(\eta_j)U^{-1}(\xi_j)|| \leq K\varepsilon .$$

References

[1] KURZWEIL, J.: *Nichtabsolut konvergente Integrale*. Teubner, Leipzig 1980, Teubner-Texte zur Mathematik, 26.

[2] KURZWEIL, J.: *The integral as a Limit of Integral Sums*. Jahrbuch Überblicke Mathematik 1984, 105-136, Bibliographisches Institut AG 1984.

[3] SCHLESINGER, L.: *Einführung in die Theorie der gewöhnlichen Differentialgleichungen auf funktionentheoretischer Grundlage*. Berlin 1922.

[4] GANTMACHER, F. R.: *Theory of Matrices*. Moskva 1966 (Russian).

[5] DOLLARD, J. D. and FRIEDMAN, CH. N.: *Product Integration with Applications to Differential Equations*. Univ. Press, Cambridge 1979.

[6] SAKS, S.: *Theory of the Integral*. Monografie matematyczne VII. G.E. Stechert, New York - Warszawa 1937.

ON THE ZEROS OF SOME SPECIAL FUNCTIONS: DIFFERENTIAL EQUATIONS AND NICHOLSON-TYPE FORMULAS

M. E. MULDOON
Department of Mathematics, York University
North York, Ontario M3J 1P3, Canada

1. Introduction. There are many results in the literature on special functions concerning the way in which a zero of a function changes with respect to one of the parameters on which the function depends. Methods based on differential equations, in particular Sturmian methods, are often useful in these discussions. Other methods are related to integral representations for the functions and seem to be provable, though not easily discoverable, by differential equations methods. Among these are methods based on Nicholson's formula [13, p.444]

$$(1)\qquad J_\nu^2(z) + Y_\nu^2(z) = \frac{8}{\pi^2}\int_0^\infty K_0(2z \sinh t)\cosh 2\nu t\, dt\ ,\quad \operatorname{Re} z > 0\ ,$$

and a companion formula

$$(2)\quad J_\nu(z)\partial Y_\nu(z)/\partial\nu - Y_\nu(z)\,\partial J_\nu(z)/\partial\nu = -\frac{4}{\pi}\int_0^\infty K_0(2z\sinh t)e^{-2\nu t}\,dt\ ,\quad \operatorname{Re} z > 0\ ,$$

from which it follows [13, p.508] that

$$(3)\qquad dc/d\nu = 2c\int_0^\infty K_0(2c\sinh t)e^{-2\nu t}\,dt\ .$$

Here J_ν and Y_ν are the usual Bessel functions, K_0 is the modified Bessel function and, in (3), $c = c(\nu,k,\alpha)$ is an x-zero of the linear combination

$$C_\nu(x) = \cos\alpha\, J_\nu(x) - \sin\alpha\, Y_\nu(x)\ .$$

Formula (1) was used by L. Lorch and P. Szego [9] to show some remarkable sign-regularity properties of the higher k-differences of the sequence $\{c(\nu,k,\alpha)\}$ in the case $|\nu| \geq \frac{1}{2}$. Beyond its obvious use to show that c increases with ν, (3) has been used to get further information about these zeros; see [10,11] for references. Á. Elbert has used (3) to show that $j_{\nu k}(=c(\nu,k,0))$ is a concave function on ν on $(-k,\infty)$. Elbert and A. Laforgia have used (3) in several recent papers. They proved, for example, that $j_{\nu k}^2$ is a convex function of ν on $(0,\infty)$ [6] and they have shown recently (personal communication) that $d^3\, j_{\nu k}/d\nu^3 > 0$, $0 < \nu < \infty$.

2. Other Nicholson-type formulas. The usefulness of (1), (2) and (3) suggests the desirability of having similar formulas for other special functions. L. Durand [3,4] has given results analogous to (1) for some of the classical orthogonal polynomials. The simplest of these, for Hermite functions, is [3, p.371]

(4) $$e^{-x^2}[H_\lambda^2(x) + G_\lambda^2(x)] = \frac{2^{\lambda+1}\Gamma(\lambda+1)}{\pi}\int_0^\infty e^{-(2\lambda+1)t + x^2\tanh t}(\cosh t \sinh t)^{-1/2}\,dt.$$

Durand does not use differential equations but points out [3, p.355] that, once the results are known, they can be checked by differential equations methods. In fact J.E. Wilkins, Jr. [14] (see also [12, pp.340-341]) proved (1) by showing that both sides satisfy the same third order differential equation and have the same asymptotic behaviour as $z \to +\infty$. I [10] did the same for equation (2) using a third-order nonhomogeneous equation.

More recently, I have tried to discover whether there is a natural way in which these formulas arise in a differential equations setting. I present such a setting here for Bessel functions but it is not clear to me yet whether the method applies to a general situation of which the Bessel function case would be a particular example. It turns out to be convenient to consider the more general formulas [2,7,13]

(5) $$J_\mu(z)J_\nu(z) + Y_\mu(z)Y_\nu(z) = \frac{4}{\pi^2}\int_0^\infty K_{\nu-\mu}(2z \sinh t)[e^{(\mu+\nu)t} + e^{-(\mu+\nu)t}\cos(\mu-\nu)\pi]dt\ ,$$

(6) $$J_\mu(z)J_\nu(z)+Y_\mu(z)Y_\nu(z) = \frac{4}{\pi^2}\int_0^\infty K_{\mu+\nu}(2z \sinh t)[e^{(\mu-\nu)t}\cos \nu\pi + e^{-(\mu-\nu)t}\cos \mu\pi]dt,$$

(7) $$J_\mu(z)Y_\nu(z) - J_\nu(z)Y_\mu(z) = \frac{4}{\pi^2}\sin(\mu-\nu)\pi\int_0^\infty K_{\nu-\mu}(2z \sinh t)e^{-(\nu+\mu)t}dt$$

(8) $$J_\mu(z)Y_\nu(z)-J_\nu(z)Y_\mu(z) = \frac{4}{\pi^2}\int_0^\infty K_{\nu+\mu}(2z \sinh t)[e^{(\nu-\mu)t}\sin \mu\pi - e^{(\mu-\nu)t}\sin \nu\pi]dt.$$

These are all valid for $\mathrm{Re}\ z > 0$ with $|\mathrm{Re}(\mu + \nu)| < 1$ in (6) and (8) and $|\mathrm{Re}(\nu - \mu)| < 1$ in (5) and (7).

Clearly (1) is got from (5) by setting $\mu = \nu$ while, as pointed out in [2], (2) (and hence (3)) is got from (7) by dividing by $\mu - \nu$ and letting $\mu \to \nu$. Dixon and Ferrar [2, p.142] find an analogue of (2) based on a similar treatment of (8).

The corresponding analogue of (3) is

$$dc/d\nu = -(2c/\pi)\int_0^\infty K_{2\nu}(2c \sinh t)[2t \sin \nu\pi - \pi \cos \nu\pi]dt\ ,\quad c > 0\ ,\quad |\nu| < \frac{1}{2}\ ,$$

but this is both more complicated and has a smaller range of validity then (3).

3. A differential equations proof of (6). The proofs of (5), (7) and (8) are quite similar to that which we will give for (6). We may clearly suppose that μ and ν are real and that z is real and positive and we write $z = e^\theta$ so that (6) becomes

(9) $$J_\mu(e^\theta)J_\nu(e^\theta) + Y_\mu(e^\theta)Y_\nu(e^\theta) = \frac{4}{\pi^2}\int_0^\infty K_{\mu+\nu}(2e^\theta \sinh t)[e^{(\mu-\nu)t}\cos \nu\pi + e^{-(\mu-\nu)t}\cos \mu\pi]dt\ .$$

The functions $J_\nu(e^\theta)$, $Y_\nu(e^\theta)$ satisfy [13, p.99]

$$d^2y/d\theta^2 + (e^{2\theta} - \nu^2)y = 0$$

so that the left-hand side of (9) is a solution of [13, p.146]

$$L_\theta u \equiv (D_\theta^2 - b^2)(D_\theta^2 - a^2)u + 4e^{2\theta}(D_\theta + 1)(D_\theta + 2)u = 0 \tag{10}$$

where $a = \mu + \nu$, $b = \mu - \nu$. There is a standard method [1; 8, Ch. 8] for finding an integral representation

$$u(\theta) = \int_\alpha^\beta k(\theta,t)\ v(t)\ dt \tag{11}$$

for a solution of (10). We try to find a linear differential operator

$$M_t = \sum_{k=0}^{m} m_k(t)\ D_t^k$$

and a function $\kappa(\theta,t)$ such that

$$L_\theta\ k(\theta,t) = M_t\ \kappa(\theta,t)\ . \tag{12}$$

We then determine $v(t)$ as a solution of $\overline{M}_t\ v = 0$ where $\overline{M}_t$ is the adjoint of M_t, i.e.

$$\overline{M}_t\ v = \sum_{k=0}^{m} (-1)^k\ D_t^k[m_k(t)v]\ .$$

Then (11) is a solution of $L_\theta\ u = 0$ provided α and β are chosen so that

$$\left[\sum_{k=1}^{m} \sum_{\ell=0}^{k-1} (-1)^\ell\ (m_k\ v)^{(\ell)}\ \kappa^{(k-\ell-1)}\right]_\alpha^\beta = 0\ . \tag{13}$$

(The differentiations in (13) are with respect to t.) Most of the standard applications of the method are to second order equations and with $\kappa = k$ and its success depends on being able to solve the equation $\overline{M}_t\ v = 0$. In the present case, if we choose

$$k(\theta,t) = K_a(2e^\theta \sinh t) \tag{14}$$

we have the convenient "factorization"

$$L_\theta\ k(\theta,t) = (D_t^2 - b^2)(D_\theta^2 - a^2)\ k(\theta,t) \tag{15}$$

which is of the form (12) with

$$\kappa(\theta,t) = (D_\theta^2 - a^2)\ K_a(2e^\theta \sinh t) = 4e^{2\theta} \sinh^2 t\ K_a(2e^\theta \sinh t)$$

and $M_t = \overline{M}_t = D_t^2 - b^2$. Thus we get $v(t) = c_1\ e^{(\mu-\nu)t} + c_2\ e^{-(\mu-\nu)t}$ and it is easily shown that (13) holds if we choose $\alpha = 0$, $\beta = \infty$. To determine c_1 and c_2 we use [13, Ch.7]

$$J_\mu(x)\ J_\nu(x) + Y_\mu(x)\ Y_\nu(x) = \frac{2}{\pi x} \cos\frac{(\mu-\nu)\pi}{2} + \frac{1}{\pi x^2}(\mu^2-\nu^2)\ \sin\frac{(\mu-\nu)\pi}{2} + O(x^{-3})\ ,\ x \to \infty\ ,$$

and, using [12, Ch.9],

$$\int_0^\infty K_a(2x \sinh t)e^{\lambda t}\, dt = \frac{\pi}{4\cos(\pi a/2)}\, x^{-1} + \frac{a\lambda\pi}{8\sin(\pi a/2)}\, x^{-2} + O(x^{-3}) \;,\; x \to \infty \,.$$

Then, by comparing coefficients, we get $c_1 = \cos \nu\pi$, $c_2 = \cos \mu\pi$ so (9), and hence (6), is proved.

The key to the success of the method in the present case is the factorization (15) arising from the choice (14) for the kernel $k(\theta,t)$. The choice of a function of the form $f(2e^\theta \sinh t)$ may be motivated by the fact that for a polynomial P we have $P(D_\theta)\, f(2e^\theta \sinh t) = P(\tanh t\, D_t)\, f(2e^\theta \sinh t)$. We see from (10) that $L_\theta\, f(2e^\theta \sinh t)$ can be expected to take on a relatively simple form if we choose f to satisfy

$$(D_\theta^2 - a^2)\, f(2e^\theta \sinh t) = 4e^{2\theta} \sinh^2 t\, f(2e^\theta \sinh t)$$

But this is the modified Bessel equation satisfied by $f = K_a$.

4. <u>Another Nicholson-type formula</u>. Here we give a differential equations proof of

$$(16)\quad J_\nu^2(x) + Y_\nu^2(x) = \frac{4}{\pi^2}\frac{\Gamma(\nu)}{\Gamma(2\nu)}(4x)^\nu \int_0^\infty K_\nu(2x \sinh t)(\cosh t)^{2\nu}(\sinh t)^\nu\, dt$$

$$= \frac{4}{\pi^2}\frac{\Gamma(\nu)}{\Gamma(2\nu)}\left(\frac{x}{2}\right)^\nu \int_0^\infty K_\nu(xu)(u^2+4)^{\nu-1/2}u^\nu\, du \;,\; x > 0 \;,\; \nu > -\tfrac{1}{2} \,,$$

the special case $n = 0$ of [3, p.368, (42)]. It is convenient to write this formula in the form

$$(17)\quad j_\alpha^2(x) + y_\alpha^2(x) = \int_0^\infty k_\alpha(xt)\,(t^{\alpha/2}+4)^{-(\alpha+4)/(2\alpha+4)}\, t^{\alpha/2-1}\, dt \;,\; x > 0 \;,\; \alpha > 0$$

where we have adopted an *ad hoc* notation for the generalized Airy functions: j_α, y_α are appropriately normalized solutions of

$$(18)\qquad y'' + x^\alpha y = 0$$

where $\alpha = -2 - 1/\nu$, while k_α is a suitable solution vanishing at $+\infty$ of

$$(19)\qquad y'' - x^\alpha y = 0 \,.$$

In the special case $\alpha = 1$ $(\nu = -1/3)$, (17) becomes

$$\mathrm{Ai}^2(-x) + \mathrm{Bi}^2(-x) = \frac{24(2/3)^{1/6}}{\sqrt{\pi}\,\Gamma(1/6)} \int_0^\infty t^{-1/2}(t^3+4)^{-5/6}\,\mathrm{Ai}(xt)\, dt \;,\; x > 0 \,.$$

In order to prove (17) we note that, using (18) and [13, p.145] its left-hand side satisfies

$$L_x u \equiv (D_x^3 + 4x^\alpha D_x + 2\alpha x^{\alpha-1})u = 0$$

and, using (19), we find that

$$L_x\, k_\alpha(xt) = M_t\, x^{\alpha-1}\, k_\alpha(xt)$$

where

$$M_t = (t^{\alpha+3} + 4t)D_t + \alpha(t^{\alpha+2} + 2)$$

and

$$\bar{M}_t = -(t^{\alpha+3} + 4t)D_t + (2\alpha - 4 - 3t^{\alpha+2}) .$$

Now $\bar{M}_t\, v = 0$ has the general solution

$$v(t) = ct^{\alpha/2-1}\,(t^{\alpha+2} + 4)^{-(\alpha+4)/(2\alpha+4)} .$$

We note that the condition (13) is also satisfied with $\alpha = 0$, $\beta = \infty$ leading to (17) apart from a constant factor. To evaluate the constant we return to the form (16) and use

$$J_\nu^2(x) + Y_\nu^2(x) = \frac{2}{\pi x} + O(x^{-2}) , \quad x \to \infty$$

and, using [12, Ch.7],

$$x^\nu \int_0^\infty K_\nu(xu)(u^2 + 4)^{\nu-1/2}\, u^\nu\, du = \frac{1}{x}\, 2^{\nu-1}\, \pi\Gamma(2\nu)/\Gamma(\nu) + O(x^{-2}) , \quad x \to \infty .$$

5. Zeros of generalized Airy functions. M.S.P. Eastham (private communication) raised the question of showing that the smallest positive zero x_α of a solution of (18), satisfying $y(0) = 0$, decreases as α increases, $0 < \alpha < \infty$. This, and more, has been proved by A. Laforgia and the author (to be published) using results (due to Elbert and Laforgia) based on (3) and the well-known connection between (18) and the Bessel equation. It would be nice to show this using (18) directly. The Sturm comparison theorem is not applicable in any obvious way because x^α is not monotonically increasing in α for each x in an interval $(0,b)$, $b > 1$. This raises the question of whether one can find an analogue of (3) (other than the awkward formula got by transforming (3) itself) for $dx_\alpha/d\alpha$. What we need in effect is a result that bears the same relation to (17) as (3) does to (1). One way to approach this problem would be to find an integral representation for $j_\alpha\, y_\beta - j_\beta\, y_\alpha$ which satisfies a known fourth order differential equation.

A perhaps more tractable problem would be to find the appropriate generalization of (4) for

$$e^{-x^2}[H_\lambda(x)\, G_\mu(x) - G_\lambda(x)\, H_\mu(x)] .$$

This would give, in particular, a formula for the derivative with respect to λ of a zero of a Hermite function.

References

1. H. Bateman, *The solution of linear differential equations by means of definite integrals*, Trans. Cambridge Philos. Soc. 21 (1909), pp. 171-196.
2. A.L. Dixon and W.L. Ferrar, *Infinite integrals in the theory of Bessel functions*, Quart. J. Math. Oxford 1 (1930), pp. 122-145.
3. L. Durand, *Nicholson-type integrals for products of Gegenbauer functions and related topics*, in Theory and Application of Special Functions (R. Askey, ed.), Academic Press, New York, 1975, pp. 353-374.
4. L. Durand, *Product formulas and Nicholson-type integrals for Jacobi functions. I: Summary of results*, SIAM J. Math. Anal. 9 (1978), pp. 76-86.
5. Á. Elbert, *Concavity of the zeros of Bessel functions*, Studia Sci. Math. Hungar. 12 (1977), pp. 81-88.
6. Á. Elbert and A. Laforgia, *On the square of the zeros of Bessel functions*, SIAM J. Math. Anal. 15 (1984), pp. 206-212.
7. G.H. Hardy, *Some formulae in the theory of Bessel functions*, Proc. London Math. Soc. 23 (1925), pp. lxi-lxiii.
8. E.L. Ince, *Ordinary Differential Equations*, Longmans, London, 1927; reprinted Dover, New York, 1956.
9. L. Lorch and P. Szego, *Higher monotonicity properties of certain Sturm-Liouville functions*, Acta Math. 109 (1963), pp. 55-73.
10. M.E. Muldoon, *A differential equations proof of a Nicholson-type formula*, Z. Angew. Math. Mech. 61 (1981), pp. 598-599.
11. M.E. Muldoon, *The variation with respect to order of zeros of Bessel functions*, Rend. Sem. Mat. Univ. Politec. Torino 39 (1981), pp. 15-25.
12. F.W.J. Olver, *Asymptotics and Special Functions*, Academic Press, New York and London, 1974.
13. G.N. Watson, *A treatise on the Theory of Bessel Functions*, 2nd ed., Cambridge University Press, 1944.
14. J.E. Wilkins, Jr., *Nicholson's integral for* $J_n^2(z) + Y_n^2(z)$, Bull. Amer. Math. Soc. 54 (1948), pp. 232-234.

SURJECTIVITY AND BOUNDARY VALUE PROBLEMS

V. ŠEDA
Faculty of Mathematics and Physics, Comenius University
Mlynská dolina, 842 15 Bratislava, Czechoslovakia

In the paper we shall deal with an initial and a boundary problem for the functional differential equation with deviating argument $x'(t) = f[t, x_{\omega(t)}]$ in a Banach space whereby the functions of the state space are defined in the interval $(-\infty, 0]$ as well as with the generalized boundary value problem for a system of differential equations in R^n. The main tool for proving the existence of a solution to these problems will be some theorems on surjectivity of an operator.

1. Surjectivity of an operator.

Let $(E, |.|)$ be a real Banach space, $\phi \neq X \subset E$ and $S : X \to E$. We recall that S is *compact* if S is continuous and maps bounded sets into relatively compact sets. Similarly $T : X \to E$ is said to be a *condensing* map if T is continuous, bounded (i.e. maps bounded sets into bounded sets) and for every bounded set $A \subset X$ which is not relatively compact we have $\alpha(T(A)) < \alpha(A)$ where α is the Kuratowski measure of noncompactness. A simple example of a condensing map is one of the form $U + V$ where $U : X \to E$ is a strict contraction and $V : X \to E$ is a compact map.

Let $G \neq \phi$ be an open subset of E and denote by $\overline{G}$ the closure of G. Let $T : \overline{G} \to E$ be a condensing map, $a \in E$. If the set $\tilde{A} = \{x \in G : x - T(x) = a\}$ is compact (possibly empty), then the degree $\deg(I - T, G, a)$ is defined in the sense of Nussbaum [6] whereby I is the identity. Notice that $\tilde{A}$ will certainly be compact if G is bounded and T is such that $x - T(x) \neq a$ for all $x \in \partial G$ (boundary of G) ([6], p. 744). If T is compact, then the degree above agrees with the classical Leray-Schauder degree.

Denote B the real Banach space of all continuous functions $x : [0, \infty) \to E$ such that there exists $\lim x(t) = x(\infty)$ ($\in E$) for $t \to \infty$. The norm in B is defined by $\| |x| \|_2 = \sup\{|x(t)| : 0 \le t < \infty\}$ for each $x \in B$. Let, further, $U(r) = \{x \in E : |x| < r\}$. Using the degree theory for condensing perturbations of identity, the topological principle in [8], p. 241, can be generalized as follows (for proof, see [9], [10]).

Theorem 1. Let $g : E \to B$ be a continuous mapping. Denote by $g(x,t)$

the value of $g(x) \in B$ at the point $t \in [0,\infty]$ ($g(x,\infty) = \lim g(x,t)$ for $t \to \infty$). Assume that

(i) $\nu(x) = \inf\{|g(x,t)| : 0 \le t \le \infty\} \to \infty$ for $|x| \to \infty$;

(ii) the mapping $I - g(.,t)$ is condensing for each $t \in [0,\infty]$;

(iii) for each $y \in E$ there is an $r_0 > 0$ such that

$$\deg(g(.,0) - y, U(r_0),0) \neq 0 ;$$

(iv) $g(x,.)$ is continuous in t, uniformly in $x \in \overline{U(r)}$ for each $r > 0$. Then for each $t \in [0,\infty]$

$$g(E,t) = E.$$

Proof. Let $y \in E$, $t_0 \in [0,\infty]$. By (i), there is an $r_0 > 0$, $|y| < r_0$, such that $y \notin g(\partial U(r_0),t)$ for each $t \in [0,\infty]$. Hence the mapping $G : \overline{U(r_0)} \times [0,\infty] \to E$ defined by $G(x,t) = x - g(x,t) + y$ is continuous and $G(x,t) \neq x$ for $x \in \partial U(r_0)$, $t \in [0,\infty]$. By (ii), $G(.,t)$ is a condensing map for $t \in [0,\infty]$ and (iv) implies that $G(x,.)$ is continuous in t, uniformly in $x \in \overline{U(r_0)}$. Hence, by Corollary 2 in [6], p.745, and (iii), for each t_0, $0 \le t_0 < \infty$,

$$\deg(I - G(.,t_0), U(r_0),0) = \deg(I - G(.,0), U(r_0),0) =$$
$$= \deg(g(.,0) - y, U(r_0),0) \neq 0.$$

As to the set $S = \{x \in U(r_0) : g(x,t_0) - y = 0\}$, either it is not compact or in case it is compact we can use Proposition 5 from [6], p. 744, and hence, in both cases it is nonempty.

Corollary 2 as well as Proposition 5 from [6] can be applied to the case $t_0 = \infty$, too, since then $t = \operatorname{tg} \frac{\pi}{2} s$ maps $[0,1]$ continuously on $[0,\infty]$ and instead of the function $G(x,t)$ we consider $G_1(x,s) = G(x, \operatorname{tg} \frac{\pi}{2} s)$, $x \in \overline{U(r_0)}$, $s \in [0,1]$.

Remark. Clearly the assumption (iii) is satisfied if $g(x,0) = x$ for each $x \in E$.

On the basis of the Schauder theorem on domain invariance ([2], p. 72) the following result can be proved ([10]).

Theorem 2. Let $T : E \to E$ be such that

(a) $\lim_{|x|\to\infty} |T(x)| = \infty$;

(b) $I - T$ is compact;

(c) T is locally one-to-one, i.e. for each point $x_0 \in E$ there is a neighbourhood N of this point such that $T|_N$ is one-to-one. Then $T(E)=E$.

Proof. The assumptions (b), (c) imply that T is an open mapping, i.e. it maps open sets onto open sets. Hence T(E) is an open subset of E. Let $\{y_n\} \subset T(E)$ be a convergent sequence and $y_0 = \lim_{n\to\infty} y_n$. Then we can find a sequence $\{x_n\}$ such that $T(x_n) = y_n$. Assumption (a) is equivalent to the statement that the inverse image of a bounded set at the mapping T is a bounded set. Hence the sequence $\{x_n\}$ is bounded together with the sequence $\{y_n\}$. By (b), there is a subsequence $\{x_m\}$ of $\{x_n\}$ and a point $x_0 \in E$ such that $x_m - y_m = x_m - T(x_m) \to x_0$ as $m \to \infty$. Then $\lim_{m\to\infty} x_m = y_0 + x_0$, and by continuity of T, $T(x_0 + y_0) = y_0$. Thus $y_0 \in T(E)$ and T(E) is closed. As E is connected, T(E) = E.

Corollary 1. Let $T : E \to E$ be such that

(a) $\lim_{|x|\to\infty} |T(x)| = \infty$;

(b) $I - T$ is compact;

(c) T is one-to-one.

Then T is a homeomorphism of E onto E and there is a compact mapping $T_1 : E \to E$ such that $T^{-1} = I - T_1$ where T^{-1} is the inverse mapping to T.

Proof. By Theorem 2 and its proof we have that T(E) = E and the mapping T^{-1} is continuous. Hence T is a homeomorphism. For T^{-1} we have the identity $I - T^{-1} = (T - I) \circ T^{-1}$. By (a), T^{-1} is a bounded mapping and thus, by (b), $I - T^{-1} = T_1$ is compact.

If $E = R^n$, then Theorem 1 is true without assuming assumptions (ii),(iv) and in Theorem 2 instead of the assumption (b) it suffices to assume the continuity of T. Choosing properly the mapping $g : R^n \to B$ (B now means the Banach space of all continuous functions $x : [0,1] \to R^n$ with the supnorm, $|.|$ is the euclidean norm in R^n and (.,.) the scalar product in this space) we get the following

Corollary 2. Let $T : R^n \to R^n$ be a continuous mapping such that

(i) $\lim_{|x|\to\infty} |T(x)| = \infty$;

(ii) either there is an $x_0 \in R^n$ such that
$T(x) - x_0 = k(x - x_0)$ implies $k \geq 0$ for each $x \in R^n$, $x \neq x_0$,
or
there is an $r_1 > 0$ such that $(x, T(x)) \geq 0$ for all $x \in R^n$, $|x| \geq r_1$

or

T is locally one-to-one.

Then

$T(R^n) = R^n$.

Proof. a. Consider the first case that there is an $x_0 \in R^n$ such that

(α) $T(x) - x_0 = k(x - x_0)$ implies $k \geq 0$ for each $x \in R^n$, $x \neq x_0$.

Without loss of generality we may assume that $x_0 = 0$. Let the mapping $g : R^n \to B$ be defined by

$$g(x,t) = tT(0) \quad \text{for } x = 0,\ 0 \leq t \leq 1,$$

$$g(x,t) = [(1 - t)|x| + t|T(x)|].[|(1 - t)x + tT(x)|]^{-1}.$$
$$[(1 - t)x + tT(x)] \quad \text{for } x \neq 0,\ 0 \leq t < 1,$$

$$g(x,t) = T(x) \quad \text{for } x \neq 0,\ t = 1.$$

By (α) the mapping g is well defined. Further $g(x,.)$ is continuous in $[0,1]$ for each $x \in R^n$ and thus, g maps R^n into B. Clearly

(β) $g(x,0) = x$, $g(x,1) = T(x)$ for each $x \in R^n$.

Now we prove that g is continuous. Let $x \neq 0$ be an arbitrary but fixed point from R^n and y be a point sufficiently close to x. Then

$$|g(x,t) - g(y,t)| \leq \left|\frac{(1 - t)x + tT(x)}{|(1 - t)x + tT(x)|} - \frac{(1 - t)y + tT(y)}{|(1 - t)y + tT(y)|}\right| .$$

$$. [(1 - t)|x| + t|T(x)|] +$$

$$+ |(1 - t)(|x| - |y|) + t(|T(x)| - |T(y)|)|, \quad 0 \leq t \leq 1.$$

Clearly the second term on the right-hand side is less or equal to

(γ) $(1 - t)|x - y| + t|T(x) - T(y)|$, $0 \leq t \leq 1$.

As to the first term, there is a constant $k > 0$ such that this term is less or equal to

$$k|(1 - t)y + tT(y)|^{-1}.|[(1 - t)x + tT(x)] . |(1 - t)y + tT(y)| -$$
$$- [(1 - t)y + tT(y)].|(1 - t)x + tT(x)|| \leq$$
$$\leq k|(1 - t)y + tT(y)|^{-1} . |[(1 - t)(x - y) + t(T(x) - T(y))] .$$
$$. |(1 - t)y + tT(y)| + [(1 - t)y + tT(y)] . [|(1 - t)y + tT(y)| -$$
$$- |(1 - t)x + tT(x)|]|.$$

Hence the first term is less or equal to

(δ) $2k[(1 - t)|x - y| + t|T(x) - T(y)|]$, $0 \leq t \leq 1$.

The inequalities (γ) and (δ) give

$$|g(x,t) - g(y,t)| \le (2k + 1)[(1 - t)|x - y| + t|T(x) - T(y)|],$$
$$0 \le t \le 1,$$

which proves the continuity of g at $x \neq 0$. In a similar way it can be shown that g is continuous at 0.

Now we derive properties (i), (iii) of g from Theorem 1 and this will complete the proof of this part of Corollary 2. As $|g(x,t)| = (1 - t)|x| + t|T(x)| \ge \min(|x|,|T(x)|)$, clearly (i) is satisfied. (iii) follows from (β).

b. Suppose that there is an $r_1 > 0$ such that

($\varkappa$) $(x,T(x)) \ge 0$ for all $x \in R^n$, $|x| \ge r_1$.

Consider the mapping g which is defined for each $x \in R^n$, $0 \le t \le 1$, by

$$g(x,t) = (1 - t)x + tT(x).$$

Clearly $g_2 : R^n \to B$ and g is continuous. Further g satisfies (β). By ($\varkappa$), $|g(x,t)|^2 \ge (1 - t)^2|x|^2 + t^2|T(x)|^2 \ge \frac{1}{2}[(1 - t)|x| + t|T(x)|]^2$. Hence g satisfies assumption (i) as well as (iii) of Theorem 1. By this theorem the result follows.

c. The statement of Corollary 2 in case that T is locally one-to-one follows directly from Theorem 2.

2. Functional Differential Equations With Deviating Argument

First we formulate the initial-value problem for these equations which includes the problem from [12],[4] and is related to one in [1], [3]. For details and proofs, see [9]. We shall employ the notations:

$(E,|.|)$ is a real Banach space.

The state space C is the Banach space of all continuous and bounded mappings $x : (-\infty,0] \to E$ with the sup-norm $\|.\|$.

$\psi : [0,\infty) \to (0,\infty)$ is a nondecreasing continuous function.

The deviation $\omega : [0,\infty) \to R$ is a continuous mapping such that $\omega(0) = 0$.

$f : [0,\infty) \times C \to E$ is a continuous mapping.

$a^+ = \max(a,0)$ for each $a \in R$, sgn $0 = 0$, sgn $a = 1$ for each $a > 0$.

Finally, if $x : (-\infty,\infty) \to E$ is a continuous mapping which is bounded in $(-\infty,0]$ and $u \in R$, then x_u is the function defined by

$$x_u(s) = x(u + s) \text{ for all } s,\ -\infty < s \le 0.$$

Clearly $x_u \in C$.

The initial-value problem in the case that $h \in C$ is uniformly continuous in $(-\infty,0]$

(1) $\quad x'(t) = f[t,x_{\omega(t)}]$

(2) $\quad x_0 = h$

means the problem to find a function x which is continuous in $(-\infty,\infty)$, $x(t) = h(t)$ for all $t \in (\infty,0]$, x is differentiable in $[0,\infty)$ and it satisfies (1) at each point from $[0,\infty)$. Since ω, f are continuous and h is uniformly continuous, the problem (1),(2) is equivalent to the problem: To find a continuous solution of the integral equation

$$x(t) = h(0) + \int_0^t f[s,x_{\omega(s)}]ds \quad (0 \le t < \infty) \tag{3}$$

which satisfies (2).

Consider the following assumptions:

(A1) The function $\int_0^t |f(s,0)|ds$ is ψ-bounded in $[0,\infty)$, i.e. $|\int_0^t |f(s,0)| ds| / \psi(t)$ $(0 \le t < \infty)$ is bounded.

(A2) There exists a nonnegative, locally integrable in $[0,\infty)$ real function n such that
$|f(t,z_1) - f(t,z_2)| \le n(t) \|z_1 - z_2\|$
for every $z_1, z_2 \in C$ and $t \in [0,\infty)$.

(A3) The function $\int_0^t n(s)ds$ is ψ-bounded in $[0,\infty)$.

(A4) There exists a q, $0 \le q < 1$, such that
$\int_0^t n(s) \operatorname{sgn} \omega^+(s)\psi[\omega^+(s)]ds \le q\psi(t) \quad (0 \le t < \infty)$.

(A5) There is a $K > 0$ such that $\int_0^t |f(s,0)|ds \le K$ for all t, $0 \le t < \infty$.

(A6) There is a q, $0 \le q < 1$, such that $\int_0^t n(s)ds \le q$, $0 \le t < \infty$.

The existence of a unique ψ-bounded solution to (1),(2) is guaranteed by

Lemma 1. If the assumptions (A1)-(A4) are satisfied, then there exists a unique ψ-bounded in $[0,\infty)$ solution x(t) of (1),(2), i.e. $|x(t)| / \psi(t)$ is bounded in $[0,\infty)$.

Proof. Let D be the vector space of all continuous mappings $x : (-\infty,\infty) \to E$ which are bounded in $(-\infty,0]$ and ψ-bounded in $[0,\infty)$, $D_h = \{x \in D : x(t) = h(t), -\infty < t \le 0\}$. Let F be the Banach space of all continuous and ψ-bounded mappings $x : [0,\infty) \to E$ with the norm $\|x\|_1 = \sup_{0 \le t < \infty} |x(t)|/\psi(t)$. Then in view of the assumptions of the lemma the mapping T defined by

$$T(x)(t) = h(t), \quad -\infty < t \le 0,$$
$$T(x)(t) = h(0) + \int_0^t f[s,x_{\omega(s)}]ds, \quad 0 \le t < \infty,$$

maps D_h into D_h or considering only the restriction of functions from D_h to $[0,\infty)$, $T : G \to G$ where $G = \{x \in F : x(0) = h(0)\}$ is a closed subset of F. By (A2) and (A4) $|T(x)(t) - T(y)(t)|/\psi(t) \le \le \int_0^t n(s)\|x_{\omega(s)} - y_{\omega(s)}\|ds/\psi(t) \le \|x - y\|_1 \int_0^t n(s)\ \mathrm{sgn}\ \omega^+(s).$ $\psi[\omega^+(s)]ds/\psi(t) \le q\|x - y\|_1$. The Banach fixed point theorem gives the result.

By considering the bounded solutions of the problem (1), (2) we can prove

Lemma 2. If the assumptions (A1)-(A4) are satisfied and ψ is bounded, then for the unique bounded solution x(t) of (1),(2) there exists $\lim_{t\to\infty} x(t) = c\ (\in E)$.

Proof. By (3) and (A2), for $0 \le t_1 < t_2 < \infty$ we have $|x(t_2) - x(t_1)| \le \int_{t_1}^{t_2} |f(s,0)|ds + \int_{t_1}^{t_2} n(s)\|x_{\omega(s)}\|ds$. In view of (A1), (A3) and the boundedness of ψ, by the Cauchy-Bolzano criterion the result follows.

Denote this unique bounded solution of (1),(2) as x(t,h). Then the continuity of the bounded solution of (1),(2) in h is proved in

Lemma 3. Suppose that (A2),(A5) and (A6) are satisfied. Then for any $h_1, h_2 \in C$, h_1, h_2 are uniformly continuous in $(-\infty,0]$ and $h_1(0) = h_2(0) = 0$

$$\|x_t(.,h_2) - x_t(.,h_1)\| \le \|h_2 - h_1\|v(t), \qquad 0 \le t < \infty,$$

where v(t) is the unique real bounded continuous solution of

$$(4) \qquad v(t) = 1 + \int_0^t n(s)\ v[\omega^+(s)]ds, \qquad 0 \le t < \infty.$$

Proof. Denote $u(t) = \|x_t(.,h_2) - x_t(.,h_1)\|$, $0 \le t < \infty$. By (3) and (A2) it follows that

$$|x(t,h_2) - x(t,h_1)| \le |h_2(0) - h_1(0)| + \int_0^t n(s)u[\omega^+(s)]ds, \qquad 0 \le t < \infty,$$

and hence $u(t) \le \|h_2 - h_1\| + \int_0^t n(s)u[\omega^+(s)]ds$, $0 \le t < \infty$. Since u is bounded and continuous, by the generalized Gronwall lemma the result follows.

Lemma 4. Assume that (A2),(A5) and (A6) are satisfied. Let $h \in C$, $h(0) = 0$ and let h be uniformly continuous in $(-\infty,0]$. Let $\{z_k\}$, $z_k \in E$, $k = 1,2,\dots$, be a sequence with $\lim_{k\to\infty} |z_k| = \infty$. Denote $m_k = \inf\{|x(t,h + z_k)| : 0 \le t < \infty\}$. Then

$\lim_{k\to\infty} m_k = \infty.$

Proof. It is similar to that of Lemma 3 in [8], p. 240.

By using Theorem 1 where $g(x_0,t) = x(t,h + x_0)$ the following *boundary value problem* for (1) can be solved. An arbitrary point $x_1 \in E$ and an initial function $h \in C$, h is uniformly continuous in $(-\infty,0]$, $h(0) = 0$, are given. To find a point $x_0 \in E$ such that

(5) $$\lim_{t\to\infty} x(t,h + x_0) = x_1 .$$

Theorem 3. Assume that (A2'),(A5),(A6) as well as the assumption:

(A7) There exists a q_1, $0 \le q_1 < 1$, such that for the bounded continuous solution $v(t)$ of the equation (4) the inequality $v(t) \le 1 + q_1$, $0 \le t < \infty$, is true,

are satisfied. Let $x_1 \in E$ and let $h \in C$, h be uniformly continuous in $(-\infty,0]$, $h(0) = 0$. Then there exists exactly one $x_0 \in E$ such that (5) is true.

Proof. Define a mapping $g : E \to B$ in this way. Given an $x_0 \in E$, let $g(x_0,t) = x(t,h + x_0)$ for $0 \le t < \infty$ and let $g(x_0,\infty) = \lim_{t\to\infty} x(t,h + x_0)$. Lemma 1 and 2 guarantee that g is well defined. By Lemma 3 g is continuous and Lemma 4 implies that the condition (i) in Theorem 1 is satisfied. Clearly (iii) in that theorem holds. Let $r > 0$, $t_1 < t$, $|x_2| \le r$. Then $|g(x_2,t) - g(x_2,t_1)| \le \int_{t_1}^{t} |f(s,0)|ds + \int_{t_1}^{t} n(s)\|x_{\omega^+(s)}(.,h + x_2)\|ds \le \int_{t_1}^{t} |f(s,0)|ds + \int_{t_1}^{t} n(s)\,[M_1 + (\|h\| + r)K_1]ds$, where $M_1 = \sup_{0\le s<\infty} \|x_{\omega^+(s)}(.,0)\|$, $K_1 = \sup_{0\le s<\infty} v(\omega^+(s))$. If $t < t_1$, we get a similar inequality. This implies that (iv) is satisfied.

Consider the mapping $U = I - g(.,t)$ for a fixed $t \in [0,\infty]$. Then by Lemma 3 and (A7) $|U(x_0) - U(y_0)| \le \int_0^t n(s)\|x_{\omega^+(s)} - y_{\omega^+(s)}\|ds \le |x_0 - y_0|(v(t) - 1) \le q_1|x_0 - y_0|$. Hence U is a strict contraction and thus a condensing mapping. By Theorem 1, $g(E,t) = E$ for each $t \in [0,\infty]$. Since U is a strict contraction, $|g(x_0,t) - g(y_0,t)| \ge (1 - q_1)|x_0 - y_0|$ which implies that $g(.,t)$ is a homeomorphic mapping of E onto itself.

Remarks. 1. In case $E = R^n$, Theorem 3 is valid without assuming (A7). Of course uniqueness of x_0 need not be true.

2. Theorem 3 extends the main result from [8], p. 239,

and in the case $\omega(t) = t$, $0 \le t < \infty$, is stronger than Theorem I in [11], p.3.

3. Generalized Boundary Value Problem for Differential Systems

The generalized boundary value problem for a differential system

(6) $\qquad x' = f(t,x), \quad t \in i, \; x \in R^n$,

and a given continuous mapping T (not necessarily linear) of the space $C(i,R^n)$ of all continuous n-dimensional vector functions defined in i into R^n can be defined as a problem of finding a solution x(t) of the system (6) on the interval i for which T(x) is a given vector r in R^n, i.e.

(7) $\qquad T(x) = r$.

The topology in $C(i,R^n)$ is given in two different cases. If $i = [a,b]$ is a compact interval, then we consider the topology of uniform convergence, while in case i is a noncompact interval, e.g. $i = (a,\infty)$, then we use the topology of locally uniform convergence.

Theorem 4. Let $f = f(t,x)$ be a continuous function on $i \times R^n$ and let the equation (6) have the following properties:

(a) There is a point $t_0 \in i$ such that for each vector $x_0 \in R^n$ there exists a unique solution x(t) on i to the initial-value problem (6),

(8) $\qquad x(t_0) = x_0$

and either:

(b) For each solution x of (6),(8) the following implication is true:

$$\text{If } T(x) = kx(t_0),\; x(t_0) \neq 0, \text{ then } k \ge 0 ,$$

or:

(c) The problem (6),(7) has at most one solution for each $r \in R^n$.

Then in the case (a),(b) a sufficient condition and in the case (a),(c) a necessary and sufficient condition that there exist at least one solution of the problem (6),(7) for each $r \in R^n$ is that the following *compactness condition* be satisfied:

(d) If $\{x_k\}$ is a sequence of solutions of (6) on the interval i such that $\{T(x_k)\}$ is bounded, then there is a subsequence $\{x_{k(l)}\}$ such that $\{x_{k(l)}\}$ is converging in $C(i,R^n)$.

The proof is based on Corollary 2 and the Kamke convergence lemma.

References

[1] ANGELOV,V.G., BAJNOV,D.D., *On the Existence and Uniqueness of a Bounded Solution to Functional Differential Equations of Neutral Type in a Banach Space* (In Russian). Arch. Math. 2, Scripta Fac. Sci. Nat. UJEP Brunensis XVII: 65-72 (1981).

[2] DEIMLING,K., *Nichtlineare Gleichungen und Abbildungsgrade*, Springer-Verlag, Berlin 1974.

[3] HALE,J.K., *Theory of Functional Differential Equations*, Appl. Math. Sci., Vol. 3, Springer-Verlag, New York 1977.

[4] HALE,J.K., *Retarded Equations With Infinite Delays*, Lecture Notes in Mathematics, Vol. 730, Springer-Verlag, Berlin, 157-193 (1979).

[5] HARTMAN,Ph., *Ordinary Differential Equations*, John Wiley, New York 1964.

[6] NUSSBAUM,R.D., *Degree Theory for Local Condensing Maps*, J. Math. Anal. Appl. 37, 741-766 (1972).

[7] OPIAL,Z., *Linear Problems for Systems of Nonlinear Differential Equations*, J. Differential Equations 3, 580-594 (1967).

[8] SMÍTALOVÁ,K., *On a Problem Concerning a Functional Differential Equation*, Math. Slovaca 30, 239-242 (1980).

[9] ŠEDA,V., *Functional Differential Equations With Deviating Argument* (Preprint).

[10] ŠEDA,V., *On Surjectivity of an Operator* (Preprint).

[11] ŠVEC,M., *Some Properties of Functional Differential Equations*, Bolletino U.M.I. (4) 11, Suppl. Fasc. 3, 467-477 (1975).

[12] WEBB,G.F., *Accretive Operators and Existence for Nonlinear Functional Differential Equations*, J. Differential Equations 14, 57-69 (1973).

SOME PROBLEMS CONCERNING THE EQUIVALENCES OF TWO SYSTEMS OF DIFFERENTIAL EQUATIONS

M. ŠVEC
Faculty of Mathematics and Physics, Comenius University
Mlynská dolina, 842 15 Bratislava, Czechoslovakia

Consider two systems

(1) $\dot{x}(t) = f(t,x_t)$

(2) $\dot{y}(t) \in f(t,y_t) + g(t,y_t)$

where $t \in J = \langle 0,\infty)$, $x : R \to R^n$, $y : R \to R^n$, $x_t = x(t + s)$, $y_t = y(t + s)$, $s \in (-\infty,0\rangle$. Denote by $C = C(-\infty,0;R^n)$ the space of all functions $\varphi : (-\infty,0\rangle \to R^n$ which are bounded and continuous with sup norm $\|.\|$. Then $f : JxC \to R^n$, $g : JxC \to$ {the set of all nonempty subsets of R^n}. Further properties of f and g will be given later. However, we will still assume that f and g are such that the existence of the solutions of (1) and (2) is guaranteed on J. $|.|$ is the vector norm in R^n. If $A \subset R^n$, then $|A| \doteq \sup\{|a| : a \in A\}$.

Our aim is to estabilish the conditions which give the possibility of pairing of the solutions x(t) of (1) and y(t) of (2) in such a way that we will be able to say something about the asymptotic behaviour of the difference $y(t) - x(t) = z(t)$. Assume that x(t) is given. Then, proceeding formally, substituting y(t) by z(t) + x(t) in (2), we get

(3) $\dot{z}(t) \in -f(t,x_t) + f(t,z_t+ x_t) + g(t,z_t+ x_t)$

We have to prove the existence of such solution z(t) to the functional differential inclusion (3) that $\lim z(t) = 0$ as $t \to \infty$ (the case of asymptotic equivalence) or that $z(t) \in L_p(J)$, $p \geq 1$ (the case of p-integral equivalence). There are many methods how to do it, e.g. use the viability theory, method of fixed point, method of Liapunov function.

First we will use the viability theory.

Theorem 1. a) Let be $f : JxC \to R^n$ continuous and let it satisfy the Lipschitz condition

(4) $|f(t,\varphi_1) - f(t,\varphi_2)| \leq L(t)\|\varphi_1- \varphi_2\|$, $L(t) \in L_1(J)$

for each (t,φ_1), $(t,\varphi_2) \in JxC$.

b) Let g be an upper semicontinuous map from JxC to the nonempty compact convex subsets of R^n and let

$|g(t,\varphi)| \leq G_0(t,\|\varphi\|)$ a.e. on J

where $G_0(t,u) : J\times J \to J$ is monotone nondecreasing in u for each fixed $t \in J$ and is integrable on J for each fixed $u \in J$.

c) Let $x : R \to R^n$ be a bounded solution of (1).

d) Let there exist a solution $u : J \to J$ to the differential equation

(5) $\dot{u}(t) = -L(t)u - G_0(t,u + \|x_t\|) \doteq -G(t,u), \ u(0) > 0$

e) Let be

$K(t) \doteq \{x \in R^n : |x| \le u(t)\}, \quad t \in J$

f) Let be

$\forall t \in J, \ \mathcal{K}(t) \doteq \{\varphi \in C : \varphi(0) \in K(t)\}$

g) Let be the image of the graph $(\mathcal{K})$ by the map

$F(t,\varphi) \doteq -f(t,x_t) + f(t,\varphi + x_t) + g(t,\varphi + x_t)$

relatively compact.

h) Let for

$\forall t \in J, \forall \varphi$ such that $\varphi(0) \in K(t), \forall x \in K(t)$

$F(t,\varphi) \cap DK(t,\varphi(0))(1) \neq \phi$

where $DK(t,\varphi(0))$ is the contingent derivative of K at $(t,\varphi(0))$. Then for each $\varphi \in \mathcal{K}(0)$ there exists a solution z(t) to the functional inclusion such that

(6) for almost all $t \in J$, $\dot{z}(t) \in F(t,z_t)$

$(z)_0 = \varphi$

which is viable in the sense that

(7) $\forall t \in J, \ z(t) \in K(t) \qquad (|z(t)| \le u(t))$

Remark 1. Evidently, if $\lim u(t) = 0$ as $t \to \infty$, then also $\lim z(t) = 0$ as $t \to \infty$ and if $u(t) \in L_p(J)$, $p \ge 1$, then also the restriction $z(t)|_J \in L_p(J)$ holds.

Remark 2. It follows from the properties of f and g that F is an upper semicontinuous map from $J\times C$ to the nonempty compact convex subsets of R^n and

(8) $|F(t,\varphi)| \le L(t)\|\varphi\| + G_0(t,\|\varphi\| + \|x_t\|) \doteq G(t,\|\varphi\|)$

Evidently, $G : J\times J \to J$ is nondecreasing in u for each fixed $t \in J$ and integrable in t for each fixed $u \in J$.

Remark 3. In our case the basic space is R^n. The set valued map K defined by e) is upper semicontinuous and therefore its graph is closed.

The proof of the Theorem 1. follows immediatly from the time dependent Viability Theorem. (See e.g. [1].)

Remark 4. The most important condition is the condition h) which is necessary in our case because R^n has a finite dimension. This follows from the viability theory. If $t > 0$ and $|\varphi(0)| < u(t)$ then $(t,\varphi(0)) \in \text{int}(\text{graph}(K))$ and therefore the contingent cone $T_{\text{graph}(K)}(t,\varphi(0)) = R^{n+1}$. Evidently, in this case the condition h) is satisfied.

As to what concerns the existence of the solution u(t) from the condition d) we have the following lemma.

Lemma 1. Let be satisfied a) from the Theorem 1. Let i) $G_0(t,c) \in L_1(J)$ for each $c \geq 0$; ii) $\liminf_{c\to\infty}(c^{-1}\,{}_0\!\int^{\infty} G_0(s,c)ds) = 0$. Then there exists a solution $u : J \to J$ of the equation (5) such that $\lim u(t) = 0$ as $t \to \infty$. If, moreover, iii) $tG_0(t,c) \in L_1(J)$ for each $c \geq 0$, then this solution $u(t) \in L_p(J)$, $p \geq 1$.

The proof of this Lemma 1 can be made via Schauder fixed point theorem.

Theorem 2. Let be satisfied a) and b) from the Theorem 1. c') Let $y : R \to R^n$ be a bounded solution to the functional differential inclusion (2).

d') Let $u : J \to J$ be a solution of the equation

(9) $\quad \dot{u}(t) = -L(t)u - G_0(t,\|y_t\|) \doteq -G_1(t,u),\ u(0) > 0$

e') Let be

$K_1 \doteq \{x \in R^n : |x| \leq u(t)\},\quad t \in J$

f') Let be

$\forall t \in J,\ \mathcal{K}_1 \doteq \{\varphi \in C : \varphi(0) \in K_1(t)\}$

g') Let be the image of the graph $(\mathcal{K}_1)$ by the map

$F_1(t,\varphi) \doteq f(t,\varphi + y_t) - f(t,y_t) - g(t,y_t)$

relatively compact.

h') Let for

$\forall t \in J, \forall\varphi$ such that $\varphi(0) \in K_1(t), \forall x \in K_1(t)$

$F_1(t,\varphi) \cap DK_1(t,\varphi(0))(1) \neq \phi$

where $DK_1(t,\varphi(0))$ is the contingent derivative of K_1 at $(t,\varphi(0))$. Then for each $\varphi \in \mathcal{K}_1(0)$ there exists a solution z(t) to the functional differential inclusion such that

(1o) $\quad$ for almost all $t \in J$, $\dot{z}(t) \in F_1(t,z_t)$

$(z)_0 = \varphi$

which is viable in the sense that

$t \in J,\ z(t) \in K_1(t) \qquad (|z(t)| \leq u(t))$

The similar remarks as Remark 1 - 4 hold also in this case. The proof of the Theorem 2 follows also immediatly from the time dependent Viability Theorem.

Lemma 2. Let be satisfied a) from the Theorem 1 and i) from Lemma 1. Then

$$u(t) = \exp(-\int_0^t L(s)ds) \int_t^{\infty} \exp(\int_0^s L(v)dv) G_0(s, \|y_s\|)ds$$

is a solution of (9) such that $\lim u(t) = 0$ as $t \to \infty$. If, moreover, iii) from Lemma 1 holds, then also $u(t) \in L_p(J)$, $p \geq 1$ holds true.

The proof can be made immediately.

From Theorem 1, Theorem 2, Lemma 1 and Lemma 2 we get

Theorem 3. Let be satisfied all conditions of Theorem 1 and 2. Then the conditions i) and ii) from Lemma 1 guarantee the asymptotic equivalence between the set of all bounded solutions of (1) and the set of all bounded solutions of (2). If, moreover, the condition iii) from Lemma 1 is satisfied, then there exists also the p- integral equivalence, $p \geq 1$, between the above mentioned sets of solutions.

Now, we will consider the same problem of asymptotic and integral equivalences for the systems (1) and (2) by use of fixed point method. Henceforth we will assume that the following assumptions are satisfied:

(F) $|f(t,\varphi_1) - f(f,\varphi_2)| \leq L(t)w(\|\varphi_1 - \varphi_2\|)$

where $L(t) \in L_1(J)$, $w : J \to J$ is a continuous function, $\int_0^{\infty} L(t)dt = S$, $\sup_{\langle 0,u\rangle} w(r) \leq S^{-1}\alpha u$, $\alpha < 1$;

(H_1) $g(t,\varphi)$ is nonempty compact convex subset of R^n for each $(t,\varphi) \in$ $\in JxC$;

(H_2) for every fixed $t \in J$ $g(t,\varphi)$ is upper semicontinuous in φ;

(H_3) for each measurable function $z : R \to R^n$ such that $z|_{(-\infty,0\rangle} \in C$ there exists a measurable selector $v : J \to R^n$ such that

$$v(t) \in g(t,z_t) \text{ a.e. on } J$$

We set $M(z(t)) \doteq$ {all measurable selectors belonging to $z(t)$}.

(H_4) there exists a function $G_0 : JxJ \to J$ such that α) $G_0(t,u)$ is monotone nondecreasing in u for each fixed $t \in J$ and $G_0(t,u) \in L_1(J)$ for any fixed $u \in J$; β) $|g(t,\varphi)| \leq G_0(t,\|\varphi\|)$ a.e. on J; γ) $\liminf_{u \to \infty} (u^{-1} \int_0^{\infty} G_0(t,u)dt) = 0$ uniformly for $t \in J$.

Lemma 3. Let $z : R \to R^n$ be a measurable and bounded function. Then for each $v(t) \in M(z(t))$ we have $v(t) \in L_1(J)$.

Proof. It follows from (H_4).

Lemma 4. Let be satisfied (F), (H_1) - (H_4). Let be $B = \{z: R \to R^n:$ continuous and bounded} and $B_u = \{z \in B : \|z\| \leq u\}$. Let $x : R \to R^n$ be a bounded solution of (1) and let be $\varphi \in C$ given. Then the operator T defined on B by the relation : for $z \in B$ it is

$$(Tz)(t) = \{-\int_t^\infty [f(s,z_s+x_s) - f(s,x_s)]ds - \int_t^\infty v(s)ds : v(t) \in \in M(z(t)+x(t))\} \text{ for } t \in J$$

$$(Tz)_0 = \{\varphi(t) - \varphi(0) - \int_0^\infty [f(s,z_s+x_s) - f(s,x_s)]ds - \int_0^\infty v(s)ds\}, \quad \text{for } t \le 0$$

maps $B \to 2^B$, is compact and upper semicontinuous in B and there exists such $u \in J$ that T maps B_u into $cf(B_u)$. ($cf(B_u)$ is the set of all closed and convex subsets of B_u.)

Proof. Let be $z(t) \in B$. Then $\|z\| = \beta < \infty$ and by (F) we have

$$\int_0^\infty |f(s,z_s+x_s) - f(s,x_s)|ds \le \int_0^\infty L(s)w(\|z_s\|)ds \le$$

$$\le \max_{0\le\tau\le\beta} w(\tau) \int_0^\infty L(s)ds < \infty .$$

By Lemma 3 $v(t) \in M(z(t)+x(t))$ is from $L_1(J)$. Thus the operator T is well defined. Evidently, for $z(t) \in B$ $(Tz)(t)$ is a subset of B.

Let be $\|x(t)\| = \rho$. Consider the set B_u. Let be $z(t) \in B_u$ and let be $\xi(t) \in (Tz)(t)$. Then there exists such $v(t) \in M(z(t)+x(t))$ that

$$\xi(t) = -\int_t^\infty [f(s,z_s+x_s) - f(s,x_s)]ds - \int_t^\infty v(s)ds, \quad t \in J$$

$$\xi(t) = \varphi(t) - \varphi(0) - \int_0^\infty [f(s,z_s+x_s) - f(s,x_s)]ds - \int_0^\infty v(s)ds, \quad t \in (-\infty,0)$$

and

$$(*) \qquad |\xi(t)| \le \max_{0\le r\le u} w(r) \int_0^\infty L(s)ds + \int_0^\infty G_0(s,\rho+u)ds = K < \infty, \quad t \in J$$

Thus the functions $\xi(t) \in (Tz)(t)$ are uniformly bounded by the constant K and because for each $z(t) \in B_u$ we get the same constant K, we may conclude that TB_u is the set of continuous and uniformly bounded functions.

Let be $0 \le t_1 < t_2$. Then for $\xi(t) \in (Tz)(t)$, $z(t) \in B_u$ we have

$$|\xi(t_2) - \xi(t_1)| \le \int_{t_1}^{t_2} |f(s,z_s+x_s) - f(s,x_s)|ds + \int_{t_1}^{t_2} |v(s)|ds \le$$

$$\leq \max_{0\leq r\leq u} w(r) \int_{t_1}^{t_2} L(s)ds + \int_{t_1}^{t_2} G_0(s,\rho + u)ds$$

From this we conclude that all functions from TB_u are equicontinuous on J. Moreover, to each $\varepsilon > 0$ there exists $t_0(\varepsilon) > 0$ such that for $t_0(\varepsilon) \leq t_1 < t_2$ we have

$$|\xi(t_2) - \xi(t_1)| \leq \max_{0\leq r\leq u} w(r) \int_{t_0}^{\infty} L(s)ds + \int_{t_0}^{\infty} G_0(s,\rho + u)ds < \varepsilon$$

Then from this, from the uniform boundedness and from the equicontinuity of all functions of TB_u it follows that TB_u is compact in the topology of uniform convergence.

Evidently, to each bounded set $A \subset B$ there exists such $u \in J$ that $A \subset B_u$ and $TA \subset TB_u$. From this it follows that T is compact in B.

Let be $z_n(t)$, $z(t) \in B$ and let $\{z_n(t)\}$ converge to $z(t)$ in B, i.e. uniformly on R. Therefore, the set $\{z_n(t), z(t), n = 1,2,..\}$ is bounded in B. Thus there exists $u \geq 0$ such that $z_n(t) \in B_u$, $z(t) \in B_u$ and TB_u is a compact set. Let $h_n(t) \in (Tz)(t)$, $n = 1,2,...$ Evidently $h_n(t) \in TB_u$, $n = 1,2...$ The set TB_u being compact there exists a subsequence $\{h_{n_i}(t)\}$ of $\{h_n(t)\}$, which converges uniformly to a function $h(t) \in TB_u$. Then to each $h_n(t)$ there exists $v_n(t) \in M(z_n(t) + x(t))$, $n = 1,2,...$ such that

$$h_n(t) = - \int_t^{\infty} [f(s,(z_n)_s + x_s) - f(s,x_s)]ds - \int_t^{\infty} v_n(s)ds,$$

$$t \in J, \quad n = 1,2,...$$

$$h_n(t) = \varphi(t) - \varphi(0) - \int_0^{\infty} [f(s,(z_n)_s + x_s) - f(s,x_s)]ds -$$

$$- \int_0^{\infty} v_n(s)ds, \quad t \in (-\infty,0)$$

By Lemma 3 we have

$$\|v_n(t)\|_1 \leq \int_0^{\infty} G_0(s,u + \rho)ds < \infty$$

It means that the sequence $\{v_n(t)\}$ is bounded in $L_1(J)$. Furthermore, if $\{E_k\}$, $E_k \subset J$, is a nonincreasing sequence of sets such that $\bigcap_{k=1}^{\infty} E_k = \phi$, then

$$\lim_{k\to\infty} |\int_{E_k} v_n(s)ds| \leq \lim_{k\to\infty} \int_{E_k} |v_n(s)|ds \leq \lim_{k\to\infty} \int_{E_k} G_0(s,u + \rho)ds = 0$$

Then (see [2], Th. IV. 8.9) it is possible to choose from $\{v_n(t)\}$ a subsequence $\{v_{n_k}(t)\}$ which weakly converges to some $v(t) \in L_1(J)$.

Now, because $\{z_{n_k}(t)\}$ converges to $z(t)$ in B and $v_{n_k}(t) \in g(t,z_{n_k}(t))$, $k = 1,2,\ldots$, using (H_2), to given $\varepsilon > 0$ and $t \in J$ there exists $N = N(t,\varepsilon)$ such that for any $n_k \geq N$ we have

$$g(t,z_{n_k}(t)) \subset 0_\varepsilon(g(t,z(t)))$$

where $0_\varepsilon(g(t,z(t)))$ is ε-neighbourhood of the set $g(t,z(t))$. It means that for all $n_k \geq N$ $v_{n_k}(t) \in 0_\varepsilon(g(t,z(t)))$.

Consider the sequence $\{v_{n_k}(t)\}$, $n_k \geq N$. Then (see [2], Corollary V.3.14) it is possible to construct such convex combinations from v_{n_k}, $n_k \geq N$, denote them $g_m(t)$, $m = 1,2,\ldots$ that the sequence $\{g_m(t)\}$ converges to $v(t)$ in $L_1(J)$. Then by Riesz theorem there exists a subsequence $\{g_{m_i}(t)\}$ of $\{g_m(t)\}$ which converges to $v(t)$ a.e. on J. From the convexity of $0_\varepsilon(g(t,z(t)))$ and from the fact that $v_{n_k}(t) \in$ $\in 0_\varepsilon(g(t,z(t)))$ it follows that $g_{m_i}(t) \in 0_\varepsilon(g(t,z(t)))$, $i = 1,2,\ldots$ and, therefore, $v(t) \in \overline{0}_\varepsilon(g(t,z(t)))$. For $\varepsilon \to 0$ we get that $v(t) \in g(t,z(t))$.

Recall that t was a fixed point and that $g(t,z(t))$ was a compact convex subset of R^n.

Thus

$$h(t) = - \int_t^\infty [f(s,z_s + x_s) - f(s,x_s)]ds - \int_t^\infty v(s)ds$$

is well defined and $h(t) \in (Tz)(t)$ for $t \in J$. It follows from the weak convergence of $\{v_{n_k}(t)\}$ to $v(t)$ in $L_1(J)$ that the subsequence $\{h_{n_k}(t)\}$ of the sequence $\{h_n(t)\}$, i.e. for $t \in J$

$$h_{n_k}(t) = - \int_t^\infty [f(s,(z_{n_k})_s + x_s) - f(s,x_s)]ds - \int_t^\infty v_{n_k}(s)ds$$

converges to $h(t)$ a.e. on J. However, the functions $h_{n_k}(t)$ belong to the compact set TB_u. Therefore, there exists a subsequence of the sequence $\{h_{n_k}(t)\}$ which converges to a function $\overline{h}(t)$ uniformly on J. It means that $\overline{h}(t) = h(t) \in (Tz)(t)$ a.e. on J. With this we end the proof of the upper semicontinuity of the operator T.

Consider now B_u. Let be $z(t) \in B_u$, $\xi(t) \in (Tz)(t)$. Then from (*), (F) and γ) from (H_4) we get for $0 < c < \frac{1-\alpha}{2}$ the existence of such $u > 0$ that

$$|\xi(t)| \leq \alpha u + (\rho + u)c < (\alpha + 2c)u < u$$

Thus $\xi(t) \in B_u$ and $TB_u \subset B_u$. We have already proved that TB_u and also $(Tz)(t)$, $z(t) \in B_u$, are compact and, therefore, also closed. From the hypotheses (H_1), (H_3) it follows that $M(z(t))$ is nonempty and convex, therefore, $(Tz)(t)$ is also nonempty and convex. Thus T maps B_u in cf (B_u).

Lemma 5. Let be satisfied (F), $(H_1) - (H_4)$. Let B, B_u be as in Lemma 4. Let $y : R \to R^n$ be a bounded solution of (2) on J. Let $\varphi \in C$ be given. Then the operator T_1 defined on B by the relations: for $z(t) \in B$ it is

$$(T_1 z)(t) = \{- \int_t^\infty [f(s,z_s + y_s) - f(s,y_s)]ds + \int_t^\infty v(s)ds,\ v(t) \in M(y(t))\},\ t \in J$$

$$(T_1 z)_0 = \{\varphi(t) - \varphi(0) - \int_0^\infty [f(s,z_s + y_s) - f(s,y_s)]ds + \int_0^\infty v(s)ds\},\ t \le 0$$

maps $B \to 2^B$, is compact and upper semicontinuous in B and there exists such $u \in J$ that T maps B_u in $cf(B_u)$.

The proof of this Lemma can be made in the same way as the proof of Lemma 4.

From Lemma 4 and Lemma 5 follows

Theorem 4. Let be satisfied (F),$(H_1)-(H_4)$. Then between the set of all bounded solutions of (1) and the set of all bounded solutions of (2) there is the asymptotic equivalence. Moreover, if

(11) $\quad tL(t) \in L_1(J)$, $tG_0(t,c) \in L_1(J)$ for each $c \ge 0$

then there is p-integral equivalence, $p \ge 1$, between the above mentioned sets of bounded solutions of (1) and of (2).

Proof. Let be $x(t)$ a bounded solution of (1) on J and let $\varphi \in C$ be given.Then by Lemma 4 there exists a ball $B_u \subset B$ such that T maps B_u into $cf(B_u)$,T is upper semicontinuous and TB_u compact. Thus by Fan fixed point theorem T has a fixed point $z(t) \in B_u$, i.e. there exists $v(t) \in M(z(t) + x(t))$ such that

$$z(t) = - \int_t^\infty [f(s,z_s + x_s) - f(s,x_s)]ds - \int_t^\infty v(s)ds, \quad t \in J$$

$$(z)_0(t) = \varphi(t) - \varphi(0) - \int_0^\infty [f(s,z_s + x_s) - f(s,x_s)]ds - \int_0^\infty v(s)ds, t \le 0$$

Evidently, $\lim z(t) = \lim(y(t) - x(t)) = 0$ as $t \to \infty$ and $y(t) = x(t) + z(t)$ is a bounded solution of (2). Moreover, if (11) is satisfied, we get

$$|z(t)| \leq \sup_{\langle 0,\|z\|\rangle} w(r) \int_t^\infty L(s)ds + \int_t^\infty G_0(s,\|x\| + \|z\|)ds$$

Thus by Lemma 2 from [3] $z(t) \in L_p(J)$, $p \geq 1$.

Let now $y(t)$ be a bounded solution of (2) and let $\varphi \in C$ be given. Then by Lemma 5 there exists a ball $B_u \subset B$ such that the operator T_1 maps B_u into $cf(B_u)$, T_1 is upper semicontinuous and $T_1 B_u$ is compact. Thus Fan fixed point theorem gives the existence of a fixed point of T_1 in B_u, i.e. there exists $v(t) \in M(y(t))$ such that

$$z(t) = -\int_t^\infty [f(s,z_s + y_s) - f(s,y_s)]ds + \int_t^\infty v(s)ds, \quad t \in J$$

$$(z)_0(t) = \varphi(t) - \varphi(0) - \int_0^\infty [f(s,z_s + y_s) - f(s,y_s)]ds + \int_0^\infty v(s)ds, \quad t \leq 0$$

Evidently, $\lim z(t) = \lim(x(t) - y(t)) = 0$ as $t \to \infty$ and $x(t) = y(t) + z(t)$ is a bounded solution of (1). Moreover, if (11) is satisfied, then

$$|z(t)| \leq \sup_{\langle 0,\|z\|\rangle} w(r) \int_t^\infty L(s)ds + \int_t^\infty G_0(s,\|y\|)ds$$

which by Lemma 2 from [3] means that $z(t) \in L_p(J)$, $p \geq 1$.

References

[1] AUBIN,J.P., CELLINA,A., *Differential Inclusions*, Springer Verlag, 1984, A Series of Comprehensive Studies in Mathematics.

[2] DUNFORD,N., SCHWARTZ,J.T., *Linear Operators*, General Theory, 1958, Interscience Publishers, New York, London.

[3] HAŠČÁK,A., ŠVEC,M., *Integral equivalence of two systems of differential equations*, Czech. Math. J., 32 (107), 1982, 423-436.

LINEAR PERTURBATIONS OF GENERAL DISCONJUGATE EQUATIONS

W. F. TRENCH
Drexel University
Philadelphia, Pennsylvania, U.S.A.

Suppose that $p_1,\dots,p_{n-1}$, $q \in C[a,\infty]$, $p_i > 0$, and

(1) $\int^\infty p_i dt = \infty, \quad 1 \le i \le n-1,$

and define the quasi-derivatives

(2) $L_0 x = x; \quad L_r x = \frac{1}{p_r}(L_{r-1}x)', \; 1 \le r \le n$

(with $p_n = 1$). We will give conditions which imply that the equation

(3) $L_n u + q(t)u = 0$

has solutions which behave as $t \to \infty$ like solutions of the equation $L_n x = 0$.

Let $I_0 = 1$ and

$$I_j(t,s;q_j,\dots,q_i) = \int_s^t q_j(w) I_{j-1}(w,s;q_{j-1},\dots,q_i)dw, \; j \ge 1.$$

Then a principal system [2] for $L_n x = 0$ is given by

$$x_i(t) = I_{i-1}(t,a;p_1,\dots,p_{i-1}), \quad 1 \le i \le n;$$

in fact,

(4) $$L_r x_i(t) = \begin{cases} I_{i-r-1}(t,a;p_{r+1},\dots,p_{i-1}), & 0 \le r \le i-1, \\ 0, & i \le r \le n-1. \end{cases}$$

We also define

$$y_i(t) = I_{n-i}(t,a;p_{n-1},\dots,p_i), \; 1 \le i \le n,$$

and

(5) $$d_{ir}(t) = \begin{cases} L_r x_i(t), & 0 \le r \le i-1, \\ 1/I_{r-i+1}(t,a;p_r,\dots,p_i), & i \le r \le n. \end{cases}$$

We give sufficient conditions for (3) to have a solution u_i such that

(6) $L_r u_i = L_r x_i + o(d_{ir}) \quad (t \to \infty), \; 0 \le r \le n-1,$

for some given i in $\{1,\dots,n\}$. This formulation of the question is due to Fink and Kusano, and the best previous result on this question is the following special case of a theorem obtained by them in [1].

THEOREM 1. *If*

$$(7) \qquad \int^{\infty} x_i y_i |q| ds < \infty,$$

then (3) *has a solution* u_i *which satisfies* (6).

Our results require less stringent integrability conditions. We need the following lemma from [4].

LEMMA 1. *Suppose that* $Q \in C[t_0,\infty)$ *for some* $t_0 \geq a$, *that* $\int^{\infty} y_i Q dt$ *converges (perhaps conditionally), and that*

$$\sup_{\tau \geq t} \left| \int_{\tau}^{\infty} y_i Q ds \right| \leq \psi(t), \quad t \geq t_0,$$

where ψ *is nonincreasing and continuous on* $[t_0,\infty)$. *Define*

$$K(t;Q) = \int_t^{\infty} I_{n-i}(t,s;p_i,\ldots,p_{n-1}) Q(s) ds,$$

and, for $t \geq t_0$, *let*

$$J(t;Q) = K(t;Q) \quad \text{if} \quad i = 1;$$

or

$$J(t;Q) = \int_{t_0}^{t} p_1(s) K(s;Q) ds = I_1(t,t_0;p_1 K(\ ;Q)) ds$$

if $i = 2$; *or*

$$J(t;Q) = I_{i-1}(t,t_0;p_1,\ldots,p_{i-1} K(\ ;Q))$$

if $3 \leq i \leq n$.

Then

$$(8) \qquad L_n J(t;Q) = -Q(t), \; t \geq t_0,$$

and

$$|L_r J(\ ;Q)| \leq \begin{cases} \Psi(t_0) d_{ir}(t), & 0 \leq r \leq i-2, \\ & \qquad t \geq t_0; \\ 2\Psi(t) d_{ir}(t), & i-1 \leq r \leq n-1, \end{cases}$$

moreover, if $\lim_{t\to\infty} \Psi(t) = 0$, *then also*

$$L_r(J(t;Q)) = o(d_{ir}(t)), \quad 0 \leq r \leq i-2.$$

The following assumption applies throughout.

ASSUMPTION A. Let $\int^{\infty} y_1 x_1 q ds$ converge (perhaps conditionally), and suppose that

$$(9) \qquad E(t) = \int_t^{\infty} y_1 x_1 q ds = O(\varphi(t))$$

with φ nonincreasing on $[a,\infty)$, and

(10) $$\lim_{t\to\infty} \varphi(t) = 0 .$$

If $t_0 > a$, let $B(t_0)$ be the set of functions h such that $L_0h,\dots,L_{n-1}h \in C[t_0,\infty)$ and

$$L_rh = \begin{cases} O(d_{ir}), & 0 \le r \le i-2, \\ & \qquad t \ge t_0 , \\ O(\varphi d_{ir}), & i-1 \le r \le n-1 , \end{cases}$$

with norm $\| \ \|$ defined by

(11) $$\|h\| = \sup_{t\ge t_0} \max \left\{ \frac{|L_rh(t)|}{\varphi(t_0)d_{ir}(t)} \ (0\le r\le i-2), \ \frac{|L_rh(t)|}{2\varphi(t)d_{ir}(t)} \ (i-1\le r\le n-1) \right\}$$

Then Lemma 1 with $Q = qv$ and $\Psi = K\varphi$ implies the following lemma.

LEMMA 2. *If* $v \in C[t_0,\infty)$ *and*

$$\left| \int_t^\infty y_i qv ds \right| \le K\varphi(t), \quad t \ge t_0 ,$$

then

$$J(\ ;qv) \in B(t_0)$$

and

$$\|J(\ ;qv)\| \le K .$$

Now define the transformation T by

(12) $$(Th)(t) = J(t;qx_i) + J(t;qh) .$$

Lemma 2 and Assumption A imply that $J(\ ;qx_i) \in B(t_0)$ for all $t_0 > a$. We need only impose further conditions which will imply that $\int^\infty y_i qhds$ converges (perhaps conditionally) if $h \in B(t_0)$, and that

$$\left| \int_t^\infty y_i qhds \right| \le \|h\|\sigma(t;t_0)\varphi(t), \quad t \ge t_0 ,$$

where σ does not depend on h, and

(13) $$\sup_{t\ge t_0} \sigma(t;t_0) = \theta < 1$$

if t_0 is sufficiently large. Lemma 2 will then imply that T is a contraction mapping of $B(t_0)$ into itself, and therefore that there is an h_i in $B(t_0)$ such that $Th_i = h_i$. It will then follow from (8) and (12) that $u_i = x_i + h_i$ is a solution of (3). Moreover, Lemma 3 with $Q = qu_i$ will imply that

(14) $$L_ru_i - L_rx_i = \begin{cases} o(d_{ir}), & 0 \le r \le i-2 \\ O(\varphi d_{ir}), & i-1 \le r \le n-1 . \end{cases}$$

The next lemma can be obtained from (9) and integration by parts.

See [3] for the proof of the special case where $p_1 = \dots = p_n = 1$.

LEMMA 3. *Let*

(15) $$H_0 = y_i q;\ H_j(t) = \int_t^\infty p_{j-1}H_{j-1}ds,\quad 1 \le j \le i \quad (p_0 = 1).$$

Then (9) implies that

(16) $$H_j = O(\varphi/L_{j-1}x_i),\quad 1 \le j \le i,$$

and that the integrals

(17) $$\int^\infty p_j(L_jx_i)H_jds,\quad 0 \le j \le i-1,$$

all converge. Moreover, if the convergence is absolute for some $j = k$ with $0 \le k \le i-2$, then it is absolute for $k \le j \le i-1$.

THEOREM 2. *If*

(18) $$\overline{\lim_{t\to\infty}}\,(\varphi(t))^{-1}\int_t^\infty p_{i-1}|H_{i-1}|\varphi ds = A < \frac{1}{2},$$

then (3) *has a solution* u_i *which satisfies* (14).

Proof. Integration by parts yields

(19) $$\int_t^T y_iqhds = -\sum_{j=1}^{i-1} H_j(L_{j-1}h)\Big|_t^T + \int_t^T p_{i-1}H_{i-1}(L_{i-1}h)ds$$

if $h \in B(t_0)$ and $2 \le i \le n$; if $i = 1$, then the sum on the right is vacuous and (19) is trivial. (Recall (2) and (15).) Now (5),(9), (11),(18), and Lemma 3 imply that we can let $T \to \infty$ in (19) and infer (13) with

(20) $$\sigma(t;t_0) = \varphi(t_0)(\varphi(t))^{-1}\sum_{j=1}^{i-1}|H_j(t)|L_{j-1}x_i(t) + $$
$$+\ 2(\varphi(t))^{-1}\int_t^\infty p_{i-1}|H_{i-1}|\varphi ds\ .$$

From (16), the sum on the right side of (20) is bounded on $[a,\infty)$; hence, (10) and (18) imply (13) for t_0 sufficiently large. This completes the proof.

With $i = 1$, (18) reduces to

$$\overline{\lim_{t\to\infty}}\,(\varphi(t))^{-1}\int_t^\infty y_1|q|\varphi ds < \frac{1}{2},$$

which is weaker than (7), since $x_1 = 1$. The next two corollaries show that (18) is also weaker than (7) if $2 \le i \le n$.

COROLLARY 1. *If* $2 \le i \le n$ *and*

(21) $$\int^\infty p_k(L_kx_i)(L_{k-1}x_i)^{-1}\varphi dt < \infty$$

for some k *in* $\{1,\dots,i-1\}$, *then* (3) *has a solution* u_i *which satisfies* (14).

Proof. From (16),

$$p_k(L_k x_i)|H_k| \le M p_k(L_k x_i)(L_{k-1}x_i)^{-1}\varphi \tag{22}$$

for some constant M, so (21) implies that (17) with $j = k$ converges absolutely. From the closing sentence of Lemma 3, this means that

$$\int^{\infty} p_{i-1}|H_{i-1}|ds < \infty ,$$

which obviously implies (18) with $A = 0$.

COROLLARY 2. *If* $2 \le i \le n$ *and*

$$\int_t^{\infty} p_{i-1}(s)\left(\int_a^s p_{i-1}(w)dw\right)^{-1}\varphi^2(s)ds = o(\varphi(t)), \tag{23}$$

then (3) *has a solution* u_i *which satisfies* (13).

Proof. From (22) with $k = i - 1$ and (4), (23) implies (18) with $A = 0$.

THEOREM 3. *If* $1 \le i \le n - 1$ *and*

$$\overline{\lim_{t\to\infty}}\, (\varphi(t))^{-1}\int_t^{\infty}\varphi(s)p_i(s)\left(\int_a^s p_i(w)dw\right)^{-1}|H_i(s)|ds = B < \frac{1}{2}, \tag{24}$$

then (3) *has a solution which satisfies* (14).

Proof. Lemma 3 and our present assumption enable us to continue the integration by parts in (19) by one more step, to obtain

$$\int_t^{\infty} y_i qh ds = \sum_{j=1}^{i} H_j(t)L_{j-1}h(t) + \int_t^{\infty} p_i H_i(L_i h)ds.$$

Because of (5) (with $r = i$) and (11), this yields

$$\sigma(t;t_0) = \varphi(t_0)(\varphi(t))^{-1}\sum_{j=1}^{i-1}|H_j(t)|L_{j-1}x_i(t) + 2H_i(t) + $$
$$+ 2(\varphi(t))^{-1}\int_t^{\infty}\varphi(s)p_i(s)\left(\int_a^s p_i(w)dw\right)^{-1}|H_i(s)|ds.$$

Now (10) and (16) imply (20) for t_0 sufficiently large. This completes the proof.

COROLLARY 3. *If* $1 \le i \le n - 1$ *and*

$$\int_t^{\infty} p_i(s)\left(\int_a^{\infty} p_i(w)dw\right)^{-1}\varphi^2(s)ds = o(\varphi(t)), \tag{25}$$

then (3) *has a solution* u_i *which satisfies* (14).

Proof. From (16) with $j = i$, it follows that (25) implies (24) with $B = 0$.

References.

[1] A.M.Fink and T.Kusano, *Nonoscillation theorems for a class of perturbed disconjugate differential equations*, Japan J. Math. 9 (1983), 277 - 291.

[2] W.F.Trench, *Canonical forms and principal systems for general disconjugate equations*, Trans. Amer. Math. Soc. 189 (1974), 319 - 327.

[3] W.F.Trench, *Evetual disconjugacy of a linear differential equation*, Proc. Amer. Math. Soc. 89 (1983), 461 - 466.

[4] W.F.Trench, *Asymptotic theory of perturbed general disconjugate equations II*, Hiroshima Math. J. 14 (1984), 169 - 187.

[5] W.F.Trench, *Eventual disconjugacy of a linear differential equation*, Proc. Amer. Math. Soc. 89 (1983), 461 - 466.

ON OPTIMAL CONTROL OF SYSTEMS WITH INTERFACE SIDE CONDITIONS

M. TVRDÝ
Mathematical Institute, Czechoslovak Academy of Sciences
115 67 Prague 1, Czechoslovakia

Let $0 < \tau < 1$. Denote by D_n the space of functions $x : [0,1] \to R_n$ which are absolutely continuous on $[0,\tau]$ and on $(\tau,1]$ and such that their derivatives $\dot{x}$ are square integrable on $[0,1]$ ($\dot{x} \in L^2_n$). We want to establish necessary conditions for a local extremum of the functional of the type

$$F : (x,u) \in D_n \times L^2_m \to g_0\big(x(0)\big) + g_\tau\big(x(\tau+)\big) + g_1\big(x(1)\big) + \int_0^1 h\big(s,x(s),u(s)\big)\, ds \in R \tag{0.1}$$

subject to the constraints

$$x(t) - A(t)x(t) - B(t)u(t) = 0 \quad \text{a.e. on} \quad [0,1] \tag{0.2}$$

and

$$Mx(0) + Nx(\tau+) + \int_0^1 K(s)\,\dot{x}(s)\, ds = 0 \ . \tag{0.3}$$

1. Preliminaries

Throughout the paper the elements in R_n are considered to be column n-vectors. Given a $c \in R_n$, c^* denotes its transposition. Given a Banach space X , $||\cdot||_X$ and X^* denote the norm on X and the dual of X , respectively. For any $x \in X$ and $\phi \in X^*$, the value of the functional ϕ on x is denoted by $\langle x,\phi\rangle_X$. If Y is also a Banach space, then $L(X,Y)$ denotes the space of linear continuous mappings of X into Y . For $A \in L(X,Y)$, $N(A)$, $R(A)$ and A^* denote its null space, range and adjoint, respectively.

Furthermore, L^2_n denotes the space of functions $x : [0,1] \to R$ square integrable on $[0,1]$, equipped with its usual norm denoted by $||\cdot||_L$. The norm on D_n is defined by $x \in D_n \to ||x||_D = |x(0)| + |x(\tau+\ )| + ||\dot{x}||_L$. Obviously D_n is isometrically isomorphic with

$L_n^2 \times R_{2n}$. Its dual will be identified with $L_n^2 \times R_{2n}$, while

$$< x,\phi >_D = a^*x(0) + b^*x(\tau+) + < \dot{x},w >_L =$$

$$= a^*x(0) + b^*x(\tau+) + \int_0^1 w^*(s)\,\dot{x}(s)\,ds$$

for any $x \in D_n$ and $\phi = (w,a,b) \in L_n^2 \times R_n \times R_n$.

We shall keep the following assumptions.

ASSUMPTIONS. $A(t)$, $B(t)$ and $K(t)$ are square integrable on $[0,1]$ matrix valued functions of the types $n \times n$, $n \times m$ and $k \times n$, respectively, M and N are $k \times n$-matrices. The functions $g_0(x)$, $g_\tau(x)$, $g_1(x)$ and $h(t,x,u)$ are continuous and continuously differentiable with respect to x and u .

2. Lagrange Multiplier Theorem

Let us define

$$A : x \in D_n \to \begin{bmatrix} \dot{x}(t) - A(t).x(t) \\ Mx(0) + Nx(\tau+) + \int_0^1 K(s)\,\dot{x}(s)\,ds \end{bmatrix},$$

$$B : u \in L_m^2 \to \begin{bmatrix} B(t)u(t) \\ 0 \end{bmatrix}$$

and

$$T : (x,u) \in D_n \times L_m^2 \to Ax - Bu .$$

Then $A \in L(D_n, L_n^2 \times R_k)$, $B \in L(L_m^2, L_n^2 \times R_k)$ and $T \in L(D_n \times L_m^2, L_n^2 \times R_k)$ and the constraints (0.2), (0.3) may be replaced by the operator equation for $(x,u) \in D_n \times L_m^2$

$$T(x,u) = 0 . \tag{2.1}$$

The operator A is related to interface boundary value problems. It is known (cf. [1]) that under our assumptions A is normally solvable, i.e. $(f,r) \in L_n^2 \times R_k$ belongs to its range iff $< y,f >_L + \gamma r = 0$ for all $(y,\gamma) \in N(A^*)$ ($N(A^*) \subset L_n^2 \times R_k$). It was also shown in [1] that $N(A^*)$ consists of all $(y,\gamma) \in L_n^2 \times R_k$ for which there exists a $z \in D_n$ such that $z^*(t) = y^*(t) + \gamma^* K(t)$ a.e. on $[0,1]$ and

$$- \dot{z}^*(t) - z^*(t)A(t) + \gamma^*K(t)A(t) = 0 \quad \text{a.e. on} \quad [0,1] , \tag{2.2}$$

$$- z^*(0) + \gamma^*M = 0 , \quad z^*(\tau-) = 0 , \tag{2.3}$$

$$- z^*(\tau+) + \gamma^*N = 0 , \quad z^*(1) = 0 . \tag{2.4}$$

It is easy to see that $0 \leq \dim N(A) + \dim N(A^*) < \infty$. Hence we may apply Proposition 1.2 of [6] to obtain necessary and sufficient conditions for the complete controllability of the system (0.2), (0.3).

PROPOSITION. $R(T) = L^2_n \times R_k$ *iff the only couple* $(z,\gamma) \in D_n \times R_k$ *fulfilling* (2.2) - (2.4) *together with*

$$- z^*(t)B(t) + \gamma^*K(t)B(t) = 0 \quad \text{a.e. on} \quad [0,1] \tag{2.5}$$

is the trivial one: $z(t) = 0$ *on* $[0,1]$ *and* $\gamma = 0$.

Let us suppose that $R(T) = L^2_n \times R_k$ and let $(x_0,u_0) \in D_n \times L^2_m$ be such that $T(x_0,u_0) = 0$. From the abstract Lagrange Multiplier Theorem (cf. [4] 9.3, Theorem 1) we obtain that if (x_0,u_0) is a local extremum on $N(T)$ of the functional F defined by (0.1) then there exists a couple $(y,\gamma) \in L^2_n \times R_k$ such that each $(x,u) \in D_n \times L^2_m$ satisfies

$$[F'(x_0,u_0)](x,u) = \langle T(x,u),(y,\gamma) \rangle_{L^2_n \times R_k} , \tag{2.6}$$

where $F'(x_0,u_0)$ stands for the Frechet derivative of F at the point (x_0,u_0) with respect to (x,u) ($F'(x_0,u_0) \in L(D_n \times L^2_m, R)$). Inserting the explicit form (0.1) of F into (2.6), applying the integration by parts formula and taking into account that

$$(x,u) \in X \to a^*x(0) + b^*x(\tau+) + \int_0^1 w^*(s)\,\dot{x}(s)\,ds + \int_0^1 v^*(s)\,u(s)\,ds \in R$$

is the zero functional on $D_n \times L^2_m$ iff $a = b = 0$, $w(s) = 0$ and $v(s) = 0$ a.e. on $[0,1]$ we obtain the following result.

THEOREM (Lagrange Multipliers). *Let* $R(T) = L^2_n \times R_k$. *Then* $(x_0,u_0) \in D_n \times L^2_m$ *is a local extremum of* F *on* $N(T)$ *only if*

$$\dot{x}_0(t) - A(t)x_0(t) - B(t)u_0(t) = 0 \quad \text{a.e. on} \quad [0,1] , \tag{2.7}$$

$$Mx_0(0) + Nx_0(\tau+) + \int_0^1 K(s)\,\dot{x}_0(s)\,ds = 0 \tag{2.8}$$

and there exist $z \in D_n$ *and* $\gamma \in R_k$ *such that*

$$- \dot{z}^*(t) - z^*(t)A(t) + \gamma^* K(t)A(t) = \left(\frac{\partial h}{\partial x}(t,x_0(t),u_0(t))\right)^* \quad \text{a.e. on } [0,1] , \tag{2.9}$$

$$- z^*(0) + \gamma^* M = \left(\frac{\partial g_0}{\partial x}(x_0(0))\right)^* , \quad z^*(\tau-) = 0 , \tag{2.10}$$

$$- z^*(\tau+) + \gamma^* N = \frac{\partial g_\tau}{\partial x}(x_0(\tau+))\Big)^* , \quad z^*(1) = \left(\frac{\partial g_1}{\partial x}(x_0(1))\right)^* , \tag{2.11}$$

$$- z^*(t)B(t) + \gamma^* K(t)B(t) = \left(\frac{\partial h}{\partial u}(t,x_0(t),u_0(t))\right)^* , \quad \text{a.e. on } [0,1] . \tag{2.12}$$

REMARK. Related topics were treated e.g. in [2], [3], [5].

References

[1] BROWN, R. C., TVRDÝ, M. and VEJVODA, O.: *Duality theory for linear n-th order integro-differential operators with domain in* L^2_m *determined by interface side conditions.* Czech. Math. J. 32, (107) (1982), 183-196.

[2] HALANAY, A.: *Optimal control of periodic solutions.* Rev. Roum.Math. Pures et Appl., 19 (1974), 3-16.

[3] CHAN, W. L., S. K. NG : *Variational control problems for linear differential systems with Stieltjes boundary conditions.* J. Austral. Math. Soc. 20 (1978), 434-445.

[4] LUENBERGER, D. G.: *Optimization by vector space methods.* J. Wiley & Sons, New York-London-Sydney-Toronto, 1969.

[5] MARCHIÒ, C.: *(M,N,F)-controllabilità completa, Questioni di controllabilità.* Istituto U. Dini, Firenze, 1973/2, 14-26.

[6] TVRDÝ, M.: *On the controllability of linear Fredholm-Stieltjes integral operator, Functional-Differential Systems and Related Topics.* (Proc. Int. Conference, ed. M. Kisielewicz) (1983), 247-252.

Section B
PARTIAL DIFFERENTIAL EQUATIONS

A DESCRIPTION OF BLOW-UP FOR THE SOLID FUEL IGNITION MODEL

J. W. BEBERNES
Department of Mathematics, University Colorado
Boulder, CO 80309, U.S.A.

The nondimensional ignition model for a supercritical high activation energy thermal explosion of a solid fuel in a bounded container Ω can be described by

(1) $u_t - \Delta u = e^u$

(2) $u(x,0) = \phi(x) \geq 0,\ x \in \Omega,\ u(x,t) = 0,\ x \in \partial\Omega,\ t > 0$

where $\Omega = \{x \in \mathbb{R}^n: |x| \leq R\}$ and ϕ is radially decreasing, i.e., $\phi(x) \geq \phi(y) \geq 0$ whenever $|x| \leq |y| \leq R$ and $\Delta\phi + e^{\phi} \geq 0$ on Ω.

Assume $R > 0$ is such that the radially symmetric solution $u(x,t)$ blows up in finite time $T > 0$. Then by the maximum principle $u(\cdot,t)$ is radially decreasing for $t \in [0,T)$ and $u_t(x,t) \geq 0$ for all $(x,t) \in \Gamma = \Omega \times [0,T)$.

Friedman and McLeod [4] recently proved that blow-up occurs only at the origin $x = 0$ and in addition that $u(x,t)$ satisfies the following estimates: I) $u(x,t) \leq -\frac{2}{\alpha}\ln|x| + c$ for all $\alpha < 1$ and $(x,t) \in \Gamma$; II) there exists $\bar{t} < T$ such that $|\nabla u(x,t)| \leq 2e^{u(0,t)/2}$, $t \in [\bar{t},T)$, $|x| \leq R$; III) there exists $\delta > 0$ such that $u_t(x,t) \geq \delta e^{u(x,t)}$, $t \in [\frac{T}{2},T)$, $x \in [-\frac{R}{2},\frac{R}{2}]$; and iv) $-\ln(T-t) \leq u(0,t) \leq -\ln(T-t) - \ln\delta$, $t \in [\frac{T}{2},T)$, $\delta > 0$.

Since $u(x,t)$ is radially symmetric, the initial boundary value problem (1)-(2) can be reduced to a problem in one spatial dimension. Let $D = \{(r,t): 0 \leq t \leq T,\ 0 \leq r \leq R\}$. Then if $r = |x|$, $v(r,t) = u(x,t)$ satisfies:

(3) $v_t = v_{rr} + \frac{n-1}{r}v_r + e^v$

(4) $v(r,0) = \phi(r),\ v_r(0) = 0,\ v(R,t) = 0.$

To study the asymptotic behavior of v as $t \to T$, consider the following change of variables: $\tau = -\ln(T-t)$, $\eta = r(T-t)^{-1/2}$, $\theta = v + \ln(T-t) = v - \tau$ whose inverse is $t = T - e^{-\tau}$, $r = \eta e^{-\tau/2}$, $v = \theta - \ln(T-\tau)$. The domain D transforms to $D' = \{(\eta,\tau): 0 \leq \eta \leq Re^{\tau/2},\ \tau \geq -\ln T\}$ and $\theta(\eta,\tau) = v - \tau$ solves

(5) $\theta_t = \theta_{\eta\eta} + \left(\frac{n-1}{\eta} - \frac{\eta}{2}\right)\theta_\eta + e^\theta - 1$

(6) $\theta(\eta, -\ell nT) = \phi(\eta T^{1/2}) + \ell nT$

$\theta_\eta(0,\tau) = 0, \ \theta(Re^{\tau/2}, \tau) = -\tau$

The following theorem is similar to a result proven by Giga-Kohn [5].

Theorem 1. As $\tau \to +\infty$, the solution $\theta(\eta,\tau)$ tends uniformly to a function $y(\eta)$ on compact subsets of $\mathbb{R}^+$ where $y(\eta)$ is a solution of the problem:

(7) $y'' + \left(\frac{n-1}{\eta} - \frac{\eta}{2}\right)y' + e^y - 1 = 0$

(8) $y'(0) = 0, \ y(0) = \alpha \geq 0$

which is globally Lipschitz continuous and nonincreasing in η.

Thus, to describe how the blow-up occurs at (T,0) for (1)-(2), we need to analyze the solutions of the steady-state equation (7)-(8) which are globally Lipschitz and are nonincreasing on $[0,\infty)$.

Theorem 2. For $n = 1$ or 2, the only solution of (7)-(8) which is globally Lipschitz continuous and nonincreasing in η is $y(\eta) \equiv 0$.

Proof. For $n = 1$, this result was first proven by Bebernes-Troy [2]. The following proof is essentially due to D.Eberly. For $n > 2$, the proof fails. Let

$g(\eta) = \frac{\eta}{2} y'(\eta) + 1$

and $h(\eta) = y''(\eta) + \frac{n-1}{\eta} y'(\eta)$

where $y(\eta)$ is a solution of (7) - (8).

Then $g(\eta)$ satisfies

$$(9) \quad \begin{cases} g'' + (\frac{n-1}{\eta} - \frac{\eta}{2})g' + (e^y - 1)g = 0 \\ g(0) = 1, \quad g'(0) = 0 \end{cases}$$

and $h(\eta)$ satisfies

$$(10) \quad \begin{cases} h'' + (\frac{n-1}{\eta} - \frac{\eta}{2})h' + (e^y - 1)h \leq 0 \\ h(0) = 1 - e^\alpha, \ h'(0) = 0. \end{cases}$$

It is clear that $g(\eta) > 0$ on $I = [0,x_0)$ where $x_0 \in (0,\infty]$.

Set $W(\eta) = gh' - g'h$, then $W(\eta)$ satisfies

$$(11) \quad \begin{cases} W' + (\frac{n-1}{\eta} - \frac{\eta}{2})W = -e^y(y')^2 g(\eta) < 0 \\ W(0) = 0 \end{cases}$$

on I. This implies $W(\eta) \leq 0$ on I and hence $h(\eta)/g(\eta) \leq h(0)/g(0) = 1 - e^{\alpha}$ on I. Thus, we have

(12) $h(\eta) \leq (1 - e^{\alpha})g(\eta)$ on I.

We now must consider two cases. We assume now that $n = 1$ or 2.

a) If $x_0 < \infty$, then $g(x_0) = 0$ and $g' - \frac{\eta}{2}g = -\frac{\eta}{2}e^y < 0$ implies $g(\eta) < 0$ for all $\eta > \eta_0$. Thus $(\eta y')' = \eta g(\eta) - \eta e^y < 0$ and $y(\eta)$ is not globally Lipschitz on $[0,\infty)$.

b) If $x_0 = +\infty$ and $g(\eta) > \varepsilon > 0$ for all $\eta \geq 0$, then $(\eta y')' < 0$ by (12) and again $y''(\eta) < 0$. If lim inf $g(\eta) = 0$ as $\eta \to \infty$ with $g(\eta) > 0$, we observe that (11) can be solved for $h(\eta)$ to give

$$(13)\quad h(\eta) = (1 - e^{\alpha})g(\eta) - g(\eta)\int_0^{\eta} \frac{1}{g^2(s)} \frac{e^{s^2/4}}{s} \left(\int_0^s u e^{-u^2/4} e^y (y')^2 g(u)\,du \right) ds$$

By analyzing (13), we can show that $h(\eta) \to -\infty$ as $\eta \to +\infty$. Once again we have that $y''(\eta) < 0$ for η large and $y(\eta)$ cannot be globally Lipschitz on $[0,\infty)$. This completes the proof in dimensions 1 and 2.

As an immediate consequence of theorems 1 and 2, we have

<u>Theorem 3</u>. Let $n = 1$ or 2. As $t \to T^-$, $v(r,t) - \ln(T - t)^{-1} \to 0$ uniformly on $0 \leq r \leq c(T - t)^{1/2}$.

These results will appear in [3].

Several open questions remain. What can be said for $n \geq 3$? What happens outside the parabolic domain $r \leq c(T-t)^{1/2}$ as $t \to T^-$?

References

[1] J.Bebernes and D.Kassoy, *A mathematical analysis of blowup for thermal reactions - the spatially monhogeneous case*, SIAM J. Appl. Math. 40 (1981), 476-484.

[2] J.Bebernes and W.Troy, *Nonexistence for the Kassoy problem*, SIAM J. Math. Analysis, submitted.

[3] J.Bebernes, A.Bressan and D.Eberly, *A description of blow-up for the solid fuel ignition model*, submitted.

[4] A.Friedman and B.McLeod, *Blow-up of positive solutions of semilinear heat equations*, Indiana Univ. Math. J. 34 (1985), 425-447.

[5] Y.Giga and R.Kohn, *Asymptotically self-similar blow-up of semilinear heat equations*, Comm. Pure Appl. Math. 38 (1985), 297-320.

[6] D. Kassoy and J.Poland, *The thermal explosion confined by a constant temperature boundary:I. The induction period solution*, SIAM J. Appl. Math. 39 (1980), 412-430.

SPECTRAL ANALYSIS OF NON-SELF-ADJOINT ELLIPTIC OPERATORS

J. BRILLA
Institute of Applied Mathematics and Computing Technique, Comenius University
842 15 Bratislava, Czechoslovakia

1. Introduction

Many important problems of mathematical physics lead to analysis of the differential equation

$$\sum_{k=0}^{n} A_k \frac{\partial^k}{\partial t^k} u = f, \qquad \text{in } \Omega, \tag{1}$$

where A_k are symmetric positive definite elliptic operators of order 2m. When dealing with analysis of these equations we assume that Ω - the domain of definition is bounded and $\partial\Omega$ - the boundary is sufficiently smooth. We consider homogeneous boundary conditions and non-homogeneous initial conditions.

When applying Laplace transform we arrive at

$$A(p)\tilde{u} = \sum_{k=0}^{n} p^k A_k \tilde{u} = \tilde{f}^*, \tag{2}$$

where a tilde denotes the Laplace transform and $\tilde{f}^*$ includes initial conditions. The operator A(p) is a complex symmetric non-self-adjoint elliptic operator.

For analysis of equations (2) we have introduced [1 - 2] spaces of analytic functions valued in Sobolev spaces, which are isomorphic to weighted anisotropic Sobolev spaces convenient for analysis of equations (1).

Now we shall deal with spectral analysis of complex symmetric operators and show that it is possible to obtain similar results on existence of eigenvalues and completeness of sets of eigenvectors as in the case of symmetric compact operators.

2. Spectral analysis

Operators A(p) are complex symmetric operators. Thus it holds $A^*(p) = A(\bar{p})$ and

$$(Ax,\bar{x}) = (x,\bar{A}\bar{x}) \ . \tag{3}$$

When $A_k A_l \neq A_l A_k$ i.e. when operatore A_k are noncommutative

$$AA^* \neq A^*A \qquad (4)$$

and A(p) is a nonnormal operator. Thus for their analysis it is not possible to apply the spectral theory of symmetric compact operators. However it is possible to generalize some of its results.

I.C. Gokhberg and M.G. Krein [3] delt with the spectral analysis of (1) from the point of view of a nonlinear eigenvalue problem

$$\sum_{k=0}^{n} \lambda^k A_k e = 0 \ . \qquad (5)$$

When applying this approach we cannot use valuable results of the linear spectral theory.

Therefore for the problem under consideration we define a linear eigenvalue problem considering the equation

$$A(p)e(p) = \sum_{k=0}^{n} p^k A_k e(p) = \lambda(p)e(p) \ , \qquad (6)$$

where $\lambda(p)$ for which the solutions of (6) exist are eigenvalues and the corresponding solutions $e(p)$ are eigenvectors of (6). Both eigenvalues and eigenvectors are in general functions of the parameter p. Eigenvalues in the sense of (5) are values of p for which

$$\lambda(p) = 0 \qquad (7)$$

and the corresponding values of $e(p)$ are eigenvectors of (5).

For nonnegative real values of p A(p) is a symmetric positive definite elliptic operator. Thus it has discrete spectrum and a complete pairwise orthogonal set of eigenvectors. Then there exists a neighbourhood Ω_{p_1} of the positive real semiaxis p_1, where A(p) has the compact inverse $B(p) = A^{-1}(p)$ and $B_1(p) = \text{Re}\ B(p)$ and $B_2(p) = \text{Im}\ B(p)$ are positive symmetric compact operators.

The we can prove:

Theorem 1. The operator $B(p) = A^{-1}(p)$ has at least one nonzero eigenvalue and its eigenvalues and eigenvectors are solutions of the variational problem

$$\min \max \left[\, |(B\, e,\bar{e})| - |\mu|\,|(e,\bar{e})| \, \right], \quad \mu = 1/\lambda \ . \qquad (8)$$

Proof: As $B_1(p)$ and $B_2(p)$ are positive operators the trace of $B(p)$ is not equal to zero. Therefore $B(p)$ is not a quasi-nilpotent operator and has at least one nonzero eigenvalue. Further the Gateaux derivative of (8) yields the condition

$$\frac{1}{|(B\,e,\bar{e})|}[(B\,e,\bar{h})(\bar{B}\,\bar{e},e) + (B\,e,\bar{e})(\bar{B}\,\bar{e},h)] - \tag{9}$$

$$- |\mu|\frac{1}{|(e,\bar{e})|}[(e,\bar{h})(\bar{e},e) + (e,\bar{e})(\bar{e},h)] = 0$$

What is fulfilled by

$$B\,e = \mu e\,. \tag{10}$$

Analysis of the second Gateaux derivative shows that (10) is a saddle point of (8).

Theorem 2. Eigenvectors of a complex symmetric operator $B(p)$ and eigenvectors of its adjoint $B^*(p) = B(\bar{p})$ form biorthogonal systems which can be biorthonormalized.

Proof: For $\mu_k \neq \mu_l$ it holds $\mu_k(e_k,\bar{e}_l) = (Ae_k,\bar{e}_l)$ and $\mu_l(e_k,\bar{e}_l) = \mu_l(e_l,\bar{e}_k) = (Ae_l,\bar{e}_k) = (Ae_k,\bar{e}_l)$. Then

$$(\mu_k - \mu_l)(e_k,\bar{e}_l) = 0\,. \tag{11}$$

Hence for $\mu_k \neq \mu_l$ $(e_k,\bar{e}_l) = 0$ and eigenvalues $e_k,\bar{e}_l$ form biorthogonal systems.

Points p, where it holds $(e(p), \overline{e(p)}) = 0$, will be called exceptional points of the operator $B(p)$. We can prove:

Theorem 3. Symmetric complex compact operators $B(p) = A^{-1}(p)$ are semisimple with exception of exceptional points.

Proof: We shall make the proof for an eigenvelue of the multiplicity two. In this case the Jordan canonical form will be

$$\begin{aligned} Be_1 &= \mu e_1 + e_2\,, \\ Be_2 &= \mu e_2\,. \end{aligned} \tag{12}$$

After biorthogonalization $x_2 = e_2$, $x_1 = k_1 e_1 + k_2 e_2$ we arrive at

$$Bx_1 = \mu_1 x_1 + a x_2 \ ,$$
$$Bx_2 = \mu x_2 \ . \tag{1 3}$$

Multiplying the first equation (1 3) by x_2 and the second one by x_1 we arrive at

$$(Ax_1, \bar{x}_2) = a\,(x_2, \bar{x}_2) \ ,$$
$$(Ax_2, \bar{x}_1) = 0 \ , \tag{1 4}$$

what can be fulfilled only when $(x_2, \bar{x}_2) = (e_2, \bar{e}_2) = 0$. In a similar way we can prove our assertion also for eigenvalues of higher multiplicity.

This theorem holds also for complex symmetric matrices. When the eigenvector e_n belonging to the eigenvalue λ of the multiplicity n fulfil the condition $(e_n, \bar{e}_n) \neq 0$ the corresponding canonical form is diagonal and the matrix is simple. J. H. Wilkinson has shown an example of a complex symmetric matrix, which cannot be diagonalized. It is [4]

$$A = \begin{bmatrix} 2i & 1 \\ 1 & 0 \end{bmatrix} . \tag{1 5}$$

This matrix has a two-fold eigenvalue $\lambda = i$ and the eigenvector $e_2 = [1, \ -i]$, thus $(e_2, \bar{e}_2) = 0$ and according to the above results the matrix cannot be diagonalized.

Then similarly as in the case of symmetric compact operators we can construct a complete system of eigenvectors. It holds:

Theorem 4. Operators $B(p) = A^{-1}(p)$ and $A(p)$ have with exception of exceptional points a countable complete set of eigenvectors $e_1, e_2, e_3, \ldots$ biorthogonal or biorthonormal to the complex conjugate set of eigenvectors of the adjoint operators $\overline{B(p)}$ and $\overline{A(p)}$ corresponding to eigenvalues $\mu_1, \mu_2, \mu_3, \ldots$ (resp. $\lambda_k = 1/\mu_k$) with $|\mu_1| \geq |\mu_2| \geq |\mu_3| \geq \ldots$ such that for $f = Bh$ we have

$$f = \sum_k (f, \bar{e}_k) e_k = \sum_k (f, e_k) \bar{e}_k \ , \tag{1 6}$$

what corresponds to covariant and contravariant expansions of vectors, respectively. Then it holds

$$\|f\|^2 = \sum_k (f, \overline{e}_k)(\overline{f}_k, \overline{e}_k) . \tag{1.7}$$

The proof is similar to that for symmetric compact operators.

At exceptional points it is necessary to replace a basis with the eigenvector by an other biorthonormal basis of the subspace corresponding to the multiple eigenvalue.

Finally we can prove the basic theorem on analycity of eigenvalues and eigenvectors of A(p).

Theorem 5. Suppose that $A(p) = A_0 + pA_1 + p^2A_2 + \ldots + p^nA_n$, where A_k are positive definite elliptic operators. Suppose that λ is an eigenvalue of multiplicity m of the operator A(p) at p_0, where p_0 assumes real nonnegative vaues. Then there exist ordinary power series $\lambda_1(p - p_0), \ldots, \lambda_m(p - p_0)$ and power series in Hilbert space $e_1(p - p_0)$, $\ldots, e_m(p - p_0)$ all convergent in a neighbourhood of p_0, which satisfy the following conditions:

1. $e_i(p - p_0)$ is an eigenvector of A(p) belonging to the eigenvalue $\lambda_i(p - p_0)$, i.e.

$$A(p)e_i(p - p_0) = \lambda_i(p - p_0)e_i(p - p_0), \quad i = 1,\ldots,m , \tag{1.8}$$

$\lambda_i(0) = \lambda$, $i = 1,\ldots,m$ and the eigenvectors $e_i(p - p_0)$ form with eigenvectors $\overline{e}_j(p - p_0)$ of $\overline{A(p)}$ biorthonormal sets, i.e.

$$(e_i(p - p_0), \overline{e}_j(p - p_0)) = \delta_{ij}, \; i,j = 1,\ldots,m, \tag{1.9}$$

2. There exists such a neighbourhood of λ and a positive number ρ such that the spectrum of $C(p - p_0) = A(p)$ for p with $|p - p_0| < \rho$ consists exactly of the points $\lambda_1(p - p_0), \ldots, \lambda_m(p - p_0)$.

Proof can be done by a generalization of results of E.Rellich [5]. F. Rellich proved such theorem for an operator $A(\varepsilon)$ for small real values of ε. He restricted himself to orthonormal systems of eigenvectors. Then scalar product of analytic functions are analytic only at real values of the parameter ε and the Weierstrass preparation theorem can be applied only to real values of ε. Introducing of biorthonormal sets of eigenfunctions and scalar products $(f,\overline{f})$ enables to apply the Weierstrass preparations theorem also to complex values of p.

Moreover after introducing biorthonormal sets of eigenvectors it is possible to generalize the proof also for complex values of p_0. Similarly it is possible to generalize other theorems of F.Rellich.

References

[1] Brilla,J., *New functional spaces and linear nonstationary problems of mathematical physics*, Proceedings of Equadiff 5, Bratislava 1981, Teubner, Leipzig 1982, 64-71.

[2] Brilla,J., *Novye funkcional'nye prostranstva i linejnye nestacionarnye problemy matematičeskoj fiziki*, Proceedings of the 7th Soviet-Czechoslovak Conference, Yerevan State University 1982, 49-58.

[3] Gokhberg,I.C., Krein, M.G., *Vvedenie v teoriju linejnych nesamosoprjazhennych operatorov*, Nauka, Moskva, 1965.

[4] Wilkinson,J.H., *The algebraic Eigenvalue Problem*, Clarendon Press, Oxford, 1965.

[5] Rellich,F., *Perturbation Theory of Eigenvalue Problems*, Gordon and Breach, New York - London - Paris, 1969.

ON THE MOUNTAIN PASS LEMMA

KUNG-CHING CHANG
Department of Mathematics, Peking University
Beijing, China

In this paper, I propose to describe a generalized Mountain Pass Lemma (MPL, in short), which extends the original MPL due to Ambrosetti and Rabinowitz [1] in two aspects:

(a) from a Banach space to a closed convex subset,

(b) from the strong separation condition of values of functions to a weaker one.

Three applications on multiple solutions of variational inequality, semilinear elliptic BVP, and minimal surface are presented.

1. Let $\mathfrak{X}$ be a Banach space. Let C be a closed convex subset of $\mathfrak{X}$. Let Q and S be two closed subsets of C.

We say that the boundary ∂Q and S link w.r.t. C, if

(1) $\quad \partial Q \cap S = \phi,$

(2) $\quad$ for each $\phi : Q \to C$ continuous, satisfying

$$\phi|_{\partial Q} = id|_{\partial Q} ,$$

we have

$$\phi(Q) \cap S \neq \phi .$$

Suppose that $f : C \to R^1$ is a restriction of a C^1 function defined on a neighborhood of C. According to the variational inequality theory, we say $x_0 \in C$ a critical point of f w.r.t. C, if

$$\langle f'(x_0), x - x_0\rangle \geq 0 \qquad \forall\, x \in C ,$$

where $\langle\, , \,\rangle$ is the duality between $\mathfrak{X}^*$ and $\mathfrak{X}$.

For $x^* \in \mathfrak{X}^*$, and $x_1 \in \mathfrak{X}$, let us define

$$\|x^*\|_{x_1} = \text{Sup}\{\langle x^*, x - x_1\rangle \mid x \in C \text{ with } \|x - x_1\| \leq 1\} .$$

We extend the Palais Smale (P.S. in short) Condition w.r.t. C as following:

For any sequence $\{x_n\} \subset C$, such that $f(x_n)$ is bounded, and $\|-f'(x_n)\|_{x_n} \to 0$ has a convergent subsequence.

THEOREM 1. Suppose that f satisfies the P.S. Condition w.r.t. C, and that $\exists\, \alpha \in R^1$ such that

$$\text{Sup}\{f(x) \mid x \in \partial Q\} \leq \alpha ,$$

$$\text{Sup}\{f(x) \mid x \in Q\} < +\infty ,$$

and

$f(x) > \alpha, \quad \forall x \in S.$

Then one of the three possibilities occurs:

(1) α is an accumulate point of critical values.

(2) α is a critical value with uncountable K_α.

(3) $c = \inf_{A \in F} \sup_{x \in A} f(x) > \alpha$ is a critical value,

where $F = \{A = \phi(Q) \mid \phi \in C(Q,C), \text{ with } \phi|_{\partial Q} = id|_{\partial Q}\}$.

The proof depends on [6] and the following deformation lemma.

Let K be the critical set of f. $\forall a \in R^1$, denote $K_a = f^{-1}(a) \cap K$ and $f_a = \{x \in C \mid f(x) \le a\}$.

DEFORMATION LEMMA. Suppose that c is the unique critical value of f in the interval $[c,b)$ and that K_c is countable, then f_c is a strong deformation retract of $f_b \setminus K_b$.

Proof. It is a combination of the proofs given in K.C. Chang [5], Chang, Eells [7] and Z.C. Wang [19]. A pseudo gradient vector field and an associate flow were constructed in [7] for $f \in C^{2-0}$ and finite K_c, it was proved in [5]. An improvement which enables to cover our conditions, was given in [19].

Proof of Theorem 1. If non of these cases occurred, then there would exist $\varepsilon > 0$ and $\phi_0 \in C(Q,C)$ such that:

$\alpha = c, \quad f^{-1}(c, c+\varepsilon] \cap K = \phi, \; K_c$ is countable and

$\phi_0(Q) \subset f_{c+\varepsilon}$.

According to the deformation lemma, there is a continuous ϕ: $f_{c+\varepsilon} \to f_c$. Since $\phi \circ \phi_0 \in C(Q,C)$ with $\phi \circ \phi_0|_{\partial Q} = id|_{\partial Q}$, we have $(\phi \circ \phi_0)(Q) \cap S \neq \phi$. It implies

$\sup\{f(x) \mid x \in \phi \circ \phi_0(Q)\} > \alpha = c$.

This is a contradiction.

As corollaries, we have

COROLLARY 1. Suppose that $x_0 \in C$ is a local minimum, and that $\exists\, x_1 \in C$ such that $f(x_0) \ge f(x_1)$, then f has a critical point other than x_0.

In case $C = \mathfrak{X}$, this was obtained in K.C.Chang [2,4] in 1982. Obviously, it implies some results in D.G. de Figueiredo [8], D.G. de Fiqueiredo S.Solmini [9], and Pucci-Serrin [12].

COROLLARY 2. Suppose that f has two local minima, then there exists a third critical point.

2. We present three applications of Theorem 1 (or its corollaries).

(1) Variational Inequality

Let Ω be an open subset in R^3, and let g be a nonnegative measurable function defined on Ω.

THEOREM 2. The functional

$$f(u) = \int_\Omega [\frac{1}{2}(\nabla u)^2 - \frac{1}{3}u^3 + gu] \tag{1}$$

has at least two critical points w.r.t. the positive cone P in $H_0^1(\Omega)$.

THEOREM 3. Let $\phi \in H^1(\Omega)$, and let $C = \{u \in H_0^1(\Omega) | \ 0 \le u(x) \le \phi(x)$ a.e.$\}$. Assume that

$$\inf\{f(u) | \ u \in C\} < 0 . \tag{2}$$

Then f(u) has at least three critical points w.r.t. C.

Outline of the proof. It is easy to see that $u_1 = 0$ is a local minimum, and that the global minimum u_2 of f is attainable. The condition (2) implies $u_1 = u_2$. Corollary 2 implies the conclusion of Theorem 2.Similarly, Corollary 1 implies the conclusion of Theorem 2.

REMARK 1. The condition (2) is satisfied, if $\phi(X)$ is large enough.

REMARK 2. For similar considerations, see C.Q. Zhung [20] and A. Szulkin [18].

(2) A combination of the variational method and the sub - and super-solutions.

Let Ω be an open bounded domain with smooth boundary $\partial\Omega$ in R^n, and let $g \in C^\gamma(\Omega \times R^1, R^1)$, for some $o < \gamma < 1$, be a function satisfying

$$|g(x,t)| \le c(1 + |t|^\alpha)$$

for some constants $C > 0$ and $\alpha < \frac{n+2}{n-2}$ if $n \ge 3$.

THEOREM 4. Let $G(x,t) = \int_0^t g(x,\xi)d\xi$. Assume that the functional

$$f(u) = \int_\Omega [\frac{1}{2}(\nabla u)^2 - G(x,u(x))]dx$$

satisfies the P.S. condition in the space $H_0^1(\Omega)$, and that f is unbounded below. Moreover if there exists a pair of strict sub- and super-solutions of the equation

$$\begin{cases} -\Delta u = g(x,u) & \text{in } \Omega , \\ u|_{\partial\Omega} = 0 . \end{cases}$$

Then the equation has at least two distinct solutions.

For a proof, cf. K.C.Chang [2]. A considerable simplification can be found in K.C.Chang [5].

Many applications derived from this theorem, which includes the superlinear Ambrosetti Prodi type problem, a nonlinear eigenvalue problem, Amann three solution theorem, and a resonance problem. See K.C.Chang [3]. The superlinear Ambrosetti Prodi type problem was rediscussed in de Figueiredo [8] and de Figueiredo Solimini [9].

(3) Minimal surfaces

Let M be a compact oriented surface of type (p,k), and let (N,h) be a compact Riemannian manifold with nonpositive sectional curvature. If μ is a conformal structure on M compactible with its orientation, then we write (M,μ) for the associated Riemann surface.

For a map $\phi : (M,\mu) \to (N,h)$, the energy is

$$E(\phi) = \frac{1}{2} \int_M |d\phi|^2 dxdy .$$

Let $\Gamma = \{\Gamma_i\}_1^k$ be a set of disjoint oriented Jordan curves in N satisfying an irreducibility condition, which prevents the degeneracy of topological type.

THEOREM 5. If $\phi_i : (M,\mu_i) \to (N,h)$, $i = 0,1$ are homotopic admissible conformal isolated E-minima, then there is a conformal structure μ on M and an admissible conformal harmonic map $\phi : (M,\mu) \to (N,h)$ homotopic to both, which is not an E-minimum.

A special case, in which M is a borded planar domain and N is Euclidean space R^n, is due to Morse-Tompkins and Shiffman [13,14,15]. If M is a disc or an annulus and $N = R^n$, that special case has been reproved by struwe [16,17].

In proving this theorem, corollary 2 is applied. The closed convex set is the following

$$C = \mathfrak{M}^k \times \mathfrak{T}(p.k),$$

where $\mathfrak{M} = \{u \in C^\circ \cap H^{1/2}([0,2\pi], R^1) | u$ is weakly monotone, and $u(\frac{2k\pi}{3}) = \frac{2k\pi}{3}$, for $k = 0,1,2,3\}$,

and $\mathfrak{T}(p,k)$ denotes the Teichmüller space of compact oriented surface M of type (p,k). The Munford compactness theorem is applied to verify the P.S. Condition.

For details see Chang Eells [6,7].

References

[1] A. Ambrosetti, P.H. Rabinowitz, *Dual variational methods in critical point theory and applications*, J. Funct Anal. 14 (1973), 349-381.

[2] K.C. Chang, *A variant mountain pass lemma*, Scientia Sinica 26, no. 12, (1983), 1241-1255.

[3] __________, *Variational method and the sub- and super-solutions*, ibid, 1256-1265.

[4] __________, *An extension of minimax principle*. Symp. DD_3 (1982) Changchuan Jilin.

[5] __________, *Infinite dimensional Morse theory and its applications*, Lecture Notes of the 22nd Session of the Seminaire de mathematiques superieures at Montreal in 1983.

[6] K.C. Chang, J. Eells, *Harmonic maps and minimal surface coboundaries*, Proc. Lefschetz Centenary. Mexico (1984).

[7] ____________, *Unstable minimal surface coboundaries*, Preprint, April 1985 Univ. of Warwick.

[8] D.G. de Figueiredo, *On the superlinear Ambrosetti-Prodi problem*, MRC Tech. Rcp. #2522, 1983.

[9] D.G. de Figueiredo, S. Solmini, *A variational Approach to superlinear elliptic problems*, Comm. in PDE, 9 (7), (1984), 699-717.

[10] M. Morse, C.B. Tompkins, *The existence of Minimal surfaces of general critical types*, Ann. Math. 40 (1939), 443-472.

[11] ______________, *Unstable minimal surfaces of higher topological structure*, Duke Math. J. 8 (1941), 350-375.

[12] P, Pucci, J. Serrin, *A mountain pass theorem*, to appear.

[13] M. Shiffman, *The Plateau problem for minimal surfaces of arbitrary topological structure*, Amer. J. Math. 61 (1939), 853-882.

[14] ______________, *The Plateau problem for non-relative minima*, Ann. Math. 40 (1939), 834-854.

[15] ______________, *Unstable minimal surfaces with several boundaries*, Ann. Math. 43 (1942), 197-222.

[16] M. Struwe, *On a critical point theory for minimal surfaces spanning a wire in* R^n, J. reine ang. Math. 349 (1984), 1-23.

[17] ______________, *A Morse theory for annulus-type minimal surfaces*, Preprint.

[18] A. Szulkin, *Minimax principles for lower semicontinuous functions and applications to nonlinear boundary value problems*, Preprint.

[19] Z.C. Wang, *Remarks on the deformation lemma* (to appear).

[20] C.Q. Zhung, *Master Thesis at Lanzhou Univ.* 1985.

ON UNIQUENESS AND STABILITY OF STEADY-STATE CARRIER DISTRIBUTIONS IN SEMICONDUCTORS

H. GAJEWSKI
Karl-Weierstraß-Institut für Mathematik der Akademie der Wissenschaften der DDR
1086 Berlin, Mohrenstraße 39, DDR

In this paper we establish a simple smallness condition guaranteeing the basic equations for carrier distributions in semiconductors to possess a unique steady-state solution. Under this condition arbitrary perturbations of the steady state decay exponentially in time.

1. Introduction

Let G be a bounded Lipschitzian domain in R^d, $d \le 3$. Let the boundary S of G be the union of two disjoint parts S_1 and S_2, S_1 closed in S, mes $S_1 > 0$. A familiar model of carrier transport in a semiconductor device occupying G is given by the system [10,13]

$$-\Delta u = (q/\varepsilon)(f + p - n), \tag{1.1}$$

$$qn_t = \nabla.J_n - qR,\quad J_p = q\mu_n(k\nabla n - n\nabla u), \tag{1.2}$$

$$qp_t = -\nabla.J_p - qR,\quad J_p = -q\mu_p(k\nabla p + p\nabla u), \tag{1.3}$$

$$u = U_s,\ n = N_s,\ p = P_s \text{ on } R^+\times S_1,\quad \nu.\nabla u = \nu.\nabla n = \nu.\nabla p = 0 \text{ on } R^+\times S_2 \tag{1.4}$$

$$n(0,x) = n_0(x),\quad p(0,x) = p_0(x),\quad x \in G. \tag{1.5}$$

Here

u is the electrostatistic potential,
n and p are the mobile electron and hole densities,
J_n and J_p are the current densities,
f is the net density of ionized impurities,
q is the electron charge,
ε is the dielectric permitivity of the semiconductor material,
$R = (np - n_i^2)/(\tau(n+p+2n_i))$ is the recombination rate,
n_i is the intrinsic semiconductor carrier density,
τ is the electron and hole lifetime,
μ_n and μ_p are the (constant) electron and hole mobilities,
U_s, N_s and P_s are given boundary values,
ν is the outward unit normal at any point of S_2.

In the expressions for the current densities the Einstein relation $D_{n,p} = k\mu_{n,p}$ between diffusion coefficients and mobilities is used. ($k = k_B T/g$, k_b= Boltzmann constant. T = absolute temperature.)

The carrier transport equations (1.1)-(1.3) were derived by Van Roosbroeck [11] in 1950 and are now generally accepted. The first significant report on using numerical techniques to solve these equations for carriers in an operating semiconductor device structure has been published by Gummel [6] in 1964. Since then, the numerical modelling of semiconductor devices proved to be a powerful tool for device designers (see [13]).

In spite of their physical and technological relevance, the device equations received relatively little attention from the side of mathematical analysis. To our knowledge, the first matematical paper devoted to these equations appeared in 1972. In this paper Mock [7] proved the solvability of the steady-state equations associated to (1.1)-(1.5) supposing that $\mu_n = \mu_p$ and R = 0. More recently, Seidman [12], the author [3] and Gröger [5] have published more general existence theorems for steady states. All these results are based on maximum principle and compactness arguments.

As to the instationary problem (1.1)-(1.5), again Mock [8] was the first to prove a global existence and uniqueness result in a special situation. Recently, the author [2] and Gajewski&Gröger [4] could show the existence and uniqueness of global solutions under rather general assumptions. Of course, the crucial step in these papers consists in finding appropriate a-priori estimates. Such estimates are obtained by means of a physically motivated Liapunov function and an iteration technique due to Moser and Alikakos.

One of the essential open questions arising from the Van Roosbroeck equations is that of the uniqueness and stability of steady states. General answers to this question are not to be expect by physical reasons [1,10]. A special result in this direction [7] concerns the case of small perturbations of the thermal equilibrium which results from the assumption

$$U_s - k\log(N_s/n_i) = U_s + k\log(P_s/n_i) = c = \text{const. on } S_1$$

and is given by

$$N = n_i\exp((U - c)/k), \quad P = n_i\exp((c - U)/k),$$

where U is the (unique) solution of the nonlinear boundary value problem

$$-\Delta U = (q/\varepsilon)(f - 2n_i\sinh((U - c)/k)) \text{ in } G,$$

$BU = U_s$, where $Bv = \{v \text{ on } S_1,\ \nu.\nabla v \text{ on } S_2\}$ and $U_s = 0$ on S_2.

The thermal equilibrium has been shown to be globally asymptotically stable (comp. [9] for the special case $S = S_2$ and [2,4] for more general situations). In fact, it was proved in [4] that for reasonable initial values the solution $(u(t),n(t).p(t))$ of (1.1)-(1.5) converges to the corresponding thermal equilibrium (U,N,P) exponentially in time. The proof of this result heavily upon the observation that the function

$$L(t) = \int_G (kq(n(\log(n/N)-1)+N+p(\log(p/P)-1)+P)+(\varepsilon/2)|\nabla(U-u)|^2)dx$$

is monotonously decreasing.

The main purpose of the present paper is to state another kind of smallness condition implying uniqueness as well as global asymptotic stability of stationary solutions. Our smallness condition involves the essential physical parameters and can be easily checked.

2. Results

Let L_2, L_∞, H_2^1 be the usual space of functions defined on G. We use the following notations

$$|v|^2 = \int_G v^2 dx, \quad |v|_\infty = \text{vrai max } v, \quad \|v\|^2 = \int_G |\nabla v|^2 = dx ,$$

$$V = \{v \in H_2^1 / \ v = 0 \text{ on } S_1\}, \quad W = \{v \in (H_2^1 \cap L_\infty)^3 / \ v_2, v_3 \geq 0 \text{ in } G\}.$$

We assume that $f \in L_\infty$ and that the boundary values can be represented by functions $(U_s, N_s, P_s) \in W$. Let λ be the smallest eigenvalue of the problem

$$-\Delta v = \lambda v \text{ in } G, \quad Bv = 0 \text{ on } S ,$$

such that we have

$$\lambda |v|^2 \leq \|v\|^2, \quad v \in V . \tag{2.1}$$

Now we can state our results.

Theorem 1. Let $(U,N,P) \in W$ be a stationary solution of (1.1)-(1.4) such that

$$r(Q) = \frac{\lambda}{2\lambda k}\left(\frac{q}{\varepsilon}(F+Q) + \frac{1}{2\mu\tau}\left(1 + \frac{Q}{2n_i}\right)\right) < 1$$

where

$$F = |f|_\infty, \quad Q = 4(|N|_\infty + |P|_\infty), \quad \mu = \min(\mu_n, \mu_p) .$$

Then (U,N,P) is unique in W.

Remark. As to existence results for steady states $(U,N,P) \in W$ we refer to [3]. In this paper also explicit bounds for $|N|_\infty$ and $|P|_\infty$ can be found which involve only f and the boundary values.

Theorem 2. Suppose $0 \le n_0, p_0 \in L_\infty$. Let (u,n,p) be the solution of (1.1)-(1.5) and let (U,N,P) be a stationary solution satisfying the hypotheses of Theorem 1. Then for $t \ge 0$ the following estimates are valid with $a = 2k\lambda\mu(1 - r(Q))$

$$\mu_p|n(t)-N|^2 + \mu_n|p(t)-P|^2 \le e^{-at}(\mu_p|n_0-N|^2 + \mu_n|p_0-P|^2) ,$$

$$\sqrt{\lambda}|u(t)-U| \le \|u(t)-U\| \le (q/(\varepsilon\sqrt{\lambda}))(|n(t)-N| + |p(t)-P|) .$$

Remark. The existence and uniqueness of the time-dependent solution (u,n,p) is guaranteed by [4], Theorem 1.

3. Proofs

We denote by $(.,.)$ the L_2-scalar product as well as the pairing between the Hilbert space V and its dual $V^* \subset L_2$. We introduce the set

$$M = \{[N,P] \in (H_2^1 \cap L_\infty)^2,\ N,P \ge 0 \text{ on } G,\ N=N_s,\ P=P_s \text{ on } S_1\} .$$

Finally, we define an operator $A \in (M \to (V^*)^2)$ by

$$(A[N,P],[h_1,h_2]) = \mu_p((\mu_n(k\nabla N-N\nabla U),\nabla h_1) + (R,h_1)) +$$
$$+ \mu_n((\mu_p(k\nabla P+P\nabla U),\nabla h_2) + (R,h_2)) \quad \forall h_1,h_2 \in V ,$$

where $R = R(N,P)$ and $U = U(N,P)$ is the solution of the boundary value problem

$$-\Delta U = (q/\varepsilon)(f + P - N), \quad BU = U_s \quad \text{on } S .$$

The main tool for proving our results is the following monotonicity property of the operator A.

Lemma. Let $[N_j,P_j] \in M$, $j=1,2$, $N_2 \le \bar{N}$, $P_2 \le \bar{P}$ in G, $\bar{N},\bar{P}$=cons. Set $Q=4(\bar{N}+\bar{P})$. Then it holds with $m=\mu_n\mu_p k(1-r(Q))$, $N=N_1-N_2$, $P=P_1-P_2$,

$$(A[N_1,P_1] - A[N_2,P_2],[N,P]) \ge m(\|N\|^2 + \|P\|^2) .$$

Proof. Setting $U_1 = U(N_1,P_1)$, $U_2 = U(N_2,P_2)$, $U=U_1-U_2$ and using (2.1) we get

$$\|U\|^2 = (q/\varepsilon)(P - N,U) \le (q/\varepsilon)|P - N||U| \le (q/(\varepsilon\lambda))\|P - N\|\|U\|$$

and consequently

$$\|U\| = (q/(\varepsilon\sqrt{\lambda}))|P - N| \le (q/(\varepsilon\lambda))\|P - N\|,\ |U| \le (q/(\varepsilon\lambda))|P-N| \quad (3.1)$$

Thus we find

$$(k\nabla N-N_1\nabla U_1 + N_2\nabla U_2,\nabla N) + (k\nabla P + P_1\nabla U_1 - P_2\nabla U_2,\nabla P) =$$
$$= k(\|N\|^2 + \|P\|^2) - (N\nabla U_1 + N_2\nabla U,\nabla N) + (P\nabla U_1 + P_2\nabla U,\nabla P) =$$
$$= k(\|N\|^2 + \|P\|^2) + (q/(2\varepsilon))(P^2-N^2,f + P_1-N_1) + (P_2\nabla P-N_2\nabla N,\nabla U) =$$
$$= k(\|N\|^2 + \|P\|^2) + (q/(2\varepsilon))(((N-P)^2,N_1+P_1) - (N^2,f + 2P_2) +$$
$$+ (P^2,f - 2N_2) + 2(NP,N_2 + P_2)) + (P_2\nabla P - N_2\nabla N,\nabla U) \ge$$

$$\geq k(\|N\|^2 + \|P\|^2) - (q/(2\varepsilon\lambda))((F + \overline{N} + 3\overline{P})\|N\|^2 + (F + 3\overline{N} + \overline{P})\|P\|^2 +$$

$$+ 2(\overline{N}\|N\| + \overline{P}\|P\|)\|P - N\|) \geq$$

$$\geq k(1 - (q/(2\varepsilon\lambda))(F + Q))(\|N\|^2 + \|P\|^2) .$$

On the other hand, setting $a_j = \tau(N_j + P_j + 2n_i)$, we get

$$(R_1 - R_2, N) = ((1/a_1)(NP_1 + N_2P - ((N_2P_2 - n_i^2)/a_2)\tau(N + P)), N) \geq$$

$$\geq -(1/\lambda)((\overline{N}/(2a_1))(\|N\|^2 + \|P\|^2) + (Q/(16a_1))(\|N\| + \|P\|)\|N\| +$$

$$+ (1/(8\tau))(\|N\|^2 + \|P\|^2)) \geq$$

$$\geq -(1/(4\lambda\tau)((\overline{N}/n_i)(\|N\|^2 + \|P\|^2) + (Q/(16n_i)(3\|N\|^2 + \|P\|^2) =$$

$$+ (1/2)(\|N\|^2 + \|P\|^2)) .$$

Evidently, an analogous estimate holds for $(R_1 - R_2, P)$. Now the lemma is an immediate consequence from these estimates.

Proof of Theorem 1. Using the operator A we can rewrite the stationary problem as follows.

$$A[N,P] = 0, \quad [N,P] \in M . \qquad (3.2)$$

From this it becomes clear that the theorem follows easily from the lemma.

Proof of Theorem 2. We can write (1.1)-(1.5) in the compact form

$$[\mu_p n_t, \mu_n p_t] + A[n,p] = 0, \ [n(t),p(t)] \in M, \ n(0) = n_0, \ p(0) = p_0.$$

Hence, using (3.2) and the lemma, we get

$$0 \geq \tfrac{1}{2}(\mu_p|n - N|^2 + \mu_n|p - P|^2)_t + k\lambda\mu(1 - r)(\mu_p|n - N|^2 + \mu_n|p - P|^2).$$

Applying a well-known differential inequality and (3.1) we obtain the theorem.

Remark. Our lemma can also be used in order to find relaxation parameters b such that the iteration sequence $([N_j, P_j])$ defined by

$$-\Delta[h_1, h_2] = b\, A[N_j, P_j], \quad h_1, h_2 \in V, \ j = 0,1,\ldots$$

$$N_{j+1} = N_j + h_1, \ P_{j+1} = P_j + h_2, \ [N_0, P_0] \in M$$

converges to a stationary solution.

References

[1] BONČ-BRUEVICH, V.L., ZVJAGIN, I.P., MIRONOV, A.G., *Spatial electrical instability in semiconductors* (russian), Moscow 1972.

[2] GAJEWSKI, H., *On existence, uniqueness and asymptotic behavior of the basic equations for carrier transport in semiconductors*, ZAMM 65, (1985), 101-108.

[3] GAJEWSKI,H., *On the existence of steady-state carrier distributions in semiconductors*, In: Probleme und Methoden der Mathematischen Physik, Teubner-Texte zur Mathematik 63. (Ed. V. Friedrich u. a.).

[4] GAJEWSKI,H., GRÖGER,K., *On the basic equations for carrier transport in semiconductors*, J. Math. Anal. Appl., to appear.

[5] GRÖGER,K., *On steady-state carrier distributions in semiconductor devices*, to appear.

[6] GUMMEL,H.K., *A selfconsistent iterative scheme for one-dimensional steady state transistor calculations*, IEEE Trans. Electron Devices ED-11 (1964), 455-465.

[7] MOCK,M.S., *On equations describing steady-state carrier distributions in a semiconductor device*, Comm. Pure Appl. Math. 25 (1972), 781-792.

[8] MOCK,M.S., *An initial value problem from semiconductor device theory*, SIAM J. Math. Anal. 5 (1974), 597-612.

[9] MOCK,M.S., *Asymptotic behavior of solutions of transport equations for semiconductor devices*, J. Math. Anal. Appl. 49 (1975), 215-225.

[10] MOCK,M.S., *Analysis of mathematical models of semiconductor devices*, Dublin 1983.

[11] VAN ROOSBROECK,W., *Theory of the flow of electrons and holes in Germanium and other semiconductors*, Bell Syst. Tech. J 29 (1950), 560-623.

[12] SEIDMAN,T.I., *Steady state solutions of diffusion-reaction systems with electrostatic convection*, Nonlinear Analysis 4 (1980), 623-637.

[13] SELBERHERR,S., *Analysis and simulation of semiconductor devices*, Wien-New York 1984.

PARTIAL REGULARITY OF MINIMIZERS

M. GIAQUINTA
Instituto di Matematica Applicata, Universita di Firenze
Via S. Marta, Firenze, Italy

After the examples shown by E. De Giorgi, E. Giusti-M. Miranda, V.G. Mazja, J. Nečas, J. Souček, it is well known that the minimizers of variational integrals

(1) $$\mathfrak{F}[u;\Omega] = \int_\Omega F(x,u(x), Du(x))dx$$

in the vector valued case, even in simple situations, are in general *non* continuous. There is only hope to show *partial regularity* of minimizers, i.e. regularity except on a closed set hopefully small.

The study of the partial regularity of minimizers and of solutions of non linear elliptic systems starts with the works by Morrey and Giusti-Miranda in 1968, and it is the aim of this lecture to refer about some of the results obtained. I shall restrict myself to some results concerning the partial regularity of minimizers referring to [7] for a general account.

Let me start by stating the most general and recent result.

THEOREM 1. *Let Ω be a bounded open set in $\mathbf{R}^n$ and let $F(x,u,p) : \Omega \times \mathbf{R}^N \times \mathbf{R}^{nN} \to \mathbf{R}$ be a function such that*

i) $|p|^m \le F(x,u,p) \le c_0|p|^m, \quad m \ge 2$

ii) F is of class C^2 with respect to p and

$$|F_{pp}(x,u,p)| \le c_1(1 + |p|^2)^{\frac{m-2}{2}}$$

iii) $(1 + |p|^2)^{-\frac{m}{2}}F(x,u,p)$ *is Hölder-continuous in (x,u) uniformly with respect to p*

iv) F is strictly quasi-convex i.e. for all x_0, u_0, p_0 and all $\varphi \in C_0^\infty(\Omega,\mathbf{R}^N)$

Let $u \in H^{1,m}_{\mathrm{loc}}(\Omega,\mathbf{R}^N)$ be a minimizer for

$$\mathfrak{F}[u;\Omega] = \int_\Omega F(x,u,Du)dx$$

i.e. $\mathfrak{F}[u;\operatorname{supp}\varphi] \le \mathfrak{F}[u+\varphi;\operatorname{supp}\varphi]$. *Then there exists an open set Ω_0 such that $u \in C^{1,\mu}(\Omega_0,\mathbf{R}^N)$, moreover* $\operatorname{meas}(\Omega - \Omega_0) = 0$.

Theorem 1, proved in [12], is the result of a series of steps due to different authors.

Under the stronger condition of ellipticity

$$F_{p^i_\alpha p^j_\beta}\xi^\alpha_i\xi^\beta_j \geq \nu(1+|p|^2)^{\frac{m-2}{2}}|\xi|^2 \qquad \forall\xi \in \mathbf{R}^{nN}; \quad \nu > 0$$

theorem 1 was proved for $m \geq 2$ by C.Morrey and E.Giusti, for $1 < m < 2$ by L.Pepe in 1968 in the case $F = F(p)$; in the case $m = 2$, $F = F(x,u,p)$ by Giaquinta - Giusti and Ivert in 1983, in the case $m \geq 2$, $F = F(x,u,p)$ by Giaquinta - Ivert in 1984. Fro these results I refer to [7] [9] [11]. Under the weaker assumption of quasi-convexity in (2) it was proved by L. Evans [5] in the case $F = F(p)$, $m \geq 2$.

The case $1 < m < 2$ is open, and essentially open are all the questions concerning the singular set; for instance

1. what about the structure of the singular set? what about the Hausdorff dimension of the singular set?
2. are there resonable structures under which minimizers are regular? (see the interesting paper [22])
3. what about the stability or instability properties of the singular set? or what about topological properties of the set of smooth minimizers?

We have results improving theorem 1 roughly only in case of quadratic functionals if we exclude the case in which F does not depend explicitly on u. So let us consider a quadratic functional

$$(3) \qquad A(u) = \int_\Omega A^{\alpha\beta}_{ij}(x,u)D_\alpha u^i D_\beta u^j dx$$

where the coefficients $A^{\alpha\beta}_{ij}$ are smooth (for example Hölder-continuous) and satisfy the ellipticity condition

$$(4) \qquad A^{\alpha\beta}_{ij}(x,u)\xi^\alpha_i\xi^\beta_j \geq |\xi|^2 \qquad \forall\xi \in \mathbf{R}^{nN}$$

Notice that the functional A is not differentiable. Concerning the strong condition of ellipticity (4), we remark that there is not much hope to weaken it. In fact in [14] it is shown that for weak solutions of the simple quasilinear system

$$\int_\Omega A^{\alpha\beta}_{ij}(x,u)D_\alpha u^i D_\beta \varphi^j dx = 0 \qquad \forall\varphi \in H^1_0(\Omega,\mathbf{R}^N)$$

with coefficients satisfying the strict Legendre-Hadamard condition

$$A^{\alpha\beta}_{ij}(x,u)\xi^\alpha\xi^\beta\eta_i\eta_j \geq |\xi|^2|\eta|^2 \qquad \forall\xi \in \mathbf{R}^n \quad \forall\eta \in \mathbf{R}^N$$

Cacciоppoli's inequality may not be true; and Caccioppoli's inequality is indeed the starting point for the regularity theory.

THEOREM 2. (*Giaquinta - Giusti* [8]) - *Let u be a minimizer for $A(u)$. Then the Hausdorff dimension of the singular set $\Omega - \Omega_0$ is strictly less than n-2. In particular minimizers are smooth in dimension n = 2.*[1)]

Now the first natural question is whether the singularities are at most isolated in dimension n = 3, where first we can have singularities. The question is open in that generality, but it has a positive answer under the extra assumption that the coefficients split as

$$A_{ij}^{\alpha\beta}(x,u) = G^{\alpha\beta}(x)g_{ij}(u) \tag{5}$$

THEOREM 3. (*Giaquinta - Giusti* [10]) - *Let u be a bounded minimizer of*

$$\int_\Omega G^{\alpha\beta}(x)g_{ij}(u)D_\alpha u^i D_\beta u^j dx$$

where G and g are smooth symmetric definite positive matrices. Then in dimension n = 3 the singularities of u are at most isolated and in general the singular set of u has Hausdorff dimension no larger than n - 3.

THEOREM 4. (*Jost - Meier* [18]) - *Under the assumption of theorem 3 if u is a bounded minimizer with smooth boundary datum, then singularities may occur only far from the boundary.*

We recall that solutions of quasilinear elliptic systems may instead have singularities at the boundary [6].

The functional (3) (4) (5) that can be rewritten as

$$\mathcal{E}(u) = \int_\Omega G^{\alpha\beta}(x)g_{ij}(u)D_\alpha u^i D_\beta u^j \sqrt{G}\, dx \tag{6}$$

where

$$G(x) = \det(G_{\alpha\beta}(x)) \qquad (G_{\alpha\beta}(x)) = (G^{\alpha\beta}(x))^{-1}$$

represents in local coordinates the energy of a map between two Riemannian manifolds $u : M^n \to M^N$ with metric tensors respectively $G_{\alpha\beta}$, g_{ij}. Smooth stationary points are called *harmonic maps*. We refer to [2][3][17] for more information.

From the general point of view of differential geometry, theorems

---

1) Actually, under some more restrictive assumptions, in the general situation of theorem 1 minimizers are also smooth in dimension 2, see [7].

2 and 3 are limited.

In fact, while we can always localize in M^n, this is in general not possible in the target manifold M^N, except that we assume that it is covered by one chart (or, worse still, that u is continuous). In the general setting of a map from M^n into M^N, theorems 2 and 3 have been proved independently by Schoen - Uhlenbeck [19] [20].

At this point we may resonably ask whether the (bounded) minimizers of (6) may be really singular. In that respect the classical result by Eells - Sampson [4] can be read: *if the sectional curvature of M^N is non-positive then the minimizers (as well as the stationary points) of (6) are smooth.* Hildebrandt - Kaul - Widman [15] in the case of target manifold with positive sectional curvature proved: *if $u(M^n)$ is contained in a geodetic ball $B_r(q)$ which is disjoint from the cut locus of its center and has radius*

(7) $$R < \frac{\pi}{2\sqrt{k}}$$

where k is an upper bound for the sectional curvature, then the minimizers (and even the stationary points) are smooth.

In case of a map from the unit ball $B_1(0)$ of $\mathbb{R}^n$ into the standard sphere S^n of $\mathbb{R}^{n+1}$ condition (7) means that $u(B_1(0))$ is strictly contained in a hemisphere. Hildebrandt - Kaul - Widman showed that the *equator map* u^* defined by $u^*(x) = (\frac{x}{|x|},0)$ is a stationary point for $\mathcal{E}(u)$.

Then Jäger-Kaul [16] proved that u^* *is a minimizer for* $n > 6$, *while it is even unstable for* $n < 7$; more recently Baldes [1] showed that u^* is *stable* even for $n = 3$ if considered as a mapping from $B_1(0)$ into a suitable ellipsoid.

In general we have

THEOREM 5. (*Schoen-Uhlenbeck [21], Giaquinta-Souček [13]*) - *Every energy minimizing map u from a domain in some n-dimensional Riemannian manifold into the hemisphere S^N_+ is regular provided $n \le 6$, and in general its singular set has Hausdorff dimension no larger than $n - 7$.*

References

[1] Baldes,A., *Stability and Uniqueness Properties of the Equator Map from a Ball into an Ellipsoid*, Math. Z. 185 (1984), 505-516.

[2] EELLS,J., Lemaire,L., *A report on harmonic maps*, Bull. London Math. Soc. 10 (1878), 1-68.

[3] EELLS,J., Lemaire,L., *Selected topics in harmonic maps*, CBMS Regional Conference series.

[4] EELLS,J., Sampson,J.H., *Harmonic mappings of Riemannian manifolds*, Amer. J. Math. 86 (1964), 109-160.

[5] Evans,C.L., *Quasiconvexity and partial regularity in the calculus of variations*, preprint 1984.

[6] Giaquinta,M., *A counterexample to the boundary regularity of solutions to elliptic quasilinear systems*, manuscripta math. 14 (1978), 217-220.

[7] Giaquinta,M., *Multiple integrals in the Calculus of Variations and non linear elliptic systems*, Annals. Math. Studies n° 105, Princeton University Press, 1983.

[8] Giaquinta,M., Giusti,E., *On the regularity of the minima of variational integrals*, Acta Math.148 (1982), 31-46.

[9] Giaquinta,M., Giusti,E., *Differentiability of minima of non differentiable functionals*, Inventiones Math. 72 (1983), 285-298.

[10] Giaquinta,M., Giusti,E., *The singular set of the minima of certain quadratic functionals*, Ann.Sc.Norm.Sup. Pisa 11 (1984), 45-55.

[11] Giaquinta,M., Ivert,P.A., *Partial regularity for minima of variational integrals*, Arkiv föf Math.

[12] Giaquinta,M., Modica,G., *Partial regularity of minimizers of quasiconvex integrals*, Ann. Inst. H. Poincaré, Analyse non linéaire.

[13] Giaquinta,M., Souček,J., *Harmonic maps into a hemisphere*, Ann. Sc. Norm. Sup. Pisa.

[14] Giaquinta,M., Souček,J., *Cacciopoli's inequality and Legendre-Hadamard condition*,Math. Ann. 270 (1985), 105-107.

[15] Hildebrandt,S., Kaul,H, Widman,K.O., *An existence theorem for harmonic mappings of Riemannian manifolds*, Acta Math. 138 (1977), 1-16.

[16] Jäger,W., Kaul,H., *Rotationally symmetric harmonic maps from a ball into a sphere and the regularity problem for weak solutions of elliptic systems*, J. reine u. angew. Math. 343 (1983), 146-161.

[17] Jost,J., *Harmonic mappings between Riemannian manifolds*, Centre for Math. Anal., Australian National Univ., vol. 4, 1983.

[18] Jost,J., Meier,M., *Boundary regularity for minima of certain quadratic functionals*, Math. Ann. 262 (1983), 549-561.

[19] Schoen,R., Uhlenbeck,K., *A regularity theory for harmonic maps*, J. Diff. Geom. 17 (1982), 307-335.

[20] Schoen,R., Uhlenbeck,K., *Boundary regularity and miscellaneous results on harmonic maps*, J. Diff. Geom. 18 (1983), 253-268.

[21] Schoen,R., Uhlenbeck,K., *Regularity of minimizing harmonic maps into the sphere*, Inventiones math.

[22] Uhlenbeck,K, *Regularity for a class of nonlinear elliptic systems*, Acta Math. 138 (1977), 219-240.

PERIODIC SOLUTIONS OF PARTIAL DIFFERENTIAL EQUATIONS WITH HYSTERESIS

P. KREJČÍ
Mathematical Institute, Czechoslovak Academy of Sciences
115 67 Prague 1, Czechoslovakia

Introduction.

In mechanics of plastic-elastic bodies or in the theory of electromagnetic field in ferromagnetic media we are led to the consideration of hysteresis phenomena. There are various approaches to the mathematical description of hysteresis (cf. [1]). Existence results for PDE's with hysteresis nonlinearities are due to Visintin (see e.g. [5]). We give here a survey of results of [2], [3], [4], where we prove the existence of periodic solutions to the problems

$$u_{tt} - u_{xx} \pm F(u) = H(t,x) \ , \quad u(t,0) = u(t,\pi) = 0 \quad ([2]), \tag{P1}$$

$$u_t - (F(u_x))_x = H(t,x) \ , \quad u(t,0) = u(t,1) = 0 \quad ([3]), \tag{P2}$$

$$u_{tt} - (F(u_x))_x = H(t,x) \ , \quad u_x(t,0) = u_x(t,\pi) = 0 \quad ([4]), \tag{P3}$$

where F is the Ishlinskiĭ hysteresis operator and H is a given time-periodic function with an arbitrary period $\omega > 0$.

1. Ishlinskiĭ operator (cf. [1], [2]).

We first define simple hysteresis operators $v \to \ell_h(v)$, $f_h(v)$ for $h > 0$ and for piecewise monotone continuous inputs $v : [0,T] \to R^1$ as follows:

$$\ell_h(v)(t) = \begin{cases} \max\{\ell_h(v)(t_0), v(t) - h\}, & t \in [t_0,t_1] , \\ \quad \text{if } v \text{ is nondecreasing in } [t_0,t_1] \\ \min\{\ell_h(v)(t_0), v(t) + h\}, & t \in [t_0,t_1] , \\ \quad \text{if } v \text{ is nonincreasing in } [t_0,t_1] , \end{cases} \tag{1.1}$$

$$\ell_h(v)(0) = \begin{cases} 0\ , & \text{if } |v(0)| \leq h \\ v(0) - h\ , & v(0) > h \\ v(0) + h\ , & v(0) < -h\ . \end{cases}$$

$$f_h(v)(t) = v(t) - \ell_h(v)(t)\ . \tag{1.2}$$

For v , w continuous and piecewise monotone we have (see [1], p. 16)

$$|\ell_h(v)(t) - \ell_h(w)(t)| \leq \max\{|v(s) - w(s)|;\ s \in [0,t]\}\ . \tag{1.3}$$

This property enables us to define $\ell_h(v)$, $f_h(v)$ for arbitrary continuous inputs. Moreover, it follows from (1.3) that ℓ_h , f_h map continuously the space $C([0,T])$ of all continuous functions into itself.

Let us introduce the space C_ω , $\omega > 0$, of all continuous ω-periodic functions $v : R^1 \to R^1$ with sup-norm $||\cdot||$. For $v \in C_\omega$ the functions $\ell_h(v)$, $f_h(v)$ are ω-periodic for $t \geq \omega$, hence ℓ_h , f_n can be considered as continuous operators $C_\omega \to C_\omega$.

Let further $g : [0,\infty) \to [0,\infty)$ be a function of class $C^2(0,\infty)$ satisfying

(i) g *is increasing,* $g(0) = 0$, $0 < g'(0+) < \infty$.

(ii) $g(h) \leq a\cdot h^\alpha$ *for some* $a > 0$, $\alpha \in (0,1)$ *and for every* $h \geq 0$.

(iii) *Put* $\gamma(r) = \inf\{-g''(h);\ 0 < h \leq r\}$. *We require the existence of constants* $b > 0$, $\beta \in (0,\alpha]$ *such that* $\liminf_{r\to\infty} \gamma(r)\, r^{2-\beta} = b$. (1.4)

(iv) $\beta > 1/3$, $3\alpha < 4\beta$.

For $v \in C_\omega$ we define

(i) $$F(v)(t) = -\int_0^\infty f_h(v)(t)\, g''(h)\, dh$$

(1.5)

(ii) $$L(v)(t) = \int_0^\infty \ell_h(v)(t)\, (g^{-1})''(h)\, dh + (g^{-1})'(0+)\, v(t)\ ,$$

where $g(g^{-1}(h)) = g^{-1}(g(h)) = h$.

Roughly speaking ,the dependence of F on v (L on v) can be represented by a system of hysteresis loops constituted by parts of the graph of g (g^{-1}, respectively) for v nondecreasing and $-g$ ($-g^{-1}$, respectively) for v nonincreasing. The operators F, L have the following properties:

(i) $F, L : C_\omega \to C_\omega$ *are continuous*

(ii) $L = F^{-1}$

(iii) $||F(v)|| = g(||v||)$

(iv) $$g'(\max\{||v||, ||w||\})\,||v - w|| \le ||F(v) - F(w)|| \le g'(0+)\,||v - w||$$

(v) *For* v *absolutely continuous* $F(v)$, $L(v)$ *are absolutely continuous and the inequalities*

$$|(F(v))'(t)| \le g'(0+)\,|v'(t)| , \qquad |(L(v))'(t)| \le (g^{-1})'(||v||)\,|v'(t)| \tag{1.6}$$

are satisfied almost everywhere.

(vi) *For* v' *absolutely continuous we have*

$$\int_\omega^{2\omega} (F(v))'v'' \ge \frac{1}{4}\gamma(||v||)\int_0^\omega |v'|^3$$

$$\int_\omega^{2\omega} (L(v))'v'' \le -\frac{1}{4}\tilde{\gamma}(||v||)\int_0^\omega |v'|^3 ,$$

where $\tilde{\gamma}(r) = \inf\{(g^{-1})''(h);\ 0 < h \le r\}$.

For functions of two variables $u(t,x)$, $t \in R^1$, $x \in I$, such that $u(.,x) \in C_\omega$ for each $x \in I$ we define

$$F(u)(t,x) = F(u(.,x))(t) , \quad L(u)(t,x) = L(u(.,x))(t) .$$

2. Existence results.

We introduce the Banach spaces or time-periodic functions $u(t,x)$, $u : R^1 \times [0,\ell] \to R^1$, $\ell > 0$, $u(t + \omega, x) = u(t,x)$, $\omega > 0$:

$C_\omega([0,\ell])$: the space of all continuous functions with norm

$$|||u||| = \max\{|u(t,x)|;\ t \quad R^1,\ x \in [0,\ell]\} ;$$

$L^p_\omega(0,\ell)$, $1 \le p \le \infty$: the space of all measurable functions such that

$$|u|_p = \left(\int_0^\omega\int_0^\ell |u(t,x)|^p\,dx\,dt\right)^{1/p} < \infty \quad \text{for } p < \infty ,$$

$$|u|_\infty = \sup\text{ess}\,\{|u(t,x)|;\ t \quad R^1,\ x \in [0,\ell]\} < \infty ,$$

with norm $|\cdot|_p$;

$$L^2(0,\ell;\, L^1_\omega) = \{u \in L^1_\omega(0,\ell);\ |u|_{2,1} = \Big(\int_0^\ell \Big(\int_0^\omega |u(t,x)|\, dt\Big)^2 dx\Big)^{1/2} < \infty\}$$

with norm $|\cdot|_{2,1}$.

Theorem. (2.1)

(i) *Let* $H \in L^2_\omega(0,\pi)$ *be given such that* $H_{tt} \in L^{3/2}_\omega(0,\pi)$ *and let* (1.4)(i) - (iii) *hold. Then there exists at least one solution* $u \in C_\omega([0,\pi])$, $u_{tt} - u_{xx} \in L^2_\omega(0,\pi)$, $u_t \in L^3_\omega(0,\pi)$ *to the problem* (P1) *such that* (P1) *is satisfied almost everywhere.*

(ii) *Let* $H \in C_\omega([0,1])$ *be given such that* $H_t \in L^2_\omega(0,1)$ *and let* (1.4)(i) - (iii) *hold. Then there exists at least one classical solution* $u \in C_\omega([0,1])$, u_t, u_x, $(F(u_x))_x \in C_\omega([0,1])$, $u_{tt} \in L^2_\omega(0,1)$, $u_{tx} \in L^3_\omega(0,1)$, $u_{xx} \in L^\infty_\omega(0,1)$ *to the problem* (P2) .

(iii) *Let* $H \in L^1_\omega(0,\pi)$ *be given such that* $H_x \in L^{3/2}_\omega(0,\pi) \cap L^2(0,\pi;\, L^1_\omega)$, $\int_0^\omega \int_0^\pi H(t,x)\, dx\, dt = 0$, *and let* (1.4)(i) - (iv) *hold. Then there exists at least one solution* $u \in C_\omega([0,\pi])$, $u_x \in C_\omega([0,\pi])$, u_{tt}, $(F(u_x))_x \in L^1_\omega(0,\pi)$, $u_{tx} \in L^3_\omega(0,\pi)$ *to the problem* (P3) *such that* (P3) *is satisfied almost everywhere.*

Sketch of the proofs.

(i) We use the classical Galerkin method. We denote

$$w_{jk}(t,x) = e_j(t) \sin kx \quad , \ j \ \text{integer}, \ \ k \ \text{natural}, \tag{2.2}$$

where $e_j(t) = \sin \frac{2\pi j}{\omega} t$ for $j > 0$ and $\cos \frac{2\pi j}{\omega} t$ for $j \le 0$.

For $m > 1$ put

$$\underset{m}{u}(t,x) = \sum_{j=-m}^{m} \sum_{k=1}^{m} u_{jk}\, w_{jk}(t,x) , \tag{2.3}$$

where the coefficients u_{jk} are solutions of the algebraic system

$$\int_\omega^{2\omega} \int_0^\pi (\underset{m}{u}_{tt} - \underset{m}{u}_{xx} \pm F(\underset{m}{u}))w_{jk}\, dx\, dt = \int_\omega^{2\omega} \int_0^\pi H\, w_{jk}\, dx\, dt , \tag{2.4}$$

$j = -m,\ldots,m$, $k = 1,\ldots,m$.

Multiplying (2.4) by $\left(\frac{2\pi j}{\omega}\right)^3 u_{-jk}$, summing over j and k and using (1.6)(vi) we get

$$\gamma(|||u_m|||)\,|u_{m t}|_3^2 \leqq \text{const.}$$

Similarly we have $|u_{m tt} - u_{m xx}|_2 \leqq \text{const.}\left(1 + |||F(u_m)|||\right)$. Classical embedding theorems and (1.4), (1.6)(iii) yield $|||u_m||| \leqq$ const. $\left(|u_{m tt} - u_{m xx}|_2 + |u_{m t}|_3\right) \leqq \text{const.}\left(|||u_m|||^{1-\beta/2} + |||u_m|||^{\alpha} + 1\right)$, hence $|||u_m|||$, $|u_{m t}|_3$, $|u_{m tt} - u_{m xx}|_2 \leqq$ const.

These estimates imply the solvability of (2.4). On the other hand there exists a subsequence $\{u_n\}$ of $\{u_m\}$ such that $u_{n tt} - u_{n xx} \to u_{tt} - u_{xx}$ in $L^2_\omega(0,\pi)$ weak, $u_{n t} \to u_t$ in $L^3_\omega(0,\pi)$ weak and $u_n \to u$ in $C_\omega([0,\pi])$ strong. Thus, we can pass to the limit in (2.4) and the proof is complete.

(ii) We replace the problem (P2) by the following system of ordinary differential equations (space discretization):

$$\begin{aligned} &v_j' - n\left(F(n(v_{j+1} - v_j)) - F(n(v_j - v_{j-1}))\right) = h_j \, , \\ &v_0 = v_n = 0 \, , \end{aligned} \tag{2.5}$$

where $h_j(t) = n \int_{j/n}^{(j+1)/n} H(t,x)\,dx$, $j = 1,\ldots,n-1$.

Using (1.6)(iv), (vi) we derive apriori estimates for v_j independent of n which ensure the existence of periodic solutions of (2.5). We further put

$$\begin{aligned} u_n(t,x) = {} & v_j(t) + (nx - j)(v_{j+1}(t) - v_j(t)) + \tfrac{1}{2}\left[(v_{j+1}(t) - v_j(t)) \right.\\ & \left. + (nx - j)^2 (v_{j+2}(t) - 2v_{j+1}(t) + v_j(t))\right] , \end{aligned}$$

$t \in R^1$, $x \in [j/n, (j+1)/n]$, $j = 0,1,\ldots,n-1$, $v_{n+1} = -v_{n-1}$.

A straigtforward computation and the compactness argument show that there exists a subsequence of u_n which converges in suitable topologies to a solution u of (P2).

(iii) We first solve the auxiliary problem

$$L(v_t - \psi)_t - v_{xx} = \tilde{H}_x \, , \quad v(t,0) = v(t,\pi) = 0 \, ,$$

where $\psi(x) = \frac{1}{\omega}\int_0^x\int_0^\omega H(t,\xi)\,dt\,d\xi\ , \quad x\in[0,\pi]\ ,$

$$\hat{H}(t,x) = \int_0^t H(\tau,x)\,d\tau - \frac{t}{\omega}\int_0^\omega H(\tau,x)\,d\tau\ , \quad t\in R^1\ , \quad x\in[0,\pi]\ .$$

We use again the Galerkin approximation scheme of the type (2.2), (2.3), (2.4). In an analogous way we derive the following estimates:

$$\tilde{\gamma}(|||\underset{m}{v}_t - \psi|||)\,|\underset{m}{v}_{tt}|_3^2 \leq \text{const.}$$

$$|\underset{m}{v}_{xt}|_2^2 \leq (g^{-1})'(|||\underset{m}{v}_t - \psi|||)\,|\underset{m}{v}_{tt}|_2^2 + \text{const.}\,|\underset{m}{v}_t|_3\ .$$

Following (1.4) there exists $\delta > 0$ such that $1/\tilde{\gamma}(r) \leq$ const. $(r^{3-1/\beta-\delta} + 1)$ for every $r > 0$. The space $\{u\in L^1_\omega(0,\pi);\ u_t\in L^3_\omega(0,\pi)\ ,\ u_x\in L^2_\omega(0,\pi)\}$ is compactly embedded into $C_\omega([0,\pi])$, hence $|||\underset{m}{v}_t - \psi||| \leq \text{const.}(|\underset{m}{v}_{tt}|_3 + |\underset{m}{v}_{tx}|_2 + 1) \leq \text{const.}(|||\underset{m}{v}_t - \psi|||^{1-\delta/2} + 1)$.
Similarly as above we get

$$|||\underset{m}{v}_t|||\ ,\ |\underset{m}{v}_{tt}|_3\ ,\ |\underset{m}{v}_{tx}|_2\ ,\ |\underset{m}{v}_{xx}|_2 \leq \text{const.},$$

so that we can repeat the argument of (i).

The solution u of (P3) is then given by the formula

$$u(t,x) = \int_\omega^t (v_x + \hat{H})(\tau,x)\,d\tau - \frac{t}{\omega}\int_\omega^{2\omega}(v_x + \hat{H})(\tau,x)\,d\tau + \int_0^x L(v_t - \psi)(\omega,\xi)\,d\xi + \text{const.},\quad t\in R^1\ ,\quad x\in[0,\pi]\ .$$

R e f e r e n c e s

[1] KRASNOSELSKIĬ, M. A. and POKROVSKIĬ, A. V.: *Systems with hysteresis*. Nauka, Moscow, 1983 (Russian).

[2] KREJČÍ, P.: *Hysteresis and periodic solutions of nonlinear wave equations*. To appear.

[3] KREJČÍ, P.: *Periodic solutions to a parabolic equation with hysteresis*. To appear.

[4] KREJČÍ, P.: *Periodic solutions to a quasilinear wave equation with hysteresis*. To appear.

[5] VISINTIN, A.: *On the Preisach model for hysteresis*. Nonlin. Anal., Theory, Meth. & Appl. 8 (1984), pp. 977-996.

STABILITY AND BIFURCATION PROBLEMS FOR REACTION-DIFFUSION SYSTEMS WITH UNILATERAL CONDITIONS

M. KUČERA
Mathematical Institute, Czechoslovak Academy of Sciences
115 67 Prague 1, Czechoslovakia

Let us consider a reaction-diffusion system of the type

$$\begin{aligned} u_t &= d\Delta u + f(u,v) \\ v_t &= \Delta v + g(u,v) \end{aligned} \quad \text{on } \langle 0,\infty) \times \Omega \qquad \text{(RD)}$$

where f, g are real smooth functions on R^2, d is a nonnegative parameter (diffusion coefficient), Ω is a bounded domain in R^n. Suppose that $\bar{u}, \bar{v} > 0$ is an isolated solution of $f(\bar{u},\bar{v}) = g(\bar{u},\bar{v}) = 0$. Thus, $\bar{u}$, $\bar{v}$ is a stationary spatially homogeneous solution of (RD) with Neumann boundary conditions and also with the boundary conditions

$$u = \bar{u} \text{ on } (0,\infty) \times \Gamma_D, \quad \frac{\partial u}{\partial n} = 0 \text{ on } (0,\infty) \times \Gamma_N, \qquad (BC_1)$$

$$v = \bar{v} \text{ on } (0,\infty) \times \Gamma_D, \quad \frac{\partial v}{\partial n} = 0 \text{ on } (0,\infty) \times \Gamma_N, \qquad (BC_2)$$

where $\Gamma_D \cup \Gamma_N = \partial\Omega$ (the boundary of Ω). Set $b_{11} = \frac{\partial f}{\partial u}(\bar{u},\bar{v})$, $b_{12} = \frac{\partial f}{\partial v}(\bar{u},\bar{v})$, $b_{21} = \frac{\partial g}{\partial u}(\bar{u},\bar{v})$, $b_{22} = \frac{\partial g}{\partial v}(\bar{u},\bar{v})$, $B = \begin{pmatrix} b_{11}, b_{12} \\ b_{21}, b_{22} \end{pmatrix}$ and suppose

$$b_{11} > 0,\ b_{21} > 0,\ b_{12} < 0,\ b_{22} < 0,\ b_{11} + b_{22} < 0,\ \det B > 0. \qquad \text{(B)}$$

Then there exists the greatest bifurcation point d_0 of (RD), (BC) at which spatially nonhomogeneous stationary solutions bifurcate from the branch of trivial solutions $\{[d,\bar{u},\bar{v}];\ d \in R\}$; the solution $\bar{u}$, $\bar{v}$ is stable for any $d > d_0$ and unstable for any $0 < d < d_0$. (All eigenvalues of the corresponding linearized problem have negative real parts for $d > d_0$ and there exists a positive eigenvalue for $d < d_0$.) For the case of Neumann conditions and $n = 1$ (i.e. $\Omega = (0,1)$) see e.g. [8], for the general case see [2]. Notice that (B) is fulfilled in models connected with population dynamics, chemistry, morphogenesis etc. In these cases u represents a prey or an activator, v represents a predator or an inhibitor. The existence of stationary spatially nonhomogeneous solutions explains the occurence of the so called striking patterns.

The aim of this lecture is to present some results obtained by the author together with P. Drábek, M. Míková and J. Neustupa, showing how this situation can change by introducing unilateral conditions given by a cone in a suitable Hilbert space. One of the simplest examples are unilateral boundary conditions

$$\begin{aligned} & v = \bar{v} \quad \text{on} \quad (0,\infty) \times \Gamma_D , \quad v \geq \bar{v} , \quad \frac{\partial v}{\partial n} \geq 0 , \\ & (v - \bar{v})\frac{\partial v}{\partial n} = 0 \quad \text{on} \quad (0,\infty) \times \Gamma_0 , \quad \frac{\partial v}{\partial n} = 0 \quad \text{on} \quad (0,\infty) \times (\Gamma_N \setminus \Gamma_0) \end{aligned} \tag{1}$$

where Γ_0 is a given subset of Γ_N. Roughly speaking, the spatially homogeneous solution $\bar{u}$, $\bar{v}$ becomes unstable even for some $d > d_0$ and the greatest bifurcation point shifts to the right of d_0 if classical conditions are replaced by unilateral ones for v; the greatest bifurcation point shifts to the left if unilateral conditions are introduced for u.

In what follows, we shall suppose $\bar{u} = \bar{v} = 0$ without loss of generality.

1. Abstract formulation.

Let K be an arbitrary closed convex cone in the space $V = \{u \in W_2^1(\Omega);\ u = 0 \text{ on } \Gamma_D\}$ with its vertex at the origin. Consider the problem

$$\begin{aligned} & \int_\Omega [u_t(t,x)\phi(x) + d\nabla u(t,x)\nabla\phi(x) - f(u(t,x),v(t,x))\phi(x)] = 0 , \\ & \quad v(t,.) \in K , \\ & \int_\Omega \{v_t(t,x)[\psi(x) - v(t,x)] + \nabla v(t,x)\nabla[\psi(x) - v(t,x)] - \\ & - g(u(t,x),v(t,x))[\psi(x) - v(t,x)]\}\, dx \geq 0 \\ & \quad \text{for all} \quad \phi \in V , \quad \psi \in K , \text{ a.a. } t \geq 0 , \end{aligned} \tag{2}$$

where $\nabla u \cdot \nabla\phi = \sum_{i=1}^n u_{x_i}\phi_{x_i}$. By a solution we can understand a couple $u, v \in L_2(0,T;V)$ such that $u_t, v_t \in L_2(0,T;V)$.

Notice that the choice $K = \{v \in V;\ v \geq 0 \text{ on } \Gamma_0\}$ corresponds to the problem (RD), (BC_1), (1). ((2) is obtained by multiplying the equations in (RD) by test functions, integrating by parts and using (BC_1), (1).)

In general, we can define a solution of (RD) with the boundary conditions (BC_1) and unilateral conditions for v given by K as a couple u, v satisfying (2). The corresponding linearization reads

$$\int_\Omega [u_t(t,x)\phi(x) + d\nabla u(t,x)\nabla\phi(x) - (b_{11}u(t,x) + b_{12}v(t,x))\phi(x)]\, dx = 0\,, \quad v(t,.) \in K\,,$$

$$\int_\Omega \{v_t(t,x)(\psi(x) - v(t,x)) + \nabla v(t,x)\cdot\nabla[\psi(x) - v(t,x)] - (b_{21}u(t,x) + b_{22}v(t,x))(\psi(x) - v(t,x))\}\, dx \geqq 0 \tag{3}$$

for all $\phi \in V$, $\psi \in K$, a.a. $t \geqq 0$.

Analogously, we can consider (RD) with (BC_2) and unilateral conditions for u, i.e.

$$u \in K\,,$$

$$\int_\Omega [u_t(\phi - u) + d\nabla u\cdot\nabla(\phi - u) - f(u,v)(\phi - u)]\, dx \geqq 0 \tag{4'}$$

for all $\phi \in K$, a.a. $t \geqq 0$,

$$\int_\Omega [v_t\psi + \nabla v\cdot\nabla\psi - g(u,v)\psi]\, dx = 0 \quad \text{for all } \psi \in V\,, \text{ a.a. } t \geqq 0\,.$$

2. Destabilization.

EXAMPLE 1. Consider $\Gamma_D = \emptyset$ (i.e. $V = W_2^1$), $K = \{v \in V;\ v \geqq 0$ on $\Omega\}$. Then (2) corresponds to a free-boundary problem

$$\begin{aligned} &u_t = d\Delta u + f(u,v) \quad \text{on} \quad \langle 0,\infty) \times \Omega\,, \\ &v_t = \Delta v + g(u,v) \quad \text{on} \quad Q_+\,, \\ &v = 0,\ - g(u,0) \geqq 0 \quad \text{on} \quad \langle 0,\infty) \times \Omega \setminus Q_+\,, \\ &u_{x_i}\,,\ v_{x_i} \quad \text{are continuous,} \quad \frac{\partial u}{\partial n} = \frac{\partial v}{\partial n} = 0 \quad \text{on} \quad (0,\infty) \times \partial\Omega\,, \end{aligned} \tag{5}$$

where the domain $Q_+ = \{[t,x] \in \langle 0,\infty) \times \Omega;\ v(t,x) > 0\}$ is unknown. The couple $u(t,x) = \exp(b_{11}t)\cdot\xi$, $v(t,x) = 0$ satisfies the linearization of (5) (i.e. also (3)) classically for any $\xi < 0$. It follows that the trivial solution of (3) is unstable for any d, even for $d > d_0$, and even with respect to spatially homogeneous perturbations. Notice that spatially homogeneous solutions of (3) (in our special case) are solutions of the inequality

$$\begin{aligned} &U(t) \in K_c \\ &\langle U_t(t) - BU(t), \Psi - U(t) \rangle \geqq 0 \quad \text{for all} \quad \Psi \in K_c\,, \text{ a.a. } t \geqq 0\,, \end{aligned} \tag{6}$$

where $U = [u,v]$, $K_c = \{[\phi,\psi] \in R^2;\ \psi \geqq 0\}$, $\langle .,. \rangle$ is the scalar product in R^2. It is not difficult to describe all trajectories of (6) and characterize also some spatially homogeneous solution of (2) under the assumption (B). As a consequence it is possible to prove also the instability for (2) for any $d > 0$ (see [7]).

Notice that the eigenvalue problem determining the stability of the trivial solution of (RD), (BC) can be written in the vector form

$$D(d)\Delta U + BU = \lambda U \tag{7}$$

(with the boundary conditions (BC)), where $U = [u,v]$, $D(d) = \begin{pmatrix} d,0 \\ 0,1 \end{pmatrix}$, $\Delta U = [\Delta u, \Delta v]$. The eigenvalue problem with unilateral conditions corresponding to (3) reads

$$\begin{aligned} &\int_\Omega [d\nabla u\cdot\nabla\phi - (b_{11}u + b_{12}v - \lambda u)\phi]\, dx = 0 \quad \text{for all} \quad \phi \in V, \\ &v \in K, \\ &\int_\Omega [\nabla v\cdot\nabla(\psi - v) - (b_{21}u + b_{22}v - \lambda v)(\psi - v)]\, dx \geq 0 \quad \text{for all} \quad \psi \in K. \end{aligned} \tag{8}$$

Denote by $E(d,\lambda)$ the set of all solution of (7), (BC) (for given d, $\lambda \in R$). Notice that $E(d_0,0) \neq \{0\}$ because d_0 is a bifurcation point of (RD), (BC). Further, we shall suppose that

> there exists a completely continuous operator $\beta : V \to V$ (a penalty operator) satisfying $< \beta u - \beta v, u - v > \geq 0$, $\beta(tu) = t\beta u$ for all $u, v \in V$, $t \geq 0$, $\beta u = 0$ for all $u \in K$, $< \beta v, v > > 0$ for all $v \notin K$, $< \beta v, u > < 0$ for all $v \notin K$, $u \in K^o$, (P)

where $<.,.>$ is the inner product in V, K^o and ∂K is the interior and the boundary of K. This assumption is fulfilled in examples. For the cone K mentioned in Section 1 we can consider the penalty operator defined by

$$<\beta v,\psi> = - \int_{\Gamma_0} v^- \psi\, dx \quad \text{for all} \quad v, \psi \in V,$$

where v^- denotes the negative part of v.

<u>THEOREM 1.</u> *Let* (P), (B) *hold and* $E(d_0,0) \cap V \times K^o \neq \emptyset$. *Then for any* $d \in (d_0,d_1)$ *(with some* $d_1 > d_0$*) there exists a solution of* (3) *of the type* $U(t) = \exp(\lambda_d^I t)\cdot U_d^I$ *with* $\lambda_d^I > 0$, $U_d^I \in \partial K$.

<u>THEOREM 2.</u> *Let* (B) *hold and* $E(d_0,0) \cap V \times K^o \neq \emptyset$, $\dim E(d_0,0) = 1$, meas $\Gamma_D > 0$. *Then there exists a bifurcation point* $d_I > d_0$ *of* (2) *at which spatially nonhomogeneous stationary solutions bifurcate from* $\{[d,0,0];\ d \in R\}$.

<u>Proof of Theorem 1</u> is based on a modification of the method developed in [5]. We shall explain main ideas only (more precisely see [3]

cf. also [2]). It is sufficient to show that for any $d \in (d_0,d_1)$ there exists a positive eigenvalue λ_d^I of (8) with the corresponding eigenvector $U_d^I = [u_d^I, v_d^I] \in V \times \partial K$. Suppose that $\dim E(d_0,0)=1$. (The general case can be reduced to this situation - see [3].) Choose a fixed $d > d_0$ and consider the system with the penalty

$$\begin{aligned} &\int_\Omega [d\nabla u\cdot\nabla\phi - (b_{11}u + b_{12}v - \lambda u)\phi]\, dx = 0 \quad \text{for all } \phi \in V, \\ &\int_\Omega [\nabla v\cdot\nabla\psi - (b_{21}u + b_{22}v - \lambda v)\psi]\, dx + \varepsilon\langle\beta v,\psi\rangle = 0 \text{ for all } \psi \in V. \end{aligned} \tag{9}$$

It is equivalent to (7), (BC) for $\varepsilon = 0$ and its eigenvalues and eigenvectors approximate those of (8) for $\varepsilon \to +\infty$. (The last assertion can be proved by standard penalty method technique.) We shall consider only solutions of (9) satisfying the norm condition

$$||U||^2 \; (= ||u||^2 + ||v||^2) = \frac{\varepsilon}{1+\varepsilon} . \tag{10}$$

Set $C_d = \overline{\{[\lambda,U,\varepsilon] \in R \times V \times V \times R;\ ||U|| \neq 0, (9), (10) \text{ is fulfilled}\}}$ (the closure in $R \times V \times V \times R$). The main idea is to show that the greatest eigenvalue λ_d of (7), (BC) can be joined with an eigenvalue λ_d^I of (8) by a connected (in $R \times V \times V \times R$) subset C_d^+ of C_d and to prove $\lambda_d^I > 0$ on the basis of the properties of this branch C_d^+ (for any $d \in (d_0,d_1)$ with some $d_1 > d_0$). The existence of a global continuum $C_d^+ \subset C_d$ of solution of (9), (10) starting at $[\lambda_d,0,0]$ in the direction $U_d = [u_d,v_d] \in E(d,\lambda_d) \cap V \times (-K^o)$ follows from a slight generalization of a Dancer's bifurcation result [1]. (Setting $x = [U,\varepsilon]$, (9), (10) can be written as the usual bifurcation equation $x - L(\mu)x + N(\mu,x) = 0$ in the space $X = V \times V \times R$ with compact linear operators $L(\mu)$ depending continuously on $\mu \in R$ and a small compact perturbation N; "starting in the direction U_d" means that for any $\delta > 0$ there is $[\lambda,U,\varepsilon] \in C_d^+$ with $||U/||U|| - U_d|| < \delta$ in any neighbourhood of $[\lambda_d,0,0]$.) An elementary investigation of solutions of (9), (10) yields that in a small neighbourhood of $[\lambda_d,0,0]$ can be only solutions $[\lambda,U,\varepsilon]$ of (9), (10) satisfying $\lambda > \lambda_d$ and that for all solutions of (9), (10) different from $[\lambda_d,0,0]$ the following implications hold: $\lambda > \lambda_d \Longrightarrow v \notin \partial K$; $v \notin K \Longrightarrow \lambda \neq \lambda_d$. This together with the fact that C_d^+ starts in the direction $U_d \notin V \times K$ and with the connectedness of C_d^+ implies $\lambda > \lambda_d$, $v \notin K$ for any $[\lambda,U,\varepsilon] \in C_d^+$, $U = [u,v]$. It follows that C_d^+ cannot intersect an analogous branch C_d^- of solution of (9), (10) starting in the direction $-U_d \in V \times K^o$. Dancer's result (see [1], Theorem 2) states that in this case C_d^+ is unbounded. It follows that there exists a sequence $\{[\lambda_n, U_n, \varepsilon_n]\} \subset C_d^+$ with $\varepsilon_n \to +\infty$. The penalty method technique gives $\lambda_n \longrightarrow \lambda_d^I$, $U_n \to U_d^I$, where λ_d^I and U_d^I is an eigenvalue and the corresponding eigenvector of (8), $U_d^I \in V \times \partial K$. If $\lambda_d^I > 0$ was not true for all $d \in (d_0, d_1)$ with some $d_1 > d_0$ then we would obtain $\lambda_{d_n}^I \to 0$ for some $d_n \to d_0+$ because $\lambda_{d_n} \to 0-$, $\lambda_d^I \geqq \lambda_d$. We could suppose $U_{d_n}^I \to U \in \partial K$ and this would contradict the assumptions $\dim E(d_0, 0) = 1$, $E(d_0,0) \cap V \times K^o \neq \emptyset$. (Any solution U of (8) with $d = d_0$, $\lambda = 0$ lies in $E(d_0, 0)$ under the last assumption.)

P r o o f of Theorem 2 is based on the same method as that of Theorem 1. The greatest bifurcation point d_0 of the stationary problem corresponding to (RD), (BC) can be joined (raughly speaking) with a bifurcation point $d_I > d_0$ of (2) by a branch of solutions of the corresponding penalty equation (with the variable d instead of λ). See [4].

REMARK 1. If $\operatorname{meas} \Gamma_D = 0$ in Theorem 2 (the case of Neumann conditions) then the bifurcation point d_I can coincide with infinity in a certain sense (see [4]).

REMARK 2. If $K = \{v \in V;\ v \geqq 0 \text{ on } \Gamma_0\}$ then $E(d_0,0) \cap V \times K^o \neq \emptyset$ holds if there exists $U = [u,v] \in E(d_0,0)$ with $v \geqq \delta$ on Γ_0 ($\delta > 0$).

3. Stabilization.

EXAMPLE 2. Consider the cone $K = \{u \in V;\ u \geqq 0 \text{ on } \Omega\}$ with $V = W_2^1$ again. Then spatially constant solutions of the linearization of (4) are solutions of (6) with $K_c = \{\psi = [\phi, \psi] \in R^2;\ \phi \geqq 0\}$. If B has a pair of complex eigenvalues and $b_{12}, b_{22} < 0$ then any solution of (6) (coinciding with the solution of $U_t = BU$ as long as it is in $K^o \times V$) touches the line $\{[0,v];\ v \geqq 0\}$ after some time t_0 and then coincides with the solution of the type $u(t) = 0$, $v(t) = \exp(b_{22}(t-t_0)) \cdot \xi$. It follows that the trivial solution of the linearization of (4) is stable with respect to spatially homogeneous perturbations even if the trivial solution of $U_t = BU$ is unstable (more precisely see [7]). Of course, in this way we cannot obtain any information about the stability with respect to nonhonogeneous perturbations.

THEOREM 3. *Let* (B) *hold and let* $E(d_0, 0) \cap K \times V = \{0\}$, $\operatorname{meas} \Gamma_D > 0$.

Then there is no bifurcation point of (4) *at which stationary spatially nonhomogeneous solutions bifurcate from* $\{[d,0,0];\ d \in R\}$ *in* $(d_0 - \delta, +\infty)$ *(with some* $\delta > 0$ *).*

Proof. Introduce the inner product $\langle .,. \rangle$ and the operator A in V by

$$\langle u,\phi \rangle = \int_\Omega \nabla u \nabla \phi \, dx , \quad \langle Au,\phi \rangle = \int_\Omega u \, \phi \, dx \quad \text{for all} \quad u, \phi \in V .$$

The linearization of the stationary problem corresponding to (4) can be written as

$$\begin{aligned} & u \in K , \\ & \langle du - b_{11}Au - b_{12}Av, \phi - u \rangle \geqq 0 \quad \text{for all} \quad \phi \in K , \\ & v - b_{21}Au - b_{22}Av = 0 . \end{aligned} \tag{11}$$

Calculating v from the second equation in (11) and substituting to the first inequality we obtain the inequality of the type

$$\begin{aligned} & u \in K , \\ & \langle du - Tu, \phi - u \rangle \geqq 0 \quad \text{for all} \quad \phi \in K \end{aligned} \tag{12}$$

with a compact linear symmetric operator T in V. It follows that any bifurcation point d_I of our unilateral stationary problem is simultaneously an eigenvalue of the inequality (12) and therefore $d_I \leqq \max_{||u||=1,\ u \in K} \langle Tu,u \rangle$. Further, the greatest bifurcation point d_0 of (RD), (BC) is simultaneously the greatest eigenvalue of T, i.e. $d_0 = \max_{||u||=1,\ u \in V} \langle Tu,u \rangle$ and any $u \in V$ realizing this maximum is an eigenvector of T corresponding to d_0. This together with the assumption $E(d_0) \cap K \times V = \{0\}$ implies $\max_{||u||=1,\ u \in K} \langle Tu,u \rangle < d_0$. For the details see [6] where a more general case is considered.

REMARK 3. Let $K = \{u \in V;\ u \geqq 0 \ \text{on} \ \Gamma_0\}$. Then $E(d_0,0) \cap K \times V = \{0\}$ is fulfilled if there exists $U = [u,v] \in E(d_0,0)$ such that u changes its sign on Γ_0.

4. Final remarks.

It is possible to consider also more general inequalities

$$\int_\Omega u_t(\phi - u) + d\nabla u\cdot\nabla(\phi - u) - f(u,v)(\phi - u)\, dx$$
$$+ \Phi_1(\phi) - \Phi_1(u) \geq 0 \ ,$$
$$\int_\Omega v_t(\psi - v) + \nabla v\cdot\nabla(\psi - v) - g(u,v)(\psi - v)\, dx \qquad (13)$$
$$+ \Phi_2(\psi) - \Phi_2(v) \geq 0 \quad \text{for all} \quad \phi,\ \psi \in V\ ,\ \text{a.a.}\quad t \geq 0\ ,$$

where Φ_1 , Φ_2 are convex proper functionals on V . More general unilateral conditions are included in this formulation. An analogy of Theorem 3 for (13) with $\Phi_2 = 0$ is contained in [6], a destabilizing effect of such unilateral conditions (for $\Phi_1 = 0$) will be the subject of a forthcomming paper.

R e f e r e n c e s

[1] DANCER, E. N.: *On the structure of solutions of non-linear eigenvalue problems*. Ind. Univ. Math. J. 23 (1974), 1069-1076.

[2] DRÁBEK, P. and KUČERA, M.: *Eigenvalues of inequalities of reaction -diffusion type and destabilizing effect of unilateral conditions*. 36(111), 1986, Czechoslovak Math. J. 36 (111), 1986, 116-130.

[3] DRÁBEK, P. and KUČERA, M.: *Reaction-diffusion systems: Destabilizing effect of unilateral conditions*. To appear.

[4] DRÁBEK, P., KUČERA, M. and MÍKOVÁ, M.: *Bifurcation points of reaction-diffusion systems with unilateral conditions*. Czechoslovak Math. J. 35 (110), 1985, 639-660.

[5] KUČERA, M.: *Bifurcation points of variational inequalities*. Czechoslovak Math. J. 32 (107), 1982, 208-226.

[6] KUČERA, M.: *Bifurcation points of inequalities of reaction-diffusion type*. To appear.

[7] KUČERA, M. and NEUSTUPA, J.: *Destabilizing effect of unilateral conditions in reaction-diffusion systems*. To appear in Comment. Math. Univ. Carol. 27 (1986), 171-187.

[8] MIMURA, M. and NISHIURA, Y.: *Spatial patterns for an interaction--diffusion equations in morphogenesis*. J. Math. Biology 7, 243-263, (1979).

BOUNDARY INTEGRAL EQUATIONS OF ELASTICITY IN DOMAINS WITH PIECEWISE SMOOTH BOUNDARIES

V. G. MAZ'YA
Leningrad University, Petrodvoretz. Math. Mech. Faculty
Bibliotechnaya pl. 2, Leningrad, USSR

0. Introduction

In the author's papers [1-3] a method for investigation of boundary integral equations arising in problems of mechanics of continuum in domains with piecewise smooth boundaries was proposed. Traditionally, equations of the potential theory are studied directly by methods of the Fredholm and singular integral operator theories. In the case of a non-smooth boundary this way leads to difficulties that have not been overcome until now. In a sense our approach is opposite to the traditional one. It is based on the well-known fact that solutions of integral equations can be expressed in terms of solutions of some exterior and interior boundary value problems. These are studied with the help of the theory developed in [4 - 8] and, as a result, theorems on solvability of equations of the potential theory are obtained. For these equations we can get differentiability properties and asymptotics of solutions near singularities of the boundary using the same approach [3, 9 - 11]. In [3] our method of construction of the potential theory was illustrated by the example of three boundary value problems of linear isotropic elasticity, namely, the first, the second and mixed, as well as of the same problems for the Laplace operator under the hypothesis that there exist a finite number of non-intersecting smooth edges on the boundary.

In the present lecture we study the first two boundary value problems for the Lamé system in domains with boundary singularities of the type of edges, conic points and polyhedral angles. New results on the harmonic potential theory are also reported.

1. Domains and function spaces

Let $G^{(i)}$ be a domain in R^3 with compact closure and with the boundary S, $Q^{(e)} = R^3 \setminus \overline{G^{(i)}}$. We suppose that S is the union of a finite number of "faces" $\{F\}$, "edges" $\{E\}$ and "vertices" $\{Q\}$ and that openings of all angles are non-zero. Confining ourselves only to the above visual description we refer the reader to [8] for the exact definition of the class of domains to be considered. In any case it contains domains with polyhedral boundaries. We place the origin into $G^{(i)}$.

Let $\{U\}$ be a sufficiently small finite covering of $\overline{G^{(i)}}$ by open sets satisfying: a) U contains not more than a single vertex Q, b) if $\overline{U}$ does not contain vertices then $\overline{U}$ intersects not more than a single edge E. With any vertex Q and any edge E we associate real numbers β_Q and γ_E, respectively.

By means of the partition of unity subordinate to the covering $\{U\}$ we define *the space* $C^{1,\alpha}_{\beta,\gamma}(G^{(i)})$ with $0 < \alpha < 1$, $\beta = \{\beta_Q\}$, $\gamma = \{\gamma_E\}$. If $\overline{U}$ intersects no singularities of S then the $C^{1,\alpha}_{\beta,\gamma}(G^{(i)})$-norm of a function with support in U is equivalent to the norm in the usual Hölder space $C^{1,\alpha}$. In the case $U \cap E \neq \emptyset$, $\overline{U} \cap \{Q\} = \emptyset$ and $\operatorname{supp} u \subset U$ we have

$$||u||_{C^{1,\alpha}_{\beta,\gamma}(G^{(i)})} \sim \sup_{x \in G^{(i)}} |u(x)| +$$

$$\sup_{x,y \in G^{(i)}} \frac{|r_E(x)^{\gamma_E} \nabla u(x) - r_E(y)^{\gamma_E} \nabla u(y)|}{|x-y|^\alpha},$$

where $r_E(x)$ is the distance from the point x to E. If U contains the vertex Q and $\operatorname{supp} u \subset U$ then

$$||u||_{C^{1,\alpha}_{\beta,\gamma}(G^{(i)})} \sim \sup_{x \in G^{(i)}} |u(x)| +$$

$$\sup_{x,y \in G^{(i)}} \frac{|\rho_Q(x)^{\beta_Q} \prod_{\{E:Q \in \overline{E}\}} r_E(x)^{\gamma_E} \nabla u(x) - \rho_Q(y)^{\beta_Q} \prod_{\{E:Q \in \overline{E}\}} r_E(x)^{\gamma_E} \nabla u(y)|}{|x-y|^\alpha},$$

where $\rho_Q(x) = |x - Q|$.

Replacing here $G^{(i)}$ by $G^{(e)}$ and $\sup_{x \in G^{(i)}} |u(x)|$ by $\sup_{x \in G^{(e)}} (1 + |x|)|u(x)|$ we obtain the definition of *the space* $C^{1,\alpha}_{\beta,\gamma}(G^{(e)})$. By $C^{1,\alpha}_{\beta,\gamma}(S)$ we denote the space of traces on $\cup F$ of functions in $C^{1,\alpha}_{\beta,\gamma}(G^{(i)})$ or $C^{1,\alpha}_{\beta,\gamma}(G^{(e)})$.

Let us introduce another *space* $C^{0,\alpha}_{\beta,\gamma}(S)$ of functions on $\cup F$. If $U \cap E \neq \emptyset$, $\overline{U} \cap \{Q\} = \emptyset$ and $\operatorname{supp} u \subset U$ then

$$||u||_{C^{0,\alpha}_{\beta,\gamma}(S)} \sim \sup_{x \in \cup F} r_E(x)^{\gamma_E - \alpha} |u(x)| +$$

$$+ \sup_{x,y \in \cup F} \frac{|r_E(x)^{\gamma_E} u(x) - r_E(y)^{\gamma_E} u(y)|}{|x-y|^{\alpha}} .$$

If $\mathcal{U}$ contains the vertex Q and $\operatorname{supp} u \subset \mathcal{U}$ then

$$||u||_{C^{0,\alpha}_{\beta,\gamma}(S)} \sim \sup_{x \in \cup F} \rho_Q(x)^{\beta_Q} \prod_{\{E:Q \in \bar{E}\}} r_E(x)^{\gamma_E} \sum_{\{E:Q \in \bar{E}\}} \frac{1}{r_E^{\alpha}} |u(x)| +$$

$$\sup_{x,y \in \cup F} \frac{|\rho_Q(x)^{\beta_Q} \prod_{\{E:Q \in \bar{E}\}} r_E(x)^{\gamma_E} u(x) - \rho_Q(y)^{\beta_Q} \prod_{\{E:Q \in \bar{E}\}} r_E(y)^{\gamma_E} u(y)|}{|x-y|^{\alpha}} .$$

If $\bar{\mathcal{U}}$ intersects no singularities of the boundary and $\operatorname{supp} u \subset \mathcal{U}$ then the norm in $C^{0,\alpha}_{\beta,\gamma}(S)$ is equivalent to the norm in $C^{0,\alpha}(S)$.

2. Boundary value problems of elasticity

We consider *interior* and *exterior Dirichlet problems:*

$$\mu\Delta u + (\lambda + \mu) \nabla \operatorname{div} u = 0 \quad \text{in } G^{(i)} , \quad u = g \quad \text{on } \cup F ; \qquad (D^{(i)})$$

$$\left.\begin{array}{l} \mu\Delta u + (\lambda + \mu) \nabla \operatorname{div} u = 0 \quad \text{in } G^{(e)} , \quad u = g \quad \text{on } \cup F , \\ u(x) = O\big((1 + |x|)^{-1}\big) \end{array}\right\} (D^{(e)})$$

and *interior* and *exterior Neumann problems:*

$$\mu\Delta u + (\lambda + \mu) \nabla \operatorname{div} u = 0 \quad \text{in } G^{(i)} , \quad Tu = h \quad \text{on } \cup F , \qquad (N^{(i)})$$

$$\left.\begin{array}{l} \mu\Delta u + (\lambda + \mu) \nabla \operatorname{div} u = 0 \quad \text{in } G^{(e)} , \quad Tu = h \quad \text{on } \cup F , \\ u(x) = O\big((1 + |x|)^{-1}\big) , \end{array}\right\} (N^{(e)})$$

where $T = T(\partial_x, n)$ is the matrix with elements $T_{kj}(\partial_x, n) = \mu\delta_k^j \, \partial/\partial n + \lambda n_k \, \partial/\partial x_j + \mu n_j \, \partial/\partial x_k$ and $n = (n_1, n_2, n_3)$ is an outer normal to $G^{(i)}$.

We introduce a collection of reals $\{\delta_E\}$ which appears in the statement of the next lemma and will be used in the sequel. Let $\phi(z) \in (0, 2\pi)$ be the angle between the tangent half-planes to S at the point $z \in E$ from the side of $G^{(i)}$. We put $\omega_E = \inf_{z \in E} (\pi + |\pi - \phi(z)|)$. Let δ_E be the root of the equation $\sin(\omega_E \delta) + \delta \sin \omega_E = 0$ with the least positive real part; δ_E is real and decreases as ω_E increases, $1/2 < \delta_E < 1$.

We formulate a theorem on solvability of all above mentioned boundary value problems, which is sufficient to justify the boundary integral equations method.

THEOREM 1. *Let* $\{\gamma_E\}$ *and* $\{\beta_Q\}$ *satisfy*

$$0 < 1 - \gamma_E + \alpha < \delta_E \quad \textit{for all} \quad E \,, \tag{1}$$

$$|\beta_Q + \sum_{\{E: Q \in \bar{E}\}} \gamma_E - \alpha - 3/2| < \varepsilon_Q \quad \textit{for all} \quad Q \,, \tag{2}$$

where $\{\varepsilon_Q\}$ *is a collection of positive numbers in* $(0,1)$ *which depend on the tangent cone to* S *with the vertex* Q *).

Then (i) $(D^{(i)})$ *and* $(D^{(e)})$ *are uniquely solvable in* $C^{1,\alpha}_{\beta,\gamma}(G^{(i)})$ *and* $C^{1,\alpha}_{\beta,\gamma}(G^{(e)})$ *for all* $g \in C^{1,\alpha}_{\beta,\gamma}(S)$; (ii) $(N^{(e)})$ *is uniquely solvable in* $C^{1,\alpha}_{\beta,\gamma}(G^{(e)})$ *for all* $h \in C^{0,\alpha}_{\beta,\gamma}(S)$; (iii) $(N^{(i)})$ *is solvable in* $C^{1,\alpha}_{\beta,\gamma}(G^{(i)})$ *for all* $h \in C^{0,\alpha}_{\beta,\gamma}(S)$ *with zero principal vector and principal moment. Its solution is unique up to a rigid displacement.*

For $(D^{(i)})$ and $(D^{(e)})$ this result is contained in Theorem 11.5 [8]. The proof for $(N^{(i)})$ and $(N^{(e)})$ requires minor technical modifications.

3. Solution of $(D^{(i)})$ and $(D^{(e)})$ by a simple layer potential

Let $V\sigma$ be *the elastic simple layer potential* with the density σ .

THEOREM 2. *Let* $\{\gamma_E\}$ *and* $\{\beta_Q\}$ *satisfy* (1) *and* (2). *Then the operators* $C^{0,\alpha}_{\beta,\gamma}(S) \ni \sigma \to (V\sigma)|_{G^{(e)}} \in C^{1,\alpha}_{\beta,\gamma}(G^{(e)})$ *and* $C^{0,\alpha}_{\beta,\gamma}(S) \ni \sigma \to (V\sigma)|_{G^{(i)}} \in C^{1,\alpha}_{\beta,\gamma}(G^{(i)})$ *are bounded. There exists a bounded inverse* V^{-1} : $C^{1,\alpha}_{\beta,\gamma}(S) \to C^{0,\alpha}_{\beta,\gamma}(S)$.

The first part of the theorem can be checked directly. The second part follows from Theorem 2 and the identity $2\sigma = Tu^{(i)} - Tu^{(e)}$ where $u^{(i)}$ and $u^{(e)}$ are restrictions of $V\sigma$ to $G^{(i)}$ and $G^{(e)}$, respectively.

4. Solution of $(D^{(i)})$, $(D^{(e)})$, $(N^{(i)})$, $(N^{(e)})$ by a double layer potential

Let $W\tau$ be *the double layer potential* with the density τ , i.e.

$$(W\tau)(x) = \frac{1}{2\pi} \int_S \big(T(\partial_y, n_y)\Gamma(x,y)\big)^* \tau(y)\, ds_y \,,$$

---

*) The numbers ε_Q can be defined by some spectral boundary value problems in spherical domains (cf. [8]) but we shall not use it in what follows. For the problems $(D^{(i)})$ and $(D^{(e)})$ it was shown in [12] that $\varepsilon_Q > 1/2$. The validity of the last inequality for $(N^{(i)})$, $(N^{(e)})$ remains an open question.

where $*$ denotes the adjoint operator and Γ is the Kelvin-Somigliana tensor.

If $u^{(i)} = W\tau$ then τ satisfies the system of singular integral equations

$$(-1 + W)\tau = g \quad \text{on} \quad \bigcup F . \tag{3}$$

The solution of $(D^{(e)})$ can be expressed as the sum $(W\tau)(x) + \Gamma(x,0)a + \mathrm{rot}\,\Gamma(x,0)b$ where a , b are unknown constant vectors. Then the triplet (τ,a,b) satisfies the system

$$(1 + W)\tau + \Gamma(\cdot,0)a + \mathrm{rot}\,\Gamma(\cdot,0)b = g \tag{4}$$

Representing the solutions of $(N^{(i)})$ and $(N^{(e)})$ as $W\tau$ we arrive at the systems

$$(1 + W^*)\tau = h , \tag{5}$$

$$(-1 + W^*)\tau = h . \tag{6}$$

THEOREM 3. *Let* $\{\gamma_E\}$ *and* $\{\beta_Q\}$ *satisfy* (1) *and* (2). *Then* (i) *the operators* W *and* W^* *are bounded in* $C^{1,\alpha}_{\beta,\gamma}(S)$ *and* $C^{0,\alpha}_{\beta}(S)$, (ii) *if* $g \in C^{1,\alpha}_{\beta,\gamma}(S)$ *then systems* (3) *and* (4) *are uniquely solvable in* $C^{1,\alpha}_{\beta,\gamma}(S)$ *and* $C^{1,\alpha}_{\beta}(S) \times R^3 \times R^3$, *respectively.*

Let us describe a scheme of the proof of the solvability confining ourselves to system (3). Let $u^{(i)}$ and $u^{(e)}$ be the solutions of $(D^{(i)})$ and $(D^{(e)})$. Then $2g = V(Tu^{(i)} - Tu^{(e)})$. We introduce the solution v of the problem $(N^{(e)})$ where $h = 1/2\,(Tu^{(i)} - Tu^{(e)})$. Since $v = 1/2(Wv - VTv)$ on $G^{(e)}$, then $(-1 + W)v = VTv = g$ on $\bigcup F$. Consequently, the vector-function $\tau = v|_{\bigcup F}$ is a solution of (3). The inclusion $\tau \in C^{1,\alpha}_{\beta,\gamma}(S)$ follows directly from Theorem 1.

To prove the unicity of the solution of (3) it suffices to establish the solvability in $C^{0,\alpha}_{\beta,\gamma}(S)$ of the formally adjoint system (6), where $h \in C^{0,\alpha}_{\beta,\gamma}(S)$. Let $v^{(e)}$ be a solution of $(N^{(e)})$. We consider the simple layer potential $V\sigma$ which coincides with $v^{(e)}$ on $G^{(e)}$ (see Theorem 2). It remains to note that the density σ satisfies (6).

The above argument contains all essential points for the proof of the following theorem on solvability of systems (5) and (6).

THEOREM 4. *There exists a bounded inverse:* $(-1 + W^*)^{-1}$ *in* $C^{0,\alpha}_{\beta,\gamma}(S)$. *System* (4) *is solvable in* $C^{0,\alpha}_{\beta,\gamma}(S)$ *for all* h *with zero principal vector and principal moment.*

For solution of systems of integral equations under consideration theorems on increasing smoothness and changing collections $\{\gamma_E\}$ and

$\{\beta_Q\}$ hold which are similar to the theorems of the same kind for solutions of boundary value problems (cf. [8]). Before giving an example let us introduce some function spaces.

Let ℓ be an integer, $\ell \geq 1$, $0 < \alpha < 1$. If in the definition of $C^{1,\alpha}_{\beta,\gamma}(G^{(i)})$ we replace ∇ by $\nabla_\ell = \{\partial^\ell/\partial x_1^{\alpha_1}\partial x_2^{\alpha_2}\partial x_3^{\alpha_3}\}$ then we obtain the definition of *the space* $C^{\ell,\alpha}_{\beta,\gamma}(G^{(i)})$. By $C^{\ell,\alpha}_{\beta,\gamma}(S)$ we mean the space of traces on S of functions in $C^{\ell,\alpha}_{\beta,\gamma}(G^{(i)})$. Here we can replace (i) by (e).

The following assertion which completes Theorem 3 holds.

THEOREM 5. *Let* $\{\gamma_E\}$, $\{\beta_Q\}$, $\{\gamma^*_E - \ell + 1\}$ *and* $\{\beta^*_Q - \ell + 1\}$ *satisfy* (1) *and* (2). *If* $g \in C^{1,\alpha}_{\beta,\gamma}(S) \cap C^{\ell,\alpha}_{\beta^*,\gamma^*}(S)$ *and* $\tau \in C^{1,\alpha}_{\beta,\gamma}(S)$ *is a solution of any system* (3) *and* (4) *then* $\tau \in C^{\ell,\alpha}_{\beta^*,\gamma^*}(S)$.

Similar complements can be made to Theorems 2 and 4.

The potential theory for plane boundary value problems can be developed with the help of the same arguments and even more easily. If by S we mean a piecewise smooth contour without cusps and by $\{Q\}$ we denote the set of its angular points then the statements of Theorems 1 - 5 remain valid up to obvious changes.

For harmonic and hydrodinamic potentials results similar to Theorems 1 - 5 can be obtained by the same method.

5. The Fredholm radius of harmonic potentials

Here we present a formula for the Fredholm radii of the double layer harmonic potential W and its formal adjoint W^* in certain Hölder spaces. We shall suppose that S contains no vertices. So the collection $\{\beta_Q\}$ in the notation of the function spaces is omitted.

Let L be a linear operator in a Banach space. The Fredholm radius $r(L)$ of L is the largest radius of a circle $C_r = \{\lambda \in \mathbb{C} : |\lambda| < r\}$ such that for all $\lambda \in C_r$ the operator $1 - \lambda L$ is Fredholm and $\mathrm{Ind}\,(1 - \lambda L) = 0$.

We introduce the operators

$$(W\tau)(x) = \frac{1}{2\pi}\int_S \frac{\cos(x - y,\, n(y))}{|x - y|^2}\,\tau(y)\,ds(y)\,, \tag{7}$$

$$(W^*\tau)(x) = -\frac{1}{2\pi}\int_S \frac{\cos(x - y,\, n(x))}{|x - y|^2}\,\tau(y)\,ds(y)\,, \tag{8}$$

where $x \in \cup F$.

The following result is proved in [9]:

THEOREM 6. *Let* $\alpha \in (0,1)$, $0 < 1 + \alpha - \beta < 1$, $\ell \geq 0$ *and*

$$R = \min_{z \in \cup E} \left| \frac{\sin \pi(1 + \alpha - \beta)}{\sin\left((\pi - \phi(z))(1 + \alpha - \beta)\right)} \right| .$$

Let W *and* W* *be operators in* $C^{\ell,\alpha}_{\beta+\ell}(S)$ *and* $C^{\ell+1,\alpha}_{\beta+\ell}(S)$, *respectively, defined by* (7) *and* (8). *Then* $r(W) = r(W^*) = R$.

We note that $R > 1$ if and only if

$$1 + \alpha - \beta < \frac{\pi}{\pi + |\pi - \phi(z)|} \quad \text{for all } z \in \cup E .$$

The proof of Theorem 6 relies heavily on Theorem 7 where the following notation is used.

Let $G = G^{(i)} \cup G^{(e)}$ and let $u^{(i)}$ and $u^{(e)}$ be restrictions of the function u to $G^{(i)}$ and $G^{(e)}$. By $C^{\ell,\alpha}_{\beta}(G)$ we denote the space of functions u with $u^{(i)} \in C^{\ell,\alpha}_{\beta}(G^{(i)})$, $u^{(e)} \in C^{\ell,\alpha}_{\beta}(G^{(e)})$. The norm in $C^{\ell,\alpha}_{\beta}(G)$ is defined as the sum of the norms of $u^{(i)}$ and $u^{(e)}$ in $C^{\ell,\alpha}_{\beta}(G^{(i)})$ and $C^{\ell,\alpha}_{\beta}(G^{(e)})$. For $\ell \geq 1$, $0 < \beta < \ell + \alpha$ we put

$$\hat{C}^{\ell,\alpha}_{\beta}(G) = \{u \in C^{\ell,\alpha}_{\beta}(G) : \Delta u \text{ in } G ,\ u^{(i)} = u^{(e)} \text{ on } S\}$$

$$\check{C}^{\ell,\alpha}_{\beta}(G) = \{u \in C^{\ell,\alpha}_{\beta}(G) : \Delta u = 0 \text{ in } G ,\ \frac{\partial u^{(i)}}{\partial n} = \frac{\partial u^{(e)}}{\partial n} \text{ on } \cup F\} .$$

Clearly, for $\ell \geq 0$ the operators

$$L_1 : \hat{C}^{\ell+1,\alpha}_{\beta}(G) \ni u \to (1 - \lambda)\frac{\partial u^{(i)}}{\partial n} - (1 + \lambda)\frac{\partial u^{(e)}}{\partial n} \in C^{\ell,\alpha}_{\beta}(S) ,$$

$$L_2 : \check{C}^{\ell+1,\alpha}_{\beta}(G) \ni u \to (1 - \lambda)u^{(i)} - (1 + \lambda)u^{(e)} \in C^{\ell+1,\alpha}_{\beta}(S)$$

are bounded.

THEOREM 7. *Let* $\alpha \in (0,1)$, $0 < 1 + \alpha - \beta < 1$, $\ell \geq 0$.

1) *If* $|\lambda| < R$ *then* L_k , $k = 1,2$, *is Fredholm and* $\operatorname{ind}(L_k) = 0$.

2) *If* $|\lambda| = R$ *then the range of* L_k *is not closed.*

R e f e r e n c e s

[1] MAZ'YA, V. G.: *Integral equations of the potential theory in domains with piecewise smooth boundaries.* Uspehi Matem.Nauk, v.36, n 4, 1981 p. 229-230.

[2] MAZ'YA, V. G.: *On solvability of integral equations in the classical*

theory of elasticity in domains with piecewise smooth boundaries. Republican School-Conf. General Mechanics and Elasticity Theory (Telavi, 1981), Abstracts of Reports, Tbilisi, 1981, pp. 55-56.

[3] MAZ'YA, V. G.: *The potential theory for the Lamè system in domains with piecewise smooth boundaries.* Proc. of the Conference on the Partial Diff. Equations in Memoriam of I. N. Vekua, Tbilisi 1982 (to appear).

[4] MAZ'YA, V. G. and PLAMENEVSKIĬ, B. A.: L_p*-estimates of solutions of elliptic boundary value problems in domains with edges.* Trudy Moskov. Mat. Obshch. 37 (1978), 49-93; English transl. in Trans. Moscow Math. Soc. 1980, no. 1 (37).

[5] ——————— : *Estimates of the Green functions and Schauder estimates of the solutions of elliptic boundary value problems in a dihedral angle.* Sibirsk. Mat. Zh. 19 (1978), 49-93; English transl. in Siberian Math. J. 19 (1978).

[6] ——————— : *Schauder estimates of solutions of elliptic boundary value problems in domains with edges on the boundary.* Partial Differential Equations (Proc. Sem. S. L. Sobolev, 1978, no. 2), Inst. Mat. Sibirsk. Otdel. Akad. Nauk SSSR, Novosibirsk, 1978, pp. 69-102; English transl. in Amer. Math. Soc. Transl. (2) 123 (1984).

[7] ——————— : *Elliptic boundary value problems on manifolds with singularities.* Problemy Mat. Anal., vyp. 6, Izdat. Leningrad. Univ., Leningrad, 1977, pp. 85-142; English transl., to appear in J. Soviet Math.

[8] ——————— : *The first boundary value problem for the classical equations of mathematical physics in domains with piecewise smooth boundaries* (I) Zeitschrift für Analysis und ihre Anwendungen, Bd. 2 (4) 1983, S. 335-359; (II) ibid. Bd. 2 (6) 1983, S.523-551.

[9] GRAČEV, N. V. and MAZ'YA, V. G.: *On the Fredholm radius for operators of the double layer potential type on piecewise smooth boundaries.* Vestn. Leningrad Univ. Math. (to appear).

[10] ZARGARYAN, S. S. and MAZ'YA, V. G.: *On singularities of solutions of a system of integral equations in potential theory for the Zaremba problem.* Vestn. Leningrad. Univ. Math. 1983, n 1. English transl.vol. 16 (1984), p. 49-55.

[11] ZARGARYAN, S. S. and MAZ'YA, V. G.: *On the asymptotic behavior of solutions of integral equations in potential theory in a neighborhood of corner points of the contour.* Prikl. Mat. Mekh. vol. 48, n 1, 1984; English transl. in J. Appl. Math. Mech. 48 (1984).

[12] MAZ'YA, V. G. and PLAMENEVSKIĬ, B. A.: *On properties of solutions of three-dimensional problems of elasticity theory and hydrodynamics in domains with isolated singular points.* Dinamika Sploshnoj Sredy ,Novosibirsk, n 50 (1981), p. 99-120 English transl. in Amer. Math. Soc. Transl. (2), vol. 123, 1984, p. 109-123.

HIGHER REGULARITY OF WEAK SOLUTIONS OF STRONGLY NONLINEAR ELLIPTIC EQUATIONS

C. G. SIMADER
Mathematisches Institut der Universität Bayreuth
Postfach 3008, D-8580 Bayreuth, West Germany

In a bounded open set $G \subset \mathbb{R}^N$ we consider the Dirichlet problem

$$Lu := -\Delta u - \sum_{i=1}^{N} \partial_i g_i(\partial_i u) + g_0(u) = f\,, \qquad u\Big|_{\partial G} = 0\,. \tag{1}$$

If f,g are measurable functions on G, we write $(f,g) := \int_G f\cdot g$ if $f\cdot g \in L^1(G)$. As usual, a weak solution of (1) is a function

$$\text{(H.1)}\quad \begin{cases} u \in H_0^{1,2}(G) \text{ such that } g_i(\partial_i u) \in L^1(G),\ i = 1,\dots,N \\ (\partial_0 u := u) \text{ and} \\ \sum_i (\partial_i u, \partial_i \phi) + \sum_{i=1}^{N} (g_i(\partial_i u), \partial_i \phi) + (g_0(u), \phi) = (f,\phi) \qquad (2) \\ \text{holds for all } \phi \in C_0^\infty(G). \end{cases}$$

For the strong nonlinearities g_i we assume

$$\text{(H.2)}\quad \begin{cases} g_i \in C^0(\mathbb{R}) \text{ are non-decreasing and} \\ g_i(t)\cdot t \geq 0 \text{ for } t \in \mathbb{R},\ i = 0,1,\dots N\,. \end{cases}$$

Existence of those weak solutions was studied by a considerable number of authors. Observe that the nonlinearities depend on derivatives up to half the order of the equation. Existence for those problems was first proved in [3]. Let us emphasize that all our considerations hold true if $-\Delta$ is replaced by a strongly elliptic operator of order 2m and the nonlinearities may depend analogously to (1) up to the derivatives of order m. All existence-proves lead to weak solutions such that in addition to (H.1) we get

$$\text{(H.3)}\qquad g_i(\partial_i u)\cdot\partial_i u \in L^1(G) \qquad (i = 0,1,\dots N)$$

Like as in the case of linear equations two questions arise: Under what conditions are the weak solutions unique? Do the weak solutions have better regularity properties? For star-shaped domains uniqueness and stability of weak solutions of (1) was proved in [4]. This result was considerable generalized to arbitrary domains with smooth boundary and to very general operators by M. Landes [2].

Concerning higher regularity properties, surprisingly it turns out that the meanwhile classical difference quotient method perfectly works in the underlying case to gain one more order of differentiability. As far as the author knows, this method goes back to S.Agmon (see e.g. [1]). For the nonlinearities we assume

(H.4)

Assume (H.2) and $g_i \in C^1(R)$, $i = 0,1,\dots,N$.

For $i = 1,\dots,N$ we assume

i) g_i' is non-decreasing in $[0,\infty)$

ii) There exists a constant $C \geq 1$ such that

$g_i'(-t) \leq Cg_i'(t)$ for $t \in R$ ("nearly odd")

iii) Let $G_i(t) := \int_0^t g_i(s)ds$. Assume that G_i satisfies a Δ_2-condition: There exists a $K > 0$ such that

$G_i(2t) \leq KG_i(t)$ for $t \in R$.

iv) There exists $\gamma > 0$ such that

$|g_i(t)| \leq \gamma |g_k(t)|$ for $t \in R$ and $i,k = 1,\dots,N$

("isotropic")

An example is given by

$$g_i(t) := \alpha_i t|t|^{p-1}, \quad \alpha_i \in R, \; P > 1 \tag{3}$$

Then we can prove the following

<u>Theorem.</u> Assume (H.1)-(H.4). Then, for $G' \subset\subset G$ we have $u\big|_{G'} \in W^{2,2}(G')$, $g_i'(\partial_i u)\cdot(\partial_i\partial_k u)^2\big|_{G'} \in L^1(G')$,

$$g_0'(u)\cdot(\partial_k u)^2\big|_{G'} \in L^1(G') \quad (i,k = 1,\dots,N)$$

and there is a constant $K = K(G',G,g_i)$ such that

$$\sum_{i,k=1}^{N} \int_{G'} (\partial_i\partial_k u)^2 + \sum_{i,k=1}^{N} \int_{G'} g_i'(\partial_i u)\cdot(\partial_i\partial_k u)^2 +$$

$$+ \sum_{k=1}^{N} \int_{G'} g_0'(u)(\partial_k u)^2 \leq K\cdot\Big(\|f\|^2_{L^2(G)} + \|u\|^2_{H_0^{1,2}(G)}$$

$$+ \sum_{i=0}^{N} \int_G g_i(\partial_i u)\cdot\partial_i u \Big)$$

As mentioned above, the proof is done by means of the difference quotient method. We can not give details here. In any case, the proof is completely elementary although it demands are careful analysis. To see what is going on, we assume now that the weak solution is arbitrarily smooth and sketch how to get the a-priori-estimate of the theorem. Roughly spoken, one can prove with this method all those regularity properties which can be read of an a-priori-estimate like as in the theorem.

For this purpose, let $\phi \in C_0^\infty(G)$ and for $k = 1,\dots,N$ put $\partial_k\phi$ as a test function in (2) and integrate at the left hand side by parts, which leads to

$$\sum_{i=1}^{N} (\partial_i\partial_k u, \partial_i\phi) + \sum_{i=1}^{N} (g_i'(\partial_i u)\partial_i\partial_k u, \partial_i\phi) + (g_0'(u)\partial_k u, \phi) = \tag{4}$$

$$= -(f, \partial_k\phi) \ .$$

Let now $\zeta \in C_0^\infty(G)$ such that $\zeta \equiv 1$ in G'. Put $\phi := \partial_k u\cdot\zeta^2$ in (4) which gives

$$\sum_{i=1}^{N} \int(\partial_i\partial_k u)^2\cdot\zeta^2 + \sum_{i=1}^{N} \int\partial_i\partial_k u\partial_k u 2\zeta\cdot\partial_i\zeta +$$

$$+ \sum_{i=1}^{N} \int g_i'(\partial_i u)\cdot(\partial_i\partial_k u)^2\cdot\zeta^2 + \sum_{i=1}^{N} \int g_i'(\partial_i u)\partial_i\partial_k u\partial_k u 2\zeta\partial_i\zeta + \tag{5}$$

$$+ \int g_0'(u)(\partial_k u)^2\cdot\zeta^2 = -(f, \partial_k\partial_k u\zeta^2) - (f, \partial_k u 2\zeta\partial_k\zeta)$$

The first, third and fifth expression at the left of (5) are that we have to estimate. E.g. the second admits trivally the estimate for $\varepsilon > 0$

$$\Big|2\sum_{i=1}^{N} \int(\partial_i\partial_k u\zeta)\cdot(\partial_k u)\cdot\partial_i\zeta\Big| \leq$$

$$\leq \varepsilon\cdot\sum_{i=1}^{N} \int(\partial_i\partial_k u)^2\zeta^2 + \varepsilon^{-1}\cdot\sum_{i=1}^{N} \int(\partial_k u)^2(\partial_i\zeta)^2$$

analogous for the right hand side of (5). Cumbersome seems the fourth term. To estimate it, observe $g_i'(t)\geq 0$. For $\delta > 0$ we get:

$$\left|2\sum_{i=1}^{N}\int\left(\sqrt{g_i'(\partial_i u)}\cdot\partial_i\partial_k u\zeta\right)\left(\sqrt{g_i'(\partial_i u)}\,\partial_k u\,\partial_i\zeta\right)\right| \le \delta\cdot\sum_{i=1}^{N}\int g_i'(\partial_i u)\cdot(\partial_i\partial_k u)^2\cdot\zeta^2+\delta^{-1}\cdot\int g_i'(\partial_i u)\cdot(\partial_k u)^2\cdot(\partial_i\zeta)^2 \quad (6)$$

To estimate the second expression on the right hand side of (6), we make use of the following

Lemma: Assume (H.4). Then

i) There is a constant $C > 0$ such that
$G_i(t) \leq C\cdot G_k(t)$ for $t\in\mathbb{R}$ and $i,k = 1,\ldots,N$

ii) There is a constant $C' > 0$ such that
$g_i'(t)s^2 \leq C'\cdot(g_i(t)\cdot t+G_i(s))$

Remark: Property ii) is no surprise if we consider e.g. $g(t):=t|t|^{p-1}$, $p > 1$. Then, $g'(t) = p\cdot|t|^{p-1}$, $g(t)\cdot t = |t|^{p+1}$ and $G(t) = (p+1)^{-1}\cdot|t|^{p+1}$. By the inequality $a\cdot b \leq \lambda^{-1}\cdot a^{\lambda}+\lambda'^{-1}\cdot b^{\lambda'}$ for $1 < \lambda,\lambda'<\infty$, $\lambda^{-1}+\lambda'^{-1} = 1$, we get with $\lambda:= \frac{p+1}{p-1}$ and therefore $\lambda' = \frac{p+1}{2}$

$$g'(t)\cdot s^2 \leq p\cdot\frac{p-1}{p+1}\cdot g(t)\cdot t + \frac{2p}{p+1}\cdot G(s)$$

from which ii) follows. The assumptions in (H.4) (especially iii)) guarantee Lemma 2, property ii) in general.

By means of the Lemma we are now able to estimate the second expression at the right hand side of (6), first pointwise:

$$g_i'(\partial_i u)\cdot(\partial_k u)^2 \leq C'(g_i(\partial_i u)\cdot\partial_i u+G_i(\partial_k u))$$

$$\leq C'(g_i(\partial_i u)\cdot\partial_i u + C\cdot G_k(\partial_k u)).$$

If we observe (H.2) and the definition of G_k, we conclude $G_k(\partial_k u) \leq g_k(\partial_k u)\,\partial_k u$. Multiplying by $(\partial_i\zeta)^2$, integrating and combining all inequalities, we have proved the desired estimate.

References

[1] AGMON,S., *Lectures on elliptic boundary value problems,* Van Nostrand, Princeton 1965.
[2] LANDES,M., *Eindeutigkeit und Stabilität schwacher Lösungen streng nichtlinearer elliptischer Randwertprobleme,* Thesis, Bayreuth 1983.
[3] SIMADER,C.G., *Über schwache Lösungen des Dirichletproblems für streng nichlineare elliptische Differentialgleichungen,* Math. Z. 150 (1976), 1-26.
[4] SIMADER,C.G., *Remarks on uniqueness and stability of weak solutions of strongly nonlinear elliptic equations,* Bayreuther Math. Schriften 11 (1982), 67-79.

SOME REGULARITY RESULTS FOR QUASILINEAR PARABOLIC SYSTEMS

J. STARÁ, O. JOHN
Faculty of Mathematics and Physics, Charles University
Sokolovská 83, 186 00 Prague 8, Czechoslovakia

We present three results on the regularity of the weak solutions of quasilinear parabolic systems of second order. Two of them concern the relation between the regularity of the system and the Liouville property. The third one is the example of the system for which there exists a weak solution of the boundary value problem in the cylinder Q with Lipschitz continuous boundary data which develops the singularity in the interior of Q.

Denote $z = [t,x] \in R \times R^n$. Let us consider the system of m equations for m unknown functions $u = [u^1,\dots,u^m]$ of the form

$$u_t^i - D_\alpha(A_{\alpha\beta}^{ij}(z,u)D_\beta u^j) = -f^i + D_\alpha g_\alpha^i, \quad i = 1,\dots,m,$$

where we sum over j from 1 to m and over α, β from 1 to n. For the sake of brevity we rewrite it at the form

$$(1) \quad u_t - \operatorname{div}(A(z,u)Du) = -f + \operatorname{div} g.$$

1. Interior regularity. Let Q be a domain in $R \times R^n$. Suppose that the following assertions on A, f and g are satisfied:

(i) $A = A(z,u)$ is continuous on $Q \times R^m$.

(ii) $(A(z,u)\xi,\xi) > 0$ for all $[z,u] \in Q \times R^m$, $\xi \neq 0$.

(iii) $f \in L_{s,loc}(Q)$ with some $s > n/2 + 1$, $g \in L_{r,loc}(Q)$ with $r > n + 1$.

Denote by $W_{2,loc}^{0,1}(Q)$ the set of all functions belonging to the $L_{2,loc}(Q)$ together with their spatial derivatives. Recall that the function $u \in W_{2,loc}^{0,1}(Q)$ is a *weak solution of (1) in* Q if for all $\varphi \in \mathcal{D}(Q)$

$$\int_Q [(u,\varphi_t) - (ADu,D\varphi)]\,dz = \int_Q [(f,\varphi) + (g,D\varphi)]\,dz .$$

Definition 1. The system (1) is said to be *regular* if every bounded weak solution of (1) is locally Hölder contiunuous in the domain Q.

Definition 2. The system (1) has *the interior Liouville property in* Q if for each $z_0 \in Q$ every bounded weak solution of the system

$w_t - \operatorname{div}(A(z_0,w)Dw) = 0$ in all $R \times R^n$ is constant.

Theorem 1. The system (1) is regular iff it has interior Liouville property in Q.

Sketch of the proof. To have the regularity of the weak solution u in Q, it is sufficient to prove that for each $z_0 \in Q$

$$\liminf_{R \to 0+} [R^{-n-2} \int_{Q(z_0,R)} |u(z) - u_{z_0,R}|^2 dz] = 0 ,$$

where $Q(z_0,R) = (t_0-R^2,t_0) \times \{x; |x - x_0| < R\}$ and $u_{z_0,R}$ is an integral mean value of u over $Q(z_0,R)$.

Using the blowing-up technique, we obtain this form the interior Liouville property. (For the details see [1], [2].)

Remark. (The elliptic case.) J. Daněček from Brno proved in his disertation that Theorem 1 (modified for elliptic systems) remains to be true if we change the request of boundedness of the weak solution (in both the definitions 1 and 2) by the assumption that the weak solutions in question belong to the space BMO. Further, he described the nontrivial class of the systems which satisfy BMO-interior Liouville property.

2. Regularity of the Cauchy problem. Let Ω be a domain in R^n and $T > 0$. Denote $Q^+ = (0,T) \times \Omega$, $\Gamma = \{[0,x];\ x \in \Omega\}$ and $Q = Q^+ \cup \Gamma$. To the conditions (i) - (iii) we add the assumption concerning the initial function u_0, namely:

(iv) $u_0 \in W^1_{q,loc}(\Omega) \cap L_\infty(\Omega)$, $q > n$.

The function $u \in W^{0,1}_{2,loc}(Q)$ is a *weak solution of Cauchy problem for the system* (1) *in* Q *with initial function* u_0 if for each $\varphi \in C^\infty(\overline{Q})$ with supp $\varphi \subset Q$

$$\int_Q [(u,\varphi_t) - (ADu,D\varphi)]\,dz = \int_Q [(f,\varphi) + (g,D\varphi)]\,dz - \int_\Omega u_0\varphi(0,x) .$$

Definition 3. Cauchy problem is *regular* if its each bounded solution is locally Hölder continuous on Q.

Definition 4. The system (1) has *boundary Liouville property on* Γ if for each $z_0 \in \Gamma$ every bounded weak solution of Cauchy problem for the system $w_t - \operatorname{div}(A(z_0,w)Dw) = 0$ in the set $\{[t,x];\ t \geq 0,\ x \in R^n\}$ with initial function $u_0 \equiv 0$ is equal zero identically.

Theorem 2. Cauchy problem for (1) in Q with initial function u_0 is regular iff the system (1) has interior Liouville property in Q^+ and boundary Liouville property on Γ.

Sketch of the proof. We extend the coefficients and the right hand side function of (1) to the cylinder $G = Q \cup (-Q^+)$. The weak solution of the Cauchy problem for (1) in Q can be shifted and prolonged in a suitable manner to the weak solution of the extended system on the whole G. Using now the interior regularity result in G we obtain the assertion of Theorem 2 immediately. (For the details see [2].)

3. Example. Let $m = n = 3$, $Q = (0,\infty) \times B$ (B is the unit ball in R^3). We obtain the example of the system (1) for which the boundary value problem with Lipschitzian boundary data on $\Gamma = [\{0\} \times B] \cup [(0,\infty) \times \partial B]$ has a solution u which develops the singularity for some $t_0 > 0$ in two steps: a) We choose the suitable u and b) to this u we construct the system for which u is the weak solution of the boudary value problem.

In the choice of the solution we were inspired by M. Struwe [3] who constructed the example for the system $u_t^i - \Delta u^i = f^i(t,x,u,Du)$, $i = 1,\dots,m$, with f^i growing quadratically in $|Du|$. We set

$$(2)\qquad u^i(t,x) = \begin{cases} \dfrac{x_i}{|x|} & \\ \Phi * \dfrac{x_i}{|x|} & \text{if } t < 1, \quad i = 1,\dots,3\,, \end{cases}$$

where Φ is fundamental solution of the equation $w_t + \Delta w = 0$. It is easy to see that u is locally Lipschitz continuous on $R \times R^n$ with except of the half-line $p = \{[t,0]\,;\ t \geq 1\}$, where it ceases to be continuous.

To construct the system, we modify the procedure due to M. Giaquinta and J. Souček. At first we seek the system with bounded and measurable coefficients in the form

$$(3)\qquad w_t - \operatorname{div}(A(z)Dw) = 0,$$

$$A^{ij}_{\alpha\beta}(z) = \delta_{\alpha\beta}\delta_{ij} - \frac{\tilde{d}_{\alpha i}\tilde{d}_{\beta j}}{(\tilde{d},Du)}, \text{ where } \tilde{d}_{\alpha i} = D_\alpha u^i - b_{\alpha i}\,.$$

Substituting u for w into this system we obtain the conditions on $b_{\alpha i}$ under which u is a solution of the system (3):

$$(4)\qquad u_t^i = D_\alpha b_{\alpha i}, \quad i = 1,\dots,3\,.$$

Choosing reasonably the form of $b_{\alpha i}$ we get after tedious calculations an explicit form of coefficients A:

$$(5)\qquad A^{ij}_{\alpha\beta}(z) = \delta_{\alpha\beta}\delta_{ij} + d_{\alpha i}d_{\beta j}\,,$$

where

$$d_{\alpha i} = \frac{1}{\sqrt{4(a-2)}}\{-\delta_{\alpha i}(a-2) - \frac{x_i x_\alpha}{|x|^2}(6+a)\} \text{ if } t \geq 1,$$

$$d_{\alpha i} = \frac{1}{\sqrt{4(a-2)+(6+a)q(4-3q)}}\{-\delta_{\alpha i}[a-2+(6+a)\frac{q}{2}] - \frac{x_i x_\alpha}{|x|^2}(6+a)(1-\frac{3q}{2})\} \quad \text{if } t < 1 .$$

Here $a > 2$ is a real parameter, $\xi = \frac{|x|}{2\sqrt{1-t}}$ and $q = q(\xi) = \xi^{-2} - \xi^{-1}e^{-\xi^2}(\int_0^\xi e^{-\tau^2}d\tau)^{-1}$.

Theorem 3. The function u given by (2) is a weak solution of the boundary value problem for (3) with the coefficients given by (5) and the boundary function u_0= Trace u on Γ. This solution is Lipschitz continuous with except of the half-line $p = \{[t,0]; t \geq 1\}$ where it ceases to be continuous.

Remark. Rewriting $x_1 x_\alpha |x|^{-2}$ in the coefficients we can pass to the desired quasilinear system of the type (1). For the details see [4].

References

[1] JOHN,O., *The interior regularity and the Liouville property for the quasilinear parabolic systems*, Comment. Math, Univ. Carolinae 23 (1982), 685-690.

[2] JOHN,O., STARÁ,J., *On the regularity of the weak solution of Cauchy problem for nonlinear parabolic systems via Liouville property*, Comment. Math. Univ. Carolinae 25 (1984), 445-457.

[3] STRUWE,M., *A counterexample in regularity theory for parabolic systems*, Czech. Math. Journal 34 (109), 1984.

[4] STARÁ,J., JOHN,O., *A counterexample...* to appear in Comment. Math. Unive. Carolinae.

CLASSICAL BOUNDARY VALUE PROBLEMS FOR MONGE-AMPÈRE TYPE EQUATIONS

N. S. TRUDINGER
Centre for Mathematical Analysis, Australian National University
Canberra, A.C.T. Australia

This report is concerned with recent work on the solvability of classical boundary value problems for elliptic Monge-Ampère type equations with particular attention to that of the author, P-L. Lions and J.I.E. Urbas [20] on Neumann type problems. The Dirichlet problem for these equations,

$$\det D^2u = f(x,u,Du) \quad \text{in} \quad \Omega \ , \tag{1}$$

$$u = \phi \quad \text{on} \quad \partial\Omega \ , \tag{2}$$

in convex domains Ω in Euclidean n space $\mathbb{R}^n$, has received considerable attention in recent years. For the standard Monge-Ampère equation,

$$\det D^2u = f(x) \quad \text{in} \quad \Omega \ , \tag{3}$$

Pogorelev [21,22] and Cheng and Yau [7] proved the existence of a unique convex solution $u \in C^2(\Omega) \cap C^{0,1}(\bar{\Omega})$, provided Ω is a uniformly convex $C^{1,1}$ domain in $\mathbb{R}^n$ and the functions $\phi, f \in C^{1,1}(\bar{\Omega})$ with f positive in Ω. Their methods depended on establishing interior smoothness of the generalized solutions of Aleksandrov [1]. These results were extended to equations of the more general form by P-L. Lions [17,18] using a direct PDE approach. Lions' approach led to the following classical existence theorem of Trudinger and Urbas [26], which we formulate explicitly for comparison with later results. Here we assume that the function f in equation (1) belongs to the space $C^{1,1}(\bar{\Omega}\times\mathbb{R}\times\mathbb{R}^n)$, is positive and non-decreasing in z, for all $(x,z,p) \in \Omega\times\mathbb{R}\times\mathbb{R}^n$ and satisfies the following growth limitations:

$$f(x,N,p) \leq g(x)/h(p) \tag{4}$$

for all $(x,p) \in \Omega\times\mathbb{R}^n$, where N is some constant and $g \in L^1(\Omega)$, $h \in L^1_{loc}(\mathbb{R}^n)$ are positive functions such that

$$\int_\Omega g < \int_{\mathbb{R}^n} h \quad ; \tag{5}$$

$$f(x,N',p) \le K[\mathrm{dist}(x,\partial\Omega)]^\alpha \, (1+|p|^2)^{\delta/2} \tag{6}$$

for all $x \in N$, $p \in \mathbb{R}^n$ where $N' = \max_{\partial\Omega} \phi$, K, α and δ are non-negative constants such that $\delta \le n+1+\alpha$ and N is some neighbourhood of $\partial\Omega$. Then we have

Theorem 1 [26] *Let* Ω *be a uniformly convex* $C^{1,1}$ *domain in* $\mathbb{R}^n$, $\phi \in C^{1,1}(\bar\Omega)$ *and suppose that* f *satisfies the above hypotheses. Then there exists a unique convex solution* $u \in C^2(\Omega) \cap C^{0,1}(\bar\Omega)$ *of the Dirichlet problem* (1), (2).

Conditions (4) and (6) were introduced by Bakelman [2] in his treatment of generalized solutions and they are both sharp [2],[26]. For the special case of the equation of prescribed Gauss curvature,

$$\det D^2u = K(x)(1+|Du|^2)^{(n+2)/2} , \tag{7}$$

conditions (5) and (6) become respectively,

$$\int_\Omega K < \omega_n , \tag{8}$$

$$K = 0 \quad \text{on} \quad \partial\Omega. \tag{9}$$

Moreover condition (8) is necessary for a $C^{0,1}(\bar\Omega)$ solution of equation (7) to exist [9],[26] while if condition (9) is violated there exist arbitrarily smooth boundary values ϕ for which the classical Dirichlet problem (7), (2) is not solvable, [26].

The above developments shed no light on the global regularity of solutions beyond being uniformly Lipschitz in Ω. This was an open problem, in more than two dimensions, for many years and was finally settled, for *uniformly* positive f, through the contributions of Ivochkina [10], who proved global bounds for second derivatives for arbitrary $\phi \in C^{3,1}(\bar\Omega)$, $\partial\Omega \in C^{3,1}$, Krylov [14],[15] and Caffarelli, Nirenberg and Spruck [5] who independently discovered the hitherto elusive global Hölder estimates for second derivatives. As a particular consequence of this work, we can infer the following existence theorem for globally smooth solutions of the classical Dirichlet problem.

Theorem 2 *Let* Ω *be a uniformly convex* $C^{3,1}$ *domain in* $\mathbb{R}^n, \phi \in C^{3,1}(\bar\Omega)$ *and suppose that* $f \in C^{1,1}(\bar\Omega\times\mathbb{R}\times\mathbb{R}^n)$ *is positive and non-decreasing with respect to* z, *for all* $(x,z,p) \in \bar\Omega\times\mathbb{R}\times\mathbb{R}^n$ *and satisfies conditions* (4) *and* (6) *with* $\alpha=0$. *Then there exists a unique convex solution* $u \in C^{3,\gamma}(\bar{\Omega})$ *for all* $\gamma < 1$ *of the Dirichlet problem* (1),(2).

More general results are in fact formulated in [5], [12] but the condition $\delta \leq n+1$ cannot be improved [26]. The situation with regard to *oblique* boundary value problems of the form

$$\beta \cdot Du = \phi(x,u) \quad \text{on } \partial\Omega , \tag{10}$$

where $\beta \cdot \nu > 0$ on $\partial\Omega$ and ν denotes the unit inner normal to $\partial\Omega$, turned out to be more satisfactory in that condition (6) is not required for the estimation of first derivatives. For the case $\beta = \nu$, that is for the usual Neumann case,

$$\nu \cdot Du = \phi(x,u) \quad \text{on } \partial\Omega , \tag{11}$$

we proved in collaboration with Lions and Urbas in [20], the following existence theorem,

<u>Theorem 3</u> *Let* Ω *be a uniformly convex* $C^{3,1}$ *domain in* $\mathbb{R}^n$ *and* $\phi \in C^{2,1}(\bar{\Omega}\times\mathbb{R})$ *satisfy*

$$\phi_z(x,z) \geq \gamma_0 \tag{12}$$

for all $x,z, \in \partial\Omega\times\mathbb{R}$ *and some positive constant* γ_0. *Then if* $f \in C^{1,1}(\bar{\Omega}\times\mathbb{R}\times\mathbb{R}^n)$ *is positive and non-decreasing with respect to* z *for all* $(x,z,p) \in \bar{\Omega}\times\mathbb{R}\times\mathbb{R}^n$ *and satisfies condition* (5), *there exists a unique convex solution* $u \in C^{3,\gamma}(\bar{\Omega})$ *for all* $\gamma < 1$ *of the boundary value problem* (1),(11).

Further regularity of the solutions in Theorems 2 and 3 follows by virtue of the Schauder theory of linear equations [9], when $\partial\Omega$, ϕ and f are appropriately smooth. In particular when $\partial\Omega \in C^\infty$, $\phi \in C^\infty(\partial\Omega\times\mathbb{R})$ and $f \in C^\infty(\bar{\Omega}\times\mathbb{R}\times\mathbb{R}^n)$ we deduce $u \in C^\infty(\bar{\Omega})$. The proofs of Theorems 2 and 3 both depend, through the method of continuity as described for example in [9], on the establishment of global $C^{2,\alpha}(\bar{\Omega})$ estimates for solutions of related problems. However the techniques employed by us to obtain these estimates in the Neumann boundary value case differ considerably from those used for the Dirichlet problem, particularly with respect to the estimation of first and second derivatives. For the estimation of sup norms we make use of the following maximum principle which does include that of Bakelman [2,3] for the Dirichlet problem as a special case.

<u>Theorem 4</u> [20] *Let* Ω *be a* C^1 *bounded domain in* $\mathbb{R}^n$ *and* $u \in C^2(\Omega) \cap C^1(\bar{\Omega})$ *a convex solution of the boundary problem* (1),(10) *in* Ω *where* f *satisfies condition* (5),

$\beta\cdot\nu \geq 0$ *on* $\partial\Omega$ *and* ϕ *satisfies* (12). *Then we have the estimate*

$$\min\left\{N, -\sup_{\partial\Omega} \phi^{+}(x,0)/\gamma_0 - (\beta_1/\gamma_0+d)R_0\right\} \leq u \leq \sup_{\partial\Omega} \phi^{-}(x,0)/\gamma_0 \tag{13}$$

where $d = \operatorname{diam}\Omega, \beta_1 = \sup_{\partial\Omega}|\beta|$, *and* R_0 *is given by*

$$\int_\Omega g = \int_{|p|<R_0} h$$

The gradient estimation in the oblique boundary condition case is a consequence of convexity as *any* convex $C^1(\bar{\Omega})$ function satisfies an estimate

$$\sup_\Omega |Du| \leq C \tag{14}$$

where C depends on $\beta_0, \beta_1, |u|_{0;\Omega}, \sup_{\partial\Omega}|\beta\cdot Du|$ and Ω, provided $\beta\cdot\nu \geq \beta_0$ where β_0 is a positive constant and $\Omega \in C^{1,1}$[20]. In contrast, a gradient estimate for solutions of the Dirichlet problem (1),(2) holds provided $\partial\Omega \in C^{1,1}$, $\phi \in C^{1,1}(\bar{\Omega})$ and condition (6) is fulfilled [26].

In both Dirichlet and Neumann problems the global estimation in Ω of *second derivatives* is reduced to considerations at the boundary $\partial\Omega$, by means of an approach which goes back to Pogorelev [21], although its implementation in the Neumann case [20] is substantially more involved than in the Dirichlet case [9], [5]. The boundary considerations are different as the Dirichlet problem is handled through barrier constructions [10], [5], whereas in [20] we employ different techniques including a device which necessitates our restriction of the vector β to the normal vector. The consequent estimates may be formulated as follows.

Theorem 5 *Let* Ω *be a* $C^{3,1}$ *uniformly convex domain in* $\mathbb{R}^n$ *and* $u \in C^4(\Omega)\cap C^3(\bar{\Omega})$ *a convex solution of the boundary value problem* (1),(11) *where* $f \in C^{1,1}(\bar{\Omega}\times\mathbb{R}\times\mathbb{R}^n)$ *is positive and* $\phi \in C^{2,1}(\partial\Omega\times\mathbb{R})$ *satisfies* (12). *Then we have*

$$\sup_\Omega |D^2u| \leq C \tag{15}$$

where C *depends on* n,Ω,f,ϕ *and* $|u|_{1;\Omega}$ *A similar estimate holds for solutions of the Dirichlet problem* (1),(2) *provided* $\phi \in C^{3,1}(\bar{\Omega})$.

We remark that the restriction (12) can be weakened to $\phi_z \geq 0$ and the case when $f^{1/n}$ is convex with respect to p is simpler. We do not know whether one need only assume $\phi \in C^{2,1}(\bar{\Omega})$ in the Dirichlet case. Once the second derivatives of solutions of the boundary value problems (1), (2), (10) are bounded, we obtain a control on the uniform ellipticity of equation (1), and further estimation hence follows from the theory of fully nonlinear uniformly elliptic equations. In particular interior $C^{2,\alpha}(\Omega)$ estimates were derived by Calabi [4] for Monge-Ampère equations and by Evans [8] and Krylov [13] for general uniformly elliptic equations. Global $C^{2,\alpha}(\bar{\Omega})$ estimates for the Dirichlet problem then arose from combination with key boundary estimates discovered by Krylov [14] and Caffarelli, Nirenberg and Spruck [5]. Global $C^{2,\alpha}(\bar{\Omega})$ estimates for oblique boundary value problems were proved by Lions and Trudinger [19], with more general results being given by Lieberman and Trudinger [16] and Trudinger [24]. The global estimates of Krylov [14] and Trudinger [24] are also applicable to classical solutions of uniformly elliptic Hamilton-Jacobi-Bellman equations. We may in fact formulate these estimates for general second order equations of the form

$$F[u] = F(x,u,Du,D^2u) = 0 \quad \text{in } \Omega , \tag{16}$$

subject to general boundary conditions

$$G[u] = G(x,u,Du) = 0 \quad \text{on } \partial\Omega , \tag{17}$$

where either G is *oblique* so that

$$G_p \cdot \nu > 0 \tag{18}$$

for all $(x,z,p) \in \partial\Omega\times\mathbb{R}\times\mathbb{R}^n$, or G is *Dirichlet* so that

$$G(x,z,p) = z-\phi(x) \tag{19}$$

for some function $\phi \in C^{2,1}(\partial\Omega)$. Here $F \in C^{1,1}(\bar{\Omega}\times\mathbb{R}\times\mathbb{R}^n\times\mathcal{U})$, $G \in C^{1,1}(\partial\Omega\times\mathbb{R}\times\mathbb{R}^n)$ where $\mathcal{U}$ is some open convex subset of the linear space $\mathcal{S}^n$ of $n\times n$ real symmetric matrices, and F is: (i) *elliptic* so that the matrix,

$$F_r = [F_{r_{ij}}] > 0 , \tag{20}$$

for all $(x,z,p) \in \bar{\Omega}\times\mathbb{R}\times\mathbb{R}^n$, $r = [r_{ij}] \in \mathcal{U}$; and (ii) *concave* with respect to r for all $(x,z,p) \in \bar{\Omega}\times\mathbb{R}\times\mathbb{R}^n$, $r \in \mathcal{U}$. Then we have

<u>Theorem 6</u> *Let Ω be a bounded $C^{3,1}$ domain in $\mathbb{R}^n$ and $u \in C^3(\bar{\Omega})\cap C^4(\Omega)$ a solution of the boundary value problem* (16),(17) *such that $D^2u(\Omega) \subset \mathcal{U}$. Then we have*

$$[D^2u]_{\alpha;\Omega} \le C \tag{21}$$

where $\alpha < 1$ *and* C *are positive constants depending only on* $n, \Omega, |u|_{2;\Omega}$ *and the first and second derivatives of* F *and* G *(excluding* F_{rr}*), (and* $|\phi|_3$ *in the Dirichlet case).*

We remark here that the solution u in Theorem 6 need only lie in the space $C^{1,1}(\bar{\Omega})$ and the smoothness of $\partial\Omega$, G, F, ϕ can be reduced, [25]. For application to Monge-Ampère type equations the convex set U becomes the set of positive symmetric matrices.

Finally we note that the sharpness of condition (8) is strikingly demonstrated by the following result of Urbas [28] concerning extremal domains for the equation of prescribed Gauss curvature.

Theorem 7 *Let* Ω *be a uniformly convex domain in* $\mathbb{R}^n$ *and* $K \in C^{1,1}(\bar{\Omega})$ *be positive in* Ω *and satisfy*

$$\int_{\Omega} K = \omega_n \tag{22}$$

Then there exists a convex solution $u \in C^2(\Omega)$ *of equation (7) in* Ω. *Furthermore the function* u *is vertical at* $\partial\Omega$ *and is unique up to additive constants. If* K *is positive in* $\bar{\Omega}$, *then the solution* u *is bounded; if* K *vanishes on* $\partial\Omega$ *then the solution* u *approaches infinity at* $\partial\Omega$.

REFERENCES

[1] A.D. Aleksandrov, Dirichlet's problem for the equation $\mathrm{Det}\|z_{ij}\| = \varphi(z_1,\ldots,z_n z, x_1,\ldots,x_n)$, *Vestnik Leningrad Univ.* 13 (1958), 5-24, [Russian].

[2] I. Ya. Bakel'man, The Dirichlet problem for equations of Monge-Ampère type and their n-dimensional analogues, *Dokl. Akad. Nauk, SSSR* 126 (1959), 923-926, [Russian].

[3] I. Ya. Bakel'man, The Dirichlet problem for the elliptic n-dimensional Monge-Ampere equations and related problems in the theory of quasilinear equations, *Proceedings of Seminar on Monge-Ampère Equations and Related Topics*, Firenze 1980), Instituto Nazionale di Alta Matematica, Roma, (1982), 1-78.

[4] E. Calabi, Improper affine hyperspheres of convex type and a generalization of a theorem by K. Jörgens, *Michigan Math. J.* 5 (1958), 105-126.

[5] L. Caffarelli, L. Nirenberg, J. Spruck, The Dirichlet problem for nonlinear second order elliptic equations, I. Monge-Ampère equation, *Comm. Appl. Math.* 37 (1984), 369-402.

[6] L. Caffarelli, J.J. Kohn, L. Nirenberg, J. Spruck, The Dirichlet problem for nonlinear second order elliptic equations II. Complex Monge-Ampère and uniformly elliptic equations, *Comm. Pure Appl. Math.* 38 (1985), 209-252.

[7] S.-Y. Cheng, S.-T. Yau. On the regularity of the solution of the n-dimensional Minkowski problem, *Comm. Pure Appl. Math.* 29 (1976), 495-516.

[8] L.C. Evans, Classical solutions of fully nonlinear, convex, second order elliptic equations, *Comm. Pure Appl. Math.* 35 (1982), 333-363.

[9] D. Gilbarg, N.S. Trudinger, *Elliptic partial differential equations of second order*, Springer-Verlag, Berlin-Heidelberg-New York-Tokyo, Second Edition, 1983.

[10] N.M. Ivochkina, Construction of a priori bounds for convex solutions of the Monge-Ampère equation by integral methods, *Ukrain. Mat. Ž.* 30 (1978), 45-53, [Russian].

[11] N.M. Ivochkina, An apriori estimate of $\|u\|_{C^2(\bar{\Omega})}$ for convex solutions of the Dirichlet problem for the Monge-Ampère equations, *Zap. Naučn. Sem. Leningrad. Otdel. Mat. Inst. Steklov. (LOMI)* 96 (1980), 69-79, [Russian]. English translation in *J. Soviet Math.* 21 (1983), 689-697.

[12] N.M. Ivochkina, Classical solvability of the Dirichlet problem for the Monge-Ampère equation, *Zap. Naučn. Sem. Leningrad. Otdel. Mat. Inst. Steklov. (LOMI)* 131 (1983), 72-79.

[13] N.V. Krylov, Boundedly inhomogeneous elliptic and parabolic equations, *Izv, Akad. Nauk. SSSR* 46 (1982), 487-523, [Russian]. English translation in *Math. USSR Izv.* 20 (1983), 459-492.

[14] N.V. Krylov, Boundedly inhomogeneous elliptic and parabolic equations in a domain, *Izv. Akad. Nauk. SSSR* 47 (1983), 75-108, [Russian].

[15] N.V. Krylov, On degenerative nonlinear elliptic equations, *Mat. Sb. (N.S.)* 120 (1983), 311-330, [Russian].

[16] G.M. Lieberman, N.S. Trudinger, Nonlinear oblique boundary value problems for nonlinear elliptic equations. Aust. Nat. Univ. Centre for Math. Anal. Research Report R24 (1984).

[17] P.-L. Lions, Sur les équations de Monge-Ampère I, *Manuscripta Math.* 41 (1983), 1-44.

[18] P.-L. Lions, Sur les équations de Monge-Ampère II, *Arch. Rational Mech. Anal.* (to appear).

[19] P.-L. Lions, N.S. Trudinger, Linear oblique derivative problems for the uniformly elliptic Hamilton-Jacobi-Bellman equation, *Math. Zeit.* (to appear).

[20] P.-L. Lions, N.S. Trudinger, J.I.E. Urbas, The Neumann problem for equations of Monge-Ampère type. Aust. Nat. Univ. Centre for Math. Anal. Research Report R16 (1985).

[21] A.V. Pogorelov, The Dirichlet problem for the n-dimensional analogue of the Monge-Ampère equation, *Dokl. Akad. Nauk. SSSR* 201 (1971), 790-793, [Russian]. English translation in *Soviet Math. Dokl.* 12 (1971), 1727-1731.

[22] A.V. Pogorelov, *The Minkowski multidimensional problem*, J. Wiley, New York, 1978.

[23] F. Schultz, A remark on fully nonlinear, concave elliptic equations, *Proc. Centre for Math. Anal.* Aust. Nat. Univ. 8, (1984), 202-207.

[24] N.S. Trudinger, Boundary value problems for fully nonlinear elliptic equations. *Proc. Centre for Math. Anal.* Aust. Nat. Univ. 8 (1984), 65-83.

[25] N.S. Trudinger, Regularity of solutions of fully nonlinear elliptic equations. *Boll. Un. Mat. Ital.*, 3-A (1984), 421-430. II Aust. Nat. Univ. Centre for Math. Anal. Research Report R38 (1984).

[26] N.S. Trudinger, J.I.E. Urbas, The Dirichlet problem for the equation of prescribed Gauss curvature, *Bull. Austral. Math. Soc.* 28 (1983), 217-231.

[27] N.S. Trudinger, J.I.E. Urbas, On second derivative estimates for equations of Monge-Ampère type, *Bull. Austral. Math. Soc.* 30 (1984), 321-334.

[28] J.I.E. Urbas, Some recent results on the equation of prescribed Gauss curvature, *Proc. Centre for Math. Anal.* Aust. Nat. Univ. 8 (1984), 215-220.

[29] J.I.E. Urbas, The equation of prescribed Gauss curvature without boundary conditions, *J. Differential Geometry* (to appear).

[30] J.I.E. Urbas, The generalized Dirichlet problem for equation of Monge-Ampère type. *Ann. L'Inst. Henri Poincaré*, (to appear).

QUALITATIVE PROPERTIES OF THE SOLUTIONS TO THE NAVIER-STOKES EQUATIONS FOR COMPRESSIBLE FLUIDS

A. VALLI
Dipartimento di Matematica, Università di Trento
38050 Povo (Trento), Italy

1. Introduction.

We want to present a new method for showing the existence of a stationary solution to the equations which describe the motion of a viscous compressible barotropic fluid.

At first it is useful to recall some known results concerning the non-stationary case. The equations of motion are

$$
(NS)\quad\begin{cases}
\rho\,[\dfrac{\partial v}{\partial t}+(v\cdot\nabla)v-f]=-\nabla[p(\rho)]+\mu\Delta v+(\zeta+\mu/3)\nabla \mathrm{div}\, v & \text{in }]0,T[\times\Omega\,,\\
\dfrac{\partial\rho}{\partial t}+\mathrm{div}(\rho v)=0 & \text{in }]0,T[\times\Omega\,,\\
v_{|\partial\Omega}=0 & \text{on }]0,T[\times\partial\Omega,\\
\int_\Omega \rho=\bar\rho\,|\Omega|>0 & (|\Omega|\equiv \mathrm{meas}(\Omega)),\\
v_{|t=0}=v_o & \text{in } \Omega\,,\\
\rho_{|t=0}=\rho_o & \text{in } \Omega\,,
\end{cases}
$$

where $\Omega\subset R^3$ is a bounded domain, with smooth boundary $\partial\Omega$; v and ρ are the velocity and the density of the fluid; p is the pressure, which is assumed to be a known function of ρ; f is the (assigned) external force field; the constants $\mu>0$ and $\zeta\geq 0$ are the viscosity coefficients; $\bar\rho>0$ is the mean density of the fluid, i.e. the total mass of fluid divided by $|\Omega|$; v_o and ρ_o are the initial velocity and density.

In the last years it has been proved that:

(i) if v_o and $\rho_o-\bar\rho$ are small enough and $f=0$, then problem (NS) has a unique global (in time) solution (Matsumura-Nishida [1]);

(ii) the preceeding result also holds for a sufficiently small $f \neq 0$; moreover, two small solutions are asymptotically equivalent as $t \to +\infty$, and consequently if f is periodic (independent of t) then there exists a periodic (stationary) solution (Valli [3]).

It must be underlined that no other method is known for showing the existence of a stationary solution, excepting when the viscosity coefficients satisfy $\zeta >> \mu$. In this case Padula [2] proved that, if f is small enough, then there exists a stationary solution. Remark, however, that in general the shear viscosity coefficient μ is larger than the bulk viscosity coefficient ζ. Moreover, from the mathematical point of view it would seem only necessary to require that μ is positive, without assumptions on the largeness of ζ.

The method that we want to present here is based on a "natural" linearization of the problem, followed by a fixed point argument. The viscosity coefficients are only required to satisfy the thermodynamic restrictions $\mu > 0$, $\zeta \geq 0$.

2. The linear problem (L).

Since we are searching for a solution in a neighbourhood of the equilibrium solution $\tilde{\rho} = \bar{\rho}$, $\tilde{v} = 0$, it is useful to introduce the new unknown

$$\sigma = \rho - \bar{\rho} .$$

The equations of motion in the stationary case thus become

$$(S)\quad \begin{cases} -\mu\,\Delta v - (\zeta+\mu/3)\,\nabla \mathrm{div}\, v + p_1 \nabla\sigma = (\sigma+\bar{\rho})[f - (v\cdot\nabla)v] + \\ \qquad\qquad + [p_1 - p'(\sigma+\bar{\rho})]\nabla\sigma & \text{in } \Omega, \\ \bar{\rho}\,\mathrm{div}\, v + \mathrm{div}(v\sigma) = 0 & \text{in } \Omega, \\ v_{|\partial\Omega} = 0 & \text{on } \partial\Omega, \\ \int_\Omega \sigma = 0 & , \end{cases}$$

where it is assumed that $p_1 \equiv p'(\bar{\rho}) > 0$.

It is easily verified that a solution of (S) exists if we find a fixed point of the map

$$\Phi : (v,\sigma) \longrightarrow (w,\eta) \quad ,$$

defined by means of the solutions of the following linear problem

$$
(L)\quad \begin{cases} -\mu\,\Delta w - (\zeta+\mu/3)\nabla \mathrm{div}\, w + p_1 \nabla\eta = (\sigma+\bar\rho)[f-(v\cdot\nabla)v] + \\ \qquad + [p_1 - p'(\sigma+\bar\rho)]\nabla\sigma \equiv F & \text{in } \Omega, \\ \bar\rho\, \mathrm{div}\, w + \mathrm{div}(v\eta) = 0 & \text{in } \Omega, \\ w_{|\partial\Omega} = 0 & \text{on } \partial\Omega, \\ \int_\Omega \eta = 0 \end{cases}
$$

3. A-priori estimates for the solution of (L).

We want to obtain a-priori estimates in Sobolev spaces of sufficiently large order, in such a way that we can control the behaviour of the non-linear terms which appear in F. We shall prove that a solution (w,η) of (L) satisfies

$$
(3.1)\qquad \|w\|_3 + \|\eta\|_2 \le c_1 \|F\|_1
$$

for $v_{|\partial\Omega} = 0$ and $\|v\|_3 \le A$ small enough. Here $\|\cdot\|_k$ is the norm in the Sobolev space $H^k(\Omega)$, and c_1 depends in a continuous way on μ, ζ and A (but it is independent of v).

(a) At first, from well-known results on Stokes problem we get

$$
(3.2)\qquad \|w\|_3 + \|\eta\|_2 \le c(\|F\|_1 + \|\mathrm{div}\, w\|_2) .
$$

Hence our aim is to estimate $\|\mathrm{div}\, w\|_2$.

(b) Multiplying $(L)_1$ by w and $(L)_2$ by $(p_1/\bar\rho)\eta$ and integrating in Ω one has

$$
(3.3)\qquad \|w\|_1 + \|\mathrm{div}\, w\|_0 \le c(\|F\|_{-1} + \|v\|_3^{1/2} \|\eta\|_0).
$$

The same argument can be used for estimating all the successive derivatives in the interior of Ω, and the tangential derivative $D_\tau \mathrm{div}\, w$ near the boundary $\partial\Omega$, obtaining in this way (in local coordinates near $\partial\Omega$)

$$
(3.4)\qquad \|D_\tau w\|_1 + \|D_\tau \mathrm{div}\, w\|_0 \le c(\|F\|_0 + \|v\|_3^{1/2} \|\eta\|_1).
$$

(c) The estimate for the normal derivative $D_n \mathrm{div}\, w$ is obtained by observing that on $\partial\Omega$

$$
\Delta w \cdot n \cong \nabla \mathrm{div}\, w \cdot n ,
$$

in the sense that their difference does not contain $D_n^2 w$.

Hence by taking the normal derivative of $(L)_2$, multiplied by $(\bar{\rho})^{-1}$ $(\zeta+4\mu/3)$, and adding it to the normal component of $(L)_1$ we get (in local coordinates near $\partial\Omega$)

$$p_1 D_n \eta + (\zeta+4\mu/3)/\bar{\rho}\, D_n \mathrm{div}(v\eta) \cong F\cdot n \, . \tag{3.5}$$

From this equation one easily gets

$$\| D_n \eta \|_0 \le c(\| F \|_0 + \| v \|_3^{1/2} \| \eta \|_1) . \tag{3.6}$$

Moreover, going back to $(L)_1$, one has

$$p_1 D_n \eta = \mu\, \Delta w\cdot n + (\zeta+\mu/3)\nabla \mathrm{div}\, w\cdot n + F\cdot n \cong (\zeta+4\mu/3) D_n \mathrm{div}\, w + $$
$$+ F\cdot n \quad ,$$

hence from (3.6)

$$\| D_n \mathrm{div}\, w \|_0 \le c(\| F \|_0 + \| v \|_3^{1/2} \| \eta \|_1) . \tag{3.7}$$

By repeating the same argument for the second order derivatives one gets

$$\| \mathrm{div}\, w \|_2 \le c(\| F \|_1 + \| v \|_3^{1/2} \| \eta \|_2) , \tag{3.8}$$

hence (3.1) holds if $\| v \|_3 \le A$ small enough.

4. Existence of the solution of (L).

Though problem (L) is linear, and we know that the a-priori estimate (3.1) holds, the existence of a solution $w \in H^3(\Omega)$, $\eta \in H^2(\Omega)$ is not obvious.

In fact, the usual elliptic approximation cannot work in this case. More precisely, if we add $-\varepsilon\, \Delta \eta_\varepsilon$ to $(L)_2$, we must also require a boundary condition (say, Dirichlet or Neumann) on η_ε. But the limit function η is free on $\partial\Omega$. Hence the sequence η_ε can only converge in $L^2(\Omega)$ (Dirichlet condition), or in $H^1(\Omega)$ (Neumann condition), and cannot converge in $H^2(\Omega)$!

Moreover, if $v \neq 0$ (L) is <u>not</u> an elliptic system in the sense of Agmon-Douglis-Nirenberg (if $v = 0$ (L) is the Stokes system). Hence the usual regularization procedures do not work.

One can proceed in the following way. By adapting the method of Padula [2] to problem (L), one defines

$$\pi \equiv (p_1/\mu)\eta - (\zeta/\mu + 1/3)\,\mathrm{div}\, w \tag{4.1}$$

and (L) is transformed into

$$(L') \quad \begin{cases} -\Delta w + \nabla\pi = F/\mu & \text{in } \Omega, \\ \mathrm{div}\, w = (\zeta/\mu + 1/3)^{-1}(p_1\eta/\mu - \pi) & \text{in } \Omega, \\ w_{|\partial\Omega} = 0 & \text{on } \partial\Omega, \end{cases}$$

$$(L'') \quad \begin{cases} \bar\rho(\zeta/\mu + 1/3)^{-1} p_1\eta/\mu + \mathrm{div}(v\eta) = \bar\rho(\zeta/\mu + 1/3)^{-1}\pi & \text{in } \Omega, \\ \int_\Omega \eta = 0 \end{cases}$$

These equations can be solved via a fixed point argument if $\zeta >> \mu$. Hence the a-priori estimates (3.1) and the continuity method give the result for any pair of viscosity coefficients satisfying $\mu > 0$ and $\zeta \geqq 0$.

5. Existence of a solution of (S).

We prove at last the existence of a fixed point for the map

$$\Phi : (v,\sigma) \longrightarrow (w,\eta).$$

Taking

$$K \equiv \{(v,\sigma) \in H^3(\Omega) \times H^2(\Omega) \mid \ v_{|\partial\Omega} = 0,\ \int_\Omega \sigma = 0,\ \| v \|_3 + \| \sigma \|_2 \leq A \},$$

by using (3.1) one sees that

$$\| w \|_3 + \| \eta \|_2 \leq c_1 \| F \|_1 \leq c\,[(\| \sigma \|_2 + 1)(\| f \|_1 + \| v \|_2^2) + \| \sigma \|_2^2\,] \leq c(A+1)(\| f \|_1 + A^2) \quad .$$

Choosing $A^2 \equiv \| f \|_1 << 1$, one has

$$\| w \|_3 + \| \eta \|_2 \leq A,$$

hence $\Phi(K) \subset K$. The set K is convex and compact in $X \equiv H^2(\Omega) \times H^1(\Omega)$, and it is easily seen that the map Φ is continuous in X. The existence of a fixed point is now a consequence of Schauder's theorem.

References.

[1] A. Matsumura - T. Nishida, Initial boundary value problems for the equations of motion of general fluids, in "Computing methods in applied sciences and engineering, V", ed. R. Glowinski - J.L. Lions, North-Holland Publishing Company, Amsterdam-New York-Oxford, 1982. (See also Preprint University of Wisconsin, MRC Technical Summary Report n° 2237 (1981)).

[2] M. Padula, Existence and uniqueness for viscous steady compressible motions, Arch. Rational Mech. Anal., to appear.

[3] A. Valli, Periodic and stationary solutions for compressible Navier-Stokes equations via a stability method, Ann. Scuola Norm. Sup. Pisa, (4) 10 (1983), 607-647.

Section C
NUMERICAL METHODS

ON GEL'FAND'S METHOD OF CHASING FOR SILVING MULTIPOINT BOUNDARY VALUE PROBLEMS

R. P. AGARWAL
Department of Mathematics, National University of Singapore
Singapore 0511

Recently, for multipoint boundary value problems for ordinary differential equations several constructive methods have been suggested, e.g. the method of complementary functions and the method of adjoints [1,2], the integral equations method [3,4], initial adjusting method [12,16], the method of quasilinearization [5,8] etc. Here, we shall report the formulation of another practical shooting method, namely the method of chasing for nth order ordinary linear differential equation

$$x^{(n)} + \sum_{i=1}^{n} p_i(t)\, x^{(n-i)} = f(t) \tag{1}$$

subject to linearly independent multipoint boundary conditions

$$\sum_{k=0}^{n-1} c_{ik}\, x^{(k)}(a_i) = A_i, \qquad 1 \leq i \leq n \tag{2}$$

where $a_1 \leq a_2 \leq \ldots \leq a_n$ $(a_1 < a_n)$. This method is originally developed for second order differential equations by Gel'fand and Lokutsiyevskii and first appeared in english literature only recently [9]. Na [11] has briefly described the method and given different formulations for the different particular cases of (1), (2). The general systems derived here include the systems given by Na [11] as special cases. The power of the method is illustrated by solving known Holt's problem.

Since the boundary conditions (2) are assumed to be linearly independent, at the point a_i at least one of the c_{ik}, $0 \leq k \leq n-1$ is not zero. Let $c_{ij} \neq 0$ then, at this point a_i the boundary condition (2) can be rewritten as

$$x^{(j)}(a_i) = \sum_{\substack{k=0\\k\neq j}}^{n-1} d_{ik}\, x^{(k)}(a_i) + \alpha_i, \quad i \leq i \leq n \tag{3}$$

where $d_{ik} = -\dfrac{c_{ik}}{c_{ij}}$; $0 \leq k \leq n-1$, $k = j$ and $\alpha_i = \dfrac{A_i}{c_{ij}}$

In the differential equation (1), we begin with the assumption that $p_1(t) \equiv 0$, so that

$$x^{(n)} = - \sum_{i=2}^{n} p_i(t)x^{(n-i)} + f(t). \tag{4}$$

Now, for the boundary condition (3) we assume that the solution x(t) of (4) satisfies (n-1)th order linear differential equation

$$x^{(j)}(t) = \sum_{\substack{k=0\\k\neq j}}^{n-1} d_{ik}(t)\, x^{(k)}(t) + \alpha_i(t) \tag{5}$$

where the n functions $d_{ik}(t)$; $0 \leq k \leq n-1$, $k \neq j$ and $\alpha_i(t)$ are to be determined.

Differentiating (5) once, we get

$$x^{(j+1)}(t) = \sum_{\substack{k=0\\k\neq j}}^{n-1} [d_{ik}(t)x^{(k+1)}(t) + d'_{ik}(t)x^{(k)}(t)] + \alpha'_i(t). \tag{6}$$

Next, we shall use (5) to eliminate the term $x^{(n-1)}(t)$ from (6), however it depends on a particular value of j and we need to consider four different cases :

(i) $j = 0$, $n \geq 3$: From (5), we have

$$x^{(n-1)}(t) = \frac{1}{d_{i,n-1}(t)}[x(t) - \sum_{k=1}^{n-2} d_{ik}(t)x^{(k)}(t) - \alpha_i(t)]. \tag{7}$$

Using (7) in (6) and rearranging the terms, we get

$$\begin{aligned} x^{(n)}(t) = &- \frac{[d_{i,n-2}(t) + d'_{i,n-1}(t)]}{d^2_{i,n-1}(t)}\, x(t) \\ &+ [\frac{1}{d_{i,n-1}(t)} + \frac{d_{i,n-2}(t) + d'_{i,n-1}(t)}{d^2_{i,n-1}(t)}\, d_{i1}(t) - \frac{d'_{i1}(t)}{d_{i,n-1}(t)}]x'(t) \\ &+ \sum_{k=2}^{n-2} [\frac{d_{i,n-2}(t) + d'_{i,n-1}(t)}{d^2_{i,n-1}(t)}\, d_{ik}(t) - \frac{d_{i,k-1}(t) + d'_{ik}(t)}{d_{i,n-1}(t)}]x^{(k)}(t) \\ &+ [\frac{d_{i,n-2}(t) + d'_{i,n-1}(t)}{d^2_{i,n-1}(t)}\, \alpha_i(t) - \frac{\alpha'_i(t)}{d_{i,n-1}(t)}]. \end{aligned} \tag{8}$$

Comparing (4) and (8), we find the system of n differential equations

$$d'_{i,n-1}(t) = -d_{i,n-2}(t) + p_n(t)d^2_{i,n-1}(t)$$

$$d'_{ki}(t) = p_{n-k}(t)d_{i,n-1}(t) - d_{i,k-1}(t) + p_n(t)d_{i,n-1}(t)d_{ik}(t);\ k=n-2,n-3,\dots,2$$

$$d'_{i1}(t) = 1 + p_n(t)d_{i,n-1}(t)d_{i1}(t) + p_{n-1}(t)d_{i,n-1}(t) \tag{9}$$

$$\alpha'_i(t) = -f(t)\, d_{i,n-1}(t) + p_n(t)\, d_{i,n-1}(t)\alpha_i(t).$$

We also desire that this solution x(t) must satisfy the boundary condition (3). For this, we compare (3) and (5) at the point a_i and find

$$d_{ik}(a_i) = d_{ik}, \quad 1 < k < n-1 \tag{10}$$

$$\alpha_i(a_i) = \alpha_i.$$

In the rest we proceed as for the case $j = 0$ and obtain the following systems

(ii) $1 < j < n-3$

$$d'_{i,n-1}(t) = -d_{i,n-2}(t) - d_{i,j-1}(t)\, d_{i,n-1}(t) + p_{n-j}(t)d^2_{i,n-1}(t)$$

$$d'_{ik}(t) = -d_{i,k-1}(t) - d_{i,j-1}(t)\, d_{ik}(t) + (p_{n-k}(t) + p_{n-j}(t)d_{ik}(t))d_{i,n-1}(t)$$

$$k = n-2,\ n-3,\dots,\ 1;\ k \neq j,\ j+1$$

$$d'_{i,j+1}(t) = 1 - d_{i,j-1}(t)d_{i,j+1}(t) + (p_{n-j-1}(t) + p_{n-j}(t)d_{i,j+1}(t))d_{i,n-1}(t)$$

$$d'_{io}(t) = -d_{i,j-1}(t)d_{io}(t) + (p_n(t) + p_{n-j}(t)d_{io}(t))d_{i,n-1}(t) \tag{11}$$

$$\alpha'_i(t) = -d_{i,j-1}(t)\alpha_i(t) + (p_{n-j}(t)\alpha_i(t) - f(t))d_{i,n-1}(t)$$

$$d_{ik}(a_i) = d_{ik};\ 0 < k < n-1,\ k = j$$

$$\alpha_i(a_i) = \alpha. \tag{12}$$

(iii) $j = n-2$

$$d'_{i,n-1}(t) = 1 - d_{i,n-1}(t)d_{i,n-3}(t) + p_2(t)d^2_{i,n-1}(t)$$

$$d'_{ik}(t) = -d_{i,k-1}(t) + (p_{n-k}(t) + p_2(t)d_{ik}(t))d_{i,n-1}(t) - d_{i,n-3}(t)d_{ik}(t),$$

$$1 < k < n-3$$

$$d'_{i0}(t) = -d_{i,n-3}(t)d_{i,0}(t) + (p_n(t) + p_2(t)d_{i,0}(t))d_{i,n-1}(t) \tag{13}$$

$$\alpha'_i(t) = -d_{i,n-3}(t)\alpha_i(t) + (-f(t) + p_2(t)\alpha_i(t))d_{i,n-1}(t)$$

$$d_{ik}(a_i) = d_{ik} \ ; \ 0 < k < n-1, \ k \neq n-2$$

$$\alpha_i(a_i) = \alpha_i. \tag{14}$$

(iv) $j = n-1$

$$d'_{ik}(t) = -d_{i,k-1}(t) - d_{i,n-2}(t)d_{ik}(t) - p_{n-k}(t), \quad 1 < k < n-2$$

$$d'_{i0}(t) = -d_{i,n-2}(t)d_{i0}(t) - p_0(t) \tag{15}$$

$$\alpha'_i(t) = -d_{i,n-2}(t)\alpha_i(t) + f(t)$$

$$d_{ik}(a_i) = d_{ik}; \quad 0 < k < n-2 \tag{16}$$

$$\alpha_i(a_i) = \alpha_i.$$

For the particular value of j, we integrate the above appropriate system from the point a_i to a_n and collect the values of $d_{ik}(a_n)$; $0 < k < n-1$, $k \neq j$ and $\alpha_i(a_n)$. Thus, (5) provides a new boundary relation at the point a_n

$$x^{(j)}(a_n) = \sum_{\substack{k=0 \\ k\neq j}}^{n-1} d_{ik}(a_n)x^{(k)}(a_n) + \alpha_i(a_n). \tag{17}$$

Let N be the number of different boundary points i.e. $a_1 < a_2 < \ldots < a_N = a_n$ $(n > N > 2)$ and $m(a_j)$ represents the number of boundary relations (3) prescribed at the point a_j and hence $\sum_{j=1}^{N} m(a_j) = n$. Thus, in (3) we have $m(a_n)$ boundary relations at the point a_n and to find $x^{(j)}(a_n)$, $0 < j < n-1$ we need $n-m(a_n)$ more new relations (17) i.e. we need to integrate $n-m(a_n)$ appropriate differential systems.

Finally, from the obtained values of $\dot{x}^{(j)}(a_n)$, $0 < j < n-1$ we integrate backward differential equation (4) and obtain the required solution.

With the help of the following guidelines unnecessary computation can be avoided : (a) $m(a_n) = \max_{1\leq j\leq N} m(a_j)$, otherwise the role of the point a_n with the point a_j where $m(a_j)$ is maximum can be interchanged. (b) We need to integrate $n-m(a_n)$ times but not necessarily different differential systems, specially because differential system does not change as long as in (3) j is same. In fact, we can have at most n different differential systems.

For the case $p_1(t) \not\equiv 0$, we rewrite the differential equation (1) as

$$[P(t)x^{(n-1)}]' = -\sum_{i=2}^{n} P(t)p_i(t)x^{(n-i)} + P(t)f(t) \tag{18}$$

where $P(t) = \exp(\int_{a_1}^{t} p_1(s)ds)$.

Assumption that the solution of (18) should satisfy (n-1)th order linear differential equation

$$d_{ij}(t)x^{(j)}(t) = \sum_{\substack{k=0\\k\neq j}}^{n-1} d_{ik}(t)x^{(k)}(t) + \alpha_i(t) \tag{19}$$

with $d_{i,n-1}(t) = P(t)$ brings the problem in the realm of the foregoing analysis.

Example. The two point boundary value problem

$$x'' = (2m + 1 + t^2)x \tag{20}$$

$$x(0) = \beta, \; x(\infty) = 0 \tag{21}$$

where m and β are specified constants, known as Holt's problem [10] is a typical example where usual shooting methods fail [10,13,14,15]. Faced with this difficulty Holt [10] used a finite difference method, whereas Osborne [13] used multiple shooting method and Roberts and Shipman [14,15] used a multipoint approach.

For this problem the solution representation (5) reduces to

$$x(t) = d_{01}(t)x'(t) + \alpha_0(t) \tag{22}$$

and the case (iii) provides the differential system to be integrated

$$\begin{aligned} d_0'(t) &= 1 - (2m + 1 + t^2)d_{01}^2(t) \\ \alpha_0'(t) &= -(2m + 1 + t^2)d_{01}(t)\alpha_0(t) \end{aligned} \tag{23}$$

together with the initial conditions

$$d_{01}(0) = 0, \ \alpha_0(0) = \beta. \tag{24}$$

We use fourth order Runge-Kutta method with step size 0.01 and obtain $d_{01}(t)$, $\alpha_0(t)$ at $t = 18.01$. These values are used to calculate $x'(18.01)$ from (22). The differential equation (20) is integrated backward with the given $x(18.01) = 0$ and the obtained value of $x'(18.01)$ using fourth order Runge-Kutta method with the same step size. The value $t = 18.01$ has been chosen in view of restricted Computer capabilities.

The solution thus obtained has been presented in Tables 1-3 for different choices of m and β. These tables also contain solutions of the problem obtained earlier in [10,13,14,15]. For further details of the method and its applications see [6,7].

References

1. R. P. Agarwal, J. Comp. Appl. Math. 5(1979), 17-24.
2. R. P. Agarwal, J. Optimization Theory and Appl. 36(1982), 139-144.
3. R. P. Agarwal, Nonlinear Analysis : TMA, 7(1983), 259-270..
4. R. P. Agarwal and S. L. Loi, Nonlinear Analysis : TMA, 8(1984), 381-391.
5. R. P. Agarwal, J. Math. Anal. Appl. 107(1985), 317-330.
6. R. P. Agarwal and R. C. Gupta, BIT. 24(1984), 342-346.
7. R. P. Agarwal and R. C. Gupta, Method of chasing for multipoint boundary value problems, Appl. Math. Comp. (to appear).
8. R. E. Bellman and R. E. Kalaba, "Quasilinearization and Nonlinear Boundary Value Problems", American Elsevier, New York, 1965.
9. I. S. Berezin and N. P. Zhidkov, Method of Chasing, in "Computing Methods" (O. M. Blum and A. D. Booth, trans.), Vol. II, Pergamon, Oxford, 1965.

10. J. F. Holt, Comm. Asso. Comp. Machinery 7(1964), 366-373.

11. T. Y. Na, "Computational Methods in Engineering Boundary Value Problems" Academic Press, New York, 1979.

12. T. Ojika and W. Welsh, Intern. J. Comput. Math. 8(1980), 329-344.

13. M. R. Osborne, J. Math. Anal. Appl. 27 (1969), 417-433.

14. S. M. Roberts and J. S. Shipman, J. Optimization Theory and Applications 7(1971), 301-318.

15. S. M. Roberts and J. S. Shipman "Two-Point Boundary Value Problems : Shooting Methods" Elsevier, New York, 1972.

16. W. Welsh and T. Ojika, J. Comp. Appl. Math. 6(1980), 133-143.

Table 1.m = 0,β = 1

t	Present Solution	Complementary Functions [8]	Finite Difference	Solution by Osborne [6]	Roberts and Shipman [7]
0	0.9999876 E 00	0.10000000 E 01	0.100000 E 01	0.1000 E 01	0.10000000 E 01
1	0.2593404 E 00	0.15729920 E 00	0.157300 E 00	0.2593 E 00	0.15729921 E 00
2	0.3456397 E-01	0.46777349 E-02	0.467778 E-02	0.3455 E-01	0.46777350 E-02
3	0.1988532 E-02	0.22090497 E-04	0.220908 E-04	0.1987 E-02	0.22090497 E-04
4	0.4595871 E-04	0.15417257 E-07	0.154175 E-07	0.4590 E-04	0.15417259 E-07
5	0.4125652 E-06	0.15366706 E-11	0.153749 E-11	0.4188 E-06	0.15374602 E-11
6	0.1413020 E-08	-0.73163560 E-15	0.215201 E-16	0.1409 E-08	0.21519753 E-16
7	0.1827268 E-11	-0.75311525 E-15	0.418390 E-22	0.1821 E-11	0.41838334 E-22
8	0.8863389 E-15	-0.75315520 E-15	0.112244 E-28	0.8825 E-15	0.11224343 E-28
9	0.1605597 E-18		0.413703 E-36	0.1597 E-18	0.41370659 E-36
10	0.1082885 E-22		0.208844 E-44	0.1058 E-22	0.20895932 E-44
11	0.2713141 E-27		0.144078 E-53		0.12279100 E-49
12	0.2521085 E-32		0.135609 E-63		0.13487374 E-49
13	0.8677126 E-38				0.17299316 E-60
14	0.1105113 E-43				-0.25496486 E-65
15	0.5203999 E-50				
16	0.9055032 E-50				
17	0.5818867 E-64				
18	0.4179442 E-72				

Table 2. m = 1, $= \pi^{-1/2}$

t	Present Solution	Complementary Function [8]	Finite Differences [5]
0	0.5641878 E 00	0.56418960 E 00	0.5642 E 00
1	0.8285570 E-01	0.50254543 E-01	0.5026 E-01
2	0.7226698 E-02	0.97802274 E-03	0.9782 E-03
3	0.3020138 E-03	0.33550350 E-05	0.3356 E-05
4	0.5431819 E-05	0.18221222 E-08	0.1823 E-08
5	0.3975088 E-07	0.12367523 E-12	0.1482 E-12
6	0.1146879 E-09	-0.29349128 E-13	0.1747 E-17
7	0.1279827 E-12	-0.34242684 E-13	0.2931 E-23
8	0.5456289 E-16	-0.39134491 E-13	0.6912 E-30
9	0.8813160 E-20		
10	0.5361614 E-24		
11	0.1223266 E-28		
12	0.1043287 E-33		
13	0.3317918 E-39		
14	0.3926980 E-45		
15	0.1727057 E-51		
16	0.2818780 E-58		
17	0.1705581 E-65		
18	0.1160366 E-73		

Table 3. m = 2, $\beta = 1/4$

	Present Solution	Complementary Functions [8]	Finite Differences [5]
0	0.2500006 E 00	0.25000000 E 00	0.2500 E 00
1	0.2340787 E-01	0.14197530 E-01	0.1420 E-01
2	0.1414359 E-02	0.19141103 E-03	0.1914 E-03
3	0.4411547 E-04	0.49007176 E-06	0.4901 E-06
4	0.6261059 E-06	0.20999802 E-08	0.2101 E-08
5	0.3764660 E-08	-0.36865462 E-13	0.1403 E-13
6	0.9193294 E-11	-0.72849101 E-13	0.1400 E-18
7	0.8879995 E-14	-0.98795539 E-13	0.2034 E-24
8	0.3334327 E-17	-0.12873356 E-12	0.4224 E-31
9	0.4809239 E-21		
10	0.2641945 E-25		
11	0.5493305 E-30		
12	0.4302831 E-35		
13	0.1265030 E-40		
14	0.1391954 E-46		
15	0.5719099 E-53		
16	0.8757835 E-60		
17	0.4990734 E-67		
18	0.3216635 E-75		

STABILITY AND ERROR ESTIMATES VALID FOR INFINITE TIME, FOR STRONGLY MONOTONE AND INFINITELY STIFF EVOLUTION EQUATIONS

O. AXELSSON
Department of Mathematics, University of Nijmegen
Toernooiveld, 6525 ED Nijmegen, The Netherlands

Abstract

For evolution equation with a monotone operator we derive unconditional stability and error estimates valid for all times. For the θ-method, with $\theta = 1/(2+\zeta\tau^{\nu}), 0<\nu\leq 1, \zeta>0)$, we prove an error estimate $O(\tau^{4/3})$, $\tau \to 0$, if $\nu = 1/3$, where τ is the maximal time-step for an arbitrary choice of the sequence of timesteps and with no further condition on F, and an estimate $O(\tau^2)$ under some additional conditions. The first result is an improvement over the implicit midpoint method ($\theta = \frac{1}{2}$), for which an order reduction to $O(\tau)$ may occur.

1. Introduction

Consider the evolution equation

(1.1) $\dot{u} + F(t,u) = 0,\ t > 0,\ u(0) = u_0 \in V,$

V a reflexive Banach space, where $\dot{u} = \frac{du}{dt}$ and $F(t,\cdot) : V \to V'$. Here V' is the space which is dual with respect to the innerproduct $(\cdot,\cdot)$ in a Hilbert space H, with norm $||v|| = (v,v)^{\frac{1}{2}}$.

We shall assume that F is a monotone operator, i.e.

(1.2) $(F(t,u) - F(t,v), u-v) \geq \rho(t)\ ||u-v||^2\ \forall\ u, v \in V,$

where $\rho : (0,\infty) \to R^+$, i.e. $\rho(t) \geq 0$, $t > 0$.

A typical example is the parabolic evolution equation

(1.3) $u_t = \underline{\nabla} \cdot (a(t,\underline{x},\underline{\nabla}u)\underline{\nabla}u) + g(t,u),\ t,\underline{x} \in (0,\infty) \cup \Omega\ ,\ \Omega \subset \mathbb{R}^d,$

with boundary conditions, say u = 0 on $\partial\Omega$. Here $V = [\overset{\circ}{H}{}^1(\Omega)]^d$ (a Cauchy product of the Sobolev space $\overset{\circ}{H}{}^1(\Omega)$), $H = L^2(\Omega)$ and, under certain conditions on a and g this is a parabolic problem, i.e. fulfills (1.2) with $\rho(t) > 0$.

Other important examples are conservative (hyperbolic) problems for which (1.2) is satisfied with $\rho(t) \geq 0$. In the present paper we restrict the analysis to the strongly monotone case, $\rho(t) \geq \rho_0 > 0$.

Classical techniques for the derivation of discretization error estimates for (1.3), uses a semidiscrete method for the discretization in space, namely the variational form

$$(u_t, \tilde{u}) + (F(t,u),\tilde{u}) = 0 \quad \forall \tilde{u} \in V_h \subset V,$$

where V_h is a finite element space depending on a mesh parameter h.

This semidiscrete method ("longitudal method of lines") results in a system of ordinary differential equations (ode) which is "stiff", i.e. components of the solution exist, which decay (exponentially) with largely different rates.

The system of ode can be solved by many methods for stiff ode's. The difficulty is in proving error estimates for the total error of the form $C_1h^p+C_2\tau^q$, $p,q > 0$, where τ is the time-step. Here C_1,C_2 should be independent of h,τ. Since the dimension of the ode depends on h, classical error estimates used in the numerical analysis of ode, cannot be applied. Furthermore, they provide usually only a bound growing with time t (sometimes even growing exponentially - see below). We want to derive error estimates which are valid (i.e. bounded) for all t.

We find it then convenient to consider a "transversal method of lines", i.e. first discretize the evolutionary problem (1.3), and more generally (1.1), w.r.t. time. A convenient time integration method turns out to be the implicit ("one-leg") form of the θ-method with $0 \le \theta \le 1 / 2(1+\zeta\tau^{\nu})$, $\zeta > 0$ for some ν, $0 \le \nu \le 1$. Error estimates $O((\frac{1}{2}-\theta)\tau + \tau^{(3-\nu)/2})$ valid for all t can now be derived in the strongly monotone case, where $\rho(t) \ge \rho_0 > 0$. Without further assumptions the optimal order we prove is $O(\tau^{4/3})$ for $\nu = 1/3$ and θ equal to the upper bound. With some additional assumptions we prove also the optimal order, $O(\tau^2)$.(At this point we remark that there is a marked difference in behaviour of the implicit and explicit forms of the θ-method, in particular for variable step-lengths. The latter method may not even converge.)

To illustrate the problems with proving error estimates for time-stepping methods, we consider the Euler (forward) method,

(1.4) $v(t+\tau) = v(t) - \tau\, F(t,v(t)),\ t = 0,\tau,2\tau,\ldots$

where v is the corresponding approximation to u.

(It is only for notational simplicity that we let the time step τ be constant.)

Let $e(t) = u(t) - v(t)$ be the error function. Classical error estimates, uses the two-sided Lipschitzconstant,

(1.5) $L = \sup\{||F(t,u) - F(t,v)|| \,/\, ||u-v||\},\ t > 0,\ u,v \in V_0 \subset V$

where V_0 contains all functions in a sufficiently large tube about the solution u. In the analysis of the Euler forward method we have to assume that F is two-sided Lipschitz bounded, i.e. that $L < \infty$, but for the implicit methods to be considered later, we need only a one-sided bound such as (1.2). From (1.1.) it follows

(1.6) $u(t+\tau) = u(t) - \int_0^1 F(t+\tau s,\ u(t+\tau s))ds$

and from (1.4) and (1.5) we get

(1.7) $e(t+\tau) = e(t) - \tau\{F(t,u(t)) - F(t,v(t))\} + \tau R(t,u),$

where

$$R(t,u) \equiv \int_0^1 [F(t,u(t)) - F(t+\tau s,u(t+\tau s))]ds = \int_0^1 [\mathring{u}(t+\tau s) - \mathring{u}(t)]ds$$

is the (normalized) local truncation error.

Note that

(1.8) $\sup\limits_{t>0} ||R(t,u)|| = \tau \sup\limits_{t>0} \int_0^1 ds \int_0^s ||u_t^{(2)}(t+\sigma\tau)||d\sigma \le \frac{1}{2}\tau D_2,$

where we use the notation

(1.9) $D_k = \sup\limits_{t>0} ||u_t^{(k)}(t)||,\ k = 1,2,\ldots$

and we assume that $u_t^{(k)} \in L_\infty(H)$, i.e. that $D_k < \infty$.

By (1.5) and (1.7) it follows

$$||e(t+\tau)|| \le (1+\tau L)||e(t)|| + \tau||R(t,u)||, \quad t = 0, \tau, 2\tau, \ldots$$

or, by recursion,

$$||e(t)|| \le (1+\tau L)^{t/\tau}||e(0)|| + \tau \sum_{j=1}^{t/\tau} (1+\tau L)^{j-1}||R(t-j\tau,u)||$$

or

$$(1.10) \quad ||e(t)|| \le e^{tL}||e(0)|| + \frac{1}{L}(e^{tL}-1) \max_{t\geq 0} ||R(t,u)||, \quad t = \tau, 2\tau, \ldots$$

Notice that the initial and truncation errors may grow as exp(tL).

By (1.8) we have $||R(t,u)|| \le C\tau$, where C depends only on the smoothness of the solution, and not on the Lipschitz constant L. However, in most problems of practical interest, L is large, so even for moderately large values of t, the truncation error is amplified by a large factor $\sim L^{-1}\exp(tL)$.

This is in particular true for stiff problems, in which case the bound (1.10) (and the method (1.4), even for very small time-steps satisfying $\tau L \ll 1$) is practically useless. This is in fact true for all explicit time-stepping methods.

However, we easily derive the following stability bound for solutions of the continuous problems valid.

$$||u(t)-w(t)|| \le \exp(\int_0^t - \rho(s)ds) \; ||u(0)-w(0)|| \le ||u(0)-w(0)||, \quad t > 0.$$

Here, u, w are solutions of (1.1) corresponding to different initial values, u(0) and w(0), respectively.

We now face the following problems:

(i) Can we find a numerical time-stepping method for which a similar stability bound is valid?

(ii) Can we derive discretization error estimates without a "nasty" large (exponentially growing) stiffness factor, such as the factor in (1.10)?

The answer to these problems is affirmative as was pointed out in [3] and [4] because the "backward" or implicit Euler method

$$(1.11) \quad v(t+\tau) + \tau F(t+\tau)) = v(t), \quad t = 0, \tau, 2\tau, \ldots$$

fulfills these conditions.

One finds now the error bound (if e(0) = 0)

$$(1.12) \quad ||e(t)|| \le \rho_0^{-1} \sup_{t>0} ||R(t,u)|| \le C\tau, \quad t > 0,$$

where C depends only on ρ_0 and D_2.

This method is only first order accurate.

In this report we discuss an extension of (1.12) to the class of θ-methods. The results found complement some of the results in [2].

2. Stability of the θ-method

We shall consider the implicit (also called one-leg) form of the θ-method

$$(2.1) \quad v(t+\tau) + \tau F(\bar{t},\bar{v}(t)) = v(t), \quad t = 0, \tau, 2\tau, \ldots,$$

$v(0) = u_0$, where $\bar{t} = \theta t + (1-\theta)(t+\tau) = t + (1-\theta)\tau$ and $\bar{v}(t) = \theta v(t) + (1-\theta)v(t+\tau)$, $0 \le \theta \le 1$. For θ = 0 and θ = 1 we get the Euler backward (i.e. the Rothe method (see [7]), for evolutionary partial differential equations and Euler forward methods, respectively.

When F is monotone, i.e. satisfies (1.2), it will follow that the nonlinear equation (2.1) has a unique solution $v(t+\tau)$ in V, if $\theta \le 1$.

As is wellknown the implicit form of the θ-method can be written as an Euler backward (implicit) step ($t \to \bar{t} = t + (1-\theta)\tau$).

(2.2) $v(\bar{t}) + \tau(1-\theta)F(\bar{t},v(\bar{t})) = v(t)$,

followed by an Euler forward (explicit) step ($\bar{t} \to t+\tau$)

(2.3) $v(t+\tau) + \tau\,\theta\,k(\bar{t}) = v(\bar{t})$,

where $k(\bar{t}) = F(\bar{t},v(\bar{t}))$.

(2.2) follows if we multiply (2.1) by $(1-\theta)$ and define $v(\ )$ as a linear function in each interval $[t,t+\tau]$. Then $v(\bar{t}) = \bar{v}(t)$. (2.3) follows if we subtract (2.2) from (2.1).

In practice we perform errors, such as iteration and round-off errors when solving (2.2) and also round-off errors when computing $v(t+\tau)$ from (2.3). (In the parabolic evolution equation, we also get space discretization errors, when solving (2.2).) We shall assume that these errors are $\tau r_\theta(t)$ and $\tau s_\theta(t)$, respectively, where $||r_\theta(t)|| \le C_1, ||s_\theta(t)|| \le C_2$, $t \ge 0$, and C_i, $i = 1,2$ are constants, independent of τ. We get then the <u>perturbed equations</u>

(2.4.1) $\tilde{v}(t) + \tau(1-\theta)F(\bar{t},\tilde{v}(\bar{t})) = \tilde{v}(t) + \tau r_\theta(t)$.

(2.4.b) $\tilde{v}(t+\tau) + \tau\,\theta\,\tilde{k}(\bar{t}) = \tilde{v}(\bar{t}) - \tau s_\theta(t)$,

$\tilde{k}(\bar{t}) = F(\bar{t},\tilde{v}(\bar{t}))$,

which are the equations the computed approximations $\tilde{v}$ actually satisfy.

Multiplying (2.4.a) by θ and subtracting (2.4.b), multiplied by $(1-\theta)$, we get

(2.5) $\tilde{v}(\bar{t}) = (1-\theta)\tilde{v}(t+\tau) + \theta\tilde{v}(t) + \tau\alpha(t) = \bar{\tilde{v}}(t) + \tau\alpha(t)$,

where $\alpha(t) = \theta r_\theta(t) + (1-\theta)s_\theta(t)$.

By summation of (2.4.a) and (2.4.b), we find

(2.6) $\tilde{v}(t+\tau) + \tau F(\bar{t},\tilde{v}(\bar{t})) = \tilde{v}(t) + \tau\beta(t)$,

where $\beta(t) = r_\theta(t) - s_\theta(t)$.

For the unperturbed equations we have

(2.5') $v(\bar{t}) = \bar{v}(t)$

and

(2.6') $v(t+\tau) + \tau F(\bar{t},v(\bar{t})) = v(t)$,

respectively.

Let the difference be $e(t) = \tilde{v}(t) - v(t)$.

We find then from (2.5), (2.5') and (2.6), (2.6'),

(2.7) $e(\bar{t}) = \bar{e}(t) + \tau\alpha(t)$,

(2.8) $e(t+\tau) - e(t) + \tau[F(\bar{t},\tilde{v}(\bar{t})) - F(\bar{t},v(\bar{t}))] = \tau\beta(t)$,

respectively.

We shall assume that $\rho(t) \ge \rho(t) \ge \rho_0 > 0$ in (1.2).

Taking the inner product by (2.8) with $e(\bar{t})$, we find then, by (1.2) and (2.7),

$(e(t+\tau) - e(t), \bar{e}(t) + \tau\alpha(t)) + \tau\rho_0||\bar{e}(t) + \tau\alpha(t)||^2 \le \tau(\beta, \bar{e}(t) + \tau\alpha(t))$.

By use of the arithmetic-geometric mean inequality, we find

$\tau(\beta,e(\bar{t})) \leq \frac{1}{2}\rho_0^{-1}\tau||\beta||^2 + \frac{1}{2}\tau\rho_0||e(\bar{t})||^2$, and

$(e(t+\tau) - e(t), \bar{e}(t)) + \frac{1}{2}\tau\rho_0||\bar{e}(t) + \tau\alpha(t)||^2 \leq \frac{1}{2}\rho_0^{-1}\tau||\beta||^2 - (e(t+\tau) - e(t), \tau\alpha)$.

By use of the inequality, $||a+b||^2 \geq \frac{1}{2}||a||^2 - ||b||^2$ and the arithmetic-geometric inequality once more we get

$$(2.9) \quad (e(t+\tau) - e(t), \bar{e}(t)) + \tfrac{1}{4}\tau\rho_0||\bar{e}(t)||^2 \leq \tfrac{1}{2}\rho_0^{-1}\tau||\beta||^2 + \tfrac{1}{2}\tau^{\nu}||e(t+\tau)-e(t)||^2 + \tfrac{1}{2}\tau^{2-\nu}(1+\rho_0\tau^{1+\nu})||\alpha||^2,$$

where $0 \leq \nu \leq 1$. The chosen value of ν will be specified later.

An elementary computation (see [1] shows that

$$(e(t+\tau) - e(t), \bar{e}(t)) = \tfrac{1}{2}[||e(t+\tau)||^2 + (1-2\theta)||e(t+\tau) - e(t)||^2 - ||e(t)||^2]$$

and

$$||\bar{e}(t)||^2 = (1-\theta)||e(t+\tau)||^2 + \theta||e(t)||^2 - (1-\theta)\theta||e(t+\tau) - e(t)||^2.$$

Using these identities in (2.9), we find

$$(2.10) \quad [1 + \tfrac{1}{2}\tau\rho_\theta(1-\theta)]||e(t+\tau)||^2 + [1-2\theta - \tfrac{1}{2}\tau\rho_0(1-\theta)\theta - \tau^{\nu}]||e(t+\tau) - e(t)||^2 \leq [1- \tfrac{1}{2}\tau\rho_0\theta]||e(t)||^2 + \rho_0^{-1}\tau||\beta||^2 + 2\tau^{2-\nu}||\alpha||^2,$$

where we have assumed that $\tau \leq 1$ is small enough so that $\rho_0\tau^{1+\nu} \leq 1$.

We shall now choose $\theta \leq \theta_0$, where θ_0 is the largest number ≤ 1, for which the factor of the second term of (2.10), $1 - 2\theta - \frac{1}{2}\tau\rho_0 (1-\theta)\theta - \tau^{\nu} \geq 0$.

We find then $\theta_0 = \frac{1}{2} - |O(\tau^{\nu})|$, $\tau \to 0$.

By recursion, it now follows from (2.10),

$$||e(t)||^2 \leq q^{t/\tau}||e(0)||^2 + \tau\rho_0^{-1}\sum_{j=1}^{(t/\tau)-1} q^{(t/\tau)-j-1}[1+\tfrac{1}{2}(1-\theta)\tau\rho_0]^{-1}\sup_{s>0} \gamma^2(s)$$

where

$$(2.11) \quad \gamma^2(s) = ||\beta(s)||^2 + 2\rho_0\tau^{1-\nu}||\alpha(s)||^2,$$

and

$$q = (1 - \tfrac{1}{2}\theta\tau\rho_0) / [1 + \tfrac{1}{2}(1-\theta)\tau\rho_0]$$

Since $\theta < \frac{1}{2}$, we have $q < 1$, and we find

$$||e(t)||^2 \leq q^{t/\tau}||e(0)||^2 + \rho_0^{-2}[2 + (1-\theta)\tau\rho_0] \sup_{s>0} \gamma^2(s),\ \forall t > 0.$$

Hence, the θ-method is unconditionally stable (independent of the stiffness and of τ), if $0 \leq \theta_0$.

We collect the result found in

Theorem 2.1. (Stability.) If (1.1) is strongly monotone, i.e. $\rho(t) \geq \rho_0 > 0$ in (1.2), and if $\theta \leq \theta_0$, where θ_0 is the largest number ≤ 1, for which $1-2\theta-\frac{1}{2}\tau\rho_0(1-\theta)\theta-\tau^{\nu} \geq 0$, $0 \leq \nu \leq 1$, then

$$||e(t)||^2 \leq g^{t/\tau}||e(0)||^2 + \rho_0^{-2}[2 + (1-\theta)\tau\rho_0] \sup_{s>0} \gamma^2(s),\ \forall t > 0$$

where $\gamma(s)$ satisfies (2.11).

Here $e(t) = \tilde{v}(t) - v(t)$ is the perturbation error, $\tilde{v}(t)$ is the solution of the perturbed equations (2.4.a,b), and $v(t)$ is the solution of the unperturbed θ-method (2.1)

Corollary 2.1. If $e(0) = 0$, then

$$(2.12) \quad ||e(t)|| \leq \rho_0^{-1} [2+(1-\theta)\tau\rho_0]^{\frac{1}{2}} \sup_{s>0}|\gamma(s)|,\ \forall t > 0.$$

This generalizes the stability part of (1.12) to the implicit class of θ-methods.

3. Truncation errors

It remains to consider the truncation errors for the θ-method. For the solution u of (1.1) we have

(3.1) $u(\bar{t}) = \bar{u}(t) + \tau\,\alpha_\theta(t)$,

where, by an elementary computation,

$$\alpha_\theta(t) = -(1-\theta)\tau \int_0^1 ds \int_{-\theta s}^{(1-\theta)s} \ddot{u}(t-\sigma\tau)\,d\sigma$$

Hence

(3.2) $\sup_{t>0} ||\alpha_\theta(t)|| = \tfrac{1}{2}\theta(1-\theta)\tau\, D_2.$

Similarly,

(3.3) $u(t+\tau) + \tau F(\bar{t},u(\bar{t})) = u(t) + \tau\beta_\theta(t)$,

where,

$$\begin{aligned}(3.4)\quad \beta_\theta(t) &= \tau^{-1}(u(t+\tau) - u(t) - \tau\mathring{u}(\bar{t}))\\ &= \tau \int_0^1 ds \int_{1-\theta}^{s} \ddot{u}(t+\sigma\tau)\,d\sigma\\ &= \tau \int_0^1 ds \int_{\frac{1}{2}}^{s} \ddot{u}(t+\sigma\tau)\,d\sigma + \tau \int_0^1 ds \int_{1-\theta}^{\frac{1}{2}} \ddot{u}(t+\sigma\tau)\,d\sigma\\ &= \tau \int_0^{\frac{1}{2}} ds \,[\int_0^s \ddot{u}(t+(\tfrac{1}{2}+\sigma)\tau)\,d\sigma - \ddot{u}(t+(\tfrac{1}{2}-\sigma))\tau]\,d\sigma + \tau \int_0^1 ds \int_{1-\theta}^{\frac{1}{2}} \ddot{u}(t+\sigma\tau)\,d\sigma.\end{aligned}$$

Hence, if $u_t^{(3)} \in L_\infty(H)$, i.e. $\sup_{s>0}||u_t^{(3)}(s)|| < \infty$, then

(3.5) $\sup_{t>0} ||\beta_\theta(t)|| \le \frac{1}{24}\tau^2 D_3 + \tau|\tfrac{1}{2}-\theta|D_2.$

Let the time-discretization error, $E(t) = u(t) - v(t)$. By (2.5'), (3.1) and (2.6'), (3.3) and using the estimates in section 2, we get by Corollary (2.1), for the strongly monotone case,

(3.6) $||E(t)|| \le \rho_0^{-1}[2 + (1-\theta)\tau\rho_0]^{\frac{1}{2}} \sup_{t>0} |\gamma_\theta(t)|,\ 0 \le \theta \le \theta_0,$

where $\gamma_\theta^2(t) = ||\beta_\theta(t)||^2 + 2\rho_0\tau^{1-\nu}||\alpha_\theta(t)||^2$.

Hence, by (3.2) and (3.5),

(3.7) $|\gamma_\theta(t)| = \frac{1}{24}\tau^2 D_3 + \tau|\tfrac{1}{2}-\theta|D_2 + \sqrt{\rho_0/2}\;\tau^{(3-\nu)/2}\theta(1-\theta)D_2.$

With $\theta = 1/(2+\zeta\tau^\nu) \le \theta_0$ (i.e. with ζ a large enough positive number), (3.7) implies

$|\gamma_\theta(t)| = |O(\tau^2)| + |O(\tau^{1+\nu})| + |O(\tau^{(3-\nu)/2})|,\ \tau \to 0.$

Its order is highest, namely $O(\tau^{4/3})$, if we choose $\nu = 1/3$.

We collect these results in

Theorem 3.1. (Discretization error) The discretization error of the θ-method with $\theta = 1/(2+\zeta\tau^\nu) \le \theta_0$, $\zeta > 0$, where θ_0 is defined in Theorem 2.1, satisfies

$$||E(t)|| \le \rho_0^{-1}[2 + (1-\theta)\tau\rho_0]^{\frac{1}{2}} \sup_{t>0} |\gamma_\theta(t)| = \begin{cases} |O(\tau^{1+\nu})|, & \text{if } 0 \le \nu \le \frac{1}{3} \\ |O(\tau^{(3-\nu)\frac{1}{2}})|, & \text{if } \frac{1}{3} \le \nu \le 1 \end{cases} \quad \forall\, t > 0,$$

for any solution u of a strongly monotone problem (1.1), for which $u_t^{(3)} \in L_\infty(H)$.

Its order is highest, $||E(t)|| = |O(\tau^{4/3})|$, if $\nu = 1/3$.

Remark 3.1. It follows readily from (3.4), that Theorem 3.1 remains valid if we replace the regularity requirement, $u_t^{(3)} \in H$, with the weaker requirement that $u_t^{(2)}$ is Höldercontinuous with exponent ν. In fact it suffices that $u_t^{(2)}$ is Hölder-continuous in the interior of each interval $(t,t+\tau)$.

Remark 3.2. Theorem 3.1 remains valid for any choice of timesteps τ_k, constant or variable, for which $\tau_k \leq C\tau$, for some positive constant C.

In some problems we have to adjust the timesteps to get convergence or fast enough convergence, because some derivative of u of low order can be discontinuous at certain points. It may for instance happen that F in (1.1) is discontinuous for certain values of t.

In such cases we want to adjust the timesteps so that those values of t become stepping-points. Hence the result in Theorem 3.1, although not of optimal order as we shall see, is of particular importance for cases where we have to change the time-steps in an irregular fashion.

We shall now present an optimal order, $O(\tau^2)$, result, but valid only if the timesteps are essentially constant.

Consider first the equations (2.4.a,b) for variable parameters $\theta = \theta_k$ and $\tau = \tau_k$, $k = 0,1,\ldots$. For the solution of (1.1) we get then truncation errors $R_k = R(t_k,\tau_k,\theta_k)$ and $S_k = S(t_k,\tau_k,\theta_k)$, defined by

(3.8.a) $u(\bar{t}) + \tau_k(1-\theta_k)F(\bar{t},u(\bar{t})) = u(t_k) - \tau_k R_k,$

(3.8.b) $u(t_{k+1}) + \tau_k\theta_k\, F(\bar{t},u(\bar{t})) = u(\bar{t}) + \tau_k S_k,$

where $\bar{t} = t_k + (1-\theta_k)\tau_k$ and t_k is the k'th stepping point.

Guided by a trick in [5] for the implicit midpoint method (i.e. (2.2), (2.3) with $\theta = \frac{1}{2}$) we define

(3.9) $\hat{u}(t_k) = u(t_k) - \tau_k R_k.$

Then (3.8.a,b) takes the form

(3.10.a) $u(\bar{t}) + \tau_k(1-\theta_k)F(\bar{t},u(\bar{t})) = \hat{u}(t_k),$

(3.10.b) $\hat{u}(t_{k+1}) + \tau_k\theta_k F(\bar{t},u(\bar{t}) = u(\bar{t}) + \tau_k\hat{\beta}_\theta(t_k),$

where

(3.11) $\hat{\beta}_\theta(t_k) = S_k - \tau_{k+1}R_{k+1}/\tau_k$

Multiplying (3.10.a) by θ_k and subtracting (3.10.b), multiplied by $(1-\theta_k)$, yields

(3.12.a) $u(\bar{t}) = \theta_k\hat{u}(t_k) + (1-\theta_k) + (1-\theta_k)\hat{u}(t_{k+1}) - (1-\theta_k)\hat{\beta}_\theta(t_k)$

Summation of (3.10.a) and (3.10.b) yields

(3.12.b) $\hat{u}(t_{k+1}) + \tau_k F(\bar{t},u(\bar{t})) = \hat{u}(t_k) + \tau_k\hat{\beta}_\theta t_k).$

We define $\hat{u}(\bar{t}) = u(\bar{t})$ (we let $0 < \theta_k$, so $t_k < \bar{t} < t_{k+1}$), so by (3.12.a),

(3.13) $\hat{u}(t) = \bar{\hat{u}}(t_k) + \tau_k\hat{\alpha}_\theta(t_k),$ where $\hat{\alpha}_\theta(t_k) = -(1-\theta_k)\hat{\beta}_\theta(t_k).$

Note that (3.13) and (3.12.b) have the same form as (3.1) and (3.3), respectively.

To estimate $\hat{\alpha}_\theta$ and $\hat{\beta}_\theta$, we need to estimate R_k and S_k. By (3.8.a) and (3.8.b) we find

$$(3.14)\quad R_k = (1-\theta_k)\,[\mathring{u}(\bar{t}) - \int_0^1 \mathring{u}(t_k + (1-\theta_k)\tau_k s)ds]$$
$$= (1-\theta_k)^2\tau_k \int_0^1 ds \int_s^1 \ddot{u}(t_k + (1-\theta_k)\tau_k\sigma(d\sigma)$$

and

$$(3.15)\quad S_k = \theta_k[\int_0^1 \mathring{u}(t_k+\tau_k-\theta_k\tau_k s)ds - \mathring{u}(\bar{t})]$$
$$= \theta_k^2\tau_k \int_0^1 ds \int_s^1 \ddot{u}(t_k+\tau_k-\theta_k\tau_k\sigma)d\sigma.$$

By (3.11), (3.14) and (3.15) we find

$$||\hat{\beta}_\theta(t_k)|| = \tau_k^{-1}|(\theta_k\tau_k)^2 - (1-\theta_{k+1})^2\tau_{k+1}^2|\ \tfrac{1}{2}D_2 + O(\tau_k^2)D_3$$

if $c\tau \le \tau_k \le C\tau$ for some positive constants c,C.

Hence $||\hat{\beta}_\theta(t_k)|| = O(\tau^2)$, $\tau \to 0$, $k = 0,1,\ldots$ if $(\theta_k\tau_k)^2 - (1-\theta_{k+1})^2\tau_{k+1}^2 = O(\tau^3)$ or $\tau_{k+1} = \frac{\theta_k}{1-\theta_{k+1}}\tau_k + O(\tau^2)$.

Remark 3.3. Since for stability reason, $\theta_k < \tfrac{1}{2}$, we see that if

$$(3.16)\quad \tau_{k+1} = \frac{\theta_k}{1-\theta_{k+1}}\tau_k,$$

then $\{\tau_k\}$ is a decreasing sequence. Frequently, in practice we want to choose small steps in the initial (transient period) and then larger steps, i.e. contrary to (3.16).

Consider now for simplicity the case $\tau_k = \tau$, $\theta_k = \theta = 1/(2+\zeta\tau)$, $\zeta > 0$, $k = 0,1,\ldots$ Let $\hat{E}(t) = \hat{u}(t) - v(t)$. It follows as before (see 3.6) and Theorem 2.1) that

$$||\hat{E}(t)||^2 \le q^{t/\tau}||\hat{E}(0)||^2 + \rho_0^{-2}\,[2 + (1-\theta)\tau\rho_0]\,O(\tau^2)D_3$$

We have $\hat{E}(0) = \hat{u}(0) - v(0) = \hat{u}(0) - u(0) = \tau_0 R_0$, i.e. by (3.9) and (3.14),
$||\hat{E}(0)|| = O(\tau^2)D_2$.

Similarly, by (3.1),

$$||u(t) - v(t)|| \le ||u(t) - \hat{u}(t)|| + ||\hat{u}(t) - v(t)|| = O(\tau^2)D_2 + ||\hat{E}(t)||.$$

We collect these results in

Theorem 3.2. If $\tau_{k+1} = \dfrac{\theta_k}{1-\theta_{k+1}}\tau_k + O(\tau^2)$,

$\theta_k = 1/(2+\zeta_k\tau_k)$, $\zeta > \zeta_k > 0$, $k = 0,1,\ldots$, then the θ-method (2.2), (2.3) has a discretization error $O(\tau^2)$, valid at all stepping points t_k, if $u_t^{(3)} \in L_\infty(H)$ and if (1.1) is strongly monotone.

Remark 3.4. In [2], it is proven an optimal order, $O(\tau^2)$ estimate, valid for arbitrary variable time-steps, if in addition to the assumptions in Theorem 3.1, we assume that $\nu = 1$, that $||\partial F/\partial t||$ is not large and that the Gataux derivative $\partial F/\partial u$ exists and satisfies: $||\partial F/\partial u\ u_t^{(2)}||$ is of the same order as D_3 (i.e. not large for smooth solutions). Note that for a linear problem $u_t = Au$ with constant operator A, we have $\partial F/\partial u\ u_t^{(2)} = A^3u = u_t^{(3)}$. For a more general parabolic problem, we have typically that $\sup_{t\ge t_0} ||\partial F/\partial u\ u_t^{(2)}||$ is of the order of $\sup_{t\ge t_0} ||u_t^{(3)}||$ when the solution (and its derivatives) is smooth for $t \geq t_0$, because then u has essentially components along the eigenfunctions corresponding to the smallest eigenvalues of the Jacobian $\partial F/\partial u$. In the results presented in the present paper, we have however not even assumed the existence of the Jacobian.

4. Conclusions

In [6] was shown by considering the problem $\mathring{u}(t) + \lambda(u-g(t)) = \mathring{g}(t)$, $t > 0$ for λ very large, that the accuracy of the approximate solutions obtained often are unrelated to the classical order of the method used.

For the implicit midpoint method (i.e. (2.1) with $\theta = \frac{1}{2}$), this error order reduction is easily seen to be caused by that the damping factor q in Theorem 2.1 approaches the value -1. For (almost) constant steplength this causes a cancellation effect and the global error remains $O(\tau^2)$, but for λ and/or τ variable this is not the case and the order is only $O(\tau)$ in general.

We have shown that by choosing $\theta = 1/(2+\zeta\tau^\nu)$, $\zeta > 0$, $0 < \nu < 1$, a higher order (at least $O(\tau^{4/3})$) can be achieved. This is due to the damping with a factor q, where $|q| \simeq \theta/(1-\theta)$ for λ large.
Under additional assumptions and with $\nu = 1$ we can also get an error $O(\tau^2)$. Hence the error order is never worse that for the implicit midpoint rule.

It is anticipated that a similar modification of higher order Lobatto type implicit Runge-Kutta methods can give a less severe order reduction than if they are not modified (cf [6] and [4]).

References

1. O. Axelsson, Error estimates for Galerkin methods for quasilinear parabolic and elliptic differential equations in divergence form, Numer. Math. 28, 1-14(1977).

2. O. Axelsson, Error estimates over infinite intervals of some discretizations of evolution equations, BIT 24(1984), 413-424.

3. G. Dahlquist, Error analysis for a class of methods for stiff nonlinear initial value problems, Numerical Analysis (G.A. Watson, ed.), Dundee 1975, Springer-Verlag, LNM 506, 1976.

4. R. Frank, J. Schneid and C.W. Ueberhuber, The concept of B-convergence, SIAM J. Numer. Anal. 18(1981), 753-780.

5. J. Kraaijevanger, B-convergence of the implicit midpoint rule and the trapezoidal rule, Report no. 01-1985, Institute of Applied Mathematics and Computer Science, University of Leiden, The Netherlands.

6. A. Prothero and A. Robinson, The stability and accuracy of one-step methods, Math. Comp. 28(1974), 145-162.

7. K. Rektorys, The Method of Discretization in Time and Partial Differential Equations D. Reidel Publ. Co., Dordrecht-Holland, Boston-U.S.A., 1982.

RECENT RESULTS IN THE APPROXIMATION OF FREE BOUNDARIES

F. BREZZI
Instituto di Analisi Numerica del C. N. R., Universita di Pavia
C.so Carlo Alberto, 5 - 27100 Pavia, Italy

§ 1. We present here a short survey on results recently obtained in the approximation of free boundaries. For examples of free boundary problems that are interesting in physics and engineering we refer for instance to |1|,|7|,|8|. Here we shall stay at a very abstract level, without considering, essentially, the nature of the free boundary problem under consideration nor the type of discretization which is employed. In that we are rather following |3| or the first part of |9|. For practical cases in which the following results apply we refer to |4|, |10| and |9|.

In the next section we present the framework in which the theory will be developed and in the third section we shall present some abstract results, most of them without proof. The proofs can be found in the corresponding references.

§ 2. For the sake of simplicity we shall consider the following "model situation. We are given a bounded domain D in $\mathbb{R}^n$ with piecewise Lipschitz boundary (to fix the ideas). We are also given a function u(x) in $C^\circ(\bar{D})$. The function u(x) will be the solution of our free boundary problem. The nature of the problem itself is immaterial at this stage. We assume that

(1) $\quad u(x) \geq 0 \quad \forall x \in \bar{D}$

and we assume that the continuous free boundary F is characterized by

(2) $\quad F := D \cap \partial(D^+)$

where

(3) $\quad D^+ := \{x \mid x \in \bar{D}\ ,\ u(x)>0\};$

We assume , finally that we have constructed a sequence $\{u_h(x)\}$, for $0<h\leq h_o$ of "approximating solutions", <u>which converges to</u> u(x) <u>in</u> $C^\circ(\bar{D})$: Again, the procedure employed to construct $\{u_h\}$ is irrelevant at the moment. We set

(4) $\quad E_p(h) := \|u-u_h\|_{L^p(D)} \qquad 1 \leq p \leq \infty$.

and we remark that we have already assumed

(5) $\lim_{h\to 0} E_\infty(h) = 0$

We would like to construct a "discrete free boundary" F_h as in (2) and then to estimate the distance of F_h from F in terms of $E_p(h)$, defined in (4). In order to make our life even easier, we assume that, as in (1),

(6) $u_h(x)\geq 0 \quad \forall x \in \bar{D}, \quad \forall h\leq h_0$

and we set, as a first trial,

(7) $D_h^+ := \{x | x \in \bar{D}, u_h(x)>0\}$

(8) $F_h := \bar{D} \cap \partial(D_h^+)$

Unfortunatly, elementary examples show that F_h can be very far from F even for u_h very close to u. For instance if $D=|-1,1|$ and $u(x)=(x)^+$ (that is $u(x)=0$ for $x<0$ and $u(x)=x$ for $x>0$) we have $F=\{0\}$. If now $u_h(x)=u(x)+h^s(x+1)$ the $F_h=\{-1\}$ no matter how small is h or how big is s. It should be clear now that the setting (1)...(8) does not allow the proof of any bounds on the distance of F_h from F. In the next section we shall present a few remedies that (under suitable additional assumptions) have been proposed to improve the situation.

§ 3. The first trial in this direction has been done about ten years ago in |2|. Assume that g(h) is a function of h such that

(9) $E_\infty(h)<g(h) \quad 0<h\leq h_1\leq h_0$

(10) $\lim_{h\to 0} g(h)=0$

and set

(11) $D_{h,g}^+ = \{x | x \in \bar{D}, u_h(x)>g(h)\}$.

Then we have |2|.

Theorem 1- Under the above assumptions we have, for $h<h_1$

(12) $D_{h,g}^+ \subset D^+$,

and, for all $x \in D^+$, there exists $h_2>0$ such that $x \in D_{h,g}^+$ for $h<h_2$.

The proof is immediate. We point out that, in other words, $D_{h,g}^+$ converges to D^+ "from inside". At our knowledge theorem 1 is still among the best result that one can obtain without additional information.

A tipical additional information that one can get, in many cases, is the behaviour of u(x) in D^+, near the free boundary. To fix the ideas, assume from now on that F is a smooth (say, C^1) surface. For any point $\bar{x}$ of F one can look at the restriction of u along the direction normal to F and pointing into D^+. If

(13) $u(x) \geq c(x-\bar{x})^s$ for $|x-\bar{x}| \leq a$ $(s \in \mathbb{R})$

with c,a and s independent of $\bar{x}$ we shall say that "u(x) grows like d^s near F" (d is for distance to the free boundary). For instance in the particular case of a nice obstacle problem, it has been proved in |5| that u(x) grows like d^2 near F. If one can prove a similar property for the approximations $u_h(x)$, one can use such an information to get estimates on the distance of F_h from F. This has been done in |4| for the case of a nice obstacle problem with a piecewise linear finite element approximation that satisfies the discrete maximum principle. The result in |4| is essentially that the distance of F_h from F behaves like $(E_\infty(h))^{\frac{1}{2}}$. This idea has then been extended in |10| to the one-phase Stefan problem in several dimensions, and in |11| to parabolic variational inequalities of obstacle type. The major drawback of this technique, however, is that it is often very difficult to prove growth properties for the discrete solutions $u_h(x)$, while the behaviour of u(x) itself is easier to analyze (see |8|,|7| for several growth properties proved on the continuousproblem).

A new and interesting set of results has been then obtained in |9| by combining, somehow, the two previous techniques: to change D_h^+ into a suitable $D_{h,g}^+$ and to use some growth property on u(x). In order to give the flavour of this procedure we shall present here two results in this direction. More detailed results and examples can be found in |9|.

Theorem 2 |9| With the notations and assumptions (1)...(11), if u(x) has the growth property (13) for some s>0, then there exist c>0 and h_3>0 such that for all h with $0<h\leq h_3$

(14) $\text{dist}\,(F_{h,g},F) \leq (cg(h))^{1/s}$,

where

(15) $F_{h,g} := D \cap \partial(D_{h,g}^+)$

and (14) means

(16) $\forall\, \bar{x}_h \in F_{h,g}\ \exists\, \bar{x} \in F$ such that $|\bar{x}-\bar{x}_h| \leq (cg(h))^{1/s}$.

Proof - Let $\bar{x}_h \in F_{h,g}$. This implies $u_h(\bar{x}_h)=g(h)>E_\infty(h)$, and therefore $\bar{x}_h \in D^+$. From (13) one has now that, for h small enough,

(17) $u(\bar{x}_h) \geq c|\bar{x}-\bar{x}_h|^s$

for some $\bar{x}\in F$. Since $u_h(\bar{x}_h)=g(h)$ one has from the triangle inequality

(18) $u(\bar{x}_h)\leq g(h)+E_\infty(h)<2g(h)$

and (16) follows from (17) and (18).

Theorem 3 Under the same assumptions of theorem 2, if

(19) $g(h)>E_p(h)^{sp/(1+sp)}$

then there exist $c>0$ and $h_4>0$ such that

(20) $\operatorname{meas}(D^+\Delta D^+_{h,g})\leq c(g(h))^{1/s}$

where in (20) the symbol Δ indicates as usual the symmetric difference of sets.

The proof can be found in |9|.

We would like to conclude with a somehow phylosophical remark (see |3|). In general, condition (13) implies that the global regularity of u(x) in $\bar{D}$ is, at best

(21) $u\in C^s(\bar{D})$.

If one uses, for instance, finite element methods in order to approximate u, one has, at best (see e.g. |6|):

(22) $E_\infty(h)=\|u-u_h\|_{C^\circ}\leq ch^r\|u\|_{C^r}$ $0\leq r\leq\min(k+1,s)$

where k is the degree of the polynomials. Using now theorem 2, say, with $g(h)=2E_\infty(h)$ one has from (14)

(23) $\operatorname{dist}(F_{h,g},F)\leq c(E_\infty(h))^{1/s}$

and from (23) and (22):

(24) $\operatorname{dist}(F_{h,g},F)\leq ch^{r/s}$ (at best).

One can now make the following observations.

1) Since $r\leq s$ in (24), the error cannot beat the mesh size.

2) For s "big" one need a "big" k (that is, polynomials of high degreee) so that the case r=s can be achieved in (22) (and hence in (24)).

3) For s "small" the error that one gets from (24) is surprising good. Unfortunately, for irregular u, (22) is often difficult to prove in practical cases.

References

|1| C. Baiocchi - A. Capelo: "Variational and quasivariational inequalities. Applications to free boundary problems", J. Wiley, 1984.

|2| C. Baiocchi - G. Pozzi: "Error estimates and free-boundary convergence for a finite difference discretization of a parabolic variational inequality", RAIRO Numer. Anal. 11, 4 (1977), 315-340.

|3| F. Brezzi: "Error estimates in the approximation of a free boundary", Math. Probl. in Structural Analysis, (G. Del Piero-F. Mauceri Eds.), Springer, CIS, Courses and Lect. n. 288 (1985), 17-23.

|4| F. Brezzi - L. Caffarelli: "Convergence of the discrete free boundaries for finite element approximations", RAIRO Numer. Anal., 17 (1983), 385-395.

|5| L. Caffarelli: "A remark on the Haussdorff measure of a free boundary, and the convergence of coincidence sets", Boll. U.M.I. (5), 18 A (1981), 109-113.

|6| P. Ciarlet: "The Finite Element Method for Elliptic Problems", North-Holland (1978).

|7| A. Friedman: "Variational Principles and Free Boundary Problems", Wiley, New York, (1982).

|8| E. Magenes, Editor (1980), "Free Boundary Problems", 2 vol., Proc. Sem. (Pavia, 1979), Istituto Nazionale di Alta Matematica, Roma.

|9| R.H. Nochetto: "A note on the approximation of free boundaries by finite element methods", To appear in R.A.I.R.O. Anal.Numer.

|10| P. Pietra - C. Verdi: "Convergence of the approximated free-boundary for the multidimensional one-phase Stefan problem", (to appear in Computational Mechanics).

|11| P. Pietra - C. Verdi: "Convergence of the approximate free boundary for the multidimensional one-phase Stefan problem" (to appear).

FINITE ELEMENT SOLUTION OF A NONLINEAR DIFFUSION PROBLEM WITH A MOVING BOUNDARY

L. ČERMÁK and M. ZLÁMAL
Computing Center of the Technical University in Brno
Obránců míru 21, 602 00 Brno, Czechoslovakia

In recent years two-dimensional process simulators for modelling and simulation in the design of VLSI semiconductor devices have appeared (see, e.g., Maldonado [2]). The underlying mathematical problem consists in solving numerically the following boundary value problem:

(1) $\frac{\partial u}{\partial t} = \nabla.[D(u)\nabla u]$ in $\Omega(t)$, $0 < t < T$,
$\Omega(t) = \{(x,y) | \varphi(y,t) < x < L_0,\ 0 < y < B\}$,

(2) $\frac{\partial u}{\partial n}\Big|_{\partial\Omega(t)-\Gamma(t)} = 0$, $0 < t < T$, $\Gamma(t) = \{(x,y) | x = \varphi(y,t),$
$0 < y < B\}$,

(3) $D(u)\frac{\partial u}{\partial n} = \gamma\dot{\varphi}_n u$ on $\Gamma(t)$, $0 < t < T$,

(4) $u(x,y,0) = u^*(x,y)$ in $\Omega(0)$.

Here u is the unknown concentration of an impurity, D(u) is the concentration dependent diffusion coefficient ($0 < d_0 \le D(u) \le d_0^{-1}$ $\forall u \ge 0$), $\varphi(y,t)$ is a given function ($0 \le \varphi \le \frac{1}{2} L_0$), $\frac{\partial u}{\partial n}$ is the derivative in the direction of the outward normal, γ is constant, $\dot{\varphi}_n$ is the rate of the motion of $\Gamma(t)$ in the direction of the outward normal and u^* is the given initial concentration.

If u is a sufficiently smooth solution of (1) - (4) (we remark that we do not know any result from which existence of a solution of (1) - (4) follows), then by multiplying (1) by $v \in H^1(\Omega(t))$ and integrating over $\Omega(t)$ we get

(5) $\forall t \in (0,T)$ $(\frac{\partial u}{\partial t},v)_{L^2(\Omega(t))} + a(u,t;u,v) = 0$ $\forall v \in V(t) \equiv H^1(\Omega(t))$;

here

$$a(w,t;u,v) = \int_{\Omega(t)} D(w)\nabla u.\nabla v dxdy - \gamma \int_{\Gamma(t)} \dot{\phi}_n uvdxdy.$$

We use (5) for defining the semidiscrete solution. First we construct a suitable moving triangulation of $\bar{\Omega}(t)$. We consider the one-to-one mapping of the rectangle $\bar{Q} = \langle 0,L_0\rangle \times \langle 0,B\rangle$ on $\bar{\Omega}(t)$:

(6) $x = F(\alpha,\beta,t) \equiv \varphi(\beta,t) + \alpha[1-L_0^{-1}\varphi(\beta,t)]$, $y = \beta$.

We cover $\bar{\Omega}(0)$ by triangles completed along $\Gamma(0)$ by curved elements in a manner described in Zlámal [4]. Let $P_k = (x_k,y_k)$, $k = 1,\dots,d$, be the nodes of this triangulation and let $Q_k = (\alpha_k, \beta_k)$ be their inverse immages in the mapping (6), i.e.

$$\alpha_k = \frac{x_k - \varphi(\beta_k,0)}{1-L_0^{-1}\varphi(\beta_k,0)}, \quad \beta_k = y_k .$$

The triangulation $T(t)$ of $\bar{\Omega}(t)$ is determined for $t > 0$ by the nodes $P_k(t) = (x_k(t), y_k)$, $x_k(t) = F(\alpha_k, \beta_k, t)$. The elements of $T(t)$ are again triangles or curved elements. As shape functions we use linear polynomials. We denote by $V_h(t) \subset H^1(\Omega(t))$ the set of all trial functions and by $w_k(x,y,t)$, $k = 1,\dots,d$, the basis functions of $V_h(t)$.

The semidiscrete solution is assumed in the form

$$U(x,y,t) = \sum_{k=1}^{d} U_k(t) w_k(x,y,t)$$

and determined by

$$\text{(7)} \qquad \forall t \in (0,T) \ \left(\frac{\partial}{\partial t} U, v\right)_{L^2(\Omega(t))} + a(U,t;U,v) = 0 \quad \forall v \in V_h(t),$$
$$U(x,y,0) = U^*(x,y) .$$

$U^* \in V_h(0)$ is a suitable approximation of u^*.

If we denote by $\underline{U}(t)$ the d-dimensional vector $(U_1(t),\dots, U_d(t))^T$ by $M(t)$, $R(t)$ and $K(U,t)$ the $d \times d$ matrices

$$M(t) = \{(w_j,w_k)_{L^2(\Omega(t))}\}_{j,k=1}^{d}, \quad R(t) = \{(w_j,\frac{\partial w_k}{\partial t})_{L^2(\Omega(t))}\}_{j,k=1}^{d},$$
$$K(U,t) = \{a(U,t;w_j,w_k)\}_{j,k=1}^{d} ,$$

then the matrix form of (7) is

$$\text{(8)} \qquad M(t)\dot{\underline{U}} + [R(t) + K(U,t)]\underline{U} = \underline{0} ,$$
$$\underline{U}(0) = \underline{U}^* .$$

Here $\dot{\underline{U}} = \frac{d}{dt}\underline{U}$ and the matrices M and K are standard mass and stiffness matrices, respectively. The matrix R is unsymmetric.

We discretize (8) in time. For simplicity, we use a uniform partition of $\langle 0,T\rangle$: $t_i = i\Delta t$, $i = 0,1,\dots,q$. In the sequal $\underline{U}^i$, $M^i,\dots$ means $\underline{U}(t_i)$, $M(t_i),\dots$. Now we set $t = t_{i+1}$ in (8), replace $\dot{\underline{U}}^{i+1}$ by $\Delta t^{-1}\Delta\underline{U}^i$, $\Delta\underline{U}^i = \underline{U}^{i+1} - \underline{U}^i$, and linearize the nonlinear term in (8). We get

$$\text{(9)} \qquad M^{i+1}\Delta\underline{U}^i + \Delta t[R^{i+1} + K(\tilde{U}^i,t_{i+1})]\underline{U}^{i+1} = \underline{0}, \quad \tilde{U}^i = \sum_{k=1}^{d} U_k^i w_k^{i+1},$$
$$\underline{U}^0 = \underline{U}^* .$$

For practical computations it is necessary to do one more step: to replace curved elements by triangles and to compute all matrices numerically. Also, we could apply the Crank-Nicholson approach for solving (8) or, more generally, the θ-method. We have restricted ourselves to justify the procedure defined by (9).

We consider a family $\{T_h^0\}$ of triangulations of $\bar{\Omega}^0$ from which a family $\{T_h(t)\}$ of the triangulations of $\bar{\Omega}(t)$ is constructed as described above. Let h_{K0} be the greatest side of an element $K^0 \in T_h^0$ and $h = \max_{K^0 \in T_h^0} h_{K0}$.

We consider a family $\{T_h^0\}$ such that $h \to 0$ and the minimum angle condition is satisfied. We have proved the following main results:

1. Let

$$b(w,t;v,v) \geq 0 \quad \forall v \in V(t),\ t \in (0,T\rangle$$

where

$$b(w,t;u,v) = a(w,t;u,v) - \frac{1}{2} \int_{\Gamma(t)} \dot{\phi}_n uv d\Gamma .$$

Then for Δt sufficiently small, $\Delta t \leq \Delta t_0$ where Δt_0 does not depend on h and on the index i, the matrices $M^{i+1} + \Delta t(R^{i+1} + K(\tilde{U}^i, t_{i+1}))$, $i = 0,\ldots,q-1$, of the systems (9) are regular so that U^i, $i = 1,\ldots,q$ are uniquely determined. Furthermore, the scheme (9) is unconditionally stable in the L^2-norm, i.e. for $\Delta t \leq \Delta t_0$ we have

$$(10) \qquad \max_{1\leq i\leq q} \|U^i\|_{L^2(\Omega^i)} \leq C\|U^0\|_{L^2(\Omega^0)}$$

where C does not depend on Δt and on h.

2. Let the form b be uniformly V(t)-elliptic, i.e.

$$b(w,t;v,v) \geq b_0\|v\|^2_{H^1(\Omega(t))} \qquad \forall v \in V(t),\ t \in (0,T\rangle$$

and let $\dot{\phi}_n \leq 0$. Then for Δt sufficiently small, $\Delta t \leq \Delta t_0$ where Δt_0 does not depend on h, there holds

$$(11) \qquad \max_{1\leq i\leq q} \|u^i - U^i\|_{L^2(\Omega^i)} + \{\Delta t \sum_{i=1}^{q} \|u^i - U^i\|_{H^1(\Omega^i)}\}^{1/2} \leq$$

$$\leq C(\|u^0 - U^0\|_{L^2(\Omega^0)} + h + \Delta t) .$$

The proof starts from the variational formulation

$$(12) \qquad \begin{cases} (U^{i+1} - \tilde{U}^i, v)_{L^2(\Omega^{i+1})} - \Delta t(\frac{\partial U^{i+1}}{\partial x}, G^{i+1}v)_{L^2(\Omega^{i+1})} + \Delta t a(\tilde{U}^i, t_{i+1}; U^{i+1}, v) = 0 \\ \qquad \forall v \in V_h^{i+1},\ i = 0,\ldots,q-1, \\ U^0 = U^* . \end{cases}$$

Here G is a computable function defined on each element by means of the map which maps uniquely this element on the reference one. From (12) the existence and an unconditional stability of the scheme can be proved. The error estimating is based on the Ritz approximation $\zeta \in V_h(t)$ defined by

$$d(u,t;u-\zeta,v) = 0 \quad \forall v \in V_h(t)$$

where $d(w,t;u,v) = b(w,t;u,v) - \frac{1}{2}\int_{\Gamma(t)} \phi_n uv d\Gamma$. $u - \zeta$ is estimated using a technique which in a case that the boundary does not move is essentially that of Wheeler [3], Dupont, Fairweather, Johnson [1] and Zlámal [5]. Instead of estimating $\frac{\partial}{\partial t}(u-\zeta)$ we estimate $D_t(u-\zeta)$ where the operator D_t is defined by $D_t = G\frac{\partial}{\partial x} + \frac{\partial}{\partial t}$. If the boundary does not move, $G = 0$ and $D_t = \frac{\partial}{\partial t}$.

References

[1] T. Dupond, G.Fairweather and J.P. Johnson, *Three-Level Galerkin Methods for Parabolic Equations*. SIAM J. Numer. Anal. 11 (1974), 392-410.

[2] C.D. Maldonado, ROMANS II, *A Two-Dimensional Process Simulator for Modeling and Simulation in the Design of VLSI Devices*. Applied Physics A 31 (1983), 119-138.

[3] M.F. Wheeler, *A priori L_2 Error Estimates for Galerkin Approximations to Parabolic Partial Differential Equations*. SIAM J. Numer. Anal. 10 (1973), 723-759.

[4] M. Zlámal, *Curved Elements in the Finite Element Method*. SIAM J. Numer. Anal. 10 (1973), 229-240.

[5] M. Zlámal, *Finite Element Methods for Nonlinear Parabolic Equations*. R.A.I.R.O. Anal. Numer. 11 (1977), 93-107.

ANALYSIS OF THACKER'S METHOD FOR SOLVING THE LINEARIZED SHALLOW WATER EQUATIONS

J. DESCLOUX, R. FERRO
EPFL - Department of Mathematics
CH 1015 Lausanne, Switzerland

1. INTRODUCTION.

In their simplest form, the shallow water equations read

$$\partial_t\vec{U}(x,t) = - b(x)\vec{\nabla}H(x,t) + fR\vec{U}(x,t), \quad x \in \Omega,\ t \geq o, \tag{1.1}$$

$$\partial_t H(x,t) = - \vec{\nabla}\cdot\vec{U}(x,t), \quad x \in \Omega,\ t \geq o, \tag{1.2}$$

$$\vec{U}(x,t)\cdot\vec{n}(x) = o, \quad x \in \partial\Omega, \quad t \geq o, \tag{1.3}$$

$$\vec{U}(x,o) = \vec{U}_o(x),\ H(x,o) = H_o(x), \quad x \in \Omega. \tag{1.4}$$

Here , $\Omega \subset \mathbb{R}^2$ is a bounded open domain with C^∞ boundary $\partial\Omega$ and closure $\bar{\Omega}$. $\vec{n}$ is the outwards unit normal to $\partial\Omega$. $\vec{U}(x,t) = (\vec{U}_1(x,t),U_2(x,t))$ is a two components vector related to the average horizontal velocity. $H(x,t)$ is the height of the surface of the basin. Up to a constant factor, $b(x)$ is the depth of the basin. We shall assume that $b(\cdot)$ is a $C^\infty(\bar{\Omega})$ strictly positive function. f which represents the intensity of the Coriolis forces is taken to be constant. $\vec{U}_o$ and H_o are given initial conditions. R is the $(-\Pi/2)$ rotation operator acting in $\mathbb{R}^2$, i.e. $R(x_1,x_2) = (x_2,-x_1)$. The tangential vector $\vec{t}$ at $\partial\Omega$ is given by $\vec{t} = - R\vec{n}$.

Equations (1.1),(1.2) can be easily set in the framework of the theory of semigroups. Let $\mathscr{H} = (L_2(\Omega))^3$ be the Hilbert space with scalar product

$$((\vec{u},g),(\vec{v},h))_{\mathscr{H}} = \int_\Omega \frac{1}{b(x)}\vec{u}(x)\cdot\vec{v}(x) + g(x)h(x) \tag{1.5}$$

and associate norm $||\cdot||_{\mathscr{H}}$ where $\vec{u} = (u_1,u_2)$, $\vec{v} = (v_1,v_2)$. We define the operator L with domain $\mathscr{D}(L)$ by the relations:

$$\mathscr{D}(L) = \{(\vec{u},g) \in \mathscr{H} |\ \vec{\nabla}\cdot\vec{u} \in L^2(\Omega),\ \vec{u}\cdot\vec{n} = o \text{ on } \partial\Omega,\ g \in H^1(\Omega)\} \tag{1.6}$$

$$L(\vec{u},g) = (-b\vec{\nabla}g, -\vec{\nabla}\cdot\vec{u}), \tag{1.7}$$

where $H^k(\Omega)$ is the classical Sobolev space of order k. One can verify that L is a skewadjoint operator, i.e. the adjoint of L is -L which implies in particular for $(\vec{u},g)$ and $(\vec{v},h)$ in $\mathscr{D}(L)$ the following relation:

$$(L(\vec{u},g),(\vec{v},h))_{\mathscr{H}} = - ((\vec{u},g),L(\vec{v},h))_{\mathscr{H}}. \tag{1.8}$$

It follows that L is the infinitesimal generator of a conservative group so that Problem (1.1)-(1.4) possesses a unique solution with the property:

$$\|(\vec{U}(\cdot,t),H(\cdot,t)\|^2_{\mathcal{H}} = \text{constant}; \qquad (1.9)$$

furthermore , from (1.2),(1.3), one deduces immediately the law of mass conservation

$$\int_\Omega H(x,t) = \text{constant}. \qquad (1.10)$$

Remarque 1.1: In [4], we give some results concerning the regularity of solutions of Problems (1.1)-(1.4).

The purpose of this paper is to analyse a numerical method proposed by Thacker [1], [2] for solving Problems (1.1)-(1.4); more exactly, we shall consider in fact two variants of Thacker's scheme.

2. DISCRETIZATION.

We consider a sequence $\{\mathcal{D}_h\}$ of standard triangulations of Ω, as shown in the figure. h denotes the maximum length of the sides of the triangles of $\mathcal{D}_h$. We assume that all angles of all triangles of all triangularizations are bounded from below by some positive constant. For a particular $\mathcal{D}_h$, let N be equal to three times the total number of nodes minus the number of nodes belonging to the boundary $\partial\Omega$, Ω_h will denote the interior of the union of the triangles of $\mathcal{D}_h$.

Let us consider a fixed triangularization $\mathcal{D}_h$. For each node P_k, let Λ_k be the polygon formed by the triangles containing P_k and let μ_k be the measure of Λ_k. Z_k will denote the set of indices j such that $P_j \in \partial\Lambda_k$: in the figure, $Z_2 = \{2,3,4,5\}$, $Z_6 = \{7,8,9,10,11\}$; clearly, $k \in Z_k$ if and only if $P_k \in \partial\Omega$. For some node P_k, let $P_j \in \partial\Lambda_k$, i.e. $j \in Z_k$; let $P_\alpha \in \partial\Lambda_k$ be the node preceding P_j with respect to the trigonometric orientation and let $P_\beta \in \partial\Lambda_k$ be the node following P_j; we define the vector:

$$\vec{Q}_{jk} = \overrightarrow{P_\alpha P_\beta}\ ; \qquad (2.1)$$

in the figure, we have, for example, $\vec{Q}_{8,6} = \overrightarrow{P_7P_5}$, $\vec{Q}_{5,2} = \overrightarrow{P_4P_2}$, $\vec{Q}_{22} = \overrightarrow{P_5P_3}$. Furthermore, we introduce at $P_k \in \partial\Omega$ the approximate tangent and normal vectors:

$$\vec{T}_k = \frac{1}{|\vec{Q}_{kk}|}\vec{Q}_{kk}, \quad N_k = R\vec{T}_k. \qquad (2.2)$$

For a function ϕ defined on Ω, let ϕ_0 be the continuous, piecewise linear (with respect to $\mathcal{D}_h$) function defined on Ω_h and equal to ϕ at the nodes;for a vector function $\vec{\psi}$, we define, in the same way, componentwise its interpolant $\vec{\psi}_0$. By Greens's formula, we have the identities:

$$\frac{1}{\mu_k}\int_{\Lambda_k}\vec{\nabla}\phi_o = \frac{1}{2\mu_k}\sum_{j\in Z_k}\phi(P_j)(R\vec{Q}_{jk}), \tag{2.3}$$

$$\frac{1}{\mu_k}\int_{\Lambda_k}\vec{\nabla}\cdot\vec{\phi}_o = \frac{1}{2\mu_k}\sum_{j\in Z_k}\vec{\phi}(P_j)\cdot(R\vec{Q}_{jk}); \tag{2.4}$$

clearly, the right members of (2.3),(2.4) define a natural approximation of $\vec{\nabla}\phi(P_k)$ and $\vec{\nabla}\cdot\vec{\phi}(P_k)$ respectively.

With the help of (2.3),(2.4), we now define a space semi-discretization of Problem (1.1)-(1.4). For all nodes P_k, $H_k(t)$ is an approximation of $H(P_k,t)$; for interior nodes $P_k \in \Omega$, $\vec{U}_k(t) = (U_{k1}(t),U_{k2}(t))$ is an approximation of $\vec{U}(P_k,t)$; for boundary nodes $P_k \in \partial\Omega$, $U_{Tk}(t)$ is an approximation of the tangential component of $\vec{U}(P_k,t)$, i.e. of $\vec{U}(P_k,t)\cdot\vec{t}(P_k)$. <u>Method IS</u> is then defined by the relations:

$$\frac{\mu_k}{b_k}\dot{\vec{U}}_k(t) = -\frac{1}{2}\sum_{j\in Z_k} H_j(t)(R\vec{Q}_{jk}) + \frac{\mu_k}{b_k} fR\vec{U}_k(t), \qquad P_k \in \Omega ; \tag{2.5}$$

$$\frac{\mu_k}{b_k}\dot{U}_{Tk}(t) = -\frac{1}{2}\sum_{j\in Z_k} H_j(t)(R\vec{Q}_{jk})\cdot\vec{t}_k , \qquad P_k \in \partial\Omega ; \tag{2.6}$$

$$\mu_k \dot{H}_k(t) = -\frac{1}{2}\sum_{j\in Z_k}\vec{U}_j(t)\cdot(R\vec{Q}_{jk}), \qquad P_k \in \bar{\Omega} ; \tag{2.7}$$

$$\vec{U}_k(o) = \vec{U}_o(P_k),\ P_k \in \Omega;\ U_{Tk}(o) = \vec{U}_o(P_k)\cdot\vec{t}(P_k), \quad P_k \in \partial\Omega ; \tag{2.8}$$

$$H_k(o) = H_o(P_k), \qquad P_k \in \bar{\Omega} ; \tag{2.9}$$

here the "dot" represents the time derivative, $b_k = b(P_k)$ and in (2.7),

$$\vec{U}_j(t) = U_{Tj}(t)\vec{t}_j \quad \text{if} \quad P_j \in \partial\Omega.$$

By choosing any fixed order, all the unknown function $U_{k1}(t)$, $U_{k2}(t)$, $U_{Tk}(t)$ and $H_k(t)$ can be set in a single vector $w(t)$ of dimension N. Then Problem (2.5)-(2.9) can be written in the compact form

$$D\dot{w}(t) = Aw(t); \quad w(o) = w_o ; \tag{2.10}$$

where D is a diagonal matrix with diagonal elements of the form μ_k or μ_k/b_k. Because of property (1.8) for L, one could expect that A is an antisymmetric matrix. Due to difficulties at the boundary, which seem inherent to the problems and impossible to overcome in a natural way, A is only "almost" antisymmetric. By inspection of the figure, one can show:

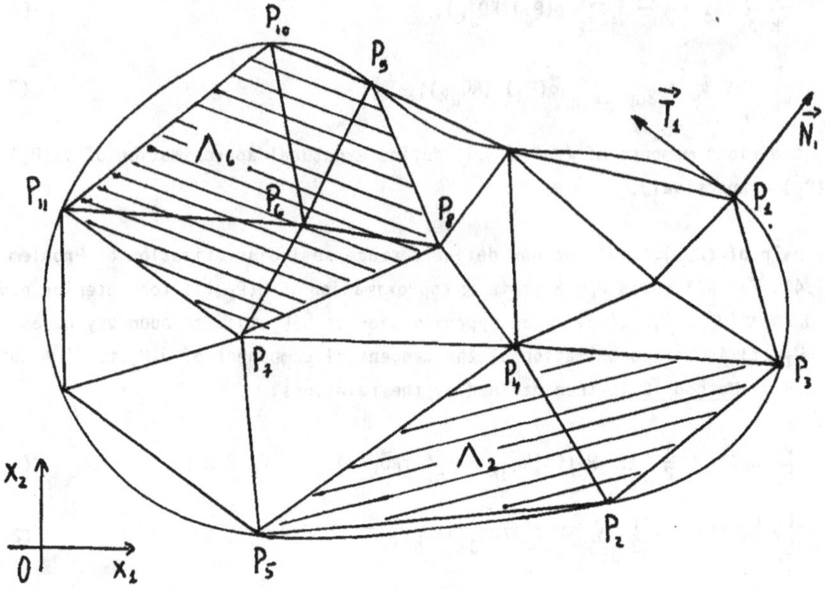

Lemma 2.1: Let P_k and P_j be two different nodes belonging to a same triangle. We suppose that at most one of them belong to the boundary $\partial\Omega$. Then

$$\vec{Q}_{jk} + \vec{Q}_{kj} = \vec{o}. \tag{2.11}$$

As easily seen from (2.5)-(2.7), A would be exactly antisymmetric if (2.11) would hold when if both P_k and P_j belong to $\partial\Omega$. There are several ways to modify Scheme (2.5)-(2.7) for obtaining the desired property of antisymmetry. One of them consists in remplacing in (2.5),(2.6) $\vec{Q}_{jk}$ by $-\vec{Q}_{kj}$; by Lemma 2.1, (2.5) is not modified whereas (2.6) becomes

$$\frac{\mu_k}{b_k} \dot{U}_{Tk}(t) = \frac{1}{2} \sum_{j \in Z_k} H_j(t)(R\vec{Q}_{jk})\cdot\vec{T}_k , \quad P_k \in \partial\Omega. \tag{2.12}$$

Method IIS is then defined by Relations (2.5),(2.12),(2.7),(2.8),(2.9) and can be written as

$$D\dot{w}(t) = Bw(t), \quad w(o) = w_o \tag{2.13}$$

where B is an antisymmetric matrix of order N.

For a vector $v \in \mathbb{R}^N$ and a matrix G of order N, let

$$\|v\| = \left(\sum_{i=1}^{N} |v_i|^2\right)^{1/2}, \quad \|G\| = \sup_{v\in\mathbb{R}^N} \frac{\|Gv\|}{\|v\|}.$$

By using the smoothness of $\partial\Omega$ and the angle property of the sequence $\{\mathcal{D}_h\}$, one can verify:

Lemma 2.2: There exist three constants c_1, c_2, c_3, independent of h such that

a) $\|D^{-1/2}(A-B)D^{-1/2}\| \le c_1$, b) $c_2 h^{-1} \le \|D^{-1/2}BD^{-1/2}\| \le c_3 h^{-1}$.

By using Lemma 2.2a, the antisymmetry of B, Relations (2.7),(2.9) in connection with the definition (2.2) of $\vec{T}_k$, one can deduce the following properties of Methods IS and IIS.

Proposition 2.1:

a) If w is solution of (2.13), then $\|D^{1/2}w(t)\|^2 = \|D^{1/2}w_o\|^2$.

b) There is a constant c, independent of t and of h, such that if w is solution of (2.10), then $\|D^{1/2}w(t)\|^2 \le e^{ct}\|D^{1/2}w_o\|^2$.

c) If w is solution of (2.10) or of (2.13), then $\sum_{P_k \in \bar{\Omega}} \mu_k H_k(t) = \sum_{P_k \in \bar{\Omega}} \mu_k H_o(P_k)$.

Remark 2.1: Since the sum is all μ_k's is equal to three times the area of Ω_k, Proposition 2.1 appears to be, up to a factor 3, the discrete counterpart of Properties (1.9),(1.10) of the exact solution.

We now turn to the time discretisation. Let $\tau > o$ be the time increment and set $t_n = n\tau$. We shall apply to (2.10) and (2.13) the two-step method, sometimes called "leap-frog" scheme. Method I is then defined by the relations:

$$w_{1/2} = w_o + \frac{\tau}{2} D^{-1}Aw_o \tag{2.14}$$

$$w_1 = w_o + \tau D^{-1}Aw_{1/2} \tag{2.15}$$

$$w_{n+1} = w_{n-1} + 2\tau D^{-1}Aw_n, \quad n = 1,2,3,\dots \tag{2.16}$$

Method II is defined by replacing in (2.14)-(2.16). A by B.

The stability analysis of Method II is trivial since $D^{-1}B$ or $D^{-1/2}BD^{-1/2}$ have a pure imaginary spectrum. By using Strang's Lemma [3] in connection with Lemma 2.2a, we easily deduce the stability of Method I from that of Method II.

Proposition 2.2: There exist a function α: $(o,1) \to \mathbb{R}$ and a constant c, both independent of h, τ and n such that if $\tau < 1/\|D^{-1/2}BD^{-1/2}\|$, we have:

a) $\|D^{-1/2}w_n\|^2 \le \alpha(\tau\|D^{-1/2}BD^{-1/2}\|)\|D^{1/2}w_o\|$ for Method II,

b) $\|D^{1/2}w_n\|^2 \le \alpha(\tau\|D^{-1/2}BD^{-1/2}\|)e^{ct_n}\|D^{1/2}w_o\|$ for Method I.

Remark 2.2: Methods I and II satisfy a law of mass conservation as Methods IS and IIS (see Proposition 2.1c).

Remark 2.3: Proposition 2.2a,b prove the stability of both Method I and II; however Method II appears to be "more" stable than Method I.

Remark 2.4: w_n can be written as a vector of order N with components U_{kn1}, U_{kn2}, U_{Tkn} and H_{kn}. If the Coriolis term $f = o$, then for Methods I and II, it is possible to compute U_{kn1}, U_{kn2}, U_{Tkn} only at even values of n and H_{kn} at odd values of n which reduces the computer time and the storage requirements by a factor 2; real Thacker's scheme, which is somewhat more difficult to analyse, keeps this property even for $f \neq o$. In fact, if $f = o$, Method I is identical to Thacker's scheme (10),(11),(11') in [2] p.683.

Remark 2.5: Thacker [2] has remarked that his scheme can be considered, to some extend, as a lumped version of a Galerkin method. In [4], we briefly analyse the effect of "lumping" on stability.

Remark 2.6: The stability condition $\tau < 1/\|D^{-1/2}BD^{-1/2}\|$ in Proposition 2.2, implies, by Lemma 2.2b that $\tau = O(h)$.

3. ERROR ESTIMATES

Our estimates will be based on the standard consistency+stability argument. Stability has already been analyzed in Section 2.

We begin with a classical study of consistency by assuming that the components of the solution of Problem (1.1)-(1.4) belong to $C^o([o,T]; C^3(\bar{\Omega}))$. We first associate to this solution a vector $u(t) \in \mathbb{R}^N$ in the following way; let w(t) the solution of the semi-discritized problem (2.10); we set: $u_i(t) = U_\ell(P_k,t)$ if $w_i(t) = U_{k\ell}$, $\ell = 1,2$;

$$u_i(t) = \vec{U}(P_k,t)\cdot\vec{t}(P_k) \text{ if } w_i(t) = U_{Tk}(t);\; u_i(t) = H(P_k,t) \text{ if } w_i(t) = H_k(t).$$

Clearly, the time discretization which is of order two, will induce errors of size $O(\tau^2)$, which, by Remark 2.6, can be written as $O(h^2)$. Let us define for $T > o$:

$$\varepsilon_I(T) = \max_{o\le t\le T}\|D^{1/2}(\dot{u}(t)-D^{-1}Au(t))\|,\quad \varepsilon_{II}(T) = \max_{o\le t\le T}\|D^{1/2}(\dot{u}(t)-D^{-1}Bu(t))\|,$$

$$R_{I(II)}(T) = \max_{o\le t_n\le T}\|D^{1/2}(u(t_n)-w_n)\| \text{ if } w_n \text{ is obtained by Method I(II).}$$

For $i = I,II$, $R_i(T)$ is the error of Method i, whereas, by using (1.1),(1.2), $\varepsilon_i(T)$ is the space consistency error.

In the following, we shall say that for a node P_k, Λ_k is <u>symmetric</u>, if for each $P_j \in \partial\Omega_k$, there exists $P_\ell \in \partial\Lambda_k$ which is symmetric to P_j with respect to P_k. Clearly if $P_k \in \partial\Omega$, Λ_k cannot be symmetric. The basic difference schemes defined by (2.3), (2.4) are of order 2, with respect to h, if Λ_k is symmetric; otherwise there are only of order 1. We shall say that the sequence of triangularizations $\{\mathcal{D}_h\}$ possesses Property G if there exists a constant c, independent of h such that for all $P_k \in \partial\Omega$ one has $|\vec{T}_k - \vec{t}(P_k)| \leq ch^2$; Property G implies a certain regularity in the distribution of the nodes on the boundary. Elementary but tidious calculations allow to establish:

Lemma 3.1: For any fixed T > o and i = I,II, we have: a) $\varepsilon_i(T) = O(h^{1/2})$; b) $\varepsilon_i(T) = O(h)$ if Property G is satisfied; c) $\varepsilon_i(T) = O(h^{3/2})$ if Λ_k is symmetric for all $P_k \in \Omega$ and if Property G is satisfied.

From Lemma 3.1 follows immediately.

Proposition 3.1: Let T > o and $\xi \in (o,1)$ be fixed numbers. For each triangularization $\mathcal{D}_h$, τ is chosen in such a way that $o < \tau < \xi/\|D^{-1/2}BD^{-1/2}\|$. Then for i = I,II:
a) $R_i(T) = O(h^{1/2})$; b) $R_i(T) = O(h)$ if Property G is satisfied; c) $R_i(T) = O(h^{3/2})$ if Λ_k is symmetric for all $P_k \in \Omega$ and if Property G is satisfied.

Remark 3.1: Suppose, that, instead of (2.2), we set $\vec{T}_k = \vec{t}(P_k)$ (exact tangent vector) Then: a) We loss the exact mass conservation property for both Methods I and II /see Remark 2.2); b) Proposition 3.1 remains valid for Method I; Proposition 3.1a remains valid for Method II.

We now turn to an error analysis under weaker regularity assumptions. We shall suppose that the components of the solution of Problem (1.1)-(1.4) belong to $C^o([o,T]; H^2(\Omega))$. For simplicity, we shall furthermore assume that Ω is convex so that $\Omega_h \subset \Omega$. The difficulty here comes from the fact that the time derivative of the solution is not a continuous function of the space variable. Let $\vec{U}_{kn} = (U_{kn1}, U_{kn2})$, H_{kn} be the approximate solution obtained by Method I or II corresponding to the exact solution $\vec{U}(P_k,t_n)$, $H(P_k,t_n)$; here we set $\vec{U}_{kn} = U_{Tkn}\vec{T}_k$ if $P_k \in \partial\Omega$. Let V_h be the space of continuous piecewise linear functions on Ω_h corresponding to $\mathcal{D}_h$. We define the function $\vec{U}_h(x,t) = (U_{h1}(x,t), U_{h2}(x,t))$, $H_h(x,t)$, for $x \in \Omega_h$ and $t = t_n$, in the following way: $U_{h\beta}(\cdot,t_n) \in V_h$, $\beta = 1,2$, $H_h(\cdot,t_n) \in V_h$; $\vec{U}_h(P_k,t_n) = \vec{U}_{kn}$, $H_h(P_k,t_n) = H_{kn}$ for all $P_k \in \bar{\Omega}$.

The main trick will consist in introducing functions $Y_h(x,t) = (Y_{h1}(x,t), Y_{h2}(x,t))$, $Z_h(x,t)$ belonging to V_h for fixed t and which are, for fixed t, Clément's approximations of the exact solution's components; for the notion of Clément's approximation, see [7]. Corresponding to the exact equation (1.2), we have for all $P_k \in \bar{\Omega}$ the following identity:

$$\mu_k Z_h(P_k(t_{n+1}) - \mu_k Z_h(P_k,t_{n-1}) + \tau \sum_{j\in Z_k} \vec{Y}(P_j,t_n)\cdot(R\vec{Q}_{jk}) \tag{3.1}$$

$$= \{\mu_k Z_h(P_k,t_{n+1}) - \mu_k Z_h(P_k,t_{n-1}) - 2\tau \int_{\Lambda_k} \dot{H}(x,t_n)\} \tag{3.2}$$

$$+ \tau \{\sum_{j\in Z_k} \vec{Y}(P_j,t_n)\cdot(R\vec{Q}_{jk}) - 2\int_{\Lambda_k} \vec{\nabla}\cdot\vec{U}(x,t_n)\}. \tag{3.3}$$

The "time" error term (3.2) can be easily estimated by using the fact that the operations of time derivative and Clément's approximation commute. The "space" error term (3.3) can be handled by remarking the following identity which is a direct consequence of (2.4):

$$\sum_{j\in Z_k} \vec{Y}(P_j,t_n)\cdot(R\vec{Q}_{jk}) - 2\int_{\Lambda_k} \vec{\nabla}\cdot\vec{Y}(x,t_n) = o.$$

Similarly to (3.1)-(3.3), we can write an equation corresponding to (1.1); it is slightly more complicated to handle because of the boundary condition (1.3) and of the presence of the function b(x). With Proposition 2.2, this allows to get error estimates between Clément's approximation and the solution of Method I or II. The final result is contained in the following proposition:

Proposition 3.2: Let $T > o$ and $\xi \in (o,1)$ be fixed numbers. We suppose: a) each component of the exact solution belong to $C^o([o,T]; H^2(\Omega))$; b) Property G is satisfied; c) for each triangularization, τ is chosen such that $o < \tau < \xi/\|D^{-1/2} BD^{-1/2}\|$. Then for both Methods I and II we have:

$$\max_{o\le t_n\le T}\{\int_{\Omega_h}(\frac{1}{b(x)}|\vec{U}(x,t_n) - \vec{U}_h(x,t_n)|^2 + |H(x,t_n) - H_h(x,t_n)|^2)\}^{1/2} = O(h).$$

Remark 3.2: In [4], we give some numerical results.

Remark 3.3: In order to compute the spectrum of the operator L defined in Section 1, one could think of using the same space discretization as in Method I or II; however this generates spurious eigenvalues. For a proper treatment of this problem, see [5], [6].

References

[1] W.C.THACKER, *Irregular Grid Finite-Difference Techniques: Simulations of Oscillations in Shallow Circular Basins,* Journal of Oceanography, Vol. 7, 1977, 284-292.

[2] W.C.THACKER, *Comparison of Finite-Element and Finite-Differences Schemes, Part I: One-Dimensional Cravity Wave Motion, Part II: Two-Dimensional Gravity Wave Motion,* Journal of Oceanography, vol. 8, 1978, 676-689.

[3] G.STRANG, *Accurate Partial Difference Methods,* Numerische Mathematik, 6, 1964, 37-46.

[4] J.DESCLOUX, R.FERRO, *On Thacker's scheme for solving the linearized shallow water equations,* Report. Departement de Mathématiques. Ecole Polytechnique Fédérale de Lausanne, 1985.

[5] M.LUSKIN, *Convergence of a Finite Element Method for the Approximation of Normal Modes of the Oceans,* Math. Comp. 33, 1979, 493-519.

[6] J.DESLOUX, M.LUSKIN, J. RAPPAZ, *Approximation of the Spectrum of Closed Operators: The Determination of Normal Modes of a Rotating Basin,* Math. Comp. 36, 1981, 137-154.

[7] Ph.CLEMENT, *Approximation by Finite Element Functions using Local Regularizations,* RAIRO 9, 1975, 77-84.

THE CONVERGENCE OF A NEW METHOD FOR CALCULATING LOWER BOUNDS TO EIGENVALUES

F. GOERISCH and J. ALBRECHT
Institut für Mathematik, Technische Universität Clausthal
Erzstraße 1, D 3392 Clausthal-Zellerfeld, West Germany

The relationship between an inclusion theorem due to N. J. Lehmann [3] and one recently proposed [2] is investigated here. The theorem due to N. J. Lehmann yields better bounds to the eigenvalues, whereas the new theorem is in general considerably easier to apply. It is shown that sequences of bounds to eigenvalues can be obtained with the use of the new theorem, and that these sequences converge to the bounds provided by Lehmann's theorem. This fact is illustrated by means of numerical results for the following eigenvalue problem:

$$\Delta^2\phi = -\lambda\Delta\phi \text{ in } \Omega, \quad \phi = \frac{\partial\phi}{\partial n} = 0 \text{ on } \partial\Omega, \quad \Omega := \{(x,y)\in\mathbb{R}^2:\ |x|<\tfrac{\pi}{2},\ |y|<\tfrac{\pi}{2}\},$$

which occurs in the calculation of buckling stresses of clamped plates under compression.

§1 The two inclusion theorems are first stated, in a version which deviates somewhat from that presented in the original papers [2], [3], but which is especially well suited for practical applications. The following assumptions and definitions are required for this purpose:

Assumptions

A1 D is a real vector space. M and N are symmetric bilinear forms on D; $M(f,f) > 0$ for all $f \in D$, $f \neq 0$.

A2 There exist sequences $(\lambda_i)_{i\in\mathbb{N}}$ and $(\phi_i)_{i\in\mathbb{N}}$ such that
$\lambda_i \in \mathbb{R}$, $\phi_i \in D$, $M(\phi_i,\phi_k) = \delta_{ik}$ for $i,k \in \mathbb{N}$,
$M(f,\phi_i) = \lambda_i N(f,\phi_i)$ for all $f \in D$, $i \in \mathbb{N}$,

$$N(f,f) = \sum_{i=1}^{\infty} \lambda_i (N(f,\phi_i))^2 \text{ for all } f \in D.$$

A3 X is a real vector space; $T: D \to X$ is a linear operator; b is a symmetric bilinear form on X. $b(f,f) \geq 0$ for all $f \in X$ and $b(Tf,Tg) = M(f,g)$ for all $f,g \in D$.

A4 $\rho \in \mathbb{R}$, $\rho > 0$; $n \in \mathbb{N}$, $v_i \in D$ for $i=1,\dots,n$.

Definitions

D1 Matrices A_0 and A_1 are defined by

$A_o := (M(v_i, v_k))_{i,k=1,\dots,n}$, $A_1 := (N(v_i, v_k))_{i,k=1,\dots,n}$.

D2 If A is a symmetric matrix of order n, with the property that $A_o - 2\rho A_1 + \rho^2 A$ is positive definite, $\mu_i(A)$ denotes the i-th smallest eigenvalue of the eigenvalue problem $(A_o - \rho A_1)z = \mu(A_o - 2\rho A_1 + \rho^2 A)z$.

The two inclusion theorems, whose relationship is to be investigated, yield inclusion intervals for the eigenvalues of the eigenvalue problem

$$M(f,\phi) = \lambda N(f,\phi) \text{ for all } f \in D. \qquad (1)$$

The theorems are as follows:

Theorem 1 (N. J. Lehmann [3])
Let $u_i \in D$ be such that $M(f,u_i) = N(f,v_i)$ for all $f \in D$, $i=1,\dots,n$; let the matrix A_2 be defined by $A_2 := (M(u_i,u_k))_{i,k=1,\dots,n}$, and let $A_o - 2\rho A_1 + \rho^2 A_2$ be positive definite. Moreover, suppose that $q \in \mathbb{N}$, $q \le n$, $\mu_q(A_2) < 0$.
The interval $[\rho - \rho(1-\mu_q(A_2))^{-1}, \rho)$ then contains at least q eigenvalues[1]) of the eigenvalue problem (1).

Theorem 2 ([2])
Let $w_i \in X$ be such that $b(Tf,w_i) = N(f,v_i)$ for all $f \in D$, $i=1,\dots,n$; let the matrix $\tilde{A}_2$ be defined by $\tilde{A}_2 := (b(w_i,w_k))_{i,k=1,\dots,n}$, and let $A_o - 2\rho A_1 + \rho^2 \tilde{A}_2$ be positive definite. Moreover, suppose that $q \in \mathbb{N}$, $q \le n$, $\mu_q(\tilde{A}_2) < 0$.
The interval $[\rho - \rho(1-\mu_q(\tilde{A}_2))^{-1}, \rho)$ then contains at least q eigenvalues[1]) of the eigenvalue problem (1).

If the assumptions of theorem 1 are satisfied, and if w_i is defined by $w_i := Tu_i$ for $i=1,\dots,n$, the assumptions of theorem 2 are also fulfilled because of $\tilde{A}_2 = A_2$. Thus, theorem 1 is an immediate consequence of theorem 2.
The importance of these theorems is due to the fact that they provide a means of calculating accurate lower bounds to the eigenvalues of problem (1). If the eigenvalues of (1) are arranged in a non-decreasing order,

$$\lambda_1 \le \lambda_2 \le \lambda_3 \le \dots ,$$

if ρ is a lower bound to the eigenvalue λ_{p+q} $(p,q \in \mathbb{N})$ and if, for example, the assumptions of theorem 2 are satisfied, then $\rho - \rho(1-\mu_q(\tilde{A}_2))^{-1}$ is a lower bound to λ_p. For an appropriate choice of the quantities involved, this bound to λ_p is very accurate, even if ρ is only a comparatively rough lower bound to λ_{p+q}.

---

1) Eigenvalues are always counted according to their multiplicity.

It is often difficult, or even impossible, to explicitly give the elements u_i required in theorem 1; by means of theorem 2, in contrast, inclusion intervals for the eigenvalues can be determined with comparative ease - provided that X, b and T have been appropriately chosen (compare §3). However, the results thus obtained cannot be better than those which would be provided by theorem 1, as is now shown:

Lemma 1

Let the assumptions of theorem 1 and 2 be fulfilled.
Then $\rho-\rho(1-\mu_q(\tilde{A}_2))^{-1} \leq \rho-\rho(1-\mu_q(A_2))^{-1}$.

Proof: Since $b(Tu_i,w_k) = N(u_i,v_k) = M(u_i,u_k)$ for $i,k=1,\ldots,n$, it follows that $b(w_i-Tu_i,w_k-Tu_k) = b(w_i,w_k) - M(u_i,u_k)$; hence, the matrix $\tilde{A}_2-A_2$ is positive semidefinite. With the use of the comparison theorem, one obtains $\mu_q(A_2) \leq \mu_q(\tilde{A}_2)$. The assertion can now be immediately deduced.

§2 On the basis of theorem 2, a sequence of inclusion intervals $[\tau_m,\rho)$ will now be constructed in such a manner that $(\tau_m)_{m\in\mathbb{N}}$ converges, and the interval $[\lim_{m\to\infty}\tau_m,\rho)$ coincides with the corresponding inclusion interval from theorem 1. For this purpose, the following additional assumptions and definitions are required.

Assumptions

A5 $\hat{w}_i \in X$ for $i=1,\ldots,n$ and $w_i^* \in X$ for $i\in\mathbb{N}$,
$b(Tf,\hat{w}_i) = N(f,v_i)$ for all $f\in D$, $i=1,\ldots,n$,
$b(Tf,w_i^*) = 0$ for all $f\in D$, $i\in\mathbb{N}$.
The matrix $(b(w_i^*,w_k^*))_{i,k=1,\ldots,m}$ is regular for all $m\in\mathbb{N}$.

A6 $X_o := \{g\in X: b(Tf,g) = 0 \text{ for all } f\in D\}$; for all $g\in X_o$ and all $\varepsilon\in\mathbb{R}$ with $\varepsilon > 0$ there exist numbers $m\in\mathbb{N}$, $c_1,\ldots,c_m\in\mathbb{R}$ such that
$b(g-\sum_{i=1}^m c_iw_i^*,g-\sum_{i=1}^m c_iw_i^*) \leq \varepsilon$.

Remark: If $b(f,f) > 0$ holds for all $f\in X$ with $f \neq 0$, that is, if $(X,b(.,.))$ is a pre-Hilbert space, the assumption A6 states precisely that the subspace spanned by $\{w_i^*: i\in\mathbb{N}\}$ is dense in X_o.

Definitions

D3 $\hat{A}_2 :=(b(\hat{w}_i,\hat{w}_k))_{i,k=1,\ldots,n}$;
$F_m :=(-b(\hat{w}_i,w_k^*))_{i=1,\ldots,n;k=1,\ldots,m}$, $G_m :=(b(w_i^*,w_k^*))_{i,k=1,\ldots,m}$,
$A_{2,m} :=\hat{A}_2-F_mG_m^{-1}F_m'$ for all $m\in\mathbb{N}$.

The inclusion intervals $[\tau_m,\rho)$ can now be given:

Theorem 3

Let $m,q\in\mathbb{N}$ with $q\le n$; let the matrix $A_0-2\rho A_1+\rho^2A_{2,m}$ be positive definite, and let $\mu_q(A_{2,m})<0$.

If τ_m is defined by $\tau_m := \rho-\rho(1-\mu_q(A_{2,m}))^{-1}$, the interval $[\tau_m,\rho)$ contains at least q eigenvalues of the eigenvalue problem (1).

Proof: Let $F_mG_m^{-1}=(d_{ik})_{i=1,\ldots,n;k=1,\ldots,m}$. The assertion follows immediately from theorem 2, if the w_i occurring there are defined by

$$w_i := \hat{w}_i+\sum_{k=1}^{m} d_{ik}w_k^* \text{ for } i=1,\ldots,n.$$

The following result concerning the convergence of the sequence $(\tau_m)_{m\in\mathbb{N}}$ is now obtained:

Theorem 4

Let the assumptions of theorem 1 be satisfied. If τ_m is defined by $\tau_m := \rho-\rho(1-\mu_q(A_{2,m}))^{-1}$ for $m\in\mathbb{N}$, then $\lim\limits_{m\to\infty}\tau_m = \rho-\rho(1-\mu_q(A_2))^{-1}$.

Proof: Let $F_mG_m^{-1} = (d_{ik}^{(m)})_{i=1,\ldots,n;k=1,\ldots,m}$ for $m\in\mathbb{N}$. Then

$$b(Tu_i-\hat{w}_i-\sum_{k=1}^{m} d_{ik}^{(m)}w_k^*,w_j^*) = 0 \qquad (2)$$

for $i=1,\ldots,n$, $j=1,\ldots,m$, $m\in\mathbb{N}$. Let $\varepsilon\in\mathbb{R}$ with $\varepsilon>0$. Since $Tu_i-\hat{w}_i\in X_0$ for $i=1,\ldots,n$, there exist numbers $l\in\mathbb{N}$, and $c_{ik}\in\mathbb{R}$ for $i=1,\ldots,n$, $k=1,\ldots,l$ such that

$$b(Tu_i-\hat{w}_i-\sum_{k=1}^{l} c_{ik}w_k^*,\ Tu_i-\hat{w}_i-\sum_{k=1}^{l} c_{ik}w_k^*) \le \varepsilon$$

for $i=1,\ldots,n$. With the use of (2), it can be shown that

$$b(Tu_i-\hat{w}_i-\sum_{k=1}^{m} d_{ik}^{(m)}w_k^*,\ Tu_i-\hat{w}_i-\sum_{k=1}^{m} d_{ik}^{(m)}w_k^*) \le \varepsilon$$

for $i=1,\ldots,n$ and all $m\in\mathbb{N}$ with $m\ge l$. By means of the Cauchy-Schwarz inequality, it follows that

$$|b(Tu_i-\hat{w}_i-\sum_{k=1}^{m} d_{ik}^{(m)}w_k^*,\ Tu_j-\hat{w}_j-\sum_{k=1}^{m} d_{jk}^{(m)}w_k^*)| \le \varepsilon$$

for $i,j=1,\ldots,n$ and all $m\in\mathbb{N}$ with $m\ge l$. Hence,

$$\lim_{m\to\infty} b(Tu_i-\hat{w}_i-\sum_{k=1}^{m} d_{ik}^{(m)}w_k^*,\ Tu_j-\hat{w}_j-\sum_{k=1}^{m} d_{jk}^{(m)}w_k^*) = 0$$

for $i,j=1,\ldots,n$. From the equation

$$(b(Tu_i-\hat{w}_i-\sum_{k=1}^{m} d_{ik}^{(m)}w_k^*,\ Tu_j-\hat{w}_j-\sum_{k=1}^{m} d_{jk}^{(m)}w_k^*))_{i,j=1,\ldots,n} = A_{2,m}-A_2$$

it follows that $A_{2,m}-A_2$ is positive semidefinite for $m\in\mathbb{N}$, and that $\lim\limits_{m\to\infty}A_{2,m} = A_2$. This gives $\lim\limits_{m\to\infty}\mu_q(A_{2,m}) = \mu_q(A_2)$, from which the assertion follows immediately.

Remark: The sequence $(\tau_m)_{m\in\mathbb{N}}$ is non-decreasing.

§3 The practical application of the results presented is now illustrated with the use of an example. - The quantities D, M, N, X, T, b occurring in the assumptions A1 and A3 are defined in the following manner ($\overset{\circ}{W}{}_2^{(2)}(\Omega)$ denotes the Sobolev space defined in [4]):

$$D := \{f\in \overset{\circ}{W}{}_2^{(2)}(\Omega):\ f(x,y) = f(y,x) = f(-x,y) \text{ for } (x,y)\in\Omega\},$$

$$M(f,g) := \int_\Omega \Delta f\Delta g\,dxdy,\quad N(f,g) := \int_\Omega \left(\frac{\partial f}{\partial x}\frac{\partial g}{\partial x} + \frac{\partial f}{\partial y}\frac{\partial g}{\partial y}\right)dxdy \text{ for } f,g\in D,$$

$$X := \{f\in L_2(\Omega):\ f(x,y) = f(y,x) = f(-x,y) \text{ for } (x,y)\in\Omega\},$$

$$Tf := -\Delta f \text{ for } f\in D,\quad b(f,g) := \int_\Omega fg\,dxdy \text{ for } f,g\in X,$$

where $\Omega := \{(x,y)\in\mathbb{R}^2:\ |x|<\frac{\pi}{2},\ |y|<\frac{\pi}{2}\}$.
In this case, the eigenvalue problem (1) is the weak form of the eigenvalue problem

$$\left.\begin{aligned}&\Delta^2\phi = -\lambda\Delta\phi \text{ in } \Omega,\ \phi = \frac{\partial\phi}{\partial n} = 0 \text{ on } \partial\Omega,\\ &\phi(x,y) = \phi(y,x) = \phi(-x,y) \text{ for } (x,y)\in\Omega.\end{aligned}\right\}\qquad (3)$$

For specifying the quantities occurring in A4 and A5, a number $r\in\mathbb{N}$ is first chosen; ρ, n, v_i, $\hat{w}_i$, w_i^* are then defined as follows:

$$\rho := 10,\ n := \tfrac{1}{2}r(r+1),$$

$$v_i(x,y) := \cos^{s+1}(x)\cos^{t+1}(y) + \cos^{s+1}(y)\cos^{t+1}(x) \text{ for } i=1,\dots,n,$$

where $s,t\in\mathbb{N}$ are determined by $r \geq s \geq t$, $i=\frac{1}{2}s(s-1)+t$,

$$\hat{w}_i := v_i \text{ for } i=1,\dots,n,$$

$$w_i^*(x,y) := \cosh(ix)\cos(iy) + \cosh(iy)\cos(ix) \text{ for } i\in\mathbb{N}.$$

The assumptions A1, A3, A4 and A5 are obviously fulfilled, and the proofs of A2 and A6 proceed in analogy with the corresponding proofs in [4], p. 472 and [1], respectively.
By means of the Cauchy-Schwarz inequality, it follows from the comparison theorem that the eigenvalues of the problem

$$-\Delta\phi = \lambda\phi \text{ in } \Omega,\ \phi = 0 \text{ on } \partial\Omega,$$
$$\phi(x,y) = \phi(y,x) = \phi(-x,y) \text{ for } (x,y)\in\Omega,$$

are lower bounds to the corresponding eigenvalues of (1); hence $\rho=10$ is a lower bound for the second eigenvalue of (1).
The first six terms of the sequence $(\tau_m)_{m\in\mathbb{N}}$ calculated for q=1 with the use of theorem 3 are compiled in the first six rows of table 1 for various values of n. Since at least one eigenvalue of (1) is contained in each of the intervals $[\tau_m,\rho)$, the numbers τ_m are lower bounds to the lowest eigenvalue of (1). An upper bound Λ to this eigenvalue, which has been determined with the use of the functions $v_1,\dots,v_n$ by

means of the Rayleigh-Ritz method, is given in the last row of table 1, for the respective value of n.

By virtue of theorem 4, the sequences $(\tau_m)_{m\in\mathbb{N}}$ converge to the corresponding bounds which would result from theorem 1. In order to apply theorem 1, however, it would be necessary to determine the exact solution u_i of the boundary value problem

$$\Delta^2 u_i = -\Delta v_i \text{ in } \Omega, \; u_i = \frac{\partial u_i}{\partial n} = 0 \text{ on } \partial\Omega,$$

which is not an easy task.

	n = 1	n = 10	n = 21
τ_1	5.049	5.057	5.057
τ_2	5.250 0	5.265 8	5.265 9
τ_3	5.283 52	5.302 42	5.302 46
τ_4	5.284 556	5.303 564	5.303 602
τ_5	5.284 582 0	5.303 587 3	5.303 625 3
τ_6	5.284 582 21	5.303 587 37	5.303 625 40
Λ	5.333 333 34	5.303 662 26	5.303 626 22

Table 1 Bounds to the lowest eigenvalue of (3)

The method based on theorem 3 has also been applied with great success to many other eigenvalue problems involving partial differential equations.

The authors gratefully acknowledge the support of this work by the Deutsche Forschungsgemeinschaft.

References

[1] Colautti, M.P.: Su un teorema di completezza connesso al metodo di Weinstein per il calcolo degli autovalori. Atti Accad. Sci. Torino, Cl. Sci. Fis. Mat. Nat. 97 (1962/63), 171 -191

[2] Goerisch, F. and H. Haunhorst: Eigenwertschranken für Eigenwertaufgaben mit partiellen Differentialgleichungen. Z. Angew. Math. Mech. 65 (1985), 129 - 135

[3] Lehmann, N.J.: Optimale Eigenwerteinschließungen. Numer. Math. 5 (1963), 246 - 272

[4] Rektorys, K.: Variational methods in mathematics, science and engineering. Dordrecht-Boston: D. Reidel Publishing Company 1977

BIFURCATION ANALYSIS OF STIMULATED BRILLOUIN SCATTERING

V. JANOVSKÝ, I. MAREK, J. NEUBERG
Faculty of Mathematics and Physics, Charles University
Malostranské nám. 25, 110 00 Prague, Czechoslovakia

1. Introduction

Our problem is motivated by the following physical effect (Stimulated Brillouin Scattering): A laser beam of a given frequency is targeted on a material sample. If the laser intensity is small then the beam penetrates without being affected. If the intensity is above a threshold then the sample acts like a mirror and reflects some energy back (Stokes' wave). This is due to stimulated pressure (acoustic) wave in the sample. The frequences of all three waves (i.e. laser and Stokes and pressure) are coupled.

Let us accept that Stimulated Brillouin Scattering can be modelled by the following initial value problem: Find complex-valued functions E_L , E_S , p (the slowly varying amplitudes of laser and Stokes and pressure waves) of time $t \geq 0$ and one spatial variable $0 \leq x \leq \ell$ such that

$$\dot{E}_L = E_L' - iE_S p \ , \quad \dot{E}_S = - E_S' - iE_L\overline{p} \ , \quad c\dot{p} = p' - p - iE_L\overline{E}_S \tag{1.1}$$

(Notation: $\dot{E} = \frac{\partial}{\partial t} E$, $E' = \frac{\partial}{\partial x} E$, $\overline{E}$ is complex conjugate, $i = (-1)^{1/2}$, c is a positive constant) with boundary condition

$$p(\ell,t) = E_S(0,t) = 0 \ , \quad E_L(\ell,t) = ae^{i\psi} \tag{1.2}$$

for $t \geq 0$; a and ψ are real parameters. At $t = 0$, an initial condition (compatible with (1.2)) is prescribed.

1.3 REMARK. The sample occupies the interval $0 \leq x \leq \ell$. Laser light of intensity a^2 is focused at the point $x = \ell$ and propagates in negative direction of the x-axis. Note that functions $E_L \equiv ae^{i\psi}$, $E_S \equiv p \equiv 0$ are a steady state solution to (1.1), (1.2). This *trivial solution* corresponds to the situation when no stimulation of pressure waves occurs which is expected if a^2 is less then the threshold. Physical significance of the model (1.1), (1.2) is discussed e.g. in [1].

1.4 OBSERVATION. If E_L, E_S, p solve (1.1) then, for each real constant γ, the functions $E_L e^{i\gamma}$, E_S, $pe^{i\gamma}$ solve (1.1), too. It implies that we can *assume* $\psi = 0$ *and* $a \geq 0$ without loss of generality.

Our aim is bifurcation analysis of trivial solution to steady state problem (1.1), (1.2) with respect to variations of parameter a. We mention dynamical stability of bifurcated solutions, too.

2. Covariance of the equations governing the steady state

Let us introduce the operator F of the steady state: Homogenising the boundary condition (1.2) by substitution $E_L := E_L + a$, we define $F = F(\mathcal{U},a)$ at $\mathcal{U} = (E_L, E_S, p)$ and at $a \in \mathbb{R}_1$ as follows:

$$F(\mathcal{U},a) = (E_L' - iE_S p,\ -E_S' - iE_L\bar{p} - ia\bar{p},\ p' - p - ia\bar{E}_S - iE_L\bar{E}_S) .$$

Then F acts (e.g.) on the *real* linear space $X = \{\mathcal{U} = (E_L,E_S,p) : E_L, E_S, p$ are complex-valued functions of x, continuously differentiable on $0 \leq x \leq \ell$, satisfying boundary condition $E_L(\ell) = E_S(0) = p(\ell) = 0\}$. The range of $F(.,a)$ is in the real linear space $Y = \{\mathcal{U} = (E_L,E_S,p) : E_L, E_S, p$ are complex-valued, continuous functions of x, $0 \leq x \leq \ell\}$. In the usual topology, there is compact imbedding of X into Y.

Let us define suitable linear transformations on Y: If $\mathcal{U} = (E_L, E_S, p) \in Y$ then

$$M_\beta \mathcal{U} = (E_L,\ e^{i\beta}E_S,\ e^{-i\beta}p) \quad \text{for each } \beta \in \mathbb{R}_1 ,$$

$$T_j \mathcal{U} = (\bar{E}_L,\ (-1)^j \bar{E}_S,\ (-1)^{j+1}\bar{p}) \quad \text{for } j = 1,2 .$$

Let $\{M_\beta,T_1,T_2\}$ denote the group generated by M_β, T_1, T_2 $(\beta \in \mathbb{R}_1)$. The covariance property of the steady state operator follows from

2.1 LEMMA. *If* $\mathcal{U} \in X$ *then* $M_\beta F(\mathcal{U},a) = F(M_\beta \mathcal{U},a)$ *and* $T_j F(\mathcal{U},a) = F(T_j \mathcal{U},a)$ *for each* $\beta \in \mathbb{R}_1$ *and* $j = 1,2$.

Proof. This can be done by a straightforward calculation.

As a direct consequence, we obtain

2.2 PROPOSITION. *If* $\Gamma \in \{M_\beta,T_1,T_2\}$ *then* $\Gamma F(\mathcal{U},a) = F(\Gamma\mathcal{U},a)$ *for each* $\mathcal{U} \in X$.

3. Bifurcation analysis

We resume the steady state problem: Given a value of $a \geq 0$, find $\mathcal{U} \in X$ such that $F(\mathcal{U},a) = 0$. Obviously, $\mathcal{U}^0 \equiv 0$ is *trivial solution* for each value of a .

Let $L(a) : X \to Y$ be Fréchet derivative of the operator $F(.,a) : X \to Y$ at $\mathcal{U}^0$. If there is a bifurcation from $\mathcal{U}^0$ at $a \in \mathbb{R}_1$ then the kernel *Ker* $L(a)$ is nontrivial. Direct calculations yield

3.1 LEMMA. *If* $a \geq 0$ *then* *Ker* $L(a)$ *is nontrivial if and only if* $a \in K$, *where* $K = \{a \geq 0 : 4a^2 - 1 \geq 0 \text{ , } \operatorname{tg} \frac{\omega \ell}{2} = -\omega \text{ , } \omega = (4a^2 - 1)^{\frac{1}{2}}\}$. *The set* K *can be arranged as an increasing sequence* $\{a_j\}_{j=1}^{\infty}$, $0 < a_1 < a_j < a_{j+1}$, $\lim_{j\to+\infty} a_j = +\infty$.

3.2 LEMMA. *If* $a_j \in K$ *then* *Ker* $L(a_j)$ *is spanned by vectors* $\xi_1 = (0,v_0,-iw_0)$, $\xi_2 = (0,-iv_0,w_0)$ *from* X , *where* v_0 *and* w_0 *are the following real functions of* x : $v_0 = e^{x/2} \sin \frac{\omega x}{2}$, $w_0 = (-1)^{j+1} e^{x/2} \sin \frac{\omega(\ell - x)}{2}$; $\omega = (4a_j^2 - 1)^{1/2}$.

Let us make some remarks on bifurcation equation: By virtue of Fredholm alternative, space Y can be decomposed as a direct sum of kernel and range of the operator $L(a_j) : X \to Y$. Thus each solution $\mathcal{U}$ to $F(\mathcal{U},a)$ can be written as $\mathcal{U} = \sum y_i \xi_i + \mathcal{U}^{\perp}$ where y_i's are real coordinates, ξ_i's span the kernel and $\mathcal{U}^{\perp}$ belongs to the range. According to Liapunov-Schmidt reduction (see e.g. [3]), if $\mathcal{U}$ and a are sufficiently close to $\mathcal{U}^0$ and a_j respectively then $\mathcal{U}$ can be *identified* with coordinates y_1 , y_2 of the projection into *Ker* $L(a_j)$. The coordinates satisfy *bifurcation equation* $H(y_1, y_2, a - a_j) = 0$, where H is germ of a mapping $H : \mathbb{R}_2 \times \mathbb{R}_1 \to \mathbb{R}_2$.

Following standard routines, Taylor expansion of H can be found. In our particular case, we have calculated that

$$H(y_1, y_2, a - a_j) = p_j(a - a_j)\begin{pmatrix} y_1 \\ y_2 \end{pmatrix} - q_j(y_1^2 + y_2^2)\begin{pmatrix} y_1 \\ y_2 \end{pmatrix} + s_j(a - a_j)^2\begin{pmatrix} y_1 \\ y_2 \end{pmatrix} + \text{terms of the 4 th order} \tag{3.3}$$

where p_j , q_j , s_j are real constants, namely $p_j = \ell + 2 \cos^2 \frac{\omega \ell}{2}$,

$q_j = (-1)^j \cos \frac{\omega \ell}{2} \int_0^{\ell} (v_0^2(\ell) - v_0^2)(v_0^2 + w_0^2)\, dx$; for ω , v_0 , w_0 see 3.2.

Both p_j and q_j are positive.

3.4 OBSERVATION. Group $\{M_\beta, T_1, T_2\}$ leaves the kernel *Ker* $L(a_j)$ invariant. The matrix representation of $\{M_\beta, T_1, T_2\}$ on *Ker* $L(a_j)$ is $O(2)$ group of 2×2 orthogonal matrices. Following Sattinger [2], the bifurcation equation is covariant under $O(2)$ symmetry group.

3.5 THEOREM. *The bifurcation equation* $H(y_1, y_2, a - a_k) = 0$ *is* $O(2)$*-equivalent to*

$$[a - a_k - (y_1^2 + y_2^2)]\begin{pmatrix} y_1 \\ y_2 \end{pmatrix} = 0$$

(for the notion of $O(2)$*-equivalence, see* [4]).

Proof. We just quote [4], Lemma 5.18 (our points 3.3 and 3.4 verify the assumptions).

We conclude that each $a_j \in K$ is a point of supercritical bifurcation. Bifurcation diagrams are $2 - D$ manifolds (symmetric, in a sense, about a-axis).

4. Stability

Let us resume the initial value problem (1.1), (1.2) where we substitute $E_L := E_L + a$: Given $a \geq 0$ and $U^{in} \in X$, find $U = (E_L, E_S, p)$ such that $\dot{U} = F(U,a)$ for $t > 0$ and $U = U^{in}$ at $t = 0$. Asymptotic stability (as $t \to +\infty$) of the above problem is under question, assuming that U^{in} is a steady state solution being subjected to a small perturbation. In [5] we have tackled this question by making use of Principle of Linearised Stability.

We just quote our results (which are not surprising anyway) : Trivial solution U^0 is stable as $0 \leq a < a_1$ (i.e. up to the first bifurcation point). Beyond this point, it looses stability. Nontrivial steady state solutions which emanate from bifurcation points $a_2, a_3, \ldots$ are unstable. On the other hand, the branch emanating from a_1 is stable.

The above facts are illustrated by the following numerical experiments.

EXAMPLE 1. Data: $\ell = c = 1$, $a = 1$ (i.e. $a < a_1$) ; U^{in} is just the trivial solution U^0 being perturbed by Gaussian "noise" (δ-correlated, dispersion = 1 , mean value = 0). Time interval: $0 \leq t \leq 0.3$.

Results are presented in Figure 1 : At each time, the (numerical) solution U is projected onto *Ker* $L(a_1)$ and then on y_1, y_2-plane.

Point S is projection of U^{in} . It is apparent that solution creeps towards the origin, i.e. U^0 . The second graph indicates velocity of the motion in time.

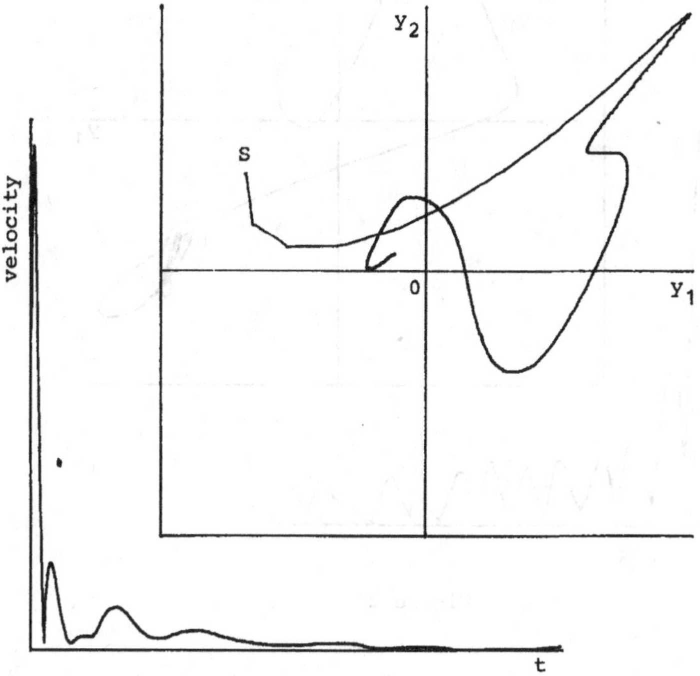

Figure 1

EXAMPLE 2. Data: $\ell = c = 1$, $a = 5$ (i.e. $a_2 < a < a_3$) ; U^{in} is a (numerical) steady state solution on the 2nd branch which is randomly perturbed as above. Time interval: $0 \leq t \leq 10$.

Legend to Figure 2 : U is projected onto *Ker* $L(a_1)$ again. S is the position of projected U^{in} . Note that projection of all steady solutions on the first branch at $a = 5$ would be a circle centred at origin O and passing through C . We observe U oscillating around a point on this circle for large t .

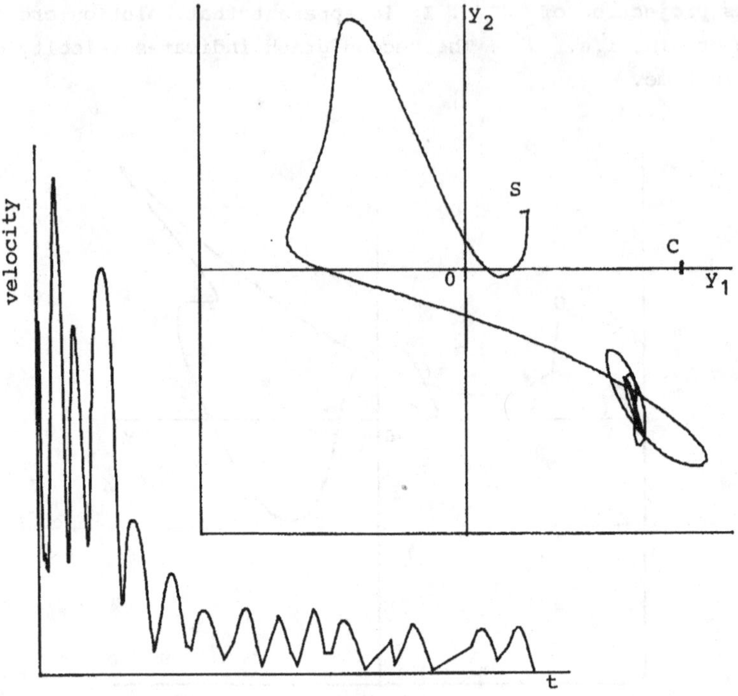

Figure 2

R e f e r e n c e s

[1] STARUNOV, V. S. and FABELINSKIJ, I. L.: *Uspekhi fiz. nauk.* 98, No 3 (1969).

[2] SATTINGER, D. H.: *Group representation theory and branch points of nonlinear functional equations.* SIAM J. Math. Anal.,8, (1977), 2, pp. 179-201.

[3] CHOW Shui-Nee and HALE, J. K.: *Methods of Bifurcation Theory.* Springer, N. Y., 1982.

[4] GOLUBITSKY, M. and SCHAEFFER, D.: *Imperfect bifurcation in the presence of symmetry.* Commun. Math. Phys., 67 (1979), pp. 205-232.

[5] HÁJEK, M. and JANOVSKÝ, V. and NEUBERG, J.: *On stability of Stimulated Brillouin Scattering.* Technical Report KNM MFF No 076/85, Charles University of Prague, 1985.

SUPERCONVERGENCE RESULTS FOR LINEAR TRIANGULAR ELEMENTS

M. KŘÍŽEK
Mathematical Institute, Czechoslovak Academy of Sciences
115 67 Prague 1, Czechoslovakia

The aim of the paper is to present several superconvergence phenomena which have been observed and analyzed when employing the standard linear elements to second order elliptic problems. We shall illustrate them in their simplest form solving the model problem:

$$\begin{aligned} -\Delta u &= f \quad \text{in} \quad \Omega \subset R^2 , \\ u &= 0 \quad \text{on} \quad \partial\Omega , \end{aligned} \tag{1}$$

where Ω is a convex polygonal domain and u is supposed to be smooth enough.

Let $\{T_h\}$ be a regular family of triangulations of $\bar{\Omega}$, i.e., Zlámal's condition on the minimal angle of triangles is fulfilled. The discrete analogue of (1) will consist in finding $u_h \in V_h$ such that

$$(\nabla u_h, \nabla v_h)_{0,\Omega} = (f, v_h)_{0,\Omega} \quad \forall v_h \in V_h , \tag{2}$$

where

$$V_h = \{v_h \in H_0^1(\Omega) \mid v_{h|T} \in P_1(T) \quad \forall T \in T_h\} .$$

It is known [15,39] that the error estimates

$$||u - u_h||_{0,p,\Omega} \le \begin{cases} C_p h^2 ||u||_{2,p,\Omega} & \text{if} \quad p \in [2,\infty) , \\ Ch^2 |\ln h| \, ||u||_{2,\infty,\Omega} & \text{if} \quad p = \infty , \end{cases} \tag{3}$$

$$||\nabla u - \nabla u_h||_{0,p,\Omega} \le Ch||u||_{2,p,\Omega} \quad \text{if} \quad p \in [2,\infty] , \tag{4}$$

are optimal. Nevertheless, we can improve the order convergence (in some norm $|||\cdot|||$ which is close to $||\cdot||_{0,p,\Omega}$) by a suitable post-processing $\sim$, and this we call the superconvergence. The post-processing $\sim$ should be easily computable and the norm $|||\cdot|||$ may be e.g. a discrete analogue of $||\cdot||_{0,p,\Omega}$, or $|||\cdot||| = ||\cdot||_{0,p,\Omega_0}$ for $\Omega_0 \subset\subset \Omega$ (i.e. $\bar{\Omega}_0 \subset \Omega$), or $|||\cdot||| = ||\cdot||_{0,p,\Omega}$, etc. We introduce several examples where $\sim$ is a restriction operator to some subset of Ω, an averaging and an integral smoothing operator. Let us emphasize that many superconvergence phenomena are very sensitive to the mesh geometry (therefore, uniform, quasiuniform or piecewise uniform triangulations are mostly employed). In this paper, we assume for brevity that each T_h is uniform, i.e., any two adjacent triangles of T_h form a parallelogram.

Let N_h be the set of nodal points of T_h. Then the use of the expansion theorem for linear elements by [32] yields (cf. (3))

$$\max_{x \in N_h} |u(x) - u_h(x)| \leq Ch^4 ||u||_{C^4(\bar{\Omega})}, \tag{5}$$

provided T_h consists of equilateral triangles. We mention that the (stiffness) matrix arising from (2), when taking the standard Courant basis functions, is the same as for the well-known 7-point finite difference scheme (see e.g. [35], p. 91)

$$\frac{2}{3}(6u_0 - u_1 - u_2 - u_3 - u_4 - u_5 - u_6) = h^2 f_0 + h^4 \Delta f_0/16$$

with the rate of convergence $0(h^4)$.

<u>Remark 1</u>. Using (1), (2), (5), and the affine one-to-one mapping F between any uniform triangulation $\hat{T}_h$ and a triangulation T_h consisting of equilateral triangles, one easily obtains an analogue of (5) for $\hat{T}_h = F^{-1}(T_h)$, indeed, but for other equation. For instance, the triangulation sketched in Fig. 1 guarantees the nodal superconvergence for the equation $-\Delta\hat{u} + \partial^2\hat{u}/\partial x\partial y = \hat{f}$.

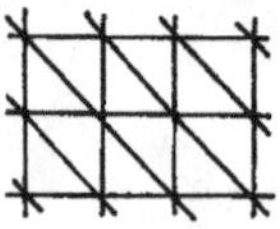

Fig. 1

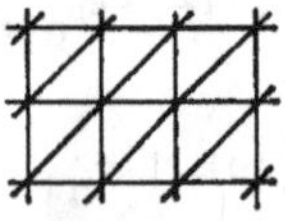

Fig. 2

<u>Remark 2</u>. A convenient combination of linear and bilinear elements may give the $0(h^4)$-superconvergence at nodes for the problem (1) on triangulations consisting of right-angled triangles. Let $\{u^i\}$ and $\{v^i\}$ be the Courant piecewise linear basis functions over the triangulation of Fig. 1 and 2, respectively, and let $\{t^i\}$ be the standard basis functions for bilinear rectangular elements. Put

$$w^i = t^i/2 + u^i/4 + v^i/4$$

and denote by W_h the linear hull of $\{w^i\}$ (dim W_h = dim V_h). Now, the matrix arising from (2), if we replace V_h by W_h, is the same as for the 9-point difference scheme over square meshes [35], p. 90; and it is thus easy to derive the rate $0(h^4)$ at nodes employing the basis $\{w^i\}$. The next table shows the values of the maximum error over all nodes for various choices of basis functions when $u(x,y) = y(y-1)\sin \pi x$ is the exact solution of (1) on the unit square $\Omega = (0,1) \times (0,1)$.

h^{-1}	v^i	$(v^i+u^i)/2$	t^i	$(t^i+v^i)/2$	w^i
4	1.2069 E-2	1.2069 E-2	1.2962 E-2	6.0703 E-4	1.6832 E-4
8	3.1027 E-3	3.1027 E-3	3.1589 E-3	1.3156 E-4	1.0307 E-5
16	7.8126 E-4	7.8126 E-4	7.8478 E-4	3.5250 E-5	6.4092 E-7
32	1.9567 E-4	1.9567 E-4	1.9589 E-4	8.7640 E-6	4.0006 E-8

Further we present superconvergence results for the gradient of $u_h \in V_h$. According to [1,26], the tangential component of ∇u_h is a superconvergent approximation to the tangential component of ∇u at midpoints of sides. Denoting by M_h the set of these midpoints, we may then define a recovery operator for both the components of the gradient by the relation (see [4,8,9,11,26,28,30,31,33,40])

$$\tilde{\nabla} u_h(x) = \frac{1}{2}(\nabla u_{h|T_1} + \nabla u_{h|T_2}) , \quad x \in M_h \cap \Omega , \tag{6}$$

where $T_1, T_2 \in \mathcal{T}_h$ are those adjacent triangles for which $x \in T_1 \cap T_2$ (note that $\nabla u_{|T_i}$ is constant). As shown in [11,30],

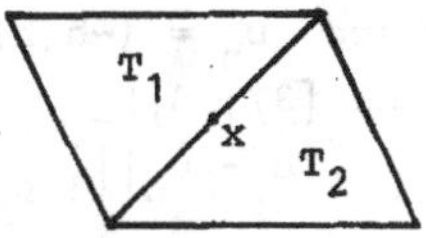

Fig. 3

$$\max_{x \in M_h \cap \Omega} ||\nabla u(x) - \tilde{\nabla} u_h(x)|| \le Ch^2 |\ln h| \; ||u||_{3,\infty,\Omega}$$

or even $O(h^2)$ for the discrete L^2-norm [26] (cf. (4)). For a three-dimensional analogue of (6), see [5].

Note that the sampling at centroids of the bilinear elements leads to the superconvergence of the gradient [24]. This is not true for the linear elements. However, a weighted averaging scheme between four elements,

$$\tilde{\nabla} u_h(x) = \frac{1}{6}(3\nabla u_{h|T} + \sum_{i=1}^{3} \nabla u_{h|T_i}) , \quad x \in C_h \cap \Omega_0 ,$$

yields [26]

$$h\Big(\sum_{x \in C_h \cap \Omega_0} ||\nabla u(x) - \tilde{\nabla} u_h(x)||^2 \Big)^{\frac{1}{2}} \le Ch^2 ||u||_{3,\Omega} .$$

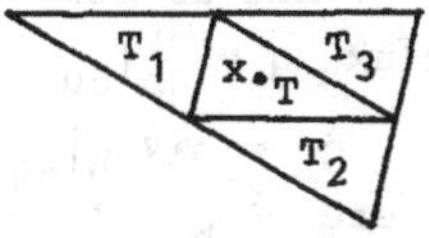

Fig. 4

Here C_h is the set of centroids of all $T \in \mathcal{T}_h$, $\Omega_0 \subset\subset \Omega$, and $T_1, T_2, T_3 \in \mathcal{T}_h$ are the triangles adjacent to that triangle $T \in \mathcal{T}_h$ for which $x \in T$. Using (6), one can define a discontinuous piecewise linear field $\tilde{\nabla} u_h$ which recovers the gradient of u even at any point of $\Omega_0 \subset\subset \Omega$ (see [36]). By the following averaging at nodes $x \in N_h$ we may determine a continuous piecewise linear field $\tilde{\nabla} u_h$ over the whole domain $\bar{\Omega}$:

$$\tilde{\nabla} u_h(x) = \begin{cases} \frac{1}{6} \sum_{T \cap \{x\} \neq \emptyset} \nabla u_{h|T} , & x \in N_h \cap \Omega , \\ 0 , & x \in Y , \\ \frac{1}{2}\Big(\sum_{i=1}^{3} \nabla u_{h|T_i} - \nabla u_{h|T_0} \Big), & x \in N_h \cap (\partial\Omega - Y) , \end{cases} \tag{7}$$

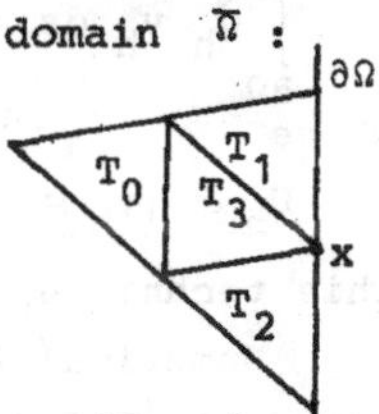

Fig. 5

where Y is the set of vertices of $\bar{\Omega}$, T_i and T_3 form a parallelogram for every $i = 0,1,2$, and $T_1 \cap T_2 \cap T_3 = \{x\}$ when $x \in N_h \cap (\partial\Omega - Y)$ - see Fig.5. In this case the global superconvergence estimate reads [23]:

$$||\nabla u - \widetilde{\nabla u}_h||_{0,p,\Omega} \leq Ch^2 |\ln h|^{1-2/p} ||u||_{3,p,\Omega} \ , \ p \in \{2,\infty\} \ . \tag{8}$$

For the generalization of the scheme (7) to elliptic systems with nonhomogeneous boundary conditions of several types, we refer to [20]. If $\partial\Omega$ is smooth then a local $O(h^{3/2})$-superconvergence in $\Omega_0 \subset\subset \Omega$ can be achieved [20,21] in the L^2-norm (T_h are not uniform near the boundary $\partial\Omega$).

Consider now triangulations as marked in Fig. 1 or 2 and the smoothing post-processing operator

$$\tilde{u}_h(x) = \tfrac{1}{4}h^{-2} \int_{D_h} u_h(x + y) \, dy \ ,$$

where $D_h = (-h,h) \times (-h,h)$. If $\Omega_0 \subset\subset \Omega$ and $\partial\Omega$ is again smooth then (see [37,38])

$$||u - \tilde{u}_h||_{1,\Omega_0} \leq Ch^{3/2} ||u||_{3,\Omega} \ ,$$

which is, in fact, a superconvergent estimate for the gradient.

Another type of an integral smoothing operator which yields a superconvergent approximation for ∇u as well as for u even on irregular meshes is presented in [3]. In [19] a least squares smoothing of ∇u_h is proposed to obtain a better approximation to ∇u . Related papers with superconvergence of linear elements further include [2,6,7,12,13,17,18, 25,29,34], see also the survey papers [10,22,27].

Let us now turn to superconvergent approximations to the boundary flux $q = \frac{\partial u}{\partial n}|_{\partial\Omega}$ (n is the outward unit normal to $\partial\Omega$). Setting

$$\tilde{q}_h = n \cdot \widetilde{\nabla u}_h|_{\partial\Omega} \ ,$$

where $\widetilde{\nabla u}_h$ is given by (7), we immediately get from (8) that

$$||q - \tilde{q}_h||_{0,\infty,\partial\Omega} \leq Ch^2 \ |\ln h| \ ||u||_{3,\infty,\Omega} \ ,$$

i.e., the continuous piecewise linear function $\tilde{q}_h$ approximates q better than the piecewise constant function $q_h = n \cdot \nabla u_h|_{\partial\Omega}$.

Another continuous piecewise linear approximation $\tilde{\tilde{q}}_h$ to the boundary flux q can be defined with the help of Green's formula

$$\int_{\partial\Omega} \tilde{\tilde{q}}_h v_h \, ds = (\nabla u_h, \nabla v_h)_{0,\Omega} - (f, v_h)_{0,\Omega} \quad \forall v_h \in U_h \ ,$$

where

$$U_h = \{v_h \in H^1(\Omega) \ | \ v_{h|T} \in P_1(T) \quad \forall T \in T_h\} \ .$$

This technique suggested by [16], p.398, is based on some ideas of [14].

Numerical tests of the presented superconvergent schemes can be found in [3,6,11,19,21,23,24,26,36].

References

[1] ANDREEV, A. B.: *Superconvergence of the gradient for linear triangle elements for elliptic and parabolic equations.* C. R. Acad. Bulgare Sci. 37 (1984), 293-296.

[2] ANDREEV, A. B., EL HATRI, M. and LAZAROV, R. D.: *Superconvergence of the gradient in the finite element method for some elliptic and parabolic problems* (Russian). Variational-Difference Methods in Math. Phys., Part 2 (Proc. Conf., Moscow, 1983), Viniti, Moscow, 1984, 13-25.

[3] BABUŠKA, I. and MILLER, A.: *The post-processing in the finite element method, Part I.* Internat. J. Numer. Methods Engrg. 20 (1984), 1085-1109.

[4] CHEN, C. M.: *Optimal points of the stresses for triangular linear element* (Chinese). Numer. Math. J. Chinese Univ. 2 (1980), 12-20.

[5] CHEN, C. M.: *Optimal points of the stresses for tetrahedron linear element* (Chinese). Natur. Sci. J. Xiangtan Univ. 3 (1980), 16-24.

[6] CHEN, C. M.: *Finite Element Method and Its Analysis in Improving Accuracy* (Chinese). Hunan Sci. and Tech. Press, Changsha, 1982.

[7] CHEN, C. M.: *Superconvergence of finite element approximations to nonlinear elliptic problems.* (Proc. China-France Sympos. on Finite Element Methods, Beijing, 1982), Science Press, Beijing, Gordon and Breach Sci. Publishers, Inc., New York, 1983, 622-640.

[8] CHEN, C. M.: *An estimate for elliptic boundary value problem and its applications to finite element method* (Chinese). Numer. Math. J. Chinese Univ. 5 (1983), 215-223.

[9] CHEN, C. M.: $W^{1,\infty}$ *-interior estimates for finite element method on regular mesh.* J. Comp. Math. 3 (1985), 1-7.

[10] CHEN, C. M.: *Superconvergence of finite element methods* (Chinese). Advances in Math. 14 (1985), 39-51.

[11] CHEN, C. M. and LIU, J.: *Superconvergence of the gradient of triangular linear element in general domain.* Preprint Xiangtan Univ., 1985, 1-19.

[12] CHEN, C. M. and THOMÉE, V.: *The lumped mass finite element method for a parabolic problem.* J. Austral. Math. Soc. Ser. B 26 (1985), 329-354.

[13] CHENG, S. J.: *Superconvergence of finite element approximation for Navier-Stokes equation.* (Proc. Conf., Bonn, 1983), Math. Schrift. No. 158, Bonn, 1984, 31-45.

[14] DOUGLAS, J., DUPONT, T. and WHEELER, M. F.: *A Galerkin procedure for approximating the flux on the boundary for elliptic and parabolic boundary value problems.* RAIRO Anal. Numér. 8 (1974), 47-59.

[15] FRIED, I.: *On the optimality of the pointwise accuracy of the finite element solution.* Internat. J. Numer. Methods Engrg. 15 (1980), 451-456.

[16] GLOWINSKI, R.: *Numerical Methods for Nonlinear Variational Problems.* Springer Series in Comp. Physics. Springer-Verlag, Berlin, New York, 1984.

[17] EL HATRI, M.: *Superconvergence of axisymmetrical boundary-value problem.* C. R. Acad. Bulgare Sci. 36 (1983), 1499-1502.

[18] EL HATRI, M.: *Superconvergence in finite element method for a degenerated boundary value problem* (to appear), 1984, 1-6.

[19] HINTON, E. and CAMPBELL, J. S.: *Local and global smoothing of discontinuous finite element functions using a least squares method.* Internat. J. Numer. Methods Engrg. 8 (1974), 461-480.

[20] HLAVÁČEK, I. and KŘÍŽEK, M.: *On a superconvergent finite element scheme for elliptic systems, I. Dirichlet boundary conditions, II. Boundary conditions of Newton's or Neumann's type* (submitted to Apl. Mat.), 1985, 1-29, 1-17.

[21] KŘÍŽEK, M. and NEITTAANMÄKI, P.: *Superconvergence phenomenon in the finite element method arising from averaging gradients.* Numer.

Math. 45 (1984), 105-116.

[22] KŘÍŽEK, M. and NEITTAANMÄKI, P.: *On superconvergence techniques.* Preprint No. 34, Univ. of Jyväskylä, 1984, 1-43.

[23] KŘÍŽEK, M. and NEITTAANMÄKI, P.: *On a global superconvergence of the gradient of linear triangular elements.* Preprint No. 85/4, Univ. Hamburg, 1985, 1-20.

[24] LASAINT, P. and ZLÁMAL, M.: *Superconvergence of the gradient of finite element solutions.* RAIRO Anal. Numér. 13 (1979), 139-166.

[25] LEVINE, N.: *Stress ampling points for linear triangles in the finite element method.* Numer. Anal. Report 10/82, Univ. of Reading, 1982.

[26] LEVINE, N.: *Superconvergent recovery of the gradient from piecewise linear finite element approximations.* Numer. Anal. Report 6/83, Univ. of Reading, 1983, 1-25.

[27] LIN, Q.: *High accuracy from the linear elements.* Proc. of the Fifth Beijing Sympos. on Differential Geometry and Differential Equations, Beijing, 1984, 1-5.

[28] LIN, Q. and LÜ, T.: *Asymptotic expansions for finite element approximation of elliptic problem on polygonal domains.* Comp. Methods in Appl. Sci. and Engrg. (Proc. Conf., Versailles, 1983), North--Holland Publishing Company, INRIA, 1984, 317-321.

[29] LIN, Q. and LÜ, T.: *Asymptotic expansions for finite element eigenvalues and finite element solution.* (Proc. Conf., Bonn, 1983), Math. Schrift. No. 158, Bonn, 1984, 1-10.

[30] LIN, Q., LÜ, T. and SHEN, S.: *Asymptotic expansion for finite element approximations.* Research Report IMS-11, Chengdu Branch of Acad. Sinica, 1983, 1-6.

[31] LIN, Q., LÜ, T. and SHEN, S.: *Maximum norm estimate, extrapolation and optimal point of stresses for the finite element methods on the strongly regular triangulations.* J. Comput. Math. 1 (1983), 376-383.

[32] LIN, Q. and WANG, J.: *Some expansions of the finite element approximation.* Research Report IMS-15, Chengdu Branch of Acad. Sinica, 1984, 1-11.

[33] LIN, Q. and XU, J. Ch.: *Linear elements with high accuracy.* J. Comp. Math. 3 (1985), 115-133.

[34] LIN, Q. and ZHU, Q. D.: *Asymptotic expansion for the derivative of finite elements.* J. Comp. Math. 2 (1984), 361-363.

[35] MICHLIN, S. G. and SMOLICKIJ, Ch. L.: *Approximation Methods for Solving Differential and Integral Equations* (Russian). Nauka, Moscow, 1965.

[36] NEITTAANMÄKI, P. and KŘÍŽEK, M.: *Superconvergence of the finite element schemes arising from the use of averaged gradients.* Accuracy Estimates and Adaptive Refinements in Finite Element Computations, (Proc. Conf., Lisbon, 1984), Lisbon, 1984, 169-178.

[37] OGANESJAN, L. A., RIVKIND, V. J. and RUCHOVEC, L. A.: *Variational--Difference Methods for the Solution of Elliptic equations* (Russian). Part I (Proc. Sem., Issue 5, Vilnius, 1973), Inst. of Phys. and Math., Vilnius, 1973, 3-389.

[38] OGANESJAN, L. A. and RUCHOVEC, L. A.: *Variational-Difference Methods for the Solution of Elliptic Equations* (Russian). Izd. Akad. Nauk Armjanskoi SSR, Jerevan, 1979.

[39] RANNACHER, R. and SCOTT, R.: *Some optimal error estimates for piecewise linear finite element approximations.* Math. Comp. 38 (1982), 437-445.

[40] ZHU, Q. D.: *Natural inner superconvergence for the finite element method.* (Proc. China-France Sympos. on Finite Element Methods, Beijing, 1982), Science Press, Beijing, Gordon and Breach Sci. Publishers, Inc., New York, 1983, 935-960.

MIXED FINITE ELEMENT IN 3D IN H(div) AND H(curl)

J. C. NEDELEC
Ecole Polytechnique, Centre de Mathématiques Appliquées
91128 Palaiseau, France

I. INTRODUCTION.

Frayes De Venbeke first introduces the mixed finite element. Then P.A. Raviart and J.M. Thomas does some mathematics on these element in 2D and others do also : F. Brezzi V. Babuska ...
In 1980 we introduce a family of some mixed finite element in 3D and we use them for solving Navier Stokes equations.
In 1984 F. Brezzi, J. Douglass and L.D. Marini introduce in 2D a new family of mixed finite element conforming in H(div). That paper was the starting point for building new families of finite element in 3D.

II. FINITE ELEMENT IN H(div).

Notations.

K is a tetrahedron

∂K its boundary

n the normal

f a face which area is $\int_f d\gamma$

a is an edge which lenght is $\int_a ds$

$\text{curl } u = \nabla \wedge u \quad u = (u_1, u_2, u_3)$

$H(\text{curl}) = \{u \in L^2(\Omega))^3 \ ; \ \text{curl } u \in (L^2(\Omega))^3 \}$

$\text{div} = \nabla . u$

$H(\text{div}) = \{u \in (L^2(\Omega))^3 \ ; \ \text{div } u \in L^2(\Omega) \}$

Spaces of polynomials.

P_k = polynomials of degree less or equal to k

$\tilde{P}_k$ = " homogeneous of degree k

$\mathcal{D}_k = (P_{k-1})^3 + \tilde{P}_{k-1}\, r \qquad r = \begin{cases} x_1 \\ x_2 \\ x_3 \end{cases}$

$S_k = \{p \in (P_k) \ ; \ (r.p) \equiv 0 \}$

$R_k = (P_{k-1})^3 \oplus S_k$

$\dim S_k = k(k+2)$

$\dim \mathcal{D}_k = \dfrac{(k+3)(k+1)\,k}{2}$

$\dim \mathcal{R}_k = \dfrac{(k+3)(k+2)\,k}{2}$

We are now able to introduce the finite element conforming in H(div).

Definition. We define the finite element by

1) K is a tetrahedron
2) $P = (P_k)^3$ is a space of polynomials
3) The set of degrees of freedom which are

$$(3.1) \qquad \int_f (p \cdot n) q \, d\gamma \; ; \; \forall q \in P_k(f) \; ;$$

$$(3.2) \qquad \int_K (p \cdot q) \, dx \; ; \; \forall q \in \mathcal{R}_{k-1} \; .$$

we have the

Theorem.

The above finite element is unisolventand conforming in H(div). The associate interpolation operator Π is such that

$$\operatorname{div} \Pi p = \Pi^* \operatorname{div} p \quad ; \quad \forall p \in H(\operatorname{div}) \quad ,$$

where Π^* is the L^2 projection on P_{k-1}.

When k = 1, the corresponding element has no interior moments and 12 degrees of freedom. Its divergence is constant.

Proposition.For a tetrahedron "regular enough" which diameter is k, we have

$$\| p - \Pi p \|_{(L^2(K))^3} \leq c \, h^{k+1} \, \| p \|_{(H^{k+1}(K))^3} \quad ;$$

$$\| D(p - \Pi p) \|_{(L^2(K))^3} \leq c \, h^{k} \, \| p \|_{(H^{k+1}(K))^3} \quad .$$

We are not going to prove this theorem. But we can recall that a finite element is said to be conforming in a functional space if the interpolate of an element of this space belong to this space.

In our case, the conformity in H(div) is equivalent to the continuity of the normal composent at each interface. This property is clearly true for our finite element since the unknowns on the face are

$$\int_f (p \cdot n) \, q \, d\gamma \quad ; \quad \forall q \in P_k(f)$$

and p.n is also $P_k(f)$.

III. FINITE ELEMENT IN H(curl).

A finite element is conforming in H(curl) if the tangential components are continue at the interface of the mesh.
We introduce the corresponding finite element.

Définition.

1) K is a tetrahedron
2) $P = (P_k)^3$ is the space of polynomials
3) The degrees of freedom are the following moments

3.1) $\int_a (p \cdot \tau)\, q\, ds \quad ; \quad \forall q \in P_k(a)$

3.2) $\int_f (p \cdot q)\, d\gamma \quad ; \quad \forall q \in \mathcal{D}_{k-1}(f)$ and tangent to the face f

3.3) $\int_K (p \cdot q)\, dx \quad ; \quad \forall q \in \mathcal{D}_{k-2}$.

We have the

Theorem.

The above finite element is unisolvent and conforming in H(curl). Moreover if Π is the corresponding interpolation operator and Π^* the interpolation operator associate to the H(div) finite element introduce previously for degree k-1 we have

$$\text{curl}\, \Pi p = \Pi^* \text{curl}\, p$$

IV. APPLICATION TO THE EQUATION OF STOKES.

The Stokes'equation is usually written in the (u,p) variable in a bounded domain Ω of R^3 as

$$\begin{cases} -\nu\, \Delta u + \text{grad}\, p = f \quad , & \text{in } \Omega \\ \text{div}\, u = 0 & \text{in } \Omega \\ u|_\Gamma = 0 & \end{cases}$$

We introduce the vector potential ϕ as

$$\begin{cases} -\Delta\phi = \text{curl}\, u \quad , & \text{in } \Omega \\ \text{div}\, \phi = 0 \quad , & \text{in } \Omega \\ \phi \wedge n|_\Gamma = 0 & \end{cases}$$

Then the Stokes equation can be written in the (ϕ,ω) variables where

$$\omega = \text{curl}\, u$$

We introduce

$$H(\text{div}^0) = \{ v \in (L^2(\Omega))^3 \ ; \ \text{div}\, v \in 0 \ , \ v.n|_\Gamma = 0 \}$$

$$\mathcal{H} = \{ \psi \in H(\text{curl}) \ ; \ \text{div}\ \psi = 0 \ ; \ \psi \wedge n|_\Gamma = 0 \}$$

Then a variational formulation of the Stokes equation is

$$\begin{cases} \nu \int_\Gamma (\text{curl } \omega . \text{curl } \psi) dx = \int_\Omega (f . \text{curl } \psi) dx \ ; \ \forall \psi \in H \\ \int_\Omega (\omega . \Pi) dx - \int_\Omega (\text{curl } \phi . \text{curl } \Pi) dx = 0 \ ; \ \forall \Pi \in H(\text{curl}) \end{cases}$$

Let C_h be a mesh covering Ω .

We can introduce some finite element spaces

$$W_h = \{ \omega_h \in H(\text{curl}) \ ; \ \omega_h|_K \in (P_k)^3 \ ; \forall K \in C_k \}$$

$$W_h^0 = \{ \omega_h \in W_h \quad ; \quad \omega_h \wedge n|_\Gamma = 0 \}$$

$$V_h = \{ v_h \in H(\text{div}) \ ; \ v_h|_K \in (P_{k-1})^3 \ ; \ \forall \ K \in C_h \}$$

$$U_h = V_h \cap H(\text{div}^0)$$

The approximate problem become then

$$\begin{cases} \nu \int_\Omega (\text{curl } w_h . v_h) dx = \int_\Omega (f . v_h) dx \quad ; \quad \forall v_h \in U_h \quad ; \\ \int_\Omega (w_h . \Pi_h) dx - \int_\Omega (u_h . \text{curl } \Pi_h) \ dx = 0 \quad ; \quad \forall \Pi_h \in W_h \quad . \end{cases}$$

We can also use a vector potential ϕ_h.
This goes like that

$$\Theta_h = \{ \theta_h \in H^1(\Omega) \quad ; \quad \theta_h|_K \in P_{k+1} \quad ; \quad \forall K \in C_h \}$$
$$\Theta_h^0 = \Theta_h \cap H_0^1(\Omega)$$

We have the

<u>Theorem</u>.

If the transgulation is regular, for every $v_h \in U_h$, there exist a unique $\psi_h \in W_h^0$ such that

$$\begin{cases} \text{curl } \psi_h = v_h \\ \int_\Omega (\psi_h \ . \ \text{grad } \theta_h) dx = 0 \ ; \ \forall \theta_h \in \Theta_h^0 \end{cases}$$

and we have also

$$\|\psi_h\|_{H(\text{curl})} \leq c \ \|v_h\|_{(L^2(\Omega))^3} \ .$$

This theorem can be use to transfer the above approximate problem in one in (ψ,ω) and also to find a local basis in the space U_h.

BIBLIOGRAPHY

F. BREZZI, On the existence, uniqueness and approximation of saddle point problems ausing from Lagrangian multipliers. RAIRO 8 : 129 - 151 (1974).

F. BREZZI, J. DOUGLASS & L.D. MARINI, Two families of mixed finite elements for second order elliptic problems. To appear in Numerische Mathematik.

P.G. CIARLET, The finite element method for elliptic problems. North Holland Amsterdam (1978).

P.G. CIARLET & P.A. RAVIART, A mixed finite element method for the biharmonic equation. Mathematical aspects in finite element method (C de Boor ed.) pp. 125 145 Academic Press New York (1974).

M. FORTIN, An analysis of the convergence of mixed finite element method. RAIRO 11 : 341 - 354 (1977).

J.C. NEDELEC, Mixed finite element in $\mathbb{R}^3$. Numerische Mathematik 35 : 315 - 341 (1980).

J.C. NEDELEC, Elements finis mixtes incompressibles pour l'equation de Stokes dans $\mathbb{R}^3$. Numerische Mathematik 39, 97 - 112 (1982).

P.A. RAVIART & J.M. THOMAS, A mixed finite element method for 2nd order elliptic problems. In Dold A Eckmann B (eds), Mathematical aspects of finite element methods, Lecture Notes 606 Springer Berlin (1977).

J.M. THOMAS, Thesis Paris (1977).

J.C. NEDELEC, A new family of mixed finite element in $\mathbb{R}^3$ (à paraître).

FREE BOUNDARY PROBLEMS FOR STOKES' FLOWS AND FINITE ELEMENT METHODS

J. A. NITSCHE
Institut für angewandte Mathematik, Albert-Ludwigs-Universität Freiburg im Breisgau, West Germany

Abstract:

In two dimensions a Stokes' flow is considered symmetric to the abscissa $\eta = 0$ and periodic with respect to ξ. On the free boundary $|\eta| = S(\xi)$ the conditions are: (i) the free boundary is a streamline, (ii) the tangential force vanishes, (iii) the normal force is proportional to the mean curvature of the boundary. By straightening the boundary, i. e. by introducing the variables $x = \xi$, $y = \eta/S(\xi)$, the problem is reduced to one in a fixed domain. The underlying differential equations are now highly nonlinear: They consist in an elliptic system coupled with an ordinary differential equation for S. The analytic properties of the solution as well as the convergence of the proposed finite element approximation are discussed.

1. In accordance to the restrictions formulated in the abstract the problem under consideration is: We ask for the free boundary $\eta = S(\xi)$, 1-periodic in ξ, such that there exists a solution pair $\underline{U} = (U_1, U_2)$ and P with the properties:

(i_1) In the domain $\Omega = \{ (\xi,\eta) \mid |\eta| < S(\xi) \}$ the system of differential equations

(1. 1) $$\sigma_{ik|k} = F_i$$

hold true with

(1. 2) $$\sigma_{ik} = U_{i|k} + U_{k|i} - P\delta_{ik}.$$

(i_2) In the domain Ω the incompressibility condition

(1. 3) $$\nabla \cdot \underline{U} = U_{1|\xi} + U_{2|\eta} = 0$$

holds true.

(ii$_1$) The free boundary $\eta = \pm S(\xi)$ is streamline, i. e.

(1.4) $$U_2 - S'U_1 = 0 \quad \text{for } \eta = \pm S(\xi) .$$

(ii$_2$) On the free boundary the shear-force vanishes, i. e.

(1.5) $$\sigma_{ik} t_i n_k = 0$$

with $\underline{t} = (t_1, t_2)$ and $\underline{n} = (n_1, n_2)$ being the tangential resp. normal unit vectors.

(ii$_3$) The normal-force is proportional to the mean curvature, i. e.

(1.6) $$\sigma_{ik} n_i n_k = \kappa H .$$

We will consider fluid motions only "not too far" from $\underline{U}^0 = (1,0)$. Together with $P^0 = 0$ and $S^0 = 1$ the triple $\{\underline{U}^0, P^0, S^0\}$ is a solution to the problem stated above with $\underline{F}^0 = \underline{0}$. - The main idea of our analysis is the "straigthening" of the free boundary, quite often used. This consists in introducing new variables

(1.7) $$x = \xi \quad , \quad y = \eta / S(\xi) .$$

Since we are looking for solutions $\{\underline{U}, P, S\}$ near to $\{\underline{U}^0, P^0, S^0\}$ we replace $\underline{U}$, P and S - depending on ξ, η - by $(1+u_1, u_2)$, p and 1+s depending on x,y . This leads to a nonlinear problem in the new variables but now in the fixed domain

(1.8) $$Q_\pm = \{ (x,y) \mid |y| < 1 \} .$$

Because of our setting all functions are assumed to be 1-periodic in x. For functions $\underline{F}$ resp. in the new variables $\underline{f}$ symmetric with respect to $y = 0$, i. e. $f_1(x,-y) = f_1(x,y)$ and $f_2(x,-y) = -f_2(x,y)$, the solution also will be symmetric to $y = 0$. Hence we can restrict ourselves to the unit square

(1.9) $$Q = \{ (x,y) \mid 0 < x, y < 1 \} .$$

The condition of symmetry implies the boundary conditions

(1.10) $$u_2(x,0) = 0 \quad , \qquad u_{1|y}(x,0) = 0 \quad .$$

By linearizing, i. e. by spltting into linar and nonlinear terms, we get from (1.1) the system

(1.11) $$\begin{aligned} \partial_x(2u_{1|x} - p) + \partial_y(u_{1|y} + u_{2|x}) &= \partial_x \Sigma_{11} + \partial_y \Sigma_{12} + f_1 \quad , \\ \partial_x(u_{1|y} + u_{2|x}) + \partial_y(2u_{2|y} - p) &= \partial_x \Sigma_{21} + \partial_y \Sigma_{22} + f_2 \quad . \end{aligned}$$

Here $\Sigma_{ik} = \Sigma_{ik}(\underline{u},p,s)$ are at least quadratic in their arguments, for example it is

$$(1.12)\qquad \Sigma_{12} = -2ys'u_{1|x} + 2(1+s)^{-1}(1+y^2s'^2)u_{2|y} - (1+s)^{-1}ys'u_{2|y} + ys'p\,.$$

In the new variables condition (1.3) becomes

$$(1.13)\qquad \begin{aligned} u_{1|x} + u_{2|y} &=: \mathbf{D} \\ &= (1+s)^{-1}(ys'u_{1|y} + su_{2|y}) \end{aligned}$$

The boundary condition (1.4) may be used as defining relation for $s = s(x)$:

$$(1.14)\qquad \begin{aligned} s' &= (1+u_1)^{-1}u_2 \\ &=: u_2 + \mathbf{P} \end{aligned}$$

(1.5) leads to a boundary condition of the type

$$(1.15)\qquad u_{1|y} + u_{2|x} = \mathbf{T}_1\,.$$

The mean curvature H of the free surface depends on the second derivative S'' resp. s''. This quantity may be computed from (1.14). In this way (1.6) leads to the second boundary condition of the type

$$(1.16)\qquad 2u_{2|y} - p + \kappa u_{2|x} = \mathbf{T}_2\,.$$

The $\mathbf{T}_i = \mathbf{T}_i(\underline{u},p,s)$ are at least quadratic in their arguments. Similar to the Σ_{ik} they depend only on the functions themselves and their first derivatives. Since s is assumed to be 1-periodic we have $\int s' = 0$. Here $\int w$ resp. later $\int\int w$ are abbreviations defined by

$$(1.17)\qquad \int w = \int_0^1 w(x,1)\,dx\,, \qquad \int\int w = \iint_Q w(x,y)\,dx\,dy$$

In view of the boundary condition (1.10) we get from (1.13) $\int\int \mathbf{D} = -\int u_2$. Therefore the quantity

$$(1.18)\qquad \begin{aligned} \gamma &= \int\int \mathbf{D} - \int \mathbf{P} \\ &=: \gamma(\underline{u},p,s) \end{aligned}$$

will be zero. Hence we may replace in (1.13) the right hand side $\mathbf{D}$ by

$$(1.19)\qquad \tilde{\mathbf{D}} = \mathbf{D} - \gamma\,.$$

In the new variables we have the

Problem:

> Given the vector $\underline{f}$ defined in Q (1.9) and 1-periodic in x. Find $\underline{u}$, p, s 1-periodic in x, fulfilling the differential equations (1.11), (1.13) in Q, and the boundary condition (1.10) on y=0 as well as (1.14), (1.15), and (1.16) on y=1.

2. The idea of proving the existence of a solution of the problem as well as deriving a ***finite element method*** in order to approximate this solution is as follows: We consider the quadruple $\mathbf{m} = \{u_1,u_2,p,s\}$ as an element of a linear space $\mathfrak{M}$ equipped with an appropriate norm. The geometric boundary condition (1.10_1) has to be imposed on u_2. Obviously u_1 as well as s are defined up to a constant only. Therefore we nomalize u_1, s according to $\int\int u_1 = 0$, $\int s = 0$. The correspondent restriction of the space $\mathfrak{M}$ will be denoted by $^*\mathfrak{M}$. Similarily we consider the octuple $\mathbf{n} = \{\Sigma_{11},\Sigma_{12},\Sigma_{21},\Sigma_{22},\tilde{D},P,T_1,T_2\}$ as an element of a linear space $\mathfrak{N}$, also equipped with a norm. By (1.12), (1.13) etc. the mapping $\mathbf{A} : \mathfrak{M} \mapsto \mathfrak{N}$ is defined. The mapping $\mathbf{B} : \mathfrak{N} \mapsto {}^*\mathfrak{M}$ which associates the solution of the boundary value problem to the right hand sides is constructed by the natural weak formulation of the problem: If $\mathbf{m} \in {}^*\mathfrak{M}$ is the solution then with any $\mu = \{\underline{v},q, r\} \in {}^*\mathfrak{M}$ the variational equations hold:

$$
(2.2)\qquad
\begin{aligned}
a(\mathbf{m},\mu) + b(\mathbf{m},\mu) &= L_1(\mathbf{n},\mu) + F(\underline{f},\mu) ,\\
b(\mu,\mathbf{m}) &= L_2(\mathbf{n},\mu) ,\\
c(\mathbf{m},\mu) - \int u_2 r' &= \int P r' .
\end{aligned}
$$

Here L_i, F, a, b, c are bilinear functionals; especially a(.,.), b(.,.), and c(.,.) are defined by

$$
(2.3)\qquad
\begin{aligned}
a(\mathbf{m},\mu) &= \int\int\left[2u_{1|x}v_{1|x} + (u_{1|y}+u_{2|x})(v_{1|y}+v_{2|x}) + 2u_{2|y}v_{2|y}\right] - \kappa\int u_{2|x}v_2 ,\\
b(\mathbf{m},\mu) &= \int\int \; q\left\{u_{1|x} + u_{2|y}\right\}\\
c(\mathbf{m},\mu) &= \int \; s' r'
\end{aligned}
$$

The standard inf-sup condition is valid for the form b(.,.), because of Korn's second inequality a(.,.) may be extended to a bounded and coercive bilinear form in the Sobolev space $H_1(Q)\times H_1(Q)$. In connection with the normalisation of u_1 and s uniqueness of the mapping **B** is guaranteed.

3. Since the mapping **A** is nonlinear we will work with Hölder-spaces: We equip the spaces $^*\mathfrak{M}$ and $\mathfrak{N}$ in the following way with norms, in these topologies they are Banach-spaces: For $\mu = \{\underline{v},q,r\} \in {}^*\mathfrak{M}$ we define

$$
(3.1)\qquad
\begin{aligned}
\|\mu\| &:= \|\mu\|_{^*\mathfrak{M}}\\
&= \sum \|v_i\|_{C_{1\cdot\lambda}(Q)} + \|q\|_{C_{0\cdot\lambda}(Q)} + \|r\|_{C_{2\cdot\lambda}(I)}
\end{aligned}
$$

Here $\|.\|_{C^{k,\lambda}(.)}$ denote the usual Hölder-norms with $\lambda \in (0,1]$, I is the unit interval.

For $v = \{\Sigma_{11},\Sigma_{12},\Sigma_{21},\Sigma_{22},\tilde{\delta},P,T_1,T_2\} \in \mathfrak{N}$ we define

$$\|v\| := \|v\|_{\mathfrak{N}} \tag{3.2}$$
$$= \sum\|\Sigma_{ik}\|_{C^{0,\lambda}(Q)} + \|\tilde{\delta}\|_{C^{0,\lambda}(I)} + \|P\|_{C^{1,\lambda}(Q)} + \sum\|T_i\|_{C^{0,\lambda}(I)}$$

Now we consider elements μ in the ball $B_\delta({}^*\mathfrak{M}) := \{\mu \mid \mu \in {}^*\mathfrak{M} \wedge \|\mu\| \le \delta\}$ with $\delta \le \delta_0 < 1$ and δ_0 fixed. Obviously the two estimates are valid:

$$\|A\mu\|_{\mathfrak{N}} \le c\delta\|\mu\|_{\mathfrak{M}} \tag{3.3}$$
$$\|A\mu^1 - A\mu^2\|_{\mathfrak{N}} \le c\delta\|\mu^1 - \mu^2\|_{\mathfrak{M}}$$

Here "c" denotes a numerical constant depending only on δ_0 which may differ at different places.

It can be shown: The mapping **B** is bounded, i. e. for $m = Bn$ the estimate

$$\|m\| \le c\|n\| + \sum\|f_i\|_{C^{0,\lambda}(Q)} \tag{3.4}$$

is valid. Thus the Banach Fixed Point Theorem leads to: For $\|f_i\|$ sufficiently small and δ chosen appropriately the mapping

$$T := B * A \tag{3.5}$$

possesses an unique fixed point in the ball $B_\delta({}^*\mathfrak{M})$. It turns out that the quantity γ (1.18) vanishes. This implies that the fixed point corresponds to the solution of the original problem.

4. Now let ${}^*\mathfrak{M}_h$ be an appropriate finite element approximation space. By restricting in (2.2) the elements $\mu = \mu_h \in {}^*\mathfrak{M}_h$ and looking for the solution $m_h \in {}^*\mathfrak{M}_h$ the mapping B_h and consequently also T_h (see (3.5)) is defined.

It can be shown: Under certain conditions concerning the approximation spaces, especially the Brezzi condition is needed, the mapping B_h is bounded, i. e. an inequality of the type (3.4) holds true. This finally leads to almost best error estimates: Let $m \in {}^*\mathfrak{M}$ and $m_h \in {}^*\mathfrak{M}_h$ be the solution of the analytic problem resp. the finite element solution then

$$\|m - m_h\| \le C\inf\{\|m - \mu_h\| \mid \mu_h \in {}^*\mathfrak{M}_h\}. \tag{4,1}$$

The proofs and the complete bibliography will appear elsewhere. Here we refer only to

Bemelmans, J. (1981a)
Gleichgewichtsfiguren zäher Flüssigkeiten mit Oberflächenspannung
Analysis 1, 241-282 (1981)

Bemelmans, J. (1981b)
Liquid Drops in a viscous Fluid under the Influence of Gravity and Surface Tension
Manuscripta math. 36, 105-123 (1981)

Bemelmans, J. and A. Friedman (1984)
Analiticity for the Navier-Stokes Equations Governed by Surface Tension on the Free Boundary
J. of Diff. Equat. 55, 135-150 (1984)

Nitsche, J. A.
Schauder Estimates for Finite Element Approximations on second Order Elliptic Boundary Value Problems
Proceedings of the Special Year in Numerical Analysis, Lecture Notes #20, Univ. of Maryland, Babuska, I., T.,-P. Liu, and J. Osborn eds., 290-343 (1981)

Schulz, F. (1982)
Über elliptische Monge-Amperesche Differentialgleichungen mit einer Bemerkung zum Weylschen Einbettungsproblem
Nachr. Akad. Wiss. Göttingen, II Math.-Phys. Klasse 1981, 93-108 (1982)

ENCLOSING METHODS FOR PERTURBED BOUNDARY VALUE PROBLEMS IN NONLINEAR DIFFERENCE EQUATIONS

J. W. SCHMIDT
Technical University of Dresden
Mommsenstr. 13, Dresden, DDR

1. In the lecture nonlinear equations $F_a(z) = 0$ are considered depending on an input parameter vector a which may be subjected to errors, shortly $a \in A$. In order to study the influence of the input $a \in A$ on the solutions z_a, by means of monotone enclosing methods intervals are constructed containing for each $a \in A$ at least one solution z_a. Such a type of methods can be developed if the operators F_a possess some monotony properties, see SCHMIDT/SCHNEIDER [1].

2. The FDM-discretization of the boundary value problem

$$u'' = 2\alpha_0 \sinh u - \varphi(t), \quad u(0) = p, \quad u(\gamma) = q \tag{2.1}$$

appearing in inner electronics is chosen as a model problem. Let the net density φ be given by

$$\varphi(t) = \varphi(t,\alpha_1,\ldots,\alpha_7) = \\ = \alpha_1 10^{10} e^{-\alpha_2 t^2} + \alpha_3 10^8 e^{-\alpha_4 t^2} + \alpha_5 10^5 + \alpha_6 10^8 e^{-\alpha_7 (t-\gamma)^2} . \tag{2.2}$$

In general the parameter vector $a = (\alpha_0,\ldots,\alpha_7)^T$ is affected with errors,

$$a = \tilde{a} \pm e, \; e = (\varepsilon_0,\ldots,\varepsilon_7)^T . \tag{2. 3}$$

This vector interval represents the set A.

Applying the common finite difference method to (2.2) ($h = \gamma/(N + 1)$ step size, $t_i = i\,h$ nodes, ζ_i approximation to $u(t_i)$) one gets the following system of equations

$$F_a(z) = F_a^+(z) + F_a^-(z) \tag{2.4}$$

with

$$(F_a^+(z))_i = -\zeta_{i-1} + 2\zeta_i - \zeta_{i+1} + \alpha_0 h^2 e^{\zeta_i} \qquad (2.5)$$

$$(F_a^-(z))_i = -\alpha_0 h^2 e^{-\zeta_i} - h^2 \varphi(t_i, \alpha_1, \ldots, \alpha_7) \quad (i = 1, \ldots, N)$$

and $\zeta_0 = p$, $\zeta_{N+1} = q$. Here the i-th component of a vector z is written as $(z)_i = \zeta_i$, and so on. Obviously, the operators F_a are offdiagonally antitone, the derivatives DF_a^+ are isotone and the derivatives DF_a^- are antitone if $\alpha_0 > 0$. These properties are essential in what follows.

3. Let R,S be finite dimensional linear spaces partially ordered by closed cones. Thus these cones are normal and regular, too. For a continuous operator

$$F : D = [y_1, x_1] \subset R \to S \qquad (3.1)$$

a mapping $\Delta F : D \times D \to L(R,S)$ is called an isotone-antitone divided difference operator if

$$F(x)-F(y) \le \Delta F(x,y)(x-y) \text{ for } y_1 \le y \le x \le x_1, \qquad (3.2)$$

$$\Delta F(x,y) \le \Delta F(u,v) \text{ for } y_1 \le v \le x \le u \le x_1 \qquad (3.3)$$

(i) For $F = F^+ + F^-$ the mapping

$$\Delta F(x,y) = DF^+(x) + DF^-(y)$$

is a divided difference operator if DF^+ is isotone and DF^- is antitone, see [10].

(ii) If, in addition, F is offdiagonally antitone

$$\Delta F(x,y) = \text{diag } DF^+(x) + \text{diag } DF^-(y)$$

is a divided difference operator, see [10].

(iii) In interval mathematics the maximal derivative

$$\Delta F(x,y) = \left(\max_{y \le z \le x} \partial_k F_i(z) \right)$$

is widely used being also a divided difference operator, see [7].

4. It is assumed that for any operator

$$F_a : D = [x_1, y_1] \subset R \to S,\ a \in A \tag{4.1}$$

an isotone-antitone divided difference operator ΔF_a exists. Because, in general, F_a and ΔF_a are not explicitly available, bounds of theirs are used. Suppose there exist mappings $U, V : D \to S$ such that

$$U(z) \leq F_a(z) \leq V(z) \text{ for } z \in D,\ a \in A\ . \tag{4.2}$$

The bounds U and V are assumed to be sharp in the following sense,

$$F_a(z) \leq 0 \text{ for all } a \in A \text{ implies } V(z) \leq 0\ , \tag{4.3}$$

$$F_a(z) \geq 0 \text{ for all } a \in A \text{ implies } U(z) \geq 0\ , \tag{4.4}$$

valid for every $z \in D$. Further, for ΔF_a let exist an upper bound $B : D \times D \to L(R,S)$ characterized by

$$\Delta F_a(x,y) \leq B(x,y) \text{ for } y_1 \leq y \leq x \leq x_1,\ a \in A, \tag{4.5}$$

$$B(x,y) \leq B(u,v) \text{ for } y_1 \leq v \leq y \leq x \leq u \leq x_1. \tag{4.6}$$

Now, the iterative process can be formulated.

Method [1]: Determine x_{n+1}, y_{n+1} such that

$$U(x_n) + B(x_n,y_n)(x_{n+1} - x_n) = 0\ , \tag{4.7}$$

$$V(y_n) + B(x_n,y_n)(y_{n+1} - y_n) = 0,\ n = 1,2,\ldots\ .$$

If ΔF_a is taken according to (i) or (ii) one gets a Newton-type method or a Jacobi-Newton-type method, respectively.

5. Monotone enclosing theorem: Let $x_1, y_1 \in R$, $y_1 \leq x_1$ be such that

$$V(y_1) \leq 0 \leq U(x_1)\ . \tag{5.1}$$

Suppose that the linear operators B(x,y) are invertible and that

$$B(x,y)^{-1} \geq 0 \text{ for } y_1 \leq y \leq x \leq x_1. \tag{5.2}$$

Then the sequence (x_n) and (y_n) are well-defined by (4.7), any of the operators F_a, $a \in A$, possesses a zero $z_a \in [y_1, x_1]$, and for such zeros the monotone enclosing

$$y_1 \leq .. \leq y_{n-1} \leq y_n \leq z_a \leq x_n \leq x_{n-1} \leq .. \leq x_1,\ n = 1,2\ldots \tag{5.3}$$

is valid.

A proof shall be skatched. The operator T,

$$T(z) = z - B(x_1,y_1)^{-1}F_a(z), \ z \in R$$

is isotone since for $y_1 \le y \le x \le x_1$ one gets

$$T(x) - T(y) = x - y - B(x_1,y_1)^{-1}\{F_a(x) - F_a(y)\},$$

$$F_a(x) - F_a(y) \le B(x,y)(x - y) \le B(x_1,y_1)(x - y)$$

implying $T(y) \le T(x)$. Further, $T(x_1) \le x_1$ and $T(y_1) \ge y_1$ hold. Thus, a fixed-point theorem of Kantorovich assures $z_a = T(z_a)$ for some vector $z_a \in [y_1,x_1]$, and hence $F_a(z_a) = 0$ follows.

Next, beinning with $y_n \le z_a \le x_n$, $F_a(z_a) = 0$, $V(y_n) \le 0 \le U(x_n)$ one gets immediately $x_{n+1} \le x_n$, further $x_{n+1} \le z_a$ because of

$$x_{n+1} - z_a = x_n - z_a - B(x_n,y_n)^{-1}\{U(x_n) - F_a(z_a)\},$$

$$U(x_n) - F_a(z_a) \le F_a(x_n) - F_a(z_a) \le B(x_n,y_n)(x_n - z_a),$$

and $U(x_{n+1}) \ge 0$ in consequence of

$$F_a(x_n) \ge U(x_n) = B(x_n,y_n)(x_n - x_{n+1}) \ge F_a(x_n) - F_a(x_{n+1}), \ a \in A.$$

Analogously $y_n \le y_{n+1} \le z_a$ and $V(y_{n+1}) \le 0$ is derived.

6. The assumption (5.2) can be weakened as follows: There esists a mapping $G \in L(S,R)$ with $\ker G = \{0\}$ and

$$G \ge 0, \ G\,B(x_1,y_1) \le I \ ,$$

see [1].

7. In the model problem (2.4),(2.5) let the input parameters be

$$\alpha_0 = 1 \pm \varepsilon_0, \ \alpha_1 = 2.8 \pm \varepsilon_1, \ \alpha_2 = 2321.385 \pm \varepsilon_2, \ \alpha_3 = 11/3 \pm \varepsilon_3,$$

$$\alpha_4 = 1121.918 \pm \varepsilon_4, \ \alpha_5 = 2 \pm \varepsilon_5, \ \alpha_6 = 20/3 \pm \varepsilon_6, \ \alpha_7 = 869.9157 \pm \varepsilon_7$$

and $\gamma = 0.25717$, $p = \operatorname{arsinh}(\varphi(0)/2)$ and $q = \operatorname{arsinh}(\varphi(\gamma)/2)$. In order to demonstrate the different influence of these parameters on the respective components of the zeros some typical examples are given computed by the Newton-type method (4.7),(i). The dimension in all cases is N = 30. Further, the notation $x^* = (\zeta_1,\dots,\zeta_{30})^T = \lim x_n$, $y^* = (\eta_1,\dots,\eta_{30})^T = \lim y_n$ is used.

Example 1: $\varepsilon_0 = 0.01$, $\varepsilon_i = 0 (i \neq 0)$ Example 2: $\varepsilon_1 = 0.01$, $\varepsilon_i = 0 (i \neq 1)$

i	η_i	ζ_i	η_i	ζ_i
5	19.943..	19.963..	19.949..	19.957
9	-12.509..	-12.489..	-12.500..	-12.498..
10	0.453..	0.473..	0.4625..	0.4646..
11	10.618..	10.638	10.6282..	10.6286..
15	12.197..	12.217..	12.20699609	12.20699609
20	13.420..	13.440..	13.43032446	13.43032446
25	18.155..	18.175..	18.16511803	18.16511803

Example 3: $\varepsilon_5 = 0.01$, $\varepsilon_i = 0 (i \neq 5)$ Example 4: $\varepsilon_6 = 0.01$, $\varepsilon_i = 0 (i \neq 6)$

i	η_i	ζ_i	η_i	ζ_i
5	19.953080..	19.953085..	19.95308310	19.95308310
9	-12.505..	-12.493..	-12.49914874	-12.49914874
10	0.413..	0.513..	0.46357969	0.46357969
11	10.602..	10.654..	10.62844138	10.62844139
15	12.201..	12.211..	12.206994..	12.206997..
20	13.428..	13.431..	13.429..	13.431..
25	18.165105..	18.165130..	18.163..	18.166..

References

[1] SCHMIDT,J.W., SCHNEIDER,H., *Enclosing methods in perturbed nonlinear operator equations*, Computing 32, 1-11 (1984).

[2] COLLATZ,L., *Aufgaben monotoner Art*, Arch. Math. (Basel)3, 366-376 (1952).

[3] NICKEL,K., *The construction of a priori bounds for the solution of a two point boundary value problem with finite elements*, Computing 23, 247-265 (1979).

[4] SCHRÖDER,H., *Operator Inequalities*, New York-London: Academic Press 1980.

[5] SPREUER,H., *A method for the computation of bounds for ordinary linear boundary value problems*, J. Math. Anal. Appl. 81, 99-133 (1981).

[6] WILDENAUER,P., *A new method for automatical computation of error bounds for nonlinear boundary value problems*, Computing 34, 131-154 (1985).

[7] ALEFELD,G., *Quadratisch konvergente Einschließung von Lösungen nichtkonvexer Gleichungssysteme*, Z. Angew. Math. Mech. 54, 335-345 (1974).

[8] ORTEGA,J.M., RHEINBOLDT,W.C., *Monotone iterations for nonlinear equations with application to Gauss-Seidel methods*, SIAM J. Numer. Anal. 4, 171-190 (1967).

[9] SCHMIDT,J.W., LEONHARDT,H., *Eingrenzung von Lösungen mit Hilfe der Regula falsi*, Computing 6, 318-329 (1970).

[10] SCHMIDT,J.W., SCHNEIDER,H., *Monoton einschließende Verfahren bei additiv zerlegbaren Gleichungen*, Z. Angew. Math. Mech, 63, 3-11 (1983).

[11] SCHNEIDER,N., *Monotone Einschließung durch Verfahren vom Regula falsi-Typ*, Computing 26, 33-44 (1981).

[12] TÖRNIG, W., *Monoton konvergente Iterationsverfahren zur Lösung nichlinearer Differenzen-Randwertprobleme*, Beiträge Numer. Math. 4, 245-257 (1975).

[13] TÖRNIG,W.,*Monoton einschließend konvergente Itartionsprozesse vom Gauss-Seidel-Typ*, Math. Meth. Appl. Sci. 2, 489-503 (1980).

[14] GROßMANN,CH.. KRÄTZSCHMAR.M., *Monotone Diskretisierung und adaptive Gittergenerierung für Zwei-Punkt-Randwertaufgaben*, Z. Angew Math. Mech. 65, T264-266 (1985).

[15] GROßMANN,CH., KRÄTZSCHMAR,M., ROOS,H.-G., *Uniformly enclosing discretization methods of high order for boundary value problems* Math. Meth. Appl. Sci. (submitted 1985).

ERROR ESTIMATES FOR FINITE ELEMENT METHODS FOR SEMILINEAR PARABOLIC PROBLEMS WITH NONSMOOTH DATA

V. THOMÉE
Department of Mathematics, Chalmers University of Technology
S-41296 Göteborg, Sweden

We shall survey some recent work on the numerical solution of the semilinear initial boundary value problem

$$(1)\qquad \begin{aligned} &u_t - \Delta u = f(u) \quad \text{in } \Omega\times I, \quad I=(0,t^*],\\ &u=0 \quad \text{on } \partial\Omega\times I,\\ &u(0)=v \quad \text{in } \Omega,\end{aligned}$$

where Ω a bounded domain in R^d with a sufficiently smooth boundary $\partial\Omega$, and f is a smooth function on R for which we assume for simplicity that f and f' are bounded. Such an assumption is normally reasonable only if the solution of (1) is known a priori to be bounded, $|u|\le B$, say, but if this is the case f may be modified if necessary for $|u|>B$ to satisfy our assumption, without changing the solution of (1).

For spatial discretization of (1), let $S_h \subset H_0^1 = H_0^1(\Omega)$ be a family of finite-dimensional spaces parametrized by a small positive parameter h and let the semidiscrete solution $u_h: \bar{I}\to S_h$ be defined by

$$(2)\qquad \begin{aligned} &(u_{h,t},\chi)+(\nabla u_h,\nabla\chi) = (f(u_h),\chi), \quad \text{for } \chi\in S_h,\ t\in I,\\ &u_h(0)=v_h \in S_h,\end{aligned}$$

where $(.,.)$ is the standard inner product in $L_2(\Omega)$.

In order to discuss the error in (2) we assume that S_h is such that the corresponding linear elliptic problem admits an $O(h^r)$ error estimate in $L_2=L_2(\Omega)$. More precisely, we assume that the elliptic projection P_1, i.e. the orthogonal projection onto S_h with respect to the Dirichlet inner product $(\nabla v,\nabla w)$, satisfies, for some $r\geq 2$ and some constant M,

$$(3)\qquad \|P_1 v-v\| \leq Mh^r\|v\|_{H^r}, \qquad \text{for } v\in H_0^1\cap H^r,$$

where $\|.\|$ denotes the norm in L_2. It is then well known that if u is sufficiently smooth on the closed interval $\bar{I}$, and if the discrete initial data v_h are suitably chosen, then

$$\|u_h(t)-u(t)\| \leq C(u,M)h^r, \qquad \text{for } t\in\bar{I}.$$

To guarantee that u is smooth enough for this result, both smoothness of v and compatiblity conditions between v and the differential equation at $\partial\Omega$ for t=0 are necessary. For instance, in the linear homogeneous case ($f\equiv 0$ in (1)) it was shown in Bramble, Schatz, Thomée and Wahlbin [3] that

$$\|u_h(t) - u(t)\| \leq Ch^r\|v\|_{H^r} \qquad \text{for } v\in D((-\Delta)^{r/2}),\ t\in I,$$

which thus requires $\Delta^j v|_{\partial\Omega}=0$ for $j<r/2$. Such requirements are not always satisfied in practice and it is therefore of interest to analyze the error for nonsmooth or incompatible data. Note that the solution of (1) will always be smooth for positive time. For the linear homogeneous equation this may be expressed by saying that the Laplacian generates an analytic semigroup $E(t)=\exp(\Delta t)$ and that $u(t)=E(t)v$ satisfies

$$(4)\qquad \|E(t)v\|_{\dot{H}^\beta} \leq Ct^{-(\beta-\alpha)/2}\|v\|_{\dot{H}^\alpha} \qquad \text{where } \|v\|_{\dot{H}^\alpha}=\|(-\Delta)^{\alpha/2}v\|.$$

For the linear homogeneous equation the nonsmooth data situation has been investigated in Blair [2], Helfrich [5], Bramble, Schatz, Thomée and Wahlbin [3] and later papers (cf. Thomée [7]). In this case, it may be shown using the smoothness property (4) that if v_h is chosen as $P_0 v$, the L_2 projection of v onto S_h. then

$$(5)\qquad \|u_h(t)-u(t)\| \leq Ch^{\alpha+\sigma}t^{-\sigma/2}\|v\|_{\dot{H}^\alpha}, \qquad \text{for } 0\leq\alpha\leq\alpha+\sigma\leq r.$$

In particular, optimal order convergence is attained for t positive even if v is only in L_2. A similar result showing $O(h^r)$ convergence for positive time without initial regularity is known also for the linear inhomogeneous problem, cf. Thomée [7].

In the semilinear situation the following result has been proved in Johnson, Larsson, Thomée and Wahlbin [6].

Theorem 1. Let u be a solution of (1) with $\|v\|\leq\rho$. Assume further that (3) is satisfied (with $r\geq 2$) and let u_h be the solution of (2) with $v_h=P_0v$. Then there exists a constant $C=C(\rho,M)$ such that

$$\|u_h(t)-u(t)\| \leq Ch^2(t^{-1}+|\log(h^2/t)|), \quad \text{for } t\in\bar{I}.$$

The above result thus shows that for r=2 the error in the semilinear case is essentially of the same order as for the linear homogeneous equation. For r>2, however, the result of Theorem 1 is weaker than the case $\alpha=0$ of (5). The reason why the above argument fails to yield higher order convergence than second is related to the lack of integrability of the right hand side of (5) for $\sigma>2$, $\alpha=0$. In spite of this, it may be shown that an analogue of (5) holds, in the sense that the convergence rate in L_2 at positive time is almost two powers of h higher than the order of regularity of the initial data (up to the optimal order $O(h^r)$).

It may be shown that Theorem 1 is, in fact, essentially sharp in the sense that an estimate of the form

$$(6) \qquad \|u_h(t_0)-u(t_0)\| \leq C(\rho,M,t_0)h^\sigma, \quad |u(x,t)| \leq B,$$

cannot hold for any $\sigma>2$ and $t_0>0$, regardless of the value of r. (Note that the requirement that u is bounded is more stringent than boundedness of $\|v\|$.) We shall sketch an example to indicate this.

Consider thus the problem

$$(7) \qquad \begin{aligned} u_t - u_{xx} &= f(u) \quad \text{for } x\in J=[0,1],\ t\in I,\\ u(0,t) &= u(1,t) = 0,\\ u(x,0) &= v(x), \end{aligned}$$

where $f(u) = u^2$ for $|u| \leq B$. Let $h=1/N$, $x_j=jh$, $J_n=(x_n,x_{n+1})$ and consider the semidiscrete analogue using the finite dimensional space

$$S_h = \{\chi\in C(J);\ \chi|_{J_n} \in \Pi_{r-1} \text{ for } n=0,\dots,N-1;\ \chi(0)=\chi(1)=0\}.$$

For the initial values we choose

$$v(x) = v_N(x) = \Psi(Nx),$$

where Ψ is a not identically vanishing function of the form

$$\Psi(x) = \Sigma_{j=1}^{r+1} \Psi_j \sin \pi jx,$$

which is orthogonal to Π_{r-1} on J. Note that v_N is then orthogonal to S_h, and also that, independently of N,

$$\|v_N\|_{L_\infty} \le \Sigma_1^{r+1} |\Psi_j| = \rho,$$

where ρ may be chosen smaller than B. The exact solution of (7) is then also smaller than B in modulus on $\bar{I}=[0,t^*]$ with t^* suitably small, independently of N. Since $v_h=v_{N,h}=P_0v_N=0$, by the construction of v_N, we have $u_h(t)\equiv 0$ on I and hence $e(t)=u_h(t)-u(t)=-u(t)$. Using comparison theorems and some Fourier series arguments, one may show for $u=u_N$ that

$$\|e(t)\| = \|u_N(t)\| \ge C/N^2 = Ch^2.$$

Hence an inequality such as (6) is not possible for $\sigma>2$. We may think of this as an example of nonlinear interaction of Fourier modes.

We shall now briefly consider the discretization of equations such as (1) and (2) with respect to the time variable. Consider thus a semilinear problem of the form

(8) $$du/dt + Au = f(u) \quad \text{for } t\in I,$$
$$u(0) = v,$$

where A is a positive definite selfadjoint linear but not necessarily bounded operator in a Hilbert space H, and where f is bounded together with its Fréchet derivative.

For the approximate solution of (8) we introduce a time step k and let $U_n\in H$ be the approximation of $u(t_n)$, $t_n=nk$, defined by a scheme of the form

(9) $$U_{n+1} = E_kU_n + kF(k,U_n), \quad n=0,1,2,\ldots$$
$$U_0 = v.$$

Here $E_k=r(kA)$ where $r(\lambda)$ is a rational function which is such that for some $p\ge 1$,

(10) $$r(\lambda) = e^{-\lambda} + O(\lambda^{p+1}) \quad \text{as } \lambda\to 0,$$

and such that

(11) $$|r(\lambda)| < 1 \quad \text{for } \lambda \ge 0.$$

Further $F(k,\phi)$ is such that (9) is consistent with (8). More precisely, assume for small k, with $\|.\|$ the norm in H,

(12) $\|F(k,\phi) - F(k,\varphi)\| \le C\|\phi - \varphi\|$

and

(13) $\|A^{-1}(F(k,\phi) - f(\phi))\| \le Ck(\|A\phi\| + 1)$ for $\phi \in D(A)$.

A simple example is provided by the linearized backward Euler method,

$$(U_{n+1}-U_n)/k + AU_{n+1} = f(U_n),$$

which is of this form with $r(\lambda) = 1/(1+\lambda)$ and $F(k,\varphi)=E_k f(\varphi)$ and which satisfies (10) with $p=1$, as well as (11), (12) and (13).

We first recall a nonsmooth data error estimate by Baker, Bramble and Thomée [1] (see also [7]) for the linear homogeneous equation, $f\equiv 0$ in (8) and the corresponding discrete scheme (9) with $F(k,\varphi)=0$:

$$\|U_n - u(t_n)\| \le Ck^p t_n^{-p}\|v\| \quad \text{for } v\in H,\ t_n\in I.$$

This result may be combined with the corresponding result for discretization in space of (1) to yield error bounds for totally discrete schemes of order $O(h^r+k^p)$ for t positive without smoothness assumptions on the initial data.

In the semilinear situation we have the following nonsmooth data error estimate by Crouzeix and Thomée [4].

<u>Theorem 2</u>. Under our present assumptions we have

$$\|U_n - u(t_n)\| \le C(\rho)k\{t_n^{-1}\log(t_{n+1}/k) + (\log(t_{n+1}/k))^2\} \text{ for } \|v\|\le\rho.$$

This result may again be combined with Theorem 1 concerning discretization in space to show an essentially $O(h^2+k)$ convergence result for the complete discretization of (1), without any other requirements for the initial data than $v\in L_2(\Omega)$.

In the same way as for the semidiscrete equation, the nonlinearity limits the order of convergence possible in the case of non-smooth data. Thus, in particular, one may show by an example that for a Runge-Kutta type method of order of accuracy $p>1$ and if $s>1$ then it is not possible to show

$$\|U_n - u(t_n)\| \le C(\rho)k^s \quad \text{for } \|v\|\le\rho,\ t_n=t>0.$$

References.

1. G.A. Baker, J.H. Bramble and V. Thomée, Single step Galerkin approximations for parabolic problems. Math. Comp. 31 (1977), 818-847.

2. J. Blair, Approximate solution of elliptic and parabolic boundary value problems, Thesis, Univ. of California, Berkeley, 1970.

3. J.H. Bramble, A.H. Schatz, V. Thomée and L.B. Wahlbin, Some convergence estimates for semilinear Galerkin type approximations for parabolic equations. SIAM J. Numer. Anal. 14 (1977), 218-241.

4. M. Crouzeix and V. Thomée, On the discretization in time of semilinear parabolic equations with non-smooth initial data. To appear.

5. H.P. Helfrich, Fehlerabschätzungen für das Galerkinverfahren zur Lösung von Evolutionsgleichungen. Manuscr. Math. 13 (1974), 219-235.

6. C. Johnson, S. Larsson, V. Thomée and L.B. Wahlbin, Error estimates for semidiscrete approximations of semilinear parabolic equations with non-smooth data. To appear.

7. V. Thomée, Galerkin Finite Element Methods for Parabolic Problems. Lecture Notes in Mathematics no. 1054, Springer-Verlag, 1984.

SINGULARITIES IN TWO- AND THREE-DIMENSIONAL ELLIPTIC PROBLEMS AND FINITE ELEMENT METHODS FOR THEIR TREATMENT

J. R. WHITEMAN
Brunel University
Uxbridge, England

1. INTRODUCTION

The effective use of finite element methods for treating elliptic boundary value problems involving singularities is well recognised. As a result considerable effort has been expended by mathematicians and engineers in developing special finite element techniques which can produce accurate approximations to the solutions of problems involving singularities.

The work of mathematicians has been mainly in the context of two-dimensional Poisson problems. It has exploited and relied on known theoretical results concerning the regularity of solutions of weak forms of problems of this type, and has produced significant finite element error estimates for this limited class of problems. Comparable progress has not been made in the finite element treatment of three-dimensional problems involving singularities, mainly on account of the lack of theoretical results for the three-dimensional case. This is particularly relevant to the case of three-dimensional re-entrant vertices.

In this paper we present a survey of the finite element treatment of singularities. This is first done in the context of a model two-dimensional Poisson problem and estimates for various norms of the error are given. Some finite element techniques for singularities are then described, taking into account their effects on convergence rates and accuracy. In problems with singularities the approximation of secondary quantities by retrieval from approximations to the solutions (primary quantities) is of great importance,and so this is also treated here.Finally Poisson problems with singularities in three dimensions are presented and the state-of-the-art for this case is contrasted with that for two dimensions.

2. POISSON PROBLEMS INVOLVING SINGULARITIES

2.1. Two Dimensional Poisson Problems.

Let $\Omega \subset \mathbf{R}^2$ be a simply connected polygonal domain with boundary $\partial\Omega$. We consider first the much studied model problem in which the scalar function $u(\underline{x})$ satisfies

$$\begin{aligned} -\Delta[u(\underline{x})] &= f(\underline{x}), & \underline{x} &\in \Omega, \\ u(\underline{x}) &= 0, & \underline{x} &\in \partial\Omega, \end{aligned} \qquad (2.1)$$

where $f \in L_2(\Omega)$. A weak form of (2.1) is defined in the usual Sobolev space $\mathring{H}^1(\Omega)$, and for this $u \in \mathring{H}^1(\Omega)$ satisfies

$$a(u,v) = F(v), \quad \forall\, v \in \mathring{H}^1(\Omega), \tag{2.2}$$

where

$$a(u,v) \equiv \int_\Omega \nabla u \nabla v \, d\underline{x}, \quad u,v \in \mathring{H}^1(\Omega), \tag{2.3}$$

and

$$F(v) \equiv \int_\Omega fv \, d\underline{x}, \quad v \in \mathring{H}^1(\Omega). \tag{2.4}$$

Problem (2.1) is treated by considering the weak form (2.2), where the bilinear form has the important properties that it is continuous, symmetric and elliptic on $\mathring{H}^1(\Omega)$, see Ciarlet [1].

For the finite element solution of (2.1) the region Ω is partitioned quasi uniformly into triangular elements Ω^e in the usual manner and the Galerkin method is applied to (2.2). Conforming trial and test functions are employed and the solution $u \in \mathring{H}^1(\Omega)$ is approximated by $u_h \in S^h$, where $S^h \subset \mathring{H}^1(\Omega)$ is a finite dimensional space of piecwise polynomial functions of degree p, $(p \geq 1)$, and u_h satisfies

$$a(u_h,v_h) = F(v_h) \;\forall\, v_h \in S^h. \tag{2.5}$$

The well known best approximation property of the Galerkin solution gives the inequality

$$\|u - u_h\|_{1,\Omega} \leq \|u - w_h\|_{1,\Omega} \;\forall\, w_h \in S^h, \tag{2.6}$$

where $\|v\|_{1,\Omega}$ is the energy norm $\|\nabla v\|_{L_2(\Omega)}$. Since (2.6) holds for all $w_h \in S^h$, we may take the interpolant $\tilde{u}_h \in S^h$ to u for w_h in (2.6) and, using approximation theory, it follows that

$$\|u - u_h\|_{1,\Omega} \leq Ch^\mu |u|_{k+1,\Omega}, \tag{2.7}$$

where $\mu = \min(p,k)$, whilst C is a constant. Throughout the paper all constants in the estimates are denoted by C.

The actual value of μ is thus dependent both on the choice of p and on the regularity of the solution u of (2.2). Under the condition that $f \in L_2(\Omega)$ the regularity of u is determined by the shape of $\partial\Omega$. If Ω is a convex polygon, then $u \in \mathring{H}^1(\Omega) \cap H^2(\Omega)$, so that $k = 1$ in (2.7) and

$$\|u - u_h\|_{1,\Omega} \leq C\, h |u|_{2,\Omega} \tag{2.8}$$

In this case, see Schatz [2],

$$\|u - u_h\|_{L_2(\Omega)} \leq C\, h^2 |u|_{2,\Omega}, \tag{2.9}$$

so that there is an O(h) convergence gain through changing from the 1-norm to the L_2-norm. The above two estimates are *optimal* in that they are the best that can be obtained by approximating from S^h a function with the regularity of u. It has also been shown, see Nitsche [3] and Ciarlet [1], that for this case the L_∞-norm of the error has $O(h^2)$ convergence.

As has been stated in Section 1, problems with boundaries having re-entrant corners, and thus containing boundary singularities, are of main interest here. We thus consider again problem (2.1), but now in the situation where Ω is a non-convex polygonal domain with interior angles α_j, $1 \le j \le M$, where

$$0 < \alpha_1 \le \alpha_2 \le \ldots \le \pi < \alpha_m \le \ldots \le \alpha_M \le 2\pi .$$

In this case the solution u of (2.2) is such that $u \in \mathring{H}^1(\Omega) - H^2(\Omega)$, and it has been shown by Grisvard [4] that over Ω u can be written as

$$u = \sum_{j=m}^{M} a_j \chi_j(r_j) u_j(r_j,\theta_j) + w, \tag{2.10}$$

where (r_j,θ_j) are local polar coordinates centred on the jth corner of $\partial\Omega$, the χ_j are smooth cut-off functions for the corners, $w \in H^2(\Omega)$ and

$$u_j(r_j,\theta_j) = r_j^{\pi/\alpha_j} \sin \frac{\pi\theta_j}{\alpha_j} .$$

The regularity of u is clearly determined by the term in the summation in (2.10) associated with the Mth corner. In fact $u \in H^{1+\pi/\alpha_M-\varepsilon}(\Omega)$ for every $\varepsilon > 0$, see also Schatz and Wahlbin [5].

Since $\alpha_M > \pi$ and $u \in H^{1+\pi/\alpha_M-\varepsilon}(\Omega)$, it follows from (2.7) that

$$\| u - u_h \|_{1,\Omega} \le C\, h^{(\pi/\alpha_M-\varepsilon)} |u|_{1+\pi/\alpha_M-\varepsilon} , \tag{2.11}$$

and, see Schatz [2], that

$$\| u - u_h \|_{L_2(\Omega)} \le C\, h^{2(\pi/\alpha_M-\varepsilon)} |u|_{1+\pi/\alpha_M-\varepsilon} . \tag{2.12}$$

Whereas the convergence gain in the changing from the 1-norm to the L_2-norm is O(h) for the case where Ω is a convex polygon, (2.8),(2.9), the gain is less for the re-entrant case.

Estimates of the type (2.11) and (2.12), being global, reflect the worst behaviour of the solution over Ω. The situation may not be so bad locally, in particular away from the corners where from (2.10) $u \in H^2$. Thus we now consider L_∞-estimates. Suppose that at the jth vertex z_j of $\partial\Omega$ the intersection of Ω with a disc centred at z_j and containing no other corner is Ω_j and that $\Omega_0 \equiv \Omega \backslash (\bigcup_{j=1} \Omega_j)$. It has been shown by Schatz and Wahlbin [5] that

$$\| u - u_h \|_{L_\infty(\Omega_0)} \le C\, h^{\min(p+1,\, 2\pi/\alpha_M)-\varepsilon} \tag{2.13}$$

$$\|u - u_h\|_{L_\infty(\Omega_M)} \le C\, h^{\pi/\alpha_M - \varepsilon} \qquad (2.14)$$

Similar estimates were discussed by Oden and O'Leary [6]. It should be emphasised that all the above estimates are based on quasi-uniform meshes and piecewise p^{th} order polynomials.

Specific examples of estimates (2.11) - (2.14) are those where the region Ω contains a slit, $\alpha_M = 2\pi$, for which

$$u \in H^{3/2-\varepsilon}(\Omega),\ \|u - u_h\|_{1,\Omega} = O(h^{1/2-\varepsilon}),\ \|u - u_h\|_{L_2(\Omega)} = O(h^{1-\varepsilon})$$

$$\|u - u_h\|_{L_\infty(\Omega_0)} = O(h^{1-\varepsilon}),\ \|u - u_h\|_{L_\infty(\Omega_M)} = O(h^{1/2-\varepsilon})$$

and where the region is L-shaped, $\alpha_M = 3\pi/2$, for which

$$u \in H^{5/3-\varepsilon}(\Omega),\ \|u - u_h\|_{1,\Omega} = O(h^{2/3-\varepsilon}),\ \|u - u_h\|_{L_2(\Omega)} \doteq O(h^{4/3-\varepsilon})$$

$$\|u - u_h\|_{L_\infty(\Omega)} = O(h^{4/3-\varepsilon}),\ \|u - u_h\|_{L_\infty(\Omega_M)} = O(h^{2/3-\varepsilon})$$

2.2 Techniques for Singularities.

The error estimates of Section 2.1 indicate the deterioration from the *optimal* state caused by the presence of the singularity. On account of the practical importance of singularities, much effort has been expended in producing special finite element techniques fro treating singularities, and a considerable literature now exists. The approaches fall mainly into three classes; augmentation of the trial ans test spaces with functions having the form of the dominant part of the singularity, use of singular elements, use of local mesh refinement. These techniques and their effects are now reviewed briefly.

Since for problems of type (2.2) with re-entrant corners the form of the singularity is known, use of this can be made by augmenting the space S^h with functions having the form of the singularity. The solution u of (2.3) is in this case approximated by $u_h \in \text{Aug}S^h$. The technique, proposed by Fix [7] and used by Barnhill and Whiteman [8] and Stephan and Whiteman [9], enables estimates as for problems with smooth solutions to be obtained. It does, however, have the disadvantage of producing a system of linear equations in which the coefficient matrix has a more complicated structure than normal.

The technique of employing *singular* elements involves in elements near the singularity the use of local functions which approximate realistically the singular behaviour. Elements of this type have been proposed by Akin [10], Blackburn [11] and Stern and Becker [12], and their use can lead to significant increase in accuracy of u_h. O'Leary [13], specifically for the Stern-Becker element, has proved that use of the element produces no improvement in the rate of convergence in the error estimate. The increase in accuracy must therefore be produced by reduction in size of the constant in the estimate.

Local mesh refinement near a singularity was originally performed on an ad-hoc basis without theoretical backing. In recent years error analysis has been produced which indicates the grading which a mesh should have near a corner in order that the effect of a singularity may be nullified. Examples of such local mesh refinement are given by Schatz and Wahlbin [5] and Babuska and Osborn [14]. Another approach is to use *adaptive* mesh refinement involving a-posteriori error estimation.

With adaptive mesh refinement the region Ω is partitioned initially and the local error in each element is estimated. If, for a particular element, this is greater than a prescribed tolerance, the element is subdivided thus causing the local refinement, see Babuska and Rheinboldt [15], [16]. Hierarchical finite elements have recently been incorporated into the technique, Craig, Zu und Zienkiewicz [17], as have multigrid methods Bank and Sherman [18] and Rivara [19].

2.3 Retrieved Quantities.

As has been stated in Section 1, for problems involving boundary singularities the approximation of secondary (retrieved) quantities is most important. Specifically the coefficients a_j in (2.10) of singular terms are of practical significance, so that ways must be found of approximating these accurately. Apart from the obvious approach of using collocation or least squares methods to fit terms to calculated results, it is often possible to exploit the mathematics of the original problem. An important case is that of a problem containing a slit, $\alpha_M = 2\pi$, and here use can be made of the "J-integral" concept to produce an integral expression for the a_M, see Destuynder et al.,[20]. This integral can be approximated using the calculated solution u_h. For piecewise linear test and trial functions on a mesh with local refinement, O(h) estimates are given in [20] for the absolute value in the error in the approximation to the singularity coefficient a_M.

Fro problem (2.2), when a singularity is present, the integrand of the "J-integral" involves derivatives of the solution u. Thus the accuracy of the approximation to the integral, and hence to the singularity coefficient, depends on the errors in the gradients of u_h.

A possibility exists here of exploiting superconvergence properties in the estimation of errors in gradients of u_h, provided local estimates can be obtained. To date the error estimates have depended on the global regularity of the solution u, see Levine [21].

2.4 Three Dimension Poisson Problems.

We consider again problems of the type (2.1), except that now $\Omega \subset \mathbf{R}^3$ is a polyhedral domain. The weak forms and the finite element method for the three-dimensional case can be described similarly, again with $\Omega \subset \mathbf{R}^3$. Singularities can in this case occur on account of re-entrant edges and vertices. The decomposition of the three dimensional weak solution corresponding to (2.10) has been shown, e.g. by Stephan [22] to have the form

$$u = \underset{(\text{vertices})}{\sum_{j=1}^{M}} a_j \chi_j u_j + \underset{(\text{edges})}{\sum_{k=1}^{N}} f_k \Xi_k v_k + w \tag{2.15}$$

where $w \in H^2(\Omega)$, $\chi_j(r_j)$ and $\Xi_k(\rho_k)$ are cut-off functions respectively for the vertices and edges, whilst the u_j and v_k are functions associated also respectively with vertices and edges. For an edge the v_k have the two dimensional form for any plane orthogonal to the edge associated with the appropriate two dimensional problem, whilst the b_k are functions of z_k.

The singular function u_j for each vertex is found by solving a Laplace-Beltrami eigenvalue problem on that part of the surface of the unit ball centred on the vertex cut off by the faces of the vertex. When the vertex is such that the eigenvalue problem is separable (has a single coordinisation), there are special cases when the problem can be solved exactly, see Walden and Kellogg [23]. When this is not so, for example for a vertex made up from three mutually orthogonal planes, Beagles and Whiteman [24], a numerical approximation to the eigenvalue must be obtained with the result that the singular function will not be known exactly.

Clearly this lack of knowledge of the exact singular functions is very important from the finite element point of view, and in particular means that the error analysis of Section 2.1 cannot in general be transferred directly to the three-dimensional singular case. All the singularity methods described in Section 2.1 are affected, although all are used in the three-dimensional context. The augmentation technique is obviously adversely affected, although Beagles and Whiteman [25] have devised the technique of *non-exact augmentation*, whereby the trial and test function spaces in the

Galerkin procedure are augmented with the non- exact singular functions. As far as we are aware no method of the "J-integral" type exists for three-dimensional Poisson problems.

The above indicates that the state-of-the-art fro treating three-dimensional singularities with finite element methods is far less advanced than that for the two-dimensional case. This arises more from limitations in the theory of three-dimensional Poisson problem than from the finite element methods themselves.

References

[1] Ciarlet P.G., *The finite Element Method for Elliptic Problems*. Nort-Holland, Amsterdam, 1979.

[2] Schatz A., *An introduction to the analysis of the error in the finite element method for second order elliptic boundary value problems*. pp. 94-139 of P.R.Turner (ed.) Numerical Analysis Lancaster 1984. Lecture Notes in Mathematics 129, Springer-Verlag, Berlin, 1985.

[3] Nitsche J.A., *L_∞-convergence of finite element approximations, mathematical aspects of finite element methods*. Lecture Notes in Mathematics 606, Springer-Verlag, Berlin, 1977.

[4] Grisvard P., *Behaviour of the solutions of an elliptic boundary value problem in a polygonal or polyhedral domain*. pp. 207-274 of B. Hubbard (ed.), Numerical Solution of Partial Differential Equations III, SYNSPADE 1975. Academic Press, New York, 1976.

[5] Schaltz A. and Wahlbin L., *Maximum norm estimates in the finite element method on plane polygonal domains*. Parts I and II. Math. Comp. 32, 73-109, 1978, and Math. Comp. 33, 465-492, 1979.

[6] Oden J.T. and O'Leary J., *Some remarks on finite element approximations of crack problems and an analysis of hybrid methods*. J. Struct. Mech. 64, 415-436, 1978.

[7] Fix G., *Higher order Rayleigh Ritz approximations*. J. Math. Mech. 18, 645-657, 1969.

[8] Barnhill R.E. and Whiteman J.R., *Error analysis of Galerkin methods for Dirichlet problems containing boundary singularities*. J. Inst. Math. Applics.15, 121-125, 1975.

[9] Stephan E. and Whiteman J.R., *Singularities of the Laplacian at corners and edges of three dimensional domains and their treatment with finite element methods*. Technical Report BICOM 81/1, Institute of Computational Mathematics, Brunel Uninersity, 1981.

[10] Akin J.E., *Generation of elements with singularities*. Int. J. Numer. Method. Eng. 10, 1249-1259, 1976.

[11] Blackburn W.S., *Calculation of stress intensity factors at crack tips using special finite elements*. pp. 327-336 of J.R. Whiteman (ed.), The Mathematics of Finite Elements and Applications. Academic Press, London, 1973.

[12] Stern M. and Becker E., *A conforming crack tip element with quadratic variation in the singular fields*. Int. J. Numer. Meth. Eng. 12, 279-288, 1978.

[13] O'Leary J.R., *An error analysis for singular finite elements*. TICOM Report 81-4, Texas Institute of Computational Mechanics, University of Texas at Austin, 1981.

[14] Babuska I. and Osborn J., *Finite element methods for the solution of problems with rough input data*. pp. 1-18 of P.Grisvard, W.Wendland and J.R.Whiteman (eds.), *Singularities and Constructive Methods for Their Treatment*. Lecture Notes in Mathematics 1121, Springer Verlag, Berlin, 1985.

[15] Babuska I. and Rheinboldt W.C., *Error estimates for adaptive finite element computations*. SIAM J. Num. Anal. 15, 736-754, 1978.

[16] Babuska I. and Rheinboldt W.C., *Reliable error estimation and mesh adaptation for the finite element method*. pp. 67-108 of J.T. Oden (ed.), Computational Methods in nonlinear Mechanics. North-Holland, Amsterdam, 1979.

[17] Craig A.W., Zhu J.Z. and Zienkiewicz O.C., *A-posterirori error estimation, adaptive mesh refinement and multigrid methods using hierarchical finite element bases*. pp. 587-594 of J.R. Whiteman (ed.), The Mathematics of Finite Elements and Applications V., MAFELAP 1984. Academic Press, London, 1985.

[18] Bank R.E. and Sherman A.H., *The use of adaptive grid refinement for badly behaved elliptic partial differential equations*. pp. 18-24 of Computers in Simulation XXII. North Holland, Amsterdam, 1980.

[19] Rivara M.C., *Dynamic implementation of the h-version of the finite element method*. pp. 595-602 of J.R. Whiteman (ed.), The Mathem. of Finite Elements and Applic. V., MAFELAP 1984, Academic Press, London, 1985.

[20] Destuynder P, Djaoua M. and Lescure S., *On numerical methods for fracture mechanics*. pp. 69-84 of P. Grisvard, W.L. Wendland and J.R. Whiteman (eds.), **Singularities and constructive methods for their treatment.** Lecture Notes in Mathematics 1121, Springer Verlag, Berlin, 1985.

[21] Levine N., *Superconvergence recovery of the gradient from piecewise linear finite element approximations*. Technical Report 6/83, Dept. of Mathematics, University of Reading, 1983.

[22] Stephan E., *A modified Fix method for the mixed boundary value problem of the Laplacian in a polyhedral domain*. Preprint Nr. 538. Fachbereich Mathematik, T.H. Darmstadt, 1980.

[23] Walden H. and Kellogg R.B., *Numerical determination of the fundamental eigenvalue for the Laplace operator on a sphercial domain*. J. Engineering Mathematics 11, 299-318, 1977.

[24] Beagles A.E. and Whiteman J.R., *Treatment of a re-entrant vertex in a three dimensional Poisson problem*. pp. 19-27 of P.Grisvard, W.H. Wendland and J.R. Whiteman (eds.), Singularities and Constructive Methods for Their Treatment. Lecture Notes in Mathematics 1121, Springer Verlag, Berlin, 1985.

[25] Beagles A.E. and Whiteman J.R., *Finite element treatment of boundary singularities by argumentation with non-exact singular functions*. Technical Report BICOM 85/1, Institute of Computational Mathematics, Brunel University, 1985.

SOME NEW CONVERGENCE RESULTS IN FINITE ELEMENT THEORIES FOR ELLIPTIC PROBLEMS

A. ŽENÍŠEK
Computing Center of the Technical University
Obránců míru 21, Brno, Czechoslovakia

THE LINEAR PROBLEM

We consider the following variational problem: Find $u \in W$ such that

$$a(u,v) = L(v) \equiv L^{\Omega}(v) + L^{\Gamma}(v) \qquad \forall v \in V \tag{1}$$

where

$$a(v,w) = \iint_{\Omega} k_{ij}\, v_{,i}\, w_{,j}\, dx, \tag{2}$$

$$L^{\Omega}(v) = \iint_{\Omega} vf\, dx, \qquad L^{\Gamma}(v) = \int_{\Gamma_2} vq\, ds, \tag{3}$$

$$V = \{v \in H^1(\Omega):\ v = 0 \text{ on } \Gamma_1;\ \text{mes}_1\Gamma_1 > 0\},\ W = z + V; \tag{4}$$

Ω is a bounded domain in E_2 with a boundary $\Gamma = \Gamma_1 \cup \Gamma_2$ ($\Gamma_1 \cap \Gamma_2 = \phi$); $z \in H^1(\Omega)$, $z = \bar{u}$ on Γ_1, $\bar{u} \in L_2(\Gamma_1)$ is a given function. In (2) and in what follows the summation convention over repeated subscripts is adopted and $v_{,i} = \partial v/\partial x_i$. The functions $k_{ij} = k_{ij}(x)$ are bounded and measurable in $\tilde{\Omega} \supset \bar{\Omega}$ and satisfy

$$k_{ij}(x)\xi_i\xi_j \geq C\xi_i\xi_i \qquad \forall x \in \tilde{\Omega} \supset \bar{\Omega} \qquad \forall \xi_i, \xi_j \in E_1, \tag{5}$$

where $C > 0$, and $f \in L_2(\Omega)$, $q \in L_2(\Gamma_2)$. Assumption (5) and Friedrichs' inequality imply that the form $a(v,w)$ is V-elliptic. Thus, according to the Lax-Milgram lemma, problem (1) has just one solution.

Problem (1) is approximated by the problem: Find $u_h \in W_h$ such that

$$a_h(u_h,v) = L_h(v) \equiv L_h^{\Omega}(v) + L_h^{\Gamma}(v) \qquad \forall v \in V_h \tag{6}$$

where V_h is a finite element approximation of V and $W_h = z_h + V_h$, z_h being a finite element approximation of z. The forms $a_h(v,w)$, $L_h(v)$ approximate the forms $a(v,w)$, $L(v)$ in the following way: The sets Ω and Γ_2 appearing in (2), (3) are substituted by Ω_h and Γ_{h2} and the obtained forms $\tilde{a}_h(v,w)$, $\tilde{L}_h(v)$ are then computed numerically.

In the Ciarlet's and Raviart's theory and its modifications (see [1],[2],[6]) the solution u of (1) is assumed to be sufficiently smooth, $u \in H^{n+1}(\Omega)$ $(n \geq 1)$, and the maximum rate of convergence is proved:

$$\|\tilde{u} - u_h\|_{1,\Omega_h} \leq Ch^n, \tag{7}$$

where $\tilde{u}$ is the Calderon's extension of u into $H^{n+1}(E_2)$.

The problem of convergence of u_h to $\tilde{u}$ (when $u \in H^1(\Omega)$ only) was solved recently in [8]. In this section we present outlines of considerations of [8].

<u>Theorem 1</u>. Let the boundary Γ of the domain Ω be piecewise of class C^{n+1}. Let Γ_h approximate Γ piecewise by arcs of degree n. Let $k_{ij}, f \in W_\infty^{(n)}(\tilde{\Omega})$ and let the quadrature formula used on the standard triangle T_0 for calculation of $a_h(v,w)$ and $L_h^\Omega(v)$ be of degree of precision 2n - 2. Let $q \in C^n(\bar{U})$, where U is a domain containing Γ_2, and let the quadrature formula used on [0,1] for calculation of $L_h^\Gamma(v)$ be of degree of precision 2n - 1. Let $\bar{u}$ be so smooth that there exists a function $z \in H^2(\Omega)$ such that $z = \bar{u}$ on Γ_1. Then

$$\|\tilde{u} - u_h\|_{1,\Omega_h} \to 0 \quad \text{if} \quad h \to 0. \tag{8}$$

<u>Proof</u>. The assumptions concerning Γ and $a_h(v,w)$ imply, according to [2] and [7], that the forms $a_h(v,w)$ are uniformly V_h-elliptic. Thus we have similarly as in [1], [2]:

$$\|\tilde{u} - u_h\|_{1,\Omega_h} \leq C \left\{ \sup_{w\in V_h} \frac{|L_h(w) - \tilde{a}_h(\tilde{u},w)|}{\|w\|_{1,\Omega_h}} + \right.$$
$$\left. + \inf_{v\in W_h} \left[\|\tilde{u} - v\|_{1,\Omega_h} + \sup_{w\in V_h} \frac{|\tilde{a}_h(v,w) - a_h(v,w)|}{\|w\|_{1,\Omega_h}} \right] \right\}. \tag{9}$$

For simplicity we prove Theorem 1 only in the case $\Gamma_1 = \Gamma$, $\bar{u} = 0$ and $n = 1$, i.e. Ω_h has a polygonal boundary and we use linear triangular finite elements. As Γ is piecewise of class C^2 each boundary triangle satisfies (for sufficiently small h) one of the following possibilities:

$$\text{a)} \quad T \subseteq T_{id}, \qquad \text{b)} \quad T_{id} \subset T, \tag{10}$$

where T_{id} is the ideal boundary triangle (see [9]) whose approximation is T.

In order to estimate the first term on the right-hand side of (9) let us define a function $\hat{w} \in V$ associated with $w \in V_h$ in the following way: $\hat{w}(P_i) = w(P_i)$ at the vertices P_i of the triangles of the triangulation $\mathcal{T}_h$ of the domain Ω_h; the function $\hat{w}$ is continuous on $\bar{\Omega}$, linear on each interior triangle of $\mathcal{T}_h$, equal to zero on $T_{id} - T$ and linear on T in the case (10a) and, finally, equal to Zlámal's ideal "linear" interpolate of w on T_{id} in the case (10b) (see [9]). Thus we can write, according to (1):

$$L_h(w) - \tilde{a}_h(\tilde{u},w) = L_h(w) - L(\hat{w}) + a(u,\hat{w}) - \tilde{a}_h(\tilde{u},w). \tag{11}$$

We have

$$L(\hat{w}) = \iint_{\Omega} f\hat{w}dx = \iint_{\Omega_h} fw\ dx + \sum_b \left\{ \iint_{T_{id}} (\hat{w} - w)f\ dx - \iint_{T-T_{id}} wf\ dx \right\}$$

where the sum is taken over all boundary triangles (10b). Expressing similarly $a(u,\hat{w}) - \tilde{a}_h(\tilde{u},w)$ we obtain from (11):

$$|L_h(w) - \tilde{a}_h(\tilde{u},w)| \le |\tilde{L}_h^{\Omega}(w) - L_h^{\Omega}(w)| +$$

$$+ \sum_b \left| \iint_{T_{id}} \left\{ (w - \hat{w})f + k_{ij}u_{,i}(\hat{w} - w)_{,j} \right\} dx \right| +$$

$$+ \sum_b \left| \iint_{T-T_{id}} (wf - k_{ij}\tilde{u}_{,i}w_{,j})dx \right|. \tag{12}$$

According to [2, Theorem 4.1.5],

$$|\tilde{L}_h^{\Omega}(w) - L_h^{\Omega}(w)| \le Ch\|w\|_{1,\Omega_h}. \tag{13}$$

Denoting the first sum in (12) briefly by S_1 we have, according to the Cauchy inequality and the boundedness of k_{ij}:

$$|S_1| \le (\|f\|_{0,\Omega} + C\|u\|_{1,\Omega}) \left\{ \sum_b \|\hat{w} - w\|^2_{1,T_{id}} \right\}^{1/2}.$$

The definition of $\hat{w}$ and the proof of [9, Theorem 2] imply

$$\|w - \hat{w}\|_{1,T_{id}} \le Ch\|w\|_{2,T_{id}} \le Ch\|w\|_{1,T}$$

because w is linear on T and (10b) holds. Thus

$$|S_1| \le Ch\|w\|_{1,\Omega_h}. \tag{14}$$

As $\mathrm{mes}(T - T_{id}) = O(h^3)$ we have

$$\sum_b \|w\|^2_{0,T-T_{id}} \le Ch^3 \sum_b \max w^2.$$

Further, similarly as [7,(39)] we can prove

$$\|w\|^2_{0,\Omega_h} \ge Ch^2 \sum_{T\in\mathcal{T}_h} \sum_{i=1}^{3} [w(P_T^i)]^2$$

where P_T^i $(i = 1,2,3)$ are the vertices of T. Using these results and the fact that $w_{,i}$ are constants on each $T \in \mathcal{T}_h$ we can easily find for the second sum S_2 on the right-hand side of (12):

$$|S_2| \cdot \|w\|^{-1}_{1,\Omega_h} \le C\left\{ \sum_b \|w\|^2_{0,T-T_{id}} \|w\|^{-2}_{0,\Omega_h} + \right.$$

$$\left. + \sum_b |w|^2_{1,T-T_{id}} |w|^{-2}_{1,\Omega_h} \right\}^{1/2} \le Ch^{1/2} . \tag{15}$$

According to (12) - (15), the first term on the right-hand side of (9) is $O(h^{1/2})$.

As to the second term on the right-hand side of (9) we can find a set $\{v_h\}$, where $v_h \in W_h$, such that

$$\|\tilde{u} - v_h\|_{1,\Omega_h} \to 0 \quad \text{if } h \to 0. \tag{16}$$

The following proof of (16) holds also in the case $\text{mes}_1 \Gamma_1 < \text{mes}_1 \Gamma$: The set $G = C^\infty(\bar{\Omega}) \cap V$ is dense in V (see [3]). Thus for every $\varepsilon > 0$ we can find $v_\varepsilon \in G$ such that $\|u - v_\varepsilon\|_{1,\Omega} < \varepsilon$. Let $\tilde{v}_\varepsilon$ and $\tilde{v}^*_\varepsilon$ be the Calderon's extensions of v_ε into $H^1(E_2)$ and $H^2(E_2)$, respectively. We have

$$\|\tilde{u} - I_h v_\varepsilon\|_{1,\Omega_h} \le \|\tilde{u} - \tilde{v}_\varepsilon\|_{1,\Omega_h} +$$

$$+ \|\tilde{v}_\varepsilon - \tilde{v}^*_\varepsilon\|_{1,\Omega_h - \Omega} + \|\tilde{v}^*_\varepsilon - I_h v_\varepsilon\|_{1,\Omega_h} \tag{17}$$

where $I_h v_\varepsilon$ is the interpolate of v_ε in W_h, i.e. the piecewise linear function which has the same function values at the vertices of $T \in \mathcal{T}_h$ as the function v_ε. The properties of the Calderon's extensions, the absolute continuity of the Lebesgue integral and the finite element interpolation theorem imply that for all $h \le h_0(\varepsilon)$ the right-hand side of (17) is bounded by $K\varepsilon$, where K does not depend on ε. As $I_h v_\varepsilon \in V_h = W_h$ relation (17) implies (16).

The set $\{v_h\}$ appearing in (16) is bounded. Thus, according to [2, Theorem 4.1.4], the third term on the right-hand side of (9) is $O(h)$. This finishes the proof in our simple case. The general case $\Gamma_1 \subset \Gamma$, $n \ge 1$ is considered in [8].

SOME NONLINEAR PROBLEMS

Let the form $a(u,v)$ appearing in (1) be now nonlinear in u, linear in v, strongly monotone and Lipschitz continuous and let $a(0,v) = 0$ for all $v \in H^1(\Omega)$. In addition, let the forms $a_h(v,w)$ be uniformly strongly monotone and uniformly Lipschitz continuous in X_h (the finite element approximations of $H^1(\Omega)$), i.e. let

$$a_h(v,v-w) - a_h(w,v-w) \ge C|v-w|^2_{1,\Omega_h} ,$$

$$|a_h(v,z) - a_h(w,z)| \le K\|v - w\|_{1,\Omega_h}\|z\|_{1,\Omega_h}$$

$$\forall v,w,z \in X_h \subset H^1(\Omega_h) \quad \forall h \in (0,h_0)$$

where the positive constants C, K do not depend on v,w,z and h. Finally, let the forms $\tilde{a}_h(v,w)$ be uniformly Lipschitz continuous. Under these assumptions the abstract error estimate has again the form (9) (see [5]).

A typical form a(u,v) satisfying all assumptions presented in this section is given by relation (2) with

$$k_{ij} = b(x,(\nabla v)^2)\delta_{ij} \tag{18}$$

where δ_{ij} is the Kronecker delta and where the function $b(x,\eta)$ has the following properties (see [4]):

a) The functions $b(x,\eta)$, $\partial b(x,\eta)/\partial x_i$, $\partial b(x,\eta)/\partial\eta$ are continuous in $\tilde{\Omega} \times [0,\infty)$, where $\tilde{\Omega} \supset \bar{\Omega}$.

b) There exist constants $c_1 > 0$, $c_2 > 0$ such that

$$c_1 \le b \le c_2, \ |\partial b/\partial x_i| \le c_2, \quad 0 \le \partial b/\partial\eta \le c_2 \quad \text{in } \tilde{\Omega} \times [0,\infty),$$

$$|\xi|(\partial b/\partial\eta)(x,\xi^2), \quad \xi^2(\partial b/\partial\eta)(x,\xi^2) \le c_2 \quad \forall x \in \tilde{\Omega}, \quad \forall \xi \in E_1.$$

The functions (18) with properties a), b) appear in many physical and technical applications.

Now we generalize the result introduced in Theorem 1:

Theorem 2. Let the form a(u,v) appearing in (1) be defined by (2) and (18) and let the function $b(x,\eta)$ have properties a), b). Let the assumptions of Theorem 1 be satisfied with n = 1. Then the solutions u and u_h of problems (1) and (6) exist and are unique and relation (8) holds. In addition, if $u \in H^2(\Omega)$ then the rate of convergence is given by (7), where n = 1.

Proof. As n = 1 we consider only linear triangular elements. We again restrict ourselves to the case $\Gamma_1 = \Gamma$, $\bar{u} = 0$. The existence and uniqueness of u and u_h is proved in [4]. The first property b) allows us to repeat (11) - (15). Thus the first term on the right-hand side of (9) is $O(h^{1/2})$. Also relation (16) remains unchanged, only the analysis of the third term on the right-hand side of (9) is different: As ∇v = const. on $T \in T_h$ for all $v \in V_h = W_h$ we can write

$$|a_h(v_h,w) - \tilde{a}_h(v_h,w)| \le \sum_{T\in T_h} |\mathrm{mes}(T)b(P_T,g_h|_T) -$$

$$- \iint_T b(x,g_h)dx|\,.\,|(\nabla v_h|_T\cdot\nabla w|_T)|$$

where v_h are the functions from (16) and $g_h = (\nabla v_h)^2$. We used one-point integration formula with the centre of gravity P_T of $T \in \mathcal{T}_h$. Using the properties of the function $b(x,\eta)$ we see, according to [2, Theorem 4.1.5], that the absolute value of the difference on the right-hand side is bounded by Ch mes(T). Thus the right-hand side of the last inequality is bounded by $Ch\|w\|_{1,\Omega_h}$ and relation (8) is valid.

The error estimate in the case $u \in H^2(\Omega)$ is derived in [5] where also more general forms a(v,w) are considered.

References

[1] CIARLET P.G., RAVIART P.A., The combined effect of curved boundaries and numerical integration in isoparametric finite element methods. In: The Mathematical Foundations of the Finite Element Method with Applications to Partial Differential Equations (A.K. Aziz, Editor), Academic Press, New York, 1972, pp. 409-474.

[2] CIARLET P.G., The Finite Element Method for Elliptic Problems. North-Holland, Amsterdam, 1978.

[3] DOKTOR P., On the density of smooth functions in certain subspaces of Sobolev space. Commentationes Mathematicae Universitatis Carolinae 14 (1973), 609-622.

[4] FEISTAUER M., On the finite element approximation of a cascade flow problem. (To appear).

[5] FEISTAUER M., ŽENÍŠEK A., Finite element methods for nonlinear elliptic problems. (To appear).

[6] ŽENÍŠEK A., Nonhomogeneous boundary conditions and curved triangular finite elements. Apl. Mat. 26 (1981), 121-141.

[7] ŽENÍŠEK A., Discrete forms of Friedrichs' inequalities in the finite element method. R.A.I.R.O. Anal. numér. 15 (1981), 265-286.

[8] ŽENÍŠEK A., How to avoid the use of Green's theorem in the Ciarlet's and Raviart's theory of variational crimes. (To appear).

[9] ZLÁMAL M., Curved elements in the finite element methods. I. SIAM J. Numer. Anal. 10 (1973), 229-240.

Section D

APPLICATIONS

MATHEMATICAL SOLUTION OF DIRECT AND INVERSE PROBLEM FOR TRANSONIC CASCADE FLOWS

P. BOLEK, J. FOŘT, K. KOZEL, J. POLÁŠEK
National Research Institute for Machine Design
11000 Prague 9 - Běchovice, Czechoslovakia

The work deals with numerical solution of direct and inverse problem of transonic cascade flows based on potential model. Governing equation of a direct problem is full potential equation, governing equation of an inverse problem is equation for Mach number in hodograph plane (Φ,ψ), Φ-velocity potential, ψ-stream function. Both equations are partial differential equations of second order, mixed elliptic-hyperbolic type. In the solution of direct problem one can consider discontinuity of the first derivatives along some curves called shock waves, in the inverse problem one must find classical solution.

Numerical solution of both problems is based on using finite difference method and Jameson's rotated difference scheme. The system of difference equations is solved iteratively using succesive line relaxation method.

The work presents results of numerical solution of transonic flows in cascade of compressor and turbine type and one example of numerical solution of inverse problem.

I: Direct problem

A steady irrational isoentropic flow is fully described by the quasilinear partial differential equation of mixed elliptic-hyperbolic type for a velocity potential:

$$(a^2-\Phi_x^2)\Phi_{xx} - 2\Phi_x\Phi_y\Phi_{xy} + (a^2-\Phi_y^2)\Phi_{yy} = 0 , \qquad (1)$$

where Φ is velocity potential and $a = a(\Phi_x^2 + \Phi_y^2)$.
We assume the existence of weak shock waves as curves of discontinuity of the first derivatives Φ_x,Φ_y. The weak solution is assumed in a class $K(\Omega)$, where Ω is a domain of solution (see [1]).

The mathematical formulation of transonic cascade flows is some combination of Dirichlet's, Neuman's and periodic boundary value problem. On the inlet boundary we prescribe a Dirichlet's condition $(\bar{w} = \bar{w}_\infty)$, on profile contour a Neuman's condition of non-permeability $(\partial\Phi \backslash \partial\bar{n} = 0)$ and on the outlet boundary also a Neuman's condition $(\bar{w} = \bar{w}_2)$, where $\bar{w}_2$ is a constant determined uniquely by the value of circulation of velocity around the one profile of the cascade γ. Potential Φ still satisfies a Kutta-Youkovski condition on the trailing edge of the profile. The value of γ, unknown in advance, is determined during iteration process of the numerical solution.

Equation (1) is possible to locally transform to the form

$$(1 - M^2)\Phi_{ss} + \Phi_{nn} = 0 \tag{2}$$

that is similar to equation (1), M-Mach number, $M = M(\Phi_s^2)$, M-given function, s - streamline direction, n - normal.

Consider (x,y) coordinate system and regular orthogonal grid. Jameson's concept of stable difference scheme is based on central difference approximation of second order for Φ_{ss} using Φ_{xx}, Φ_{xy}, Φ_{yy} in elliptic point $(1 - M^2 < 0)$ and backward approximation of first order for Φ_{ss} in hyperbolic point $(1 - M^2 < 0)$. Central approximation of second order in both cases is used for Φ_{nn} (details see [1]).

The system of difference equations is solved by a SLOR method. It is solved in one step of iteration for grid points lying on line x_i = const., succesive in the direction of flow. The relaxation parameter is chosen 1.7 for all mesh points in line x_i = const., if all this points does not lie on profile contour and if their local Mach number in computed iteration is less than 1; and equal to 1 in other cases.

II: Inverse problem

Solving inverse problem of transonic flow over an airfoil or through a cascade the following governing equation in hodograph plane has been used

$$AM_{\Phi\Phi} + BM_{\psi\psi} + CM_{\Phi}^2 + DM_{\psi}^2 = 0, \tag{3}$$

$$A = M(1 - M^2)P\frac{2}{\varkappa-1}, \quad P = 1 + \frac{\varkappa-1}{2}M^2,$$

$$B = M$$

$$C = -(1 + 3\frac{\varkappa-1}{2}M^2 + \frac{3-\varkappa}{2}M^4)P\frac{3-\varkappa}{\varkappa-1},$$

$$D = -(1 + \varkappa M^2 P^{-1})$$

M-Mach number, Φ - velocity potential, ψ - stream function. Smooth solution is considered in this case due to regularity of transformation $(x,y) \rightarrow (\Phi,\psi)$. Boundary value problem is based on eq. (3) and Dirichlet's conditions for an airfoil or combination of Dirichlet's, Neuman's and periodicity conditions for a cascade.

The details are described in [2]. Numerical solution of the problem is a similar to the solution of eq. (1). Knowing $M(\Phi,\psi)$ we find angle ϑ (oriented angle of the flow in (x,y) system)

$$\vartheta = \int_{\Phi_0}^{\Phi} P^{-\frac{\varkappa}{\varkappa-1}} M_{\psi} M^{-1} d\tau$$

and then streamline coordinates ("zero" streamlines)

$$x(\Phi,\psi) = x(\Phi_0,\psi) + \int_{\Phi_0}^{\Phi} \frac{\cos\vartheta}{q(M)}\, d\tau,\quad y(\Phi,\psi) = y_0(\Phi_0,\psi) + \int_{\Phi_0}^{\Phi} \frac{\sin\vartheta}{q(M)}\, d\tau,$$

$q = (u^2 + v^2) = F_1(M)$, F_1 - given function.

III: Numerical results

Fig. 1 shows the iso-Mach lines of transonic flows calculation for compressor cascade with upstream Mach number $M_\infty = 0.83$. We can see the typical choked fows with so called closed sonic line ($M = 1$). It means that first end of the sonic line is situated on lower profile surface and the other end is situated on the upper profile surface.

Fig. 2 shows the iso-Mach lines of transonic flows calculation for turbine cascade with upstream Mach number $M_\infty = 0.337$ and downstream Mach number $M_2 = 0.803$. Small supersonic region ($M > 1$) is situated near lower profile surface. This cascade is more cambered and therefore the problem of numerical solution of transonic flows through this cascade is very complicated. The comparisons of our numerical results and experimental data is published in [4].

Fig. 3 shows results of inverse problem for given Mach number along upper (M_h) and lower (M_d) profile surface (fig. 3a); fig. 3b showes geometry of found cascade corresponding given distribution of Mach number along profile surface and other parameters.

References

[1] Fořt J., Kozel K., *Numerical Solution of Potential Transonic Flow Past Blade Cascades*, Strojnický čas. 35 (1984), 3 (in czech).

[2] Bolek P., *Mathematical Solution of Inverse Problem of Transonic Potential Flow Past Airfoil and Through a Cascade*, Thesis ČVUT, Prague, 1982 (in czech).

[3] Jameson A., *Numerical Computation of Transonic Flows With Shock Waves*, Symposium Transonicum II, Götingen 1975, Springer Verlag, 1976.

[4] Fořt J., Kozel K., *Calculation of Transonic Flow Through Compressor and Turbine Cascades Using Relaxation Method for Full Potential Equation*, International Conference on Numerical Methods and Applications, Sofia, 1984 (Proceedings).

[5] Fořt J., Kozel K., *Numerical Solution of the Inviscid Stationary Transonic Flow Past an Isolated Airfoil and Through a Cascade*, Applmath I, Bratislava 1984 (Proceedings).

[6] Kozel K., Polášek J., *Numerical Solution of Two-Dimensional and Three-Dimensional Inviscid Transonic Flow*, Probleme und Methoden der Mathematischen Physik, 8. Tagung, 1983, in Teubner Texte zur Mathematic, band 63, Leipzig.

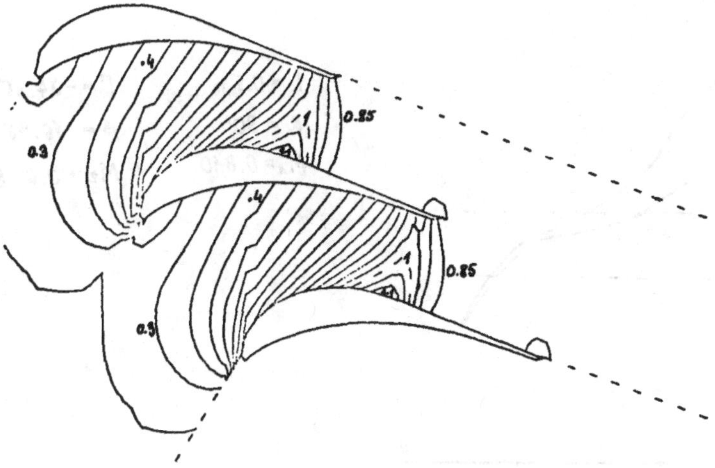

Fig. 1 : Compressor cascade. Iso-Mach lines of computed flow field, increment $\Delta M = 0.05$, $M_\infty = 0.83$

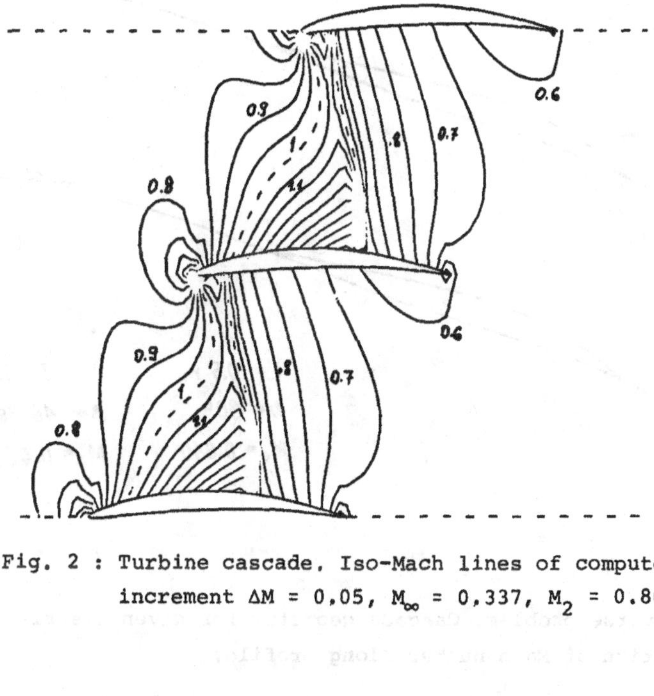

Fig. 2 : Turbine cascade. Iso-Mach lines of computed flow field, increment $\Delta M = 0.05$, $M_\infty = 0.337$, $M_2 = 0.809$

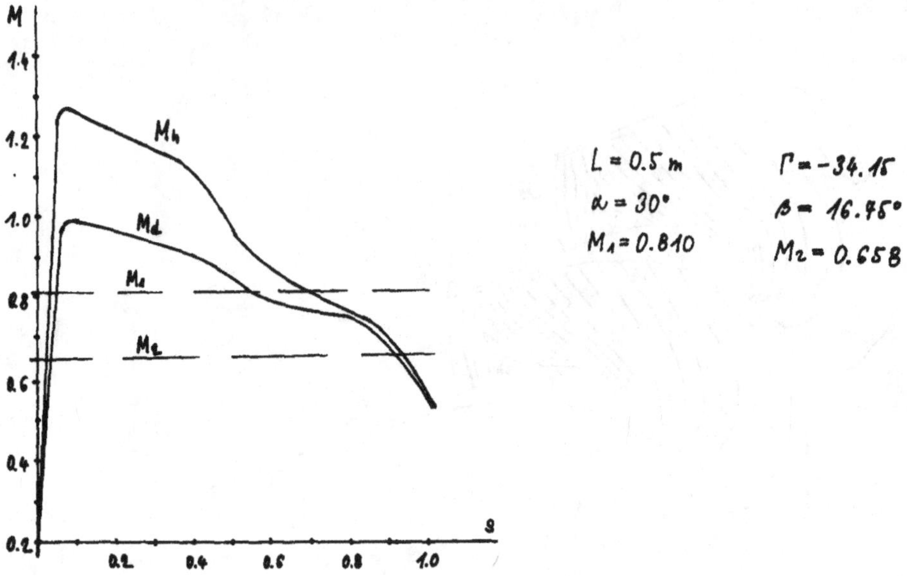

Fig. 3a: Inverse problem. Distribution of Mach number along upper and lower side of profile (M_h, M_α).

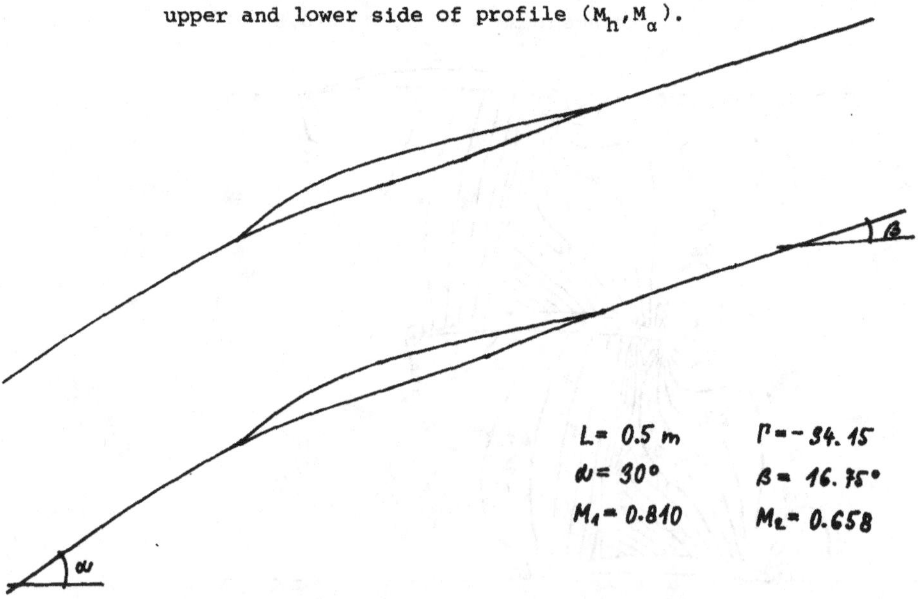

Fig. 3b: Inverse problem. Cascade geometry for given distribution of Mach number along profile.

EINIGE ANWENDUNGEN DER MEHRDIMENSIONALEN APPROXIMATIONSTHEORIE ZUR LÖSUNGSEINSCHLIEßUNG BEI RANDWERTAUFGABEN

L. COLLATZ
Inst. für Angewandte Mathematik der Universität Hamburg
Bundesstraße 55, D-2000 Hamburg 13, West Germany

Summary. In this survey lecture we summarize at first some elsewhere described methods for inclusion of solutions of linear and nonlinear boundary value problems, and apply them to certain threedimensional problems. In simple case one can check with aid of multivariate Approximation-theorie, which degree of accuracy can be reached. A numerical three-dimensional example shows this.

Abstract. In diesem Übersichtsvortrag werden zunächst auch schon andernorts beschriebene Methoden zur Einschließung der Lösungen von linearen und nichtlinearen Randwertaufgaben zusammengestellt und dann an verschiedenen dreidimensionalen Aufgaben getestet. Die Güte der erreichten Näherung kann in einfachen Fällen mit Hilfe der multivariaten Approximationstheorie beurteilt werden, wie es an einem einfachen Beispiel vorgeführt wird.

I. Einführung und Zielsetzung

Vorgelegt sei eine Operatorgleichung der Form

$$Tu = r. \tag{1.1}$$

T sei ein gegebener (evtl. nichtlinearer) Operator, der einen Bereich D eines halbgeordneten Banachraumes R_1 in einen halbgeordneten Banachraum R_2 abbildet. $r \in R_2$ ist gegeben und $u \in D$ gesucht. Häufig ist $R_1 = R_2$. Für die Praxis ist folgender Ordnungsbegriff nützlich: Für zwei in einem Bereich B des n-dimensionalen Punktraumes R^n definierte reellwertige Funktionen g,h bedeute

$$g \leq h, \text{ daß } g(x) \leq h(x) \text{ ist für alle } x \in B. \tag{1.2}$$

Dabei bezieht sich das Zeichen $\leq$ auf die klassische Ordnung reeller Zahlen. Wenn $g \leq h$ gilt, kann man das Intervall $I=[g,h]$ einführen als die Menge $p(x)$ der Funktionen $I = \{p(x),\ g \leq p \leq h\}$.

Hat man eine Näherungslösung $z(x)$ für $u(x)$ berechnet, so interessiert sich der Anwender dafür, wieviele der vom Computer ausgedruckten Dezimalen richtig sind und garantiert werden können, d.h. es soll ein genügend kleines Intervall $I=[v,w]$ angegeben werden, welches mit Sicherheit eine Lösung u von (1.1) enthält mit

$$v \leq u \leq w \tag{1.3}$$

Die in den Anwendungen auftretenden Probleme sind häufig so komplexer Natur, daß es für den Mathematiker in der Regel zu schwierig ist, ein solches Intervall anzugeben. Trotzdem wurden hier in neuerer Zeit Fortschritte erzielt, und es soll hier beschrieben werden, wie man bei verschiedenen einfachen Modellen zum Ziel kommen kann. Dabei sollen besonders dreidimensionale Aufgaben betrachtet werden.

II. Methode der Berechnung und Zusammenstellung der Grundlagen

1. Operatoren monotoner Art. Der Operator T heißt "von monotoner Art" oder "inversmonoton" (SCHRÖDER [62]), falls

$$\text{aus } Tf < Tg \text{ folgt: } f < g \text{ für alle } f,g \in D. \tag{2.1}$$

Für weitreichende Klassen linearer und nichtlinearer elliptischer und parabolischer Gleichungen und Integralgleichungen, aber auch für gewisse Fälle hyperbolischer Gleichungen wurde monotone Art bewiesen (vgl. COLLATZ [68],[81], WALTER [70], BOHL [74], SCHRÖDER [80] u.a.).

Für viele Anwendungen ist auch ein anderer Monotoniebegriff wichtig:

T heißt synton (antiton), wenn $Tf<Tg$ aus $f<g$ ($f>g$) folgt für alle $f,g\in D$.

2. Lineare und nichtlineare Approximation. Sei T ein Operator von monotoner Art; man versucht zwei Funktionen $v(x)$, $w(x)$ aufzustellen mit

$$Tv < r < Tw; \tag{2.2}$$

dann gilt die Einschließung (1.3). Für die Numerik läßt man v und w noch von Parametern a_ν, b_μ abhängen:

$$v=v(x,a)=v(x,a_1,\ldots,a_p), \text{ analog } w=w(x,b_1,\ldots,b_q)$$

und bestimmt die a_ν, b_μ so, daß das Intervall möglichst klein wird.

3. Lineare und nichtlineare Optimierung. Auf dem Computer bestimmt man die a_ν, b_μ aus der semi-infiniten Optimierung:

$$-\delta\le w(x,b) - v(x,a) \le \delta; \ Tv(x,a) \le r(x) \le Tw(x,b); \quad \delta = \text{Min.} \tag{2.3}$$

4. Iterationsverfahren. Brauchbare Schranken v,w bekommt man häufig mit Hilfe eines Iterationsverfahrens, ausgehend von passenden Startelementen v_0, w_0 wie es schon oft ausführlich dargestellt worden ist (BOHL [74], SCHRÖDER [80], COLLATZ [68] u.a.).

5. Singularitäten. Diese erfordern i.a. eine besondere Beachtung, wie es hier in III näher behandelt wird. Wenn man den Typ und die Lage der Singularitäten kennt, so kann man sie "abspalten". (vgl. TOLKSDORF [85], DOBROWOLSKI [85], WHITEMAN [85]). Hier ist noch viel Forschung nötig.

6. Schauderscher Fixpunktsatz. Der Operator T sei "monoton zerlegbar", d.h. er lasse sich als Summe eines syntonen Operators T_1 und eines antitonen Operators T_2 schreiben. Wenn der Operator T überdies vollstetig ist, so läßt sich unter gewissen Voraussetzungen mit Hilfe des genannten Satzes die Existenz von mindestens einer Lösung u im Intervall $[v_0, w_0]$ nachweisen, vgl. die in 4. genannte Literatur.

7. Multivariate Tschebyscheff Approximation. Eine gegebene stetige reellwertige Funktion $f(x)\in C(B)$ soll durch Funktionen φ einer gegebenen

Menge $W \subset C(B)$ im Tschebyscheffschen Sinne, d.h. bezüglich der Maximumnorm "möglichst gut" approximiert werden; man fragt nach der "Minimalabweichung" $\rho(f,W) = \inf_{\varphi \in W} ||\varphi - f||$.

Wenn es zwei (normalerweise endliche) Punktmengen M_1, M_2 von B gibt mit der Eigenschaft: Für kein Paar $\varphi, \hat{\varphi} \in W$ ist

$$\varphi(x) - \hat{\varphi}(x) > 0 \text{ für alle } x \in M_1; \ \varphi(x) - \hat{\varphi}(x) < 0 \text{ für alle } x \in M_2 \qquad (2.4)$$

so heißt die Vereinigung M von M_1 und M_2 eine H-Menge ("Haar-Menge") $M = M_1 \cup M_2$. Nun sei h(x) eine Funktion aus W und es gebe eine H-Menge M, für die der Fehler $\varepsilon = h - f$ auf $M \neq 0$ ist; ferner soll es keine Funktion $\varphi \in W$ geben mit $\varepsilon \cdot (h - \varphi) > 0$ auf M. Dann bestehen für ρ die Schranken (COLLATZ-KRABS [73], MEINARDUS [67])

$$m_1 = \inf_M |\varepsilon| \leq \rho(f,W) \leq ||\varepsilon|| = m_2 \qquad (2.5)$$

Dies gibt einen Anhaltspunkt für die Güte der erreichbaren Annährung für f in der Klasse W: Ist m_1 "nahe" an m_2, so hat man mit h nahezu die beste Annäherung an f erreicht; für $m_1 = m_2$ ist h eine beste Approximation Ist m_1 größer als die gewünschte Genauigkeit, so muß man zu einer anderen Funktionenklasse W übergehen (z.B. mehr Parameter verwenden).

8. Algorithmus für H-Mengen. **In (2.4) sind φ und $\hat{\varphi}$ beliebige Elemente aus W, sodaß die Nachprüfung bei nichtlinearer Approximation schwierig werden kann. Bei vorliegender Näherung $h^* \in W$ genügt es aber, h^* als φ zu nehmen und (2.4) nur für beliebiges $\hat{\varphi} \in W$ zu prüfen (Ph.Defert-J.P.Thiran, Exchange Algorithm for multivariate Polynomials, Intern. Ser. Numerical** Math.59(1982) 115–128. Dann ist $H = M_1 \cup M_2$ eine H-Menge, wenn die Ungleichungen

$$h^*(x) - \hat{\varphi}(x) > 0 \text{ für } x \in M_1; \ -[h^*(x) - \hat{\varphi}(x)] > 0 \text{ für } x \in M_2 \qquad (2.6)$$

keine Lösung $\hat{\varphi}(x) \in W$ zulassen. Falls W ein linearer Unterraum von C(B) ist, kann man zum Nachweis, daß eine H-Menge vorliegt, den Gaußschen Algorithmus benutzen (ein Beispiel in Nr. IV). Wenn eine H-Menge vorliegt, gelten die Schranken (2.5) und man kann die Güte der Näherung beurteilen.

III. Einige Testbeispiele

1. Eine nichtlineare Anfangs-Randwertaufgabe. Ein sehr einfaches Modell der Navier-Stokesschen Gleichungen ist die eindimensionale Gleichung (Burger's Gleichung) für eine Funktion u(x,t):

$$Tu = \frac{\partial u}{\partial t} + u \cdot \frac{\partial u}{\partial x} - \nu \cdot \frac{\partial^2 u}{\partial x^2} = 0 \text{ in B}$$

Fig.1

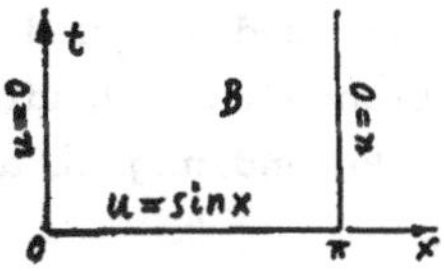

mit dem Bereich $B = \{(x,t),\ 0 < x < \pi,\ t > 0\}$

und den Anfangs- und Randbedingungen, Fig. 1:

$$u(x,0) = \sin x \text{ für } 0 \leq x \leq \pi, \ u(0,t) = u(\pi,t) = 0 \text{ für } t \geq 0.$$

Für die Näherungsfunktion $w(x,t)$ wird der Ansatz gemacht:

$$u \approx w(x,t) = \sin x + \sum_{j=1}^{n} a_j \varphi_j (x,t)$$

mit $\varphi_1 = t \sin x$, $\varphi_2 = tx(\pi-x)$, $\varphi_3 = tx^2(\pi-x)$, $\varphi_4 = t^2x(\pi-x)$, $\varphi_5 = \frac{t}{2}\sin(2x)$.
Bei Diskresisierung mit den Schrittweiten 1/32, (bzw. 1/20) in x-, bzw. t-Richtung erhält man im Intervall $0 \le t \le 0.2$ die Fehlerschranken

	$\nu=0.1$	$\lvert w-u \rvert \le$	$\nu=0.5$
n=2	0.1013		0.0999
n=3	0.0450		0.0420
n=4	0.0408		0.0377
n=5	0.0171		0.0252

Für den größeren Wert $\nu=0.5$ fallen die Fehlerschranken ungünstiger aus als für $\nu=.0.1$. Ich danke Herrn Dipl.Math. Uwe Grothkopf für die numerische Durchrechnung auf dem Computer.

2. Einschließung einer Ableitung bei einem unbeschränkten dreidimensionalen Bereich. (Elektrostatisches Potential $u(x,y,z)$ zwischen einer Kugel (Potential $u=1$) und zwei Ebenen mit dem Potential $u=0$, Fig.2). Gesucht ist die elektrische Feldstärke η im Punkte $P=(0,0,1)$, $\eta=\partial u/\partial z(P)$

Randwertaufgabe: $\overbrace{U_{xx}+U_{yy}+U_{zz}}^{\Delta U}=0$ in $(|z|<3, |r|>1)$ mit $r=(x,y,z)$

$u = 0$ auf Γ_1 und Γ_2 $(|z|=3)$ und $u=1$ auf $\Gamma_3(|r|=1)$

Ansatz für eine Näherungsfunktion

$$u \approx v = a + \frac{a_0}{|r|} + \sum_{i=1}^{m} a_i \left(\frac{1}{|r-(0,0,z_i)|} + \frac{1}{|r-(0,0,-z_i)|}\right)$$

mit $v(P)=1$; $0<z_i<1$ oder $z_i>3$)

Fig.2

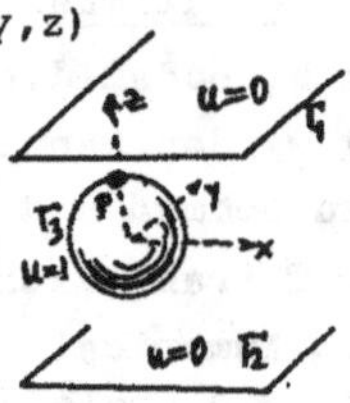

Nichtlineare Optimierung: $v(r) \ge u(t)$ auf Γ_1 Γ_2 Γ_3, $(\frac{\partial v}{\partial z})_{(P)}=\text{Min}$.

Einschließung für η (berechnet von Herrn Dipl.Math. Jörg Haarmeyer)

m=Anzahl der Parameter	untere Schranke für η	obere Schranke für η
4	- 1.358 022	- 1.357 97
8	- 1.358 022	- 1.358 016

3. Singuläre Kante bei einer dreidimensionalen Aufgabe. Modell einer Temperaturverteilung $u(x,y,z)$ in einem Raumteil B, Fig. 3 mit $\Delta u=0$ in $B = \{(x,y,z),\ |z|<1;\ |x|<1$ für $0<y<1$, und $-1<x<0$ für $-1<y\le 0.\}$ und $u=1$ auf Grund- und Deckfläche und auf den beiden an die z-Achse grenzenden Rechteckrandflächen; $u=z^2$ für $x=-1$ und $y=1$, und u linear in x, bzw. y auf den restlichen zwei Randflächenstücken, Fig. 3 (ausführliche Behandlung an anderer Stelle).

Fig.3

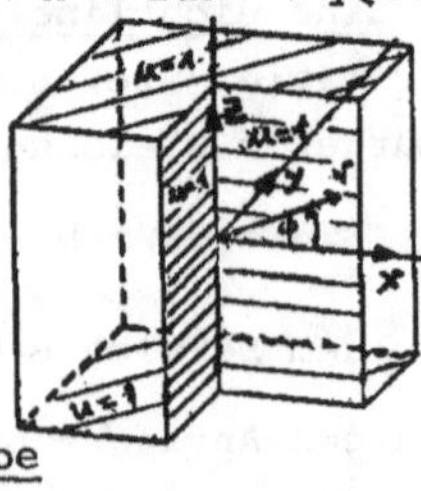

IV. H-Mengen bei einer dreidimensionalen Randwertaufgabe

Als ähnliches Beispiel für eine dreidimensionale Temperaturverteilung u(x,y,z) werde gewählt; Fig. 4

$-\Delta u=1$ in $B=\{(x,y,z)\ 0<x<2,\ 0<y<2,\ x+y<3,\ |z|<1\}$; $u=0$ auf ∂B

Ansatz für die Näherungsfunktionen v_n:

$$u \approx v_n = -\tfrac{1}{2}z^2 + \sum_{i=1}^{n} a_i v_i,$$ mit den die Potentialgleichung erfüllenden

Polynomen: $v_1=1,\ v_2=x+y,\ v_3=x\cdot y,\ v_4=x^2+y^2-2z^2,$

$v_5=x^3+y^3-3/x+y)z^2,\ v_6=xy(x+y)-(x+y)z^2,\ldots$

Mit $\varepsilon_k = v_k - u$ lautet die Optimierungsaufgabe:

$$-\delta \le \varepsilon_k \le \varepsilon \text{ in } B,\quad \Delta = \text{Min.}$$

Ich danke Herrn Dipl.-Math. Zheng Tsinghua aus Schanghai für die Durchführung der Technung auf einem Computer und für Fig. 4 und 5; man erhält:

Verwendeter Polynomgrad	Anzahl k der Parameter	δ
1	2	0.25
2	4	0.153
4	9	0.0382
6	16	0.0145
8	25	0.00493

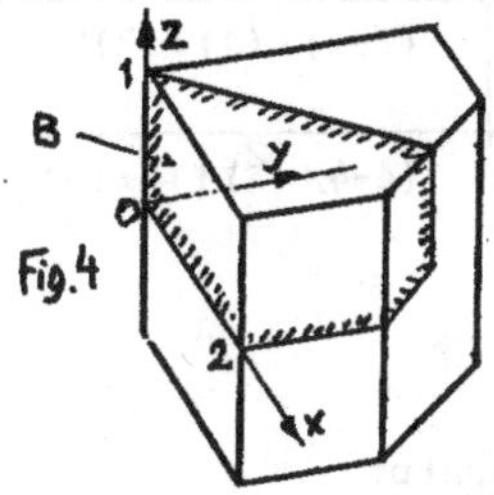

Man hat damit die Einschließung $|u-v_{25}|\le 0.00493$.

<u>2. Algorithmus zur Testung der H-Mengeneigenschaft.</u> Um ein einfaches Beispiel zu haben, werden die Punkte $P_1,P_2,\ldots,P_5$ wie in Fig. 5 ausgewählt; Die Koordinaten x,y,z der Punkte sind in dem folgenden Schema angegeben. Daneben stehen die Faktoren v_j der zugehörigen Terme a_j,v_j. Bei den P_ν bedeutet ein "+"-Zeichen (bzw. ein "-"-Zeichen, daß P_ν zu M_1 (bzw. M_2) gehört. Die erste Zeile des Schemas ist zu lesen als

$$1\cdot a_1 + 0\cdot a_2 + 0\cdot a_3 - 2\,a_4 > 0$$

Da hier lineare Approximation vorliegt, kann man wie beim Gaußschen Eliminationsverfahren Unbekannte eliminieren, indem man Ungleichungen mit positiven Faktoren multipliziert und addiert, wie es in der Spalte "Operation" der folgenden Tabelle angegeben ist. Dann stellen die Ungleichungen (9)(10) einen Widerspruch dar, die Punkte $P_1,\ldots,P_5$ bilden eine H-Menge.

Bei dem Ansatz $a_1v_1+a_2v_2$ genügen P_1,P_2,P_3 wie es im Schema der gestrichelte Rahmen andeuten soll, zum Nachweis der H-Mengeneigenschaft, in dem (1) und (7) mit $a_1>0$, $(-2+r)a_1>0$ einen Widerspruch ergeben. Bei Hinzunahme von a_3v_3 und P_4 mit (1)(8) ergeben $a_1>0$, $(-4+4s-s^2)a_1>0$ einen Widerspruch.

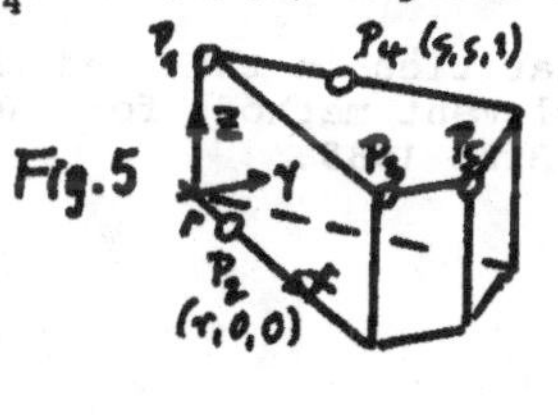

	Punkte	x	y	z	$a_1 \cdot 1$	$a_2(x+y)$	a_3xy	$a_4(x^2+y^2-2z^2)$
(1)	$P_1(+)$	0	0	1	1	0	0	-2
(2)	$P_2(-), 0<r<2$	r	0	0	-1	$-r$	0	$-r^2$
(3)	$P_3(+)$	2	0	1	1	2	0	2
(4)	$P_4(-), 0<s\leq 4/3$	s	s	1	-1	$-2s$	$-s^2$	$2-2s^2$
(5)	$P_5(+)$	2	1	1	1	3	2	3
	Operation							
(6)	$2\cdot(4)+s^2\cdot(5)$				$-2+s^2$	$-4s+3s^2$	0	$4-s^2$
(7)	$2\cdot(2)+r\cdot(3)$				$-2+r$	0		$2r-2r^2$
(8)	$(4s-3s^2)\cdot(3)+2\cdot(6)$				$-4+4s-s^2$	0		$8+8s-8s^2$
(9)	$(2-r)\cdot(1)+(7)$				0			$-4+4r-2r^2$ $=-2-2(1-r)^2<0$
(10)	$(4-4s+s^2)(1)+(8)$				0			$16s-10s^2$ $=2s(8-5s)>0$ Widerspruch.

Literatur

BOHL, E. [74] Monotonie, Lösbarkeit und Numerik bei Operatorgleichungen. Springer, 1974, 255 S.

COLLATZ, L. [52] Aufgaben monotoner Art, Arch.Math.Anal.Mech.3 (1952) 366-376.

COLLATZ, L. [68] Funktional Analysis und Numerische Mathematik, Springer, 1968, 371 S.

COLLATZ, L. [81] Anwendung von Monotoniesätzen zur Einschließung der Lösungen von Gleichungen; Jahrbuch Überblicke der Mathematik 1981, 189-225.

COLLATZ, L. [85] Inclusion of regular and singular solutions of certains types of integral equations, Intern.Ser.Num.Math. 73 (1985) 93-102.

COLLATZ, L. - W. KRABS [73] Approximationstheorie, Teubner, Stuttgart, 1973, 208 S.

DOBROWOLSKI, M. [85] On quasilinear elliptic equations in domains with conical boundary points, Bericht Nr. 8506, Juni 1985, Univers. d. Bundeswehr München, 16 S.

MEINARDUS, G. [67] Approximation of functions, Theory and numerical methods, Springer, 1967, 198 p.

SCHRÖDER, J. [62] Invers-monotone Operatoren, Arch.Rat.Mech.Anal. 10 (1962), 276-295.

SCHRÖDER, J. [80] Operator inequalities, Acd.Press (1980), 367 p.

TOLKSDORF, P. [85] On the Dirichletproblem for quasilinear equations in domains with conical boundary points, erscheint in Comm.Diff. Equ.

WALTER, W. [70] Differential and Integral Inequalities, Springer (1970) 352 p.

WHITEMAN, J.R. [85] Singularities in two- and threedimensional elliptic problems and finite element methods for their treatment, erscheint in Proc. Equadiff 6, Brno 1985.

A POSTERIORI ESTIMATIONS OF APPROXIMATE SOLUTIONS FOR SOME TYPES OF BOUNDARY VALUE PROBLEMS

R. KODNÁR
Institute of Applied Mathematics, Comenius University
Mlynská dolina, 842 15 Bratislava, Czechoslovakia

1. Motive

When any approximate method is employed, it is of importance to know an estimation of the error involved in the approximate solution. In general, there exist two kinds of such estimations (i) a priori and (ii) a posteriori. The a priori assessments are obtained from the qualitative properties of the problem. They ussually possess an asymptotic character, are pessimistic and are used chiefly in theoretical considerations. The a posteriori assessments are carried out on the basis of already constructed approximate solution. Among the ways of construction of a posteriori estimations, a significant role is played by s.c. counter-direction methods. They are based on the following simple considerations:
Let A be a linear positively definite operator in a real Hilbert space H. Then the problem to find a generalized solution of the equation

$$Au=f, \quad f\in H \tag{1}$$

and minimization of the functional

$$F(u)=[u,u]_A-2(u,f)_H \tag{2}$$

are equivalent [1]. In (2) $[\ ,\ ]_A$ denotes a scalar product in the space H_A of the generalized solution to equation (1). It holds

$$\min_{H_A} F(u)=-\|u_0\|_A^2 ,$$

$$\|u_n-u_0\|_A^2=F(u_n)+\|u_0\|_A^2 ,$$

where u_n is an approximate solution constructed by the variational method and $\|\ \|_A$ is the norm in H_A. Usually the numbers d can be constructed greater than $\|u_0\|_A^2$ but close to it (lower bound estimation of $F(u_0)$). Then with their aid we get

$$\|u_n-u_0\|_A \leq (F(u_n)+d)^{1/2}. \tag{3}$$

Such a construction of numbers d is possible, for example, in the following case:
Let us assume that there exists a functional $F_1(u)$ such that

$$\inf_{H_A} F_1(u)=\|u_0\|_A^2 .$$

Then it is possible to select $d=F_1(u)$, $u\in H_A$. With the selection of u being suitable, it can be achieved that d is very near to $\|u_0\|_A^2$.

Inequality (3) explicitly gives the error estimation of the approximate solution. In the case of equations with worse operators than those mentioned in (1), it is not always possible to obtain an estimation in the form of (3), but one gets estimations of some other types. In literature several constructions of the lower estimations d are described. But the majority of them is applicable for the single special problems only.

Functions in the whole text are real.

2. Basic notions

Let Q be a bounded domain in R^m with the boundary ∂Q which satisfies several conditions of smoothness. Let on $\overline{Q}$ be given a boundary value problem such that $u_0\in V$ is its weak solution if

$$\forall v\in V:\ ((v,u_0))=\langle v,f\rangle+Z(v,h)-((v,w)). \tag{4}$$

Nearby $((,))$ is a bounded bilinear form on $H^k(Q)$, $k\geq 1$, $V\subset H^k(Q)$, $\langle\ ,\rangle$ is a duality on V, $w\in H^k(Q)$,

$$Z(v,h)=\sum_{p=1}\ \sum_{l=1}\ \int_{Q_p} \frac{\partial^{t_{pl}} v}{\partial n^{t_{pl}}}\, h_{pl}\, dS,$$

$$\bigcup_1^r \partial Q_p = Q,$$

$f\in V^*$, $h_{pl}\in L_2(\partial Q_p)$, n is a direction an outward normal to Q. Let

$$\forall v\in V:\ ((v,v))\geq a^2\|v\|_V^2\ ,\ a\ R^1, \tag{5}$$

$$\forall u,v\in V:\ ((u,v))=((u,v)). \tag{6}$$

<u>Theorem 1</u>. Let

$$\forall u\in V, \forall v\in H^k(Q):\ ((u-v,u-v))\geq 0. \tag{7}$$

Let $v_1\in H^k(Q)$ be such that

$$\forall u\in V:\ ((u,v_1))=\langle u,f\rangle+Z(u,h)-((u,w)). \tag{8}$$

Then

$$\forall v_n \in V:\ a^2\|u_0-v_n\|_V^2=((v_n,v_n))+2\langle v_n,f\rangle-2Z(v_n,h)+ \\ +2((v_n,w))+((v_1,v_1)), \tag{9}$$

holds.

Proof. Immediately follows from (5),(6),(7),(8).

If the informations on regularity of solution of the starting boundary value problem are avalaible construction of the lower estimation may be simplified. Consider a linear boundary value problem

$$Au=f \quad \text{in } Q, \qquad B_1u=0 \quad \text{on } \partial Q, \tag{10}$$

where A is a differential operator of 2k-th order. Denote

$$K=\{u\in C^{(2k-1)}(\overline{Q})\cap C^{(2k)}(Q),\ Au\in L_2(Q)\},$$

$$D_A=\{u\,|\,u\in K,\ u \text{ fulfills all the boundary conditions from (10)}\}.$$

Suppose that A,f,Q and the boundary conditions in (10) are such, that the solution u_0 of the problem (10) belongs to D_A. Let $((,))_1$ be a symmetrical bilinear form such that

$$\forall u\in K:\ ((u,u))_1\geqq 0,\ ((u,u))_1=0\Leftrightarrow u=0, \tag{11}$$

$$\forall u\in K,\forall v\in D_A:\ ((u,v))_1=(Au,v)_{L_2}. \tag{12}$$

Remark 1. It can be shown that such a form exists for the most of the boundary value problem with Laplace and also with biharmonic operator. Form $((,))_1$ to the given problem is not uniquely defined.

Theorem 2. u_0 minimizes in D_A the functional

$$F_1(u)=((u,u))_1-2(u,f)_{L_2}.$$

Let $v\in K$ be such that $Av=f$. Then

$$F_1(u_0)\geqq -((v,v))_1.$$

Proof. $\forall u\in D_A,\forall v\in K:\ ((v-u,v-u))_1\geqq 0$. Then

$$((u,u))_1-2(u,f)_{L_2}\geqq -((v,v))_1-2(u,f)_{L_2}+2((u,v))_1.$$

From it follows

$$\forall u\in D_A,\forall v\in K:\ F_1(u)\geqq -((v,v))_1-2(Av-f,u)_{L_2}.$$

Denote $J(v)=((v,v))$. $J(v)$ is a functional defined on $H^k(Q)$. Let the assumptions of Theorem 1 be fulfilled. Then for $v\in H^k(Q)$ fulfilling (8) there is $J(v)\geqq((u_0,u_0))$. Let

$$D_J=\{v\,|\,v\in H^k(Q) \text{ and fulfil (8)}\}.$$

Then $((u_0,u_0))=\min_{D_J} J(v)$.

Lemma 1. The minimizing sequence for the functional $J(v)$ converges to the solution u_0 of the equation (4) in the following sense

$((u_n-u_0,u_n-u_0)) \longrightarrow 0.$

Proof. Denote the minimizing sequence $\{u_n\}_1^\infty$. Then $D_J \ni z_n = u_n - u_0$. It holds $\forall u \in V$: $((z_n,u))=0$. From that

$$((u_n-u_0,u_n-u_0))=((u_n,u_n))-2((z_n,u_0))-((u_0,u_0))=$$
$$=((u_n,u_n))-((u_0,u_0)).$$

Remark 2. Procedure formulated by Theorem 1,2 is a generalization of Trefftz method.

3. The construction of minimizing sequence

Denote

$$U=\{v \in H^k(Q) \mid \forall u \in V:\ ((u,v))=\langle u,f\rangle+Z(u,h)-((u,w))\}.$$

Lemma 2. The set U is convex and closed.

Proof. By direct verifying.

Lemma 3. The functional J(v) is convex on $H^k(Q)$ and its minimum on U is attained at u_0.

Proof. Convexity follows from differentiability.

Corollary 1. Relations

$$\forall v \in U:\ J(u) \leqq J(v),$$
$$\forall v \in U:\ J'(u,v-u) \geqq 0$$

are equivalent. The given problem can be solved by means of variational inequalities. The obtained minimizing sequence converges by given way to the solution u_0.

Thus we get further counter-direction methods to variational method of the solution of the primary problem.

4. Several special cases

Let A be a linear, positively definite operator $A \in (H,H)$. Let be given further Hilbert space H_1 with scalar product $(\,,)_1$. Let

$$A=T^*BT, \tag{13}$$

where T is operator $T \in (D_T \subset H,H_1)$, B is positively definite operator $B \in (H_1,H_1)$. $T^* \in (D_{T^*} \subset H_1,H)$ is operator adjoint to T. Assume that

$$D_T \supset D_A,\ D_{T^*} \supset BTD_A.$$

Operator T^*BT is thus defined at least on D_A. Let the problem $Au=f$ have solution $u_0 \in D_A$, while $f=T^*g$, where $g \in D_{T^*}$. Denote

$$w_0=Tu_0,\ w=Tu,\ u \in D_T,$$

$$G(w)=(Bw,w)_1-2(g,w)_1\ ,\ w\in H_1.$$

<u>Theorem 3</u>. w_0 minimizes G on $TD_T\subset H_1$. If $T^*v=f$, then

$$G(w_0)=-\frac{1}{a}(v,v)_1, \tag{14}$$

where a is a constant from positively definiteness of operator B.

<u>Remark 3</u>. In case $A=\Delta\Delta$ and Dirichlet boundary conditions we get from Theorem 3 the principle of the method of unharmonic residue [1] .

<u>Corollary 2</u>. $U_1=\{v\in D_{T^*}\,|\,T^*v=f\}$ is convex and closed set. $(v,v)_1$ is a convex functional on H_1. If there is $w_0\in U_1$ and $a\geq 1$ (from (14)), then w_0 minimizes functional $(v,v)_1$ on U_1.

<u>Example</u>. Consider a boundary value problem

$$\Delta\Delta u=f,\quad u\Big|_{\partial Q}=\frac{\partial u}{\partial n}\Big|_{\partial Q}=0,\ f\in L_2(Q). \tag{15}$$

Let the problem (15) have the solution $u_0\in C^{(1)}(\overline{Q})\cap C^{(4)}(Q)$. Denote:

$H=L_2(Q)=H_1$,

$D_A=\{u\,|\,u\in C^{(4)}(Q)\cap C^{(1)}(\overline{Q}),\ \Delta\Delta u\in H$, u fulfil the boundary conditions from (15)$\}$,

$D_T=\{u\,|\,u\in C^{(1)}(\overline{Q})\cap C^{(4)}(Q)$, u fulfil the boundary conditions from (15)$\}$,

$D_{T^*}=\{v\,|\,v\in C^{(1)}(\overline{Q})\cap C^{(2)}(Q),\ \Delta v\in H\}$.

Define

$$\forall u,v\in D_{T^*}:\ (u,v)_1=\sum_{|i|\leq 2}\int_Q D^iuD^iv\ dQ=(u,v)_{H^2}\ .$$

Then there is in (13) $T=T^*=\Delta$, B is an identic operator. On lower estimations of minimum of the functional

$$u\in D_A:\ (\Delta\Delta u,u)_{L_2}-2(u,f)_{L_2}$$

Theorem 3 and Corollary 2 can be used.

<u>Remark 4</u>. Problem (15) is a mathematical model of clamped plate. Similarly the mathematical models of further kinds of boundary of plate and web we may investigated [2] .

<u>Remark 5</u>. Procedures from sections 2,3,4 may also be used for non-linear problem of the special type:

$u_0\in V$ is called the solution of the problem if

$f(x,u_0+w)\in L_2(Q)$,

$$\forall v\in V:\ ((v,u_0))+\int_Q f(x,u_0(x)+w(x))v(x)\ dQ=Z(v,h)-((v,w)).$$

At the same time it is supposed that function $f:\ \overline{Q}\times R^1\to R^1$ is continuous and for fixed $x\in Q$

$$f(x,r_1)\leq f(x,r_2)\quad \forall r_1,r_2\in R^1,\quad r_1\leq r_2$$

holds.

All the rest notations are the same as in (4).

5. Slobodyanskii procedure

In [3] Slobodyanskii proposed procedure to get lower bound assessment. Generalization of this procedure for further, even non-linear problems, is described in [4].

REFERENCES

[1] Michlin, S.G.: Variationsmethoden der Mathematischen Physik. Berlin: Akademie Verlag 1962.

[2] Kodnár, R.: A posteriori estimation of approximate solution for mathematical models of plate and web. (in preparation).

[3] Slobodyanskii M.G.: On corvension of the problem of minimum of a functional into the problem of maximum. Doklady Akad. Nauk SSSR 91, Nr.4 (1953) (Russian).

[4] Djubek, J., R.Kodnár, M.Škaloud: Limit state of the plate elements of steel structures. Basel-Boston-Stuttgart: Birkhäuser Verlag 1984, 298p.

NONLINEAR DYNAMICS SYSTEMS - BIFURCATIONS, CONTINUATION METHODS, PERIODIC SOLUTIONS

M. KUBÍČEK and M. HOLODNIOK
Department of Mathematics and Computer Centre, Prague Institute of Chemical Technology
166 28 Praha 6, Czechoslovakia

1. INTRODUCTION

Let us consider a system of autonomous nonlinear ordinary differential equations (nonlinear dynamic system)

$$y' = f(y,p) \quad , \qquad (1)$$

where $' = d/dt$, $y \in R^n$, $p \in R^m$ are parameters, $f : R^n \times R^m \to R^n$, $f \in C^1$. Steady state solutions of (1) are defined by

$$f(y,p) = 0 \quad . \qquad (2)$$

A set $S(f) \in R^n \times R^m$, $S(f) : (y,p)$, $f(y,p) = 0$, is called "solution diagram" [19], sometimes also "bifurcation diagram" [10]. Solution diagram is mostly considered for one parameter, i.e. p_1 only, while values of remaining parameters $p_2, \dots, p_m$ are fixed. Continuation algorithms have been developed in the last ten years for an automatic computation of such solution diagrams [e.g., 14, 24, 9, 10, 11]. Stability of steady state solutions can be determined on the basis of eigenvalues λ of the Jacobian matrix $J = \{\partial f / \partial y\}$. If an eigenvalue λ crosses the imaginary axis in complex plane (by varying some parameter, e.g., p_1) a bifurcation occurs in generic cases. Several review papers surveying numerical methods for location of bifurcation points appeared recently [22, 20, 19, 9]. Four iterative algorithms for the evaluation of Hopf bifurcation points have been published in [5] .

Main purpose of this paper is to discuss periodic behaviour observed in two typical mathematical models of chemical and engineering systems, review computational methods for continuation and bifurcation of periodic solutions.

The first model is well known Lorenz model [21] of the flow in the layer of liquid heated from below (the Rayleigh - Bénard problem). The system

$$y_1' = -\sigma y_1 + \sigma y_2 \quad , \quad y_2' = -y_1 y_3 + r y_1 - y_2 \quad , \quad y_3' = y_1 y_2 - b y_3 \qquad (3)$$

is obtained by a reduction of the system of Navier Stokes equations and the equation describing heat transfer. The dimensionless parameters $p = (r, \sigma, b)$ correspond to : σ - Prandtl number, r - reduced Rayleigh number, b is related to a wave-number of the convective structure . A detailed description of behaviour of the model can be found in the Sparrow's book [27] , a structure of periodic solutions was discussed in [7] .

The second model describes behaviour of two well mixed reaction cells with linear diffusion coupling and the "Brusselator" reaction kinetic scheme. The model is used as a standard model system for the discussion of dissipative structures in nonlinear chemical systems [23] . It can be written in the form

$$\begin{aligned} y_1' &= A - (B + 1)y_1 + y_1^2 y_2 + D(y_3 - y_1) \\ y_2' &= By_1 - y_1^2 y_2 + D(y_4 - y_2)/\rho \\ y_3' &= A - (B + 1)y_3 + y_3^2 y_4 + D(y_1 - y_3) \\ y_4' &= By_3 - y_3^2 y_4 + D(y_2 - y_4)/\rho \end{aligned} \qquad (4)$$

Here $p = (D,B,A,\rho)$, A and B are constant concentrations, D and D/ρ define the intensity of mass exchange between the cells, y_1 , y_2 and y_3 , y_4 are dimensionless concentrations of reaction intermediates in the first and second cell, respectivelly.

2. CONTINUATION OF PERIODIC SOLUTIONS

We shall present here a short description of an algorithm for the continuation (and computation) of periodic solutions based on the shooting method together with a continuation along the arclength of the solution locus. Detailed description of the algorithm is presented in [4] .

A periodic solution with the period T fulfils

$$y(T) - y(0) = 0 \qquad (5)$$

Considering shooting method we choose initial conditions

$$y(0) = x \qquad (6)$$

$x \in R^n$, and the value of the period T . Then the system (1) can be numerically integrated for fixed p from $t = 0$ to $t = T$, the results of integration

$$y(T) = \Psi(x,T,p) \qquad (7)$$

are dependent on the choice of x , T and p . Inserting (7) into (5) we obtain a system of n nonlinear equations

$$F(x,T,p) = \Psi(x,T,p) - x = 0 \qquad (8)$$

with $n + 1$ unknowns x , T and m parameters p . We have to fix one variable except T (or add some "normalization" equation). Let us fix x_k for some k , in such a way that x_k actually exists on the trajectory of the k-th component of the wanted periodic solution $y_k(t)$, $t \in [0,T]$. To continue periodic solutions in dependence on one parameter, say p_1 , we can use standard continuation algorithm DERPAR [14, 19] for continuation of solutions of n equations (8) for n unknowns $x_1,\dots,x_{k-1}, x_{k+1},\dots, x_n$, T and one parameter p_1 . This continuation algorithm requires an evaluation of the functions F in (8) and of the Jacobi matrix $\partial F/\partial x$, $\partial F/\partial T$, $\partial F/\partial p_1$. Elements of the Jacobi matrix can be determined on the basis of variational differential equations for variational variables

$$V(t) = \partial y/\partial x \quad , \quad q(t) = \partial y/\partial p_1 \qquad (9)$$

V is n by n matrix and q is n by 1 , i.e.,

$$V' = JV \quad , \quad V(0) = I \quad , \quad q' = Jq + \partial f/\partial p_1 \quad , \quad q(0) = 0 \quad . \qquad (10)$$

The elements of the Jacobi matrix of the system (8) are then defined as

$$\partial F/\partial x = V(T) - I \quad , \quad \partial F/\partial T = f(y(T),p) \quad , \quad \partial F/\partial p_1 = q(T) \quad . \qquad (11)$$

The continuation routine can proceed until the fixed value of x_k "disappears" from the course of the periodic solution. To avoid this disappearance, the algorithm exchanges x_k adaptively.

Several solution diagrams obtained by the continuation algorithm are presented in Figs 1 - 3 .

The stability of the computed periodic solution can be determined on the basis of characteristic multipliers, i.e., of eigenvalues μ of the monodromy matrix

$$M = \partial \Psi/\partial x = V(T) \quad . \qquad (12)$$

One multiplier is always equal to 1 because (1) is autonomous. If all remaining multipliers lie inside the unit circle, the periodic solution is stable, if at least one of them lies outside, then the periodic solution is unstable.

The use of the above described continuation algorithm is limited by the applicability of the shooting method. If the initial value problems are unstable, i.e., there are multipliers of the order 10^5 or higher, the integration and thus the simple shooting method usually fails. In such cases the multiple shooting method can be successfully used as, e.g., for the Hodgkin - Huxley model of the conduction of the nervous impulse, where $\mu \sim 10^9$ [6].

3. BIFURCATIONS OF PERIODIC SOLUTIONS

Bifurcation of periodic solutions occurs when a multiplier crosses the unit circle when varying a parameter. It can happen in three qualitatively different ways, i.e., when $\mu = 1$; $\mu = -1$; $|\mu| = 1$, $\mu^s \neq 1$, s = 1, 2, 3, 4. The cases $\mu^3 = 1$ and $\mu^4 = 1$ are of special interest [e.g., 8].

3.1 LIMIT POINTS AND SYMMETRY BREAKING BIFURCATION POINTS ($\mu = 1$)

The monodromy matrix M has $\mu = 1$ as an eigenvalue (of multiplicity two) and, therefore, the Jacobi matrix $\partial F/\partial x$ has two zero eigenvalues. It means that no unique dependence of the periodic solutions on a parameter exists in the neighbourhood of this point. The bifurcation point can be either limit (turning) point (cf., e.g., point denoted L.P. in the Fig. 1) or symmetry breaking bifurcation point when there exists an inherent symmetry in the system (cf. point denoted SB in the fig. 1). Both bifurcation points can be determined by using shooting (or multiple shooting) method and methods for steady-state bifurcations. Either methods which use evaluation of the determinant of a matrix [13, 18, 19] or method without evaluation of the determinant [e.g., 1, 26] can be used.

3.2 PERIOD - DOUBLING BIFURCATIONS

When the monodromy matrix has $\mu = -1$ as an eigenvalue, then the so called

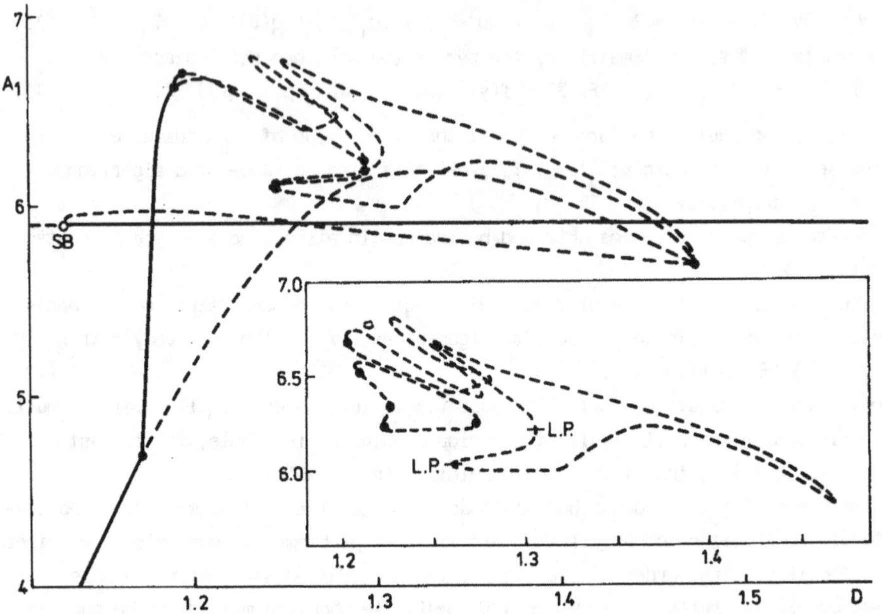

FIG. 1 : Solution diagram of periodic solutions of (4), A = 2, B = 5.9, ρ = 0.1 [25]. A_1 - amplitude of y_1 . ● - period-doubling bifurcation point, o - symmetry breaking bifurcation point. —— stable, ---- unstable.

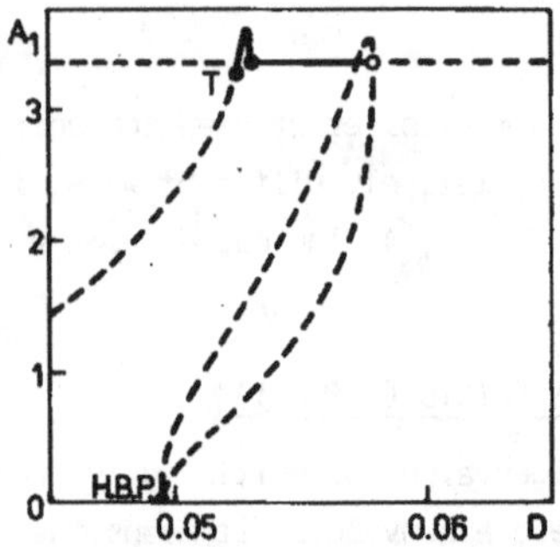

FIG. 2 : Solution diagram of periodic solutions of (4), A = 2, B = 5.5, ρ = 0.1 [25]. T - tori bifurcation point, H.B.P. - two mutually symmetric Hopf bifurcation points. Further see legende to Fig. 1.

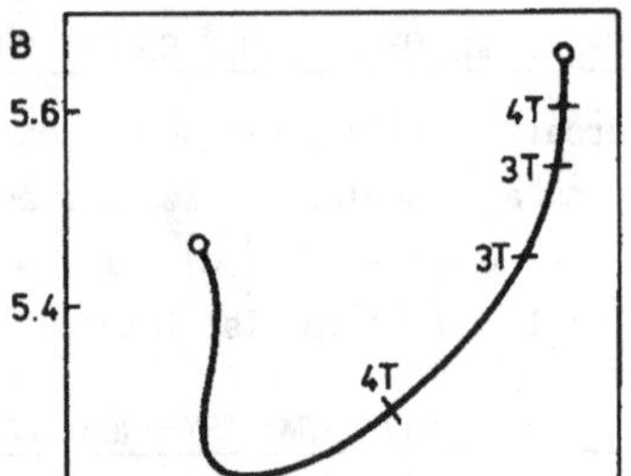

FIG. 4 : Bifurcation diagram of periodic solutions of (4), tori bifurcation points. A = 2, ρ = 0.1. o - point of higher degeneration.

period-doubling bifurcation occurs, i.e., a branch of periodic solutions with approximately double period (asymptotically) branches off the original branch of periodic solutions (cf. points denoted ● in the Figs 1 - 3).

<u>DETERMINATION OF PERIOD - DOUBLING BIFURCATION POINTS (μ =.-1).</u> Four iteration algorithms for computation of period-doubling bifurcation points have been publi-

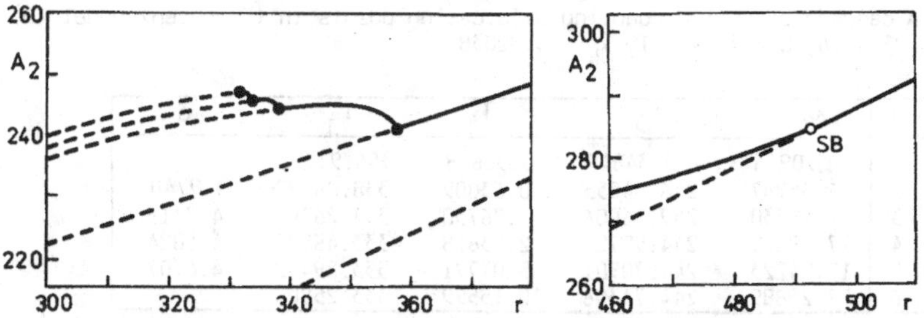

FIG. 3 : Solution diagram of periodic solutions of (3), $\sigma = 4$, $b = 16$ [7]. A_2 - amplitude of y_2 . ● - period doubling bifurcation point, o - symmetry breaking bifurcation point, —— stable, ---- unstable.

shed and are compared in [3]. We shall summarize very briefly two of them.

Let the characteristic polynomial of the monodromy matrix M be

$$P(\mu) = (-1)^n \det(M - \mu I) = \mu^n + a_1\mu^{n-1} + \ldots + a_{n-1}\mu + a_n \quad . \qquad (13)$$

The coefficients a_j can be computed by using standard software. $\mu = -1$ is the root of (13) if

$$F_{n+1}(x,T,p) = 1 + \sum_{i=1}^{n} (-1)^i a_i = 0 \quad . \qquad (14)$$

As a result we obtain $n + 1$ nonlinear equations (8) (14) for $n + 1$ unknowns $x_1,\ldots, x_{k-1}, x_{k+1},\ldots, x_n, T, p_1$. Newton method is used to solve this system.

$P(\mu)$ must have one root equal to unity, therefore, it can be decomposed into the form

$$P(\mu) = (\mu + 1)(\mu - 1)(\mu^{n-2} + b_1\mu^{n-3} + \ldots + b_{n-2}) + C\mu + D \quad . \qquad (15)$$

The coefficients $b_1, \ldots, b_{n-2}, C, D$ can be evaluated recurrently. If we determine the periodic solution where $C = 0$ and $D = 0$, we have a period doubling bifurcation point. We can use

$$F_{n+1}(x,T,p) = D = 0 \qquad (16)$$

instead of (14) and solve the system (8) (16) again by Newton method ($C = 0$ automatically for the solution).

A number of period-doubling bifurcation points has been successfully computed in this way. Some of them are reported in Figs 1 - 3 .

CASCADE OF PERIOD - DOUBLING BIFURCATIONS. FEIGENBAUM SEQUENCE. High accuracy of computed period-doubling bifurcation points enables to test the validity of Feigenbaum´s results also for more complicated and continuous dynamic systems. Several period-doubling bifurcation points of the Lorenz model (3) are presented in the Table 1 Results in the Table correspond to a cascade of period-doubling bifurcations, cf. Fig. 3. The values of the parameter r ($= p_1$) at the individual bifurcation points form a Feigenbaum sequence $\{r_j\}$ [2] . The values

TABLE 1 : A cascade of period-doubling bifurcation points in the Lorenz model (3), $\sigma = 16$, $b = 4$, $k = 1$, $x_k = 3.82038$.

j	x_2	x_3	T_j	r_j	δ_j
1	20.90946	273.34849	0.30618	356.93391	
2	16.85987	246.64055	0.63009	338.06197	4.9740
3	21.19530	259.36006	1.26750	334.26789	4.7313
4	17.29002	244.99724	2.53818	333.46599	4.6824
5	17.24223	244.70901	5.07771	333.29472	4.6707
6	17.25889	244.74356	10.15599	333.25806	

$$\delta_j = (r_{j+1} - r_j)/(r_j - r_{j-1}) \quad (17)$$

are presented in the Table, too. We can observe a very good convergence to a limit, which is approximately $\delta^* \sim 4.6692$ [2] .

DIRECTION OF EMANATING BRANCHES. Let us have a period-doubling bifurcation point (x^*,T^*,p^*) , determined, e.g., by the algorithms described above. Let us seek periodic solutions with the period approximately equal to $2T^*$ in the neighbourhood of (x^*,T^*,p^*) . Therefore, we define a nonlinear system

$$G(x,T,p) = \Psi[\Psi(x,T,p),\ T,\ p] - x = 0 \quad (18)$$

for the unknowns $x_1,\ldots, x_{k-1}, x_k,\ldots, x_n$, T and the parameters p . The system (18) has a bifurcation (crossection) point at (x^*,T^*,p^*) . There are two branches which intersect at this point. One branch is a branch of "composed" periodic solutions obtained by a composition of two original periodic solutions on a known branch. The second branch is the bifurcated branch of solutions with a double period. Directions of the branches can be evaluated by the algorithm described in [17] . Let us note that we need second derivatives of G (computed,e.g., by finite differences). Directions of branches resulting for the first period-doubling bifurcation point from the Table 1 are presented in the Table 2 together with starting points used for the continuation of the bifurcated branch. More detailed description will be presented in [16] .

TABLE 2 : Directions of branches emanating from the first bifurcation point in the Table 1.

Direction on original branch	Direction on bifurcated branch	Starting point ($\Delta x_2 = 0.1$)
$\frac{dx_2}{dr} = 0.1262$	$\frac{dx_3}{dx_2} = 2.9431$	$x_2 = 21.0095$ (+) * $x_3 = 273.643$ (+)
$\frac{dx_3}{dr} = 1.0042$	$\frac{dT}{dx_2} = 0.56E-7$	$T = 0.30617$ (+) (period = 2T)
$\frac{dT}{dr} = -0.43E-3$	$\frac{dr}{dx_2} = -0.17E-3$	$r = 356.93$ (-)

* sign of the change of individual variables for starting continuation (direction parameters).

3.3 TORI BIFURCATIONS

The monodromy matrix M has the eigenvalues

$$\mu_{1,2} = a \pm ib \quad , \quad a^2 + b^2 = 1 \quad , \quad \mu_{1,2}^s \neq 1 \quad , \quad s = 1, 2, 3, 4 \quad . \qquad (19)$$

Decomposition of the characteristic polynomial (13) gives

$$P(\mu) = (\mu^2 - \omega\mu + 1)(\mu^{n-2} + b_1 \mu^{n-3} + \ldots + b_{n-2}) + C\mu + D \quad , \qquad (20)$$

where $\omega = 2a$ and $(\mu^2 - \omega\mu + 1) = (\mu - \mu_1)(\mu - \mu_2)$. The coefficients $b_1, \ldots, b_{n-2}$, C, D can be again evaluated recurrently similarly as above. Two additional equations

$$F_{n+1}(x,T,p,\omega) = C = 0 \quad , \quad F_{n+2}(x,T,p,\omega) = D = 0 \qquad (21)$$

have to be fulfilled at the tori bifurcation point. As a result we have $n + 2$ equations (7) (21) for $n + 2$ unknowns $x_1, \ldots, x_{k-1}, x_{k+1}, \ldots, x_n$, T, p, ω . The Newton method can be used to solve this nonlinear system.

A modified method makes use of the fact that $\mu = 1$ is an eigenvalue of M , i.e., the decomposition of $P(\mu)$ is in the form

$$P(\mu) = (\mu^3 - (2a + 1)\mu^2 + (2a + 1)\mu - 1)(\mu^{n-3} + b_1\mu^{n-2} + \ldots + b_{n-3}) + $$
$$+ C\mu^2 + D\mu + E \quad . \qquad (22)$$

Coefficients $b_1, \ldots, b_{n-3}$, C, D, E can be evaluated recurrently and the Newton method is used for the solution of the $n + 2$ by $n + 2$ nonlinear system (8) (21). $E = 0$ automatically at the resulting tori bifurcation point.

Resulting tori bifurcation points for the model (4) are shown in the Table 3, cf. Fig. 2 for $B = 5.5$. The parameter $D (= p_1)$ has been considered as a bifurcation parameter. If we continue tori bifurcation points in dependence on another parameter of the problem, here, e.g., $B (= p_2)$, we obtain so called bifurcation diagram [19] . Results of one such continuation are presented in Fig. 4 . The points where $\mu^3 = 1$ or $\mu^4 = 1$ are denoted 3T or 4T , respectively. The curve ends at the point where $\mu_1 = \mu_2 = \mu_3 = 1$.

4. DISCUSSION AND REMARKS

The algorithm for the continuation of periodic solutions can be used also for parabolic partial differential equations when these are transformed into a set of ordinary differential equations by using a semidiscretization (method of lines) [12] .

TABLE 3 : Resulting tori bifurcation points for (4), $A = 2$, $\rho = 0.1$, $k = 1$, $x_1 = 2$.

B	x_2	x_3	x_4	T	a	D
5.3	2.43727	1.90236	2.49646	7.15820	- 0.01694	0.048626
5.4	2.45219	1.92098	2.50011	7.62984	- 0.36244	0.051152
5.5	2.47887	1.92659	2.52457	8.32408	- 0.54139	0.052495
5.6	2.51478	1.92259	2.56374	9.19639	- 0.12304	0.053030

An algorithm for evaluation of Hopf bifurcation points in parabolic equations has been published recently [15] . The algorithms for evaluation of period-doubling and tori bifurcation points can be easily used for most autonomous dynamic systems of lower order, say $n \ll 20$. Of course, the use of the algorithms is limited by the applicability of the shooting method (stability). Simple modifications of the algorithms can be used also for a nonautonomous system with a time-periodic right hand side.

REFERENCES

1. Becker K.H., Seydel R. : Lect. Notes in Math. 878, Springer Verlag, Berlin 1981, p. 99.
2. Feigenbaum M.J. : J. Stat. Phys. 6, 669 (1979).
3. Holodniok M., Kubíček M. : Computation of period doubling bifurcation points in O.D.E. Preprint, Tech. Univ. München, M-8406 (1984).
4. Holodniok M., Kubíček M. : J. Comput. Phys. 55, 254 (1984).
5. Holodniok M., Kubíček M. : Appl. Math. Comput. 15, 261 (1984).
6. Holodniok M., Kubíček M. : Continuation of periodic solutions in ordinary differential equations with application to the Hodgkin-Huxley model, Inter. Symp. of Numerical Analysis, Madrid, 17. - 19. 9. 1985 .
7. Holodniok M., Kubíček M., Marek M. : Stable and unstable periodic solutions in the Lorenz model. Preprint, Tech. Univ. München, M-9217 (1982).
8. Iooss G., Joseph D.D. : Elementary stability and bifurcation theory. Springer Verlag, New York 1981.
9. Jepson A.D., Keller H.B. : in [20] , p. 219.
10. Keller H.B. : in "Applications of Bifurcation Theory", Ed. by P. Rabinowitz, Academic Press, New York, 1977, p. 359.
11. Keller H.B. : in "Recent Advances in Numerical Analysis", Ed. by C. de Boor, G.H. Golub, Academic Press, New York 1978, p. 73.
12. Knedlík P., Holodniok M., Kubíček M., Marek M. : Periodic solutions in reaction--diffusion problems, 7th CHISA Congress, Prague 1984.
13. Kubíček M. : Appl. Math. Comput. 1, 341 (1975).
14. Kubíček M. : ACM TOMS 2, 98 (1976).
15. Kubíček M., Holodniok M. : Chem. Eng. Sci. 39, 593 (1984).
16. Kubíček M., Klíč A., Holodniok M. : in preparation.
17. Kubíček M., Klíč A. : Appl. Math. Comput. 13, 125 (1983).
18. Kubíček M., Marek M. : Appl. Math. Comput. 5, 253 (1979).
19. Kubíček M., Marek M. : Computational methods in bifurcation theory and dissipative structures. Springer Verlag, New York 1983.
20. Küpper T., Mittelmann H.D., Weber H., Eds. : Numerical methods for bifurcation problems. Birkhäuser, Basel 1984.
21. Lorenz E.N. : J. Athmosph. Sci. 20, 130 (1963).
22. Mittelmann H.D., Weber H., Eds. : Bifurcation problems and their numerical solution. Birkhäuser, Basel 1980.
23. Nicolis G., Prigogine I. : Self-organization in nonequilibrium systems. J. Wiley, New York 1977.
24. Rheinboldt W.C., Burkardt J.V. : ACM TOMS 9, 215 (1983).
25. Schreiber I., Holodniok M., Kubíček M., Marek M. : J. Stat. Phys., to be published
26. Seydel R. : Numer. Math. 32, 51 (1979).
27. Sparrow C. : The Lorenz Equations : Bifurcations, Chaos and Strange Attractors. Springer Verlag, Berlin 1982.

THE ROTHE METHOD FOR NONLINEAR HYPERBOLIC PROBLEMS

E. MARTENSEN
Mathematisches Institut II, Universität Karlsruhe
7500 Karlsruhe 1, West Germany

The ROTHE method or the horizontal method of lines, if it is applied to parabolic as well as to hyperbolic evolution problems, reduces these problems to a sequence of elliptic problems. That from a former point of view, such an approach has appeared more natural in the case of parabolic than of hyperbolic problems, may serve as an explanation for the considerable delay of time in studying the method for both classes of problems. So after ROTHE [11] has introduced his method in the early thirties of our century, numerous parabolic differential equation problems, linear as well as nonlinear ones, have been treated by it successfully; the names of LADYSHENSKAJA, REKTORYS, NEČAS, and KAČUR may stand here for many others (references, for instance, may be seen from the book of REKTORYS [10]). On the other hand, efforts for applying the ROTHE method to hyperbolic problems firstly have been started during the last decade. Results have been given mainly for certain linear problems of mathematical physics, so as for the wave equation [1,2,5], the continuity equation [3], and the MAXWELL equations [4]; recently the vibrating string problem with discontinuous data has been completely solved by the ROTHE method [6]. Further linear hyperbolic problems have been investigated by REKTORYS [10]. With regard to nonlinear hyperbolic problems, however, one is standing at the very beginning. First results of MUNZ [8,9] concerning the quasilinear scalar conservation equation, especially have shown the ROTHE method as a suitable tool for approximation of shocks and rarefaction waves.

In the following we shall consider the CAUCHY problem for the BURGERS equation

$$u_t + \frac{1}{2}(u^2)_x = 0 \quad , \quad (x,t) \in \mathbb{R} \times (0,\infty) \quad , \tag{1}$$

where the initial values

$$u(x,0) = u_o(x) \quad , \quad x \in \mathbb{R} \quad , \tag{2}$$

are assumed to be piecewise continuous with at most a finite number of discontinuities and existing limits for $x \to \mp\infty$. The ROTHE method for a fixed chosen time step length $h > 0$ leads to the ordinary differential equation

$$u + \frac{h}{2}(u^2)' = u_o \qquad x \in \mathbb{R} \tag{3}$$

which for given $u_o(x)$, $x \in \mathbb{R}$, has to be solved successively according to the next time step. On the solutions $u(x)$, $x \in \mathbb{R}$, of (3), there are imposed piecewise continuity with at most a finite number of discontinuities and existing limits for $x \to \mp\infty$; furthermore, the square $[u(x)]^2$, $x \in \mathbb{R}$, is asked as a piecewise continuously differentiable function (as a consequence of the foregoing, for the derivative there may occur at most a finite number of discontinuities). Without mentioning in detail, the following assertions will concern to solutions of (3) having at least these properties.

Theorem 1 (Behaviour at the infinity). For a solution $u(x)$, $x \in \mathbb{R}$, of (3) it holds

$$\lim_{x\to\mp\infty} u(x) = \lim_{x\to\mp\infty} u_o(x) \quad . \tag{4}$$

Proof follows immediately from (3) in connection with the second L'HOSPITAL rule:

$$0 = \lim_{x\to\mp\infty} \frac{h[u(x)]^2}{2x} = \lim_{x\to\ \infty} \left\{\frac{h}{2}\frac{d}{dx}[u(x)]^2\right\} = \lim_{x\to\mp\infty} u_o(x) - \lim_{x\to\mp\infty} u(x) \quad .$$

Remark. The proof of Theorem 1 makes only use of the conservation property of the underlying partial differential equation (1). Thus the accordance of the limits (4) will be obtained analogously for other hyperbolic problems when they are given in conservation form. For instance, this holds for the EULER equations.

Theorem 2 (Global uniqueness). There exists at most one continuous solution $u(x)$, $x \in \mathbb{R}$, of (3).

Proof. Assuming that there exist two different continuous solutions $u(x), v(x)$, $x \in \mathbb{R}$, so the continuous function $w(x) := u(x) - v(x)$, $x \in \mathbb{R}$, does not vanish everywhere. Note that because of Theorem 1 it holds

$$\lim_{x\to\mp\infty} w(x) = \lim_{x\to\mp\infty} u(x) - \lim_{x\to\mp\infty} v(x) = 0 \tag{5}$$

Let now $x_o \in \mathbb{R}$ be a point with $w(x_o) \neq 0$. If $w(x)$ has at least one zero in the open interval $(-\infty, x_o)$, then for continuity there exists a maximum zero in this interval and we denote it by $a < x_o$; if, however, there are no zeroes in $(-\infty, x_o)$, we put $a = -\infty$. Analogously let $b > x_o$ denote the minimum zero for $w(x)$ in (x_o, ∞) or stand for ∞, respectively. Together with (5) we get

$$\lim_{x\to a} w(x) = \lim_{x\to b} w(x) = 0 \quad . \tag{6}$$

Observing the continuity and piecewise continuous differentiability of $[u(x)]^2, [v(x)]^2$, $x \in \mathbb{R}$, it follows from (3) and (6) by improper integration that

$$\int_a^b w(x)\,dx = \int_a^b \{u(x) - v(x)\}\,dx = -\frac{h}{2}\int_a^b \left\{\frac{d}{dx}[u(x)]^2 - \frac{d}{dx}[v(x)]^2\right\} dx$$

$$= -\frac{h}{2}\Big[[u(x)]^2 - [v(x)]^2\Big]_a^b = -\frac{h}{2}\Big[w(x)(u(x)+v(x))\Big]_a^b = 0$$

This is a contradiction to $w(x) \not\equiv 0$, $x \in (a,b)$.

Theorem 3 (Local uniqueness). Let $(a,b) \subseteq \mathbb{R}$ be an arbitrary finite or infinite open interval and let the above ordinary differential equation problem be formulated analogously for (a,b) instead of $\mathbb{R}$. Let further $u(x), x \in (a,b)$, be a positive continuous (negative continuous) solution of (3) which has a positive limit for $x \to a$ (negative limit for $x \to b$). Then there does not exist another continuous solution of (3) with the same limit for $x \to a$ ($x \to b$).

Proof only for the first case. Assume that there exists a continuous solution $v(x)$, $x \in (a,b)$, different from $u(x)$, $x \in (a,b)$, but with the same limit for $x \to a$. Then the difference $w(x) := u(x) - v(x)$, $x \in (a,b)$ forms a continuous function satisfying

$$\lim_{x \to a} w(x) = 0 \quad . \tag{7}$$

Next we are able to find a point $x_o \in (a,b)$ with properties

$$w(x_o) \not\equiv 0 \quad , \quad u(x_o) + v(x_o) \geqq 0 \quad . \tag{8}$$

Indeed, if $u(x) + v(x)$, $x \in (a,b)$, has no zeroes, from continuity and

$$\lim_{x \to a} \{u(x) + v(x)\} = 2 \lim_{x \to a} u(x) > 0$$

it follows that $u(x) + v(x) > 0$, $x \in (a,b)$, and so it is trivial to find $x_o \in (a,b)$ satisfying (8); if, however, $u(x) + v(x)$, $x \in (a,b)$, has a zero $x_o \in (a,b)$, so this zero immediately fulfills the second condition in (8) and the first condition follows from $v(x_o) = -u(x_o)$ as

$$w(x_o) = u(x_o) - v(x_o) = 2\,u(x_o) > 0$$

Now we denote by $a^* < x_o$ the maximum zero for $w(x)$ in the open interval (a,x_o) if there exists a zero at all, otherwise we put $a^* = a$. So in any case when observing (7), we get

$$\lim_{x \to a^*} w(x) = 0 \quad . \tag{9}$$

Then by improper integration, it follows from (3) and (9) that

$$\int_{a^*}^{x_o} w(x)\,dx = \int_{a^*}^{x_o} \{u(x) - v(x)\}\,dx = -\frac{h}{2}\int_{a^*}^{x_o} \left\{\frac{d}{dx}[u(x)]^2 - \frac{d}{dx}[v(x)]^2\right\} dx$$

$$= -\frac{h}{2}\Big[w(x)(u(x) + v(x))\Big]_{a^*}^{x_o} = -\frac{h}{2}\, w(x_o)(u(x_o) + v(x_o))$$

here because of (8), we have the contadiction, that the left hand side has the sign of $w(x_o) \neq 0$ whilst the right hand side either has the opposite sign or vanishes.

Remark 1. Theorem 3 gives a hint how to proceed for solving the differential equation (3) uniquely. So if starting at some point with a positive or negative initial value, one has to integrate to the right or to the left, respectively. On the other hand, the sign of the exact solution analogously indicates the direction of the characteristics. So it turns out that local uniqueness for the ROTHE solution is assured by integrating into the direction of characteristics.

Remark 2. As it can be seen from the example $u_o(x) = 1$, $x \in \mathbb{R}$, the sign condition in Theorem 3 plays a significant role. So the solution $u(x) = 1$, $x \in \mathbb{R}$, is the only one of (3) with limit 1 for $x \to -\infty$, but there exist an infinite number of further solutions with limit 1 for $x \to \infty$; indeed, with an arbitrary real constant C, such a solution $u(x)$, $x \in \mathbb{R}$, may be obtained as the inverse of the monotonously decreasing function

$$x(u) = -h\,\{u + \ln(u-1)\} + C \quad , \quad u \in (1,\infty) \quad .$$

We shall make use of the foregoing theorems when discussing the following four examples.

Example 1 (MUNZ [8]). If $u_o(x)$, $x \in \mathbb{R}$, is the step function with value 2 for negative or 1 for positive x, respectively, the exact solution $u(x,t), (x,t) \in \mathbb{R} \times [0,\infty)$, of the evolution problem (1) and (2) is given as a shock wave at $x = \frac{3}{2}t$ with value 2 left or 1 right of the shock, respectively. Assume that for an arbitrary time step a ROTHE solution exists which, for convenience, will be denoted by $u_o(x), x \in \mathbb{R}$; besides the general properties mentioned above let this solution be monotonously nonincreasing with lower bound 1, let it have the value 2 for $x \in (-\infty,0)$, and let it be continuous for $x \in (0,\infty)$. Note that for such solution the limits for $x \to \mp\infty$ exist and that everything holds for the given initial function. The next ROTHE step $u(x)$, $x \in \mathbb{R}$, then may be computed from (3) as a continuous solution with value 2 for $x \in (-\infty,0)$ for $x \in [0,\infty)$ the solution follows by means of the initial condition $u(0) = 2$ in connection with the lower function $u_o(x)$ and the upper function 2. This especially yields

$$1 \le u_o(x) \le u(x) \quad , \quad x \in (0,\infty) \tag{10}$$

From the differential equation (3) together with (10) it follows that $u'(x) \le 0$, $x \in (0,\infty)$; so $u(x)$, $x \in \mathbb{R}$, is monotonously nonincreasing and because of (10), it has the lower bound 1. Theorem 2 as well as the first case of Theorem 3 say that there is no further continuous solution, so the next ROTHE step is well-defined. Finally by induction, all ROTHE solutions are uniquely determined. Because of Theorem 1, for every ROTHE solution the limit 2 for $x \to -\infty$ or 1 for $x \to \infty$ is obtained, respectively.

Example 2 (MUNZ [8]). Here $u_o(x)$, $x \in \mathbb{R}$, is considered as a step function with value 1 for negative or 2 for positive x, respectively. The exact solution is a rarefaction wave with values $\frac{x}{t}$ for $t \le x \le 2t$, $0 < t < \infty$ and value 1 left or 2 right of the wave, respectively. As it turns out quite similarely to Example 1, the ROTHE method again can be carried out uniquely.

Example 3 (MARTENSEN [7]). The initial values $u_o(x)$, $x \in \mathbb{R}$, are given as -1 for negative or 1 for positive x, respectively. The exact solution is a rarefaction wave with values $\frac{x}{t}$ for $-t \le x \le t$, $0 < t < \infty$ and value -1 left or 1 right of the wave, respectively. Evidently Theorem 3 is not applicable with respect to both the infinities. If beginning with the first time step, the ROTHE solutions $u(x)$, $x \in \mathbb{R}$, are further asked to be continuous, monotonously increasing, and skew-symmetric with respect to the origin, then such solutions can be constructed successively by means of a fixed point method. Uniqueness is now assured by Theorem 2. As a secondary result it turns out that all the ROTHE solutions (contrarily to their squares) are not from each side differentiable at the origin.

Example 4 (MUNZ [9]). If $u_o(x)$, $x \in \mathbb{R}$, has the value 2 for negative or -1 for positive x, respectively, the exact solution is obtained as a shock wave at $x = \frac{1}{2}t$ with value 2 left or -1 right of the shock, respectively. For the piecewiese continuous ROTHE solution $u(x)$, $x \in \mathbb{R}$, beginning with the first time step, the further supposition is made that the square $[u(x)]^2$, $x \in \mathbb{R}$, remains continuous when passing through a discontinuity; in such a way there is made use of the conservation property governing the ROTHE differential equation (3). In particular, with a well-defined discontinuity $x^* \in (0,\infty)$, the ROTHE solution $u(x), x \in \mathbb{R}$ is obtained with constant value 2 for $x \in (-\infty,0)$, as a monotonously decreasing solution of the differential equation (3) for $x \in [0,x^*]$ satisfying the initial condition $u(0) = 2$ and the free boundary condition

$u(x^*) = 1$, and with constant value -1 for $x \in (x^*,\infty)$. Here Theorem 3 leads to local uniqueness for the left interval $(-\infty,x^*)$ as well as for the right one (x^*,∞); furthermore by means of Theorem 3, this ROTHE solution turns out to be the only one with exactly one discontinuity whilst a continuous solution does not exist. With regard to the complete ROTHE method, the discontinuities form a monotonously increasing sequence.

For the examples mentioned before numerical computations have been done by standard methods, where the results have shown a high accuracy in comparison with the exact solutions [7,8,9]. Recently for such nonlinear hyperbolic problems the L_1-convergence of the ROTHE method with respect to any compactum in the upper (x,t)-plane has been proved [9]. The pointwise convergence, however, remains still as an open question.

References

[1] Gerdes, W.; Martensen, E.: Das Rotheverfahren für die räumlich eindimensionale Wellengleichung. ZAMM 58 (1978) T367-T368

[2] Halter, E.: Das Rotheverfahren für das Anfangs-Randwertproblem der Wellengleichung im Außenraum. Dissertation, Karlsruhe 1979

[3] Halter, E.: The convergence of the horizontal line method for the continuity equation with discontinuous data. ZAMP 35 (1984) 715-722

[4] Martensen, E.: The convergence of the horizontal line method for Maxwell's equations. Math. Methods Appl. Sci. 1 (1979) 101-113

[5] Martensen, E.: The Rothe method for the wave equation in several space dimensions. Proc. Roy. Soc. Edinburgh 84A (1979) 1-18

[6] Martensen, E.: The Rothe method for the vibrating string containing contact discontinuities. Meth. Verf. math. Phys. 26 (1983) 47-67

[7] Martensen, E.: Approximation of a rarefaction wave by discretization in time. Applications of Mathematics in Technology, V. Boffi and H. Neunzert eds. Stuttgart: Teubner 1984, 195-211

[8] Munz, C.-D.: Über die Gewinnung physikalisch relevanter Stoßwellenlösungen mit dem Rotheverfahren. Dissertation, Karlsruhe 1983

[9] Munz, C.-D.: Approximate solution of the Riemann problem for the Burgers equation by the transversal method of lines. To appear in ZAMP

[10] Rektorys, K.: The Method of Discretization in Time and Partial Differential Equations. Dordrecht/Boston/London: Reidel Publishing Company 1982

[11] Rothe, E.: Zweidimensionale parabolische Randwertaufgeben als Grenzfall eindimensionaler Randwertaufgaben. Math.Ann. 102 (1930) 650-670

SOME SOLVED AND UNSOLVED CANONICAL PROBLEMS OF DIFFRACTION THEORY

E. MEISTER
Technical University Darmstadt
Schlosgartenstr. 7, D 6100 - Darmstadt, West Germany

1. Introduction

Mathematical diffraction theory is concerned with the following boundary value problem in case of an incoming or primary time-harmonic wave-field $\mathrm{Re}[\Phi_{pr}(\underline{x})e^{-i\omega t}]$:

Given an obstacle $\Omega \subset R^n$; $n = 2$ or 3; with boundary $\Gamma = \partial\Omega$. Find the scattered field $\Phi_{sc}(\underline{x})$ in $\Omega_a := R^n - \overline{\Omega}$, s.th.

(1.1) $(\Delta + k^2)\Phi_{sc}(\underline{x}) = 0$ for $x \in \Omega_a$

with a wave-number $k = k_1 + ik_2 \in C_{++} - \{0\}$ fulfilling a boundary condition

(1.2a) $B_1[\Phi_{sc}(\underline{x})]|_\Gamma := \Phi_{sc}(\underline{x})|_\Gamma = f(\underline{x})$ of Dirichlet-type

or

(1.2b) $B_2[\Phi_{sc}(\underline{x})]|_\Gamma := (\frac{\partial}{\partial n} + i\, p(\underline{x}))\Phi_{sc}(\underline{x})|_\Gamma = g(\underline{x})$

of $\left\{ \begin{matrix} \text{Neumann } (p \equiv 0) \\ \text{Impedance } (p \not\equiv 0) \end{matrix} \right\}$ - type.

In the case of edges E and/or vertices $V \subset \Gamma$ existing the "edge condition"

(1.3) $\Phi_{sc}(\underline{x}) = O(1)$ and $\nabla\Phi_{sc}(\underline{x}) \in L^2_{loc}(\Omega_a)$

should hold. Besides this the scattered field should be "outgoing", i.e. "Sommerfeld's radiation conditions" should hold

(1.4) $\Phi_{sc}(\underline{x}) = \vartheta(e^{-k_2 r})$, $(\frac{\partial}{\partial r} - i.k)\Phi_{sc}(\underline{x}) = \vartheta(e^{-k_2 r}/r^{\frac{n-1}{2}})$

as $r = |\underline{x}| \to \infty$

For smooth compact boundaries Γ this problem has completely been solved, e.g. by the boundary integral equation method (BEM) (c.f. e.g. COLTON-KRESS (1983) [2]) or by means of Sobolev space methods (c.f. e.g. LEIS (1985) [11]). Generalizations to piecewise smoothly bounded domains were carried out by GRISVARD (1980) [6] and COSTABEL (1984) [4], e.g.

2. The Sommerfeld Half-Plane Problem

There are a number of "canonical diffraction problems" with domains whose boundaries extend to infinity and having corners and cusps. The most famous one is the "Sommerfeld half-plane problem", the first diffraction problem having been treated in a mathematically rigorous way (1896) [15].

Applying the well-known representation formula for outgoing solutions of the Helmholtz equation (1.1) the Sommerfeld half-plane problems leads to the following integral or integro-differential equations (of the first kind) of the Wiener-Hopf type:

$$(2.1)\qquad \int_0^\infty H_0^{(1)}(k|x-x'|)\, I(x')dx' = -4i.\Phi_{pr}(x,0) \quad \text{for} \quad x \geq 0$$

in the case of the Dirichlet problem and

$$(2.2)\qquad \left(\frac{d^2}{dx^2} + k^2\right) \int_0^\infty H_0^{(1)}(k|x-x'|)\, Q(x')dx' = 4i\, \frac{\partial\Phi_{pr}}{\partial y}(x,0) \quad \text{for } x > 0$$

in the case of the Neumann problem with the unknown jumps

$$(2.3)\qquad I(x') := \frac{\partial\Phi_{sc}}{\partial y}(x',+0) - \frac{\partial\Phi_{sc}}{\partial y}(x',-0) \qquad \text{for } x' > 0$$

and

$$(2.4)\qquad Q(x') := \Phi_{sc}(x',+0) - \Phi_{sc}(x',-0) \qquad \text{for } x' \geq 0\,,$$

respectively.

The theory of such equations, but of the second kind, in $L^p(\mathbb{R}_+)$ or $W^{m,p}(\mathbb{R}_+)$-spaces for $m \in \mathbb{N}_0$, $1 \leq p \leq \infty$ has been developed by M.G. KREIN (1958/62) [9], E.Gerlach (1969) [5] and, combined with other integral operators than 1-convolutions, by G.THELEN (1985) [17].

To solve the equations (2.1) or (2.2) on the half-line, or more directly the original boundary value problem, one applies a one-dimensional Fourier transform to the scattered wave function

$$(2.5)\qquad \hat{\Phi}_{sc}(\lambda,y) := \int_{-\infty}^\infty e^{i\lambda x}\Phi_{sc}(x,y)dx, \quad \lambda \in \mathbb{R}, \quad y \lessgtr 0\,.$$

The usual, or S'-distributional Fourier transform technique leads to the following "function-theoretic Wiener-Hopf equations", in the case of a damping medium, i.e. $\operatorname{Im} k = k_2 > 0$, and an incoming plane wave:

$$(2.6)\qquad \hat{E}_-(\lambda) + \tfrac{1}{2}\,\hat{I}_+(\lambda)/\sqrt{\lambda^2 - k^2} = [i(\lambda + k\cos\theta)]^{-1}$$

and

$$(2.7)\qquad \hat{V}_-(\lambda) + \tfrac{1}{2}\,\hat{Q}_+(\lambda).\sqrt{\lambda^2 - k^2} = -k\sin\theta\,[\lambda + k\cos\theta]^{-1}\,,$$

respectively, for the Dirichlet and Neumann case with the unknown F-transforms $\hat{E}_-$, $\hat{V}_-$ being holomorphic for $\operatorname{Im}\lambda < k_2$ and I_+, Q_+ being holomorphic for $\operatorname{Im}\lambda > -k_2\cos\theta$. The equations (2.6) and (2.7) are equivalent to "non-normal Riemann boundary value problems on a line" parallel to the real λ-axis.

The well-known steps of factorization of $\gamma(\lambda) := \sqrt{\lambda^2 - k^2}$ into

$\gamma_+(\lambda).\gamma_-(\lambda)$, the multiplication of (2.6) and (2.7) by γ_- and by γ_-^{-1}, respectively, then additive decomposition of $\gamma_-.[\lambda + k\cos\theta]^{-1}$ and $\gamma_-^{-1}.[\lambda + k\cos\theta]^{-1}$ in the λ-strip gives after rearrangement and application of Liouville's theorem the explicite solutions to eqs. (2.6) and (2.7) as

$$(2.8)\qquad \hat{I}_+(\lambda) = 2\sqrt{2k}\cos\theta/2.\gamma_+(\lambda)[\lambda + k\cos\theta]^{-1}$$

and

$$(2.9)\qquad \hat{Q}_+(\lambda) = -2i\sqrt{2k}\sin\theta/2.\gamma_+^{-1}(\lambda)[\lambda + k\cos\theta]^{-1}$$
$$\text{for } \operatorname{Im}\lambda > -k_2\cos\theta\ .$$

These functions being known allow to calculate $\Phi_{sc}(x,y)$ in both cases after applying an inverse F-transform and shifting the line of integration in the complex λ-plane to get all informations relevant, i.e. the edge behaviour an the far field in the geometrically different regions.

This functiontheoretic method has been applied successfully to a big number of canonical problems in microwave theory and to other diffraction problems, e.g. for systems of parallel semi-infinite plates (A.E.Heins (1948) [7]), or cascades of such (J.F. Carlson, A.E. Heins (1946/50) [1]), or cylindrical semi-infinite pipes (e.g. L.A. Vajnshtejn (1948) [18]).

The "canonical mixed Sommerfeld half-plane problems", where there are given different boundary conditions on the faces $\delta_\pm$ of the semi-infinite screen $\delta := \{(x,y) \in \mathbb{R}^2 : y = 0,\ x \geq 0\}$, may be transformed by the same Fourier technique into a 2×2-functiontheoretic system of Wiener-Hopf equations

$$(2.10)\qquad \underline{\hat{\Phi}}_-(\lambda) = \underline{\underline{K}}(\lambda)\underline{\hat{\Phi}}_+(\lambda) + \underline{\hat{r}}(\lambda) \quad \text{for } -k_2\cos\theta < \operatorname{Im}\lambda < k_2$$

with the known 2×2-function matrix

$$(2.11)\qquad \underline{\underline{K}}(\lambda) := \begin{pmatrix} \sqrt{(\lambda-k)/(\lambda+k)} & 1 \\ -1 & \sqrt{(\lambda+k)/(\lambda-k)} \end{pmatrix}$$

and the unknown 2×1-function-vectors

$$(2.12)\qquad \underline{\hat{\Phi}}_-(\lambda) := \begin{pmatrix} \sqrt{\lambda-k}\ .\ \hat{E}_-(\lambda) \\ \hat{V}_-(\lambda)/\sqrt{\lambda-k} \end{pmatrix},$$

$$\underline{\hat{\Phi}}_+(\lambda) := -\frac{1}{2}\begin{pmatrix} \sqrt{\lambda+k}\ .\ \hat{\Phi}_+(\lambda,-0) \\ \hat{\Phi}'_+(\lambda,+0)/\sqrt{\lambda+k} \end{pmatrix}.$$

The matrix $\underline{\underline{K}}(\lambda)$ - or a closely related one - has been factorized into $[\underline{\underline{K}}_-(\lambda)]^{-1}\ \underline{\underline{K}}_+(\lambda)$ only (1982/83) by A.E.Heins [8], (1981) by A.D.Rawlins [14] and (1981/85) by the present author [12], independently by different methods. Now the solution of the mixed Sommerfeld problem may be written down explicitely and gives full information on the behaviour of Φ_{sc}, $\nabla\Phi_{sc}$ as $r \to 0$ and $r \to \infty$, respectively, which is now different at the edge compared to the one-boundary-condition-problems. The corresponding mixed boundary value problems for systems of parallel semi-infinite plates or a tube are unsolved up to now due to the lack of a known explicit factorization of the 2×2-function matrices involved (c.f. e.g. the authors paper (1984/85) [12]!).

The Sommerfeld half-plane problems have been generalized to the so called "Quarter-plane Problems of Diffraction Theory" where the half-plane, i.e. the screen $\delta \subset \mathbf{R}^3$, is replaced by a screen $\Sigma \subset \mathbf{R}^3$ which is the quarter-plane $\mathbf{R}^2_{++} := \{(x,y,z) \in \mathbf{R}^3 : z = 0,\ x \geq 0,\ z \geq 0\}$ with two semi-infinite lines as edges meeting in the corner E at the origin. Like for an arbitrary plane screen $\Sigma \subset \mathbf{R}^2_{xy}$ the 2-dimensional F-transform applied to the unknown scattered field $\Phi_{sc}(\underline{x})$, $\underline{x} \in \mathbf{R}^3$, leads to the following "Two-dimensional Wiener-Hopf functional equations"

(2.13) $\gamma^{-1}(\lambda_1,\lambda_2)\hat{I}_\Sigma(\lambda_1,\lambda_2) - \hat{\Phi}_{\mathbf{R}^2\setminus\Sigma}(\lambda_1,\lambda_2,0) = -\Phi_{pr,\mathbf{R}^2\setminus\Sigma}(\lambda_1,\lambda_2,0)$

and

(2.14) $\gamma(\lambda_1,\lambda_2)\hat{Q}_\Sigma(\lambda_1,\lambda_2) - (\frac{\partial}{\partial z}\Phi)\hat{}_{\mathbf{R}^2\setminus\Sigma}(\lambda_1,\lambda_2,0) = -(\frac{\partial}{\partial z}\Phi_{pr})_{\mathbf{R}^2\setminus\Sigma}(\lambda_1,\lambda_2,0)$

where $\gamma(\lambda_1,\lambda_2) := \sqrt{\lambda_1^2 + \lambda_2^2 - k^2}$ and the indices Σ and $\mathbf{R}^2\setminus\Sigma$ refer to the 2D-F-transforms of the restrictions to Σ and $\mathbf{R}^2\setminus\Sigma$, respectively.

Up to now there exists no explicit factorization of the multiplication operator γ with respect to the complementary projectors $\hat{P}_\Sigma$, $\hat{Q}_\Sigma := I - \hat{P}_\Sigma$ in spaces $FL^p(\mathbf{R}^2)$ or $FW^{s,p}(\mathbf{R}^2)$, $s > 0$, $1 < p \leq 2$ (∞). But there exists now a very general theory for "general Wiener-Hopf or Toeplitz operators" of the form

(2.15) $P_2A|_{P_1X}\, u = v \in P_2Y$

for bijective continuous operators $A : X \to Y$ acting between two Banach-spaces X,Y with bounded projectors $P_1 \in \mathscr{L}(X)$, $P_2 \in \mathscr{L}(Y)$. This theory by F.-O.Speck (1983/85) [16] gives necessary and sufficient conditions for the general invertibility and Fredholm property of operators of type (2.15) in dependance on factorization properties of A w.r.t. (P_1,P_2).

3. Canonical Transmission Problems

Another big class of canonical diffraction problems exists given by the following specification:

Given a primary time-harmonic wave-field $\mathrm{Re}[\Phi_{pr}(\underline{x})e^{-i\omega t}]$ and a region $\Omega_1 \subset \mathbf{R}^n$, $n = 2$ or 3, and finitely many disjoint regions $\Omega_2,\dots,\Omega_N \subset \mathbf{R}^n$, s. th. $\bigcup_{j=1}^{N} \bar{\Omega}_j = \mathbf{R}^n$.. Then one looks for a scattered field $\Phi_{sc}(\underline{x})$, $x \in \mathbf{R}^n$, s. th. $\Phi_{sc}(\underline{x})|_{\Omega_j} \in C^2(\Omega_j) \cap C^1(\bar{\Omega}_j\setminus\{0\})$ and

(3.1) $(\Delta + k_j^2)\Phi_{sc}(\underline{x}) = 0$ in Ω_j, $j = 1,\ldots,N$,

fulfilling the "transmission conditions"

(3.2a) $\Phi_{sc,j}(\underline{x}) - \Phi_{sc,t}(\underline{x}) = F_{jl}(\underline{x})$

and on $\partial\Omega_j \cap \partial\Omega_l \neq \phi$

(3.2b) $\rho_j \cdot \frac{\partial \Phi_{sc,j}}{\partial n_j}(\underline{x}) + \rho_l \cdot \frac{\partial \Phi_{sc,l}}{\partial n_l}(\underline{x}) = G_{jl}(\underline{x})$

with prescribed data F_{jl}, G_{jl} from the primary field on the common boundary parts $\partial\Omega_j \cap \partial\Omega_l$.
Additionally the edge conditions $\Phi_j(\underline{x}) = \Phi(x)|_{\Omega_j} = O(1)$ and $\nabla\Phi_j \in L^2_{loc}(\Omega_j)$ and the radiation condition for $\Phi_1(x)$ as $|x| = r \to \infty$ have to hold.

Again in the case of smoothly bounded domains with compact boundaries $\partial\Omega_j$ this "transmission or interface problem" has been solved by the boundary integral method and in the case of two-dimensional polygonal domains by M.Costabel and E.Stephan (1985) [3].

In the special case of two different media (i.e. N = 2) and a plane interface (i.e. $\partial\Omega_1 = \partial\Omega_2 = Rxy$ or $= R_x^1$) the problem is elementary and gives, for a plane wave as the primary wave-function, the well-known relations from Snellius' law and the reflection and transmission coefficients explicitely. The corresponding "two-dimensional Sommerfeld half-plane problems with two media" are unsolved up to now - as far as an explicit representation is concerned - due to the unknown matrix factors of the 2X2-Wiener-Hopf function matrices involved here having two different square roots $\sqrt{\lambda^2 - k_1^2}$ and $\sqrt{\lambda^2 - k_2^2}$ to be taken into account [12].

A very important canonical transmission problem is the so-called "Dielectric Wedge Problem", i.e. the case of $\Omega_1 = R^2_{++}$ and $\Omega_2 = R^2 \setminus R^2_{++}$ in R^2 or the corresponding "Dielectric Octant Problem" in R^3-space: This has been generalized to the "Four-Quadrant-Transmission-Problem" in R^2 with the four quadrants filled with different media. Applying 2D-Fouriertransformation the restrictions of the unknown scattered field may be represented by the 1D - F - transformed Cauchy-data on the semi-infinite lines, the boundaries of the quadrants. For $\hat{\Phi}_1(\lambda_1,\lambda_2)$ one gets e.g.

(3.3) $$\hat{\Phi}_1(\Lambda_1,\lambda_2) = [i\lambda_2 . \hat{f}_1^{(1)}(\lambda_1) + g_1^{(1)}(\lambda_1) + i\lambda_1 . \hat{f}_2^{(1)}(\lambda_2) + \hat{g}_2^{(1)}(\lambda_2)] \cdot (\lambda_1^2 + \lambda_2^2 - k_1^2)^{-1} \text{ for } \mathrm{Im}\,\lambda_1, \mathrm{Im}\,\lambda_2 > -\beta_1, -\beta_2$$

with $\beta_1^2 + \beta_2^2 < (\mathrm{Jm}k_1)^2$.

Due to the transmission conditions (3.2) the total sum of all numerators of the $\hat{\Phi}_j(\lambda_1,\lambda_2)$ is a known function $Z(\lambda_1,\lambda_2)$. Dividing by the known $N(\lambda_1,\lambda_2,k^2)$ with an appropriate $k \in \mathbb{C}_+$ one arrives at the "Four-part Wiener-Hopf functional equation"

$$(3.4) \qquad \sum_{j=1}^{4} \left(1 + \frac{k^2 - k_j^2}{\lambda_1^2+\lambda_2^2-k^2}\right)\hat{P}_j\hat{\Phi}(\lambda_1,\lambda_2) = \frac{Z(\lambda_1,\lambda_2)}{N(\lambda_1,\lambda_2,k^2)}$$

holding for a pair of strips of $\mathbb{C}^2$. Here we have

$$(3.5) \qquad \hat{\Phi}_j(\lambda_1,\lambda_2) := \hat{P}_j\hat{\Phi}(\lambda_1,\lambda_2) := (F_{2 \times Q_j} : F_2^{-1}\Phi)(\lambda_1,\lambda_2)$$

It has been shown (e.g. by N.Latz (1968) [10]) that in the case of $\mathrm{Im}\, k_j > 0$ the auxiliary k may be chosen in such a way that eq. (3.4) is uniquely solvable in $FL^p(\mathbb{R}^2)$, $1 < p \leq 2$, for any $\Phi_{pr}(\underline{x}) \in L^p(\mathbb{R}^2)$. The present author has derived quite recently (1984) [13] a 4X4-system of integral equations for the Fourier-cosine transforms of the normal derivatives on the bounding semi-axis's of the four quadrants Q_j. This system is uniquely solvable in the case of $\mathrm{Im}\, k_j > 0$ and $|k_j - k_\nu|$ and $|\rho_j - \rho_\nu|$ small by Banach's fixed point theorem in the spaces $(L^q(\mathbb{R}_+))^4$ for $2 \leq q < \infty$, but the general case of four different wave numbers k_j is still unsolved.

References

[1] CARLSON,J.F., A.E.HEINS, *The reflection of electromagnetic waves by an infinite set of plates,* I: Quart.Appl. Math. 4(1946),313-329, II: 5(1947),82-88, III: 8(1950),281-291.

[2] COLTON,D., R.KREIS, *Integral Equations in Scattering Theory,* J.Wiley, New York et al. 1983.

[3] COSTABEL,M., E.STEPHAN, *A direct boundary integral equation method for transmission problems,* Journ. Math, Appl. 106(1985), 367-413.

[4] COSTABEL,M., *Starke Elliptizitat von Randintegraloperatoren erster Art.*, Habil.-srift=preprint Nr. 868, FB Mathematik TH Darmstadt, Dez. 1984.

[5] GERLACH,E., *Zur Theorie einer Klasse von Integrodifferentialgleichungen,* Dissert. TU Berlin 1969.

[6] GRISVARD,P., *Boundary value problems in non-smooth domains,* Univ. of Maryland, MD 20742 Lecture Notes 19(1980).

[7] HEINS,A.E., *The radiation and transmission properties of a pair of semi-infinite parallel plates,* I. Quart. Appl. Math. 6(1948), 157-166, II: 215-220.

[8] HEINS,A.E., *The Sommerfeld Half-Plane Problem Revusited I: The Solution of a pair of Coupled Wiener-Hopf Integral Equations,* Math. Meth. Appl. Sci. 4(1982), 74-90, II: 5(1983), 14-21.

[9] KREIN,M.G., *Integral Equations on the Half-Plane with Kernels Depending upon the Difference of the Arguments,* Amer. Math. Soc. Transl. 22(1962), 163-288.

[10] LATZ,N., *Untersuchungen uber ein skalares Ubergangswertproblem aus den Theorie der Beugung elektromagnetische Wellen an dielektrischen Keilen,* Dissert. U Saarbrucken 1968.

[11] LEIS,R., *Lectures on initial-boundary value problems in mathematical physics,* B.G.Teubner-J.Wiley, Stuttgart-New York (1985 to appear).

[12] MEISTER,E., *Some multiple-part Wiener-Hopf problems in mathematical physics,* Banach Center Public. (to appear 1985) = Preprint No. 600, FB Math. TH Darmstadt, Mai 1981.

[13] MEISTER,E., *Integral equations for the Fourier transformed boundary values for the transmission problems for right angled wedges and octants,* Math. Meth. Appl. Sci. 7.

[14] RAWLINS,A.D., *The explicit Wiener-Hopf factorization of a special matrix,* Z. Angew.Math.Mech. 61(1981),527-528.

[15] SOMMERFELD,A., *Mathematische Theorie der Diffraktion,* Math. Ann. 47(1896), 317-374.

[16] SPECK,F.-O., *General Wiener-Hopf factorization methods,* Res. Notes in Math., vol. 119, Pitman, Boston et al. 1985

[17] THELEN-ROSEMANN-NIEDRIG,G., *Zur Fredholmtheorie singularer Integro-Differentialoperatoren auf der Halbachse,* Diss. TH Darmstadt 1985.

ENTROPY COMPACTIFICATION OF THE TRANSONIC FLOW

J. NEČAS
Faculty of Mathematics and Physics, Charles University
Malostranské nám. 25, 110 00 Prague 1, Czechoslovakia

1. Introduction

Let us consider a compressible, irrotational, steady, adiabatic, isentropic and inviscid fluid in a bounded, simply connected domain $\Omega \subset \mathbf{R}^n$, $n = 2,3$, with Lipschitz boundary. The relation between the presure p and the density ρ is

(1.1) $$\frac{p}{p_0} = \left(\frac{\rho}{\rho_0}\right)^{\varkappa}, \quad 1 < \varkappa < 2,$$

where quantities with the zero index correspond to the speed $\vec{v} = 0$. If $\vec{v}$ is the velocity vector, then the condition of the irrotational flow is

(1.2) $$\operatorname{rot} \vec{v} = 0 .$$

The flow satisfies the continuity equation

(1.3) $$\operatorname{div} (\rho\vec{v}) = 0$$

and the Euler equation of motion:

(1.4) $$\vec{v} \operatorname{grad} \vec{v} = -\frac{1}{\rho} \operatorname{grad} p .$$

This implies for the potential of the velocity is satisfied the equation

(1.5) $$\operatorname{div} (\rho\nabla u) = 0 ,$$

where

(1.6) $$\rho = \rho(|\nabla u|^2) = \rho_0\left(1 - \frac{\varkappa-1}{2a_0^2}|\nabla u|^2\right)^{\frac{1}{\varkappa-1}}$$

and a is the speed of the sound. If the Mach number defined as

$$M \overset{\text{def}}{=} \frac{|\nabla u|}{a} \text{ is } < 1 \Leftrightarrow |\nabla u|^2 < \frac{2a_0^2}{1+\varkappa} ,$$

the flow is subsonic and the equation (5) is elliptic. In the opposite case the flow is supersonic and the equation (5) is hyperbolic. A flow with subsonic and supersonic regions is called transonic.

It is important to underline that the equation (5) does not contain an information about the behaviour of the entropy on the shock surfaces. The entropy condition across the shock: $|\nabla u|$ is decreasing.

This can be formulated, for example, in the form

(1.7) $\Delta u \le K < \infty$;

we shall consider in the next only physical speed, i.e. such that

(1.8) $|\nabla u| \le \frac{\sqrt{2}a_0}{\sqrt{\varkappa-1}}$.

We suppose the boundary $\partial\Omega = \Gamma_1 \cup \Gamma_2$, $u = 0$ on Γ_1 and $\rho(|\nabla u|^2)\frac{\partial u}{\partial \nu} = g$ on Γ_2.

The transonic flow problem was considered numerically by many authors. We mention here a book by R. Glowinski [1]; there are excelent numerical results by many authors: M.O. Bristeau, R.Glowinski, J. Periaux, P. Perrieer, O. Pironneau, G.Poirer, M.Feistauer, A.Jameson, K.Kozel, J.Polášek, M.Vavřincová. They used entropy conditions of the type (7), upwinding iterations and viscosity approximations. We shall do the same in the next. The entropy condition (7) is compactifying, which follows from some slight generalisation of the result by F.Murat [2]. More complete discussion of the result is in M.Feistauer, J.Nečas [3], M.Feistauer, J.Madel, J.Nečas [4], J.Nečas [5]. A justification of the finite element approximation is discussed in Ph.G.Ciarlet, J.Mandel, J.Nečas [6].

2. Formulation of the problem, compactness bv entropy

We look for a week solution to the equation (1.5), i.e. for $u \in W^{1,\infty}(\Omega)$, such that for $v \in V = \{v \in W^{1,2}(\Omega);\ v = 0 \text{ on } \Gamma_1\}$

(2.1) $\int_\Omega \rho(|\nabla u|^2)\nabla u \nabla v dx = \int_{\partial\Omega} g v dS$, $g \in L^\infty(\partial\Omega)$.

If $\Gamma_1 = 0$, we suppose $\int_{\partial\Omega} g dS = 0$.

We can give also Dirichlet data on a part $\Gamma_t \subset \Gamma_2$, where $\Gamma_t \subset \{x \in \partial\Omega;\ (\vec{v},\vec{\nu}) < 0\}$. For an illustration, let us consider a paralel flow: $\vec{v} = (u,0,0)$, $u = u(x_1)$. For $w = \frac{u}{a_0}$ let us consider $w \in W^{1,\infty}((0,1))$, $w(1) = 0$, $|w'|^2 \le \frac{2}{\varkappa-1}$, satisfying with $\mu(s) = (1 - \frac{\varkappa-1}{2}s)^{1/(\varkappa-1)}$

(2.2) $(\mu(w'^2)w')' = 0$ in $(0,1)$,

(2.3) $w'(0)\mu(w'(0)^2) = A$

(2.4) $w'' \le K,\ K > 0$.

So first $A \in [0,0.57]$ which is clear from teh Fig. 1; the general solution of (2.2), (2.3), (2.4) is sketched on the Fig. 2 and is unique, the Cauchy data in the origine being prescribed.

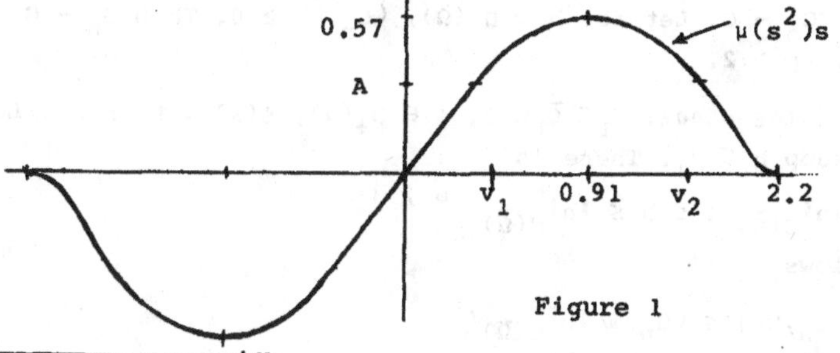

Figure 1

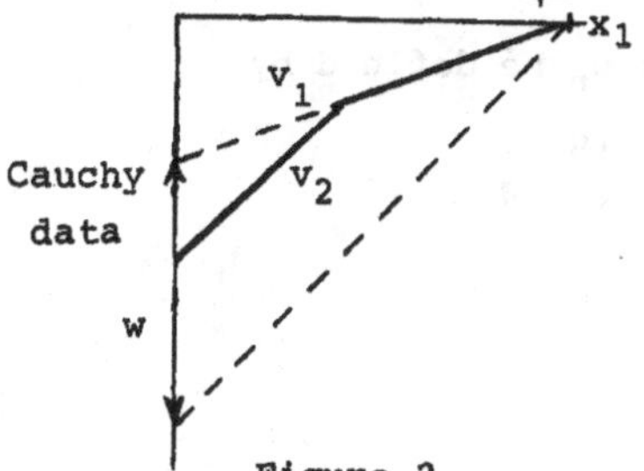

Figure 2

Let us mention that a uniqueness of the entropx solution is probably not true in more dimensions. The existence of the solution will follow from some "à posteriori" conditions given by an ideal computer.

2.5. Definition.

Let $h \in C^1([0,s_0])$, $\frac{2a_0^2}{\varkappa+1} < s_0 < \frac{2a_0^2}{\varkappa-1}$, $h(s) > 0$ in $(0,s_0]$ and let it satisfy here the monotony condition: $h(s) + 2sh'(s) > 0$. A transonic flow is called h-entropic if $\forall\varphi \in D_+(\Omega)$: $(\varphi \geq 0)$

(2.6) $$-\int_\Omega h(|\nabla u|^2)\nabla u\nabla\varphi dx \leq K\int_\Omega \varphi dx, \quad K \in R^1 .$$

(2.6′) Examples:

(1) $h(s) \equiv 1$, $s_0 < \frac{2a_0^2}{\varkappa-1}$,

$$-\int_\Omega \nabla u\nabla\varphi dx \leq K\int_\Omega \varphi dx \Leftrightarrow \Delta u \leq K, \ M = \infty ,$$

(2) $h(s) = s\rho(s)$, $s_0 < \frac{6a_0^2}{3\varkappa-1}$,
$M = \sqrt{3}$: entropy by viscosity ,

(3) $h(s) = -\rho(s)\ln\left(1 - \frac{\varkappa-1}{2a_0^2}s\right)$,

$$s_0 < \frac{2a_0^2}{\varkappa-1}\tau_0, \quad \tau_0 = [\tfrac{1}{2}\ln(1-\tau_0)](1-6\tau_0) ;$$

$M = 1.91$: Hugoniot's entropy,

(4) $h(s) = -\rho'(s)s$, $s_0 < \frac{6a_0^2}{\varkappa-1}$,
natural entropy, $M = 2.23$. □

In a formal way: the monotony condition for h and $s > \frac{2a_0^2}{\varkappa+1}$ is sufficient and necessary, for the solution satisfies the entropy condition on the shock surface.

2.7. Theorem (F.Murat). Let $\{G_n\}$ be a sequence of functionals defined on $W^{1,2}(\Omega)$, $G_n \to G$. Let for $h \in D_+(\Omega)$, $\langle G_n, h\rangle \geq 0$. Then $G_n \to G$ in $[W^{1,p}(\Omega)]'$, $\forall p > 2$.

Idea of the proof: $\Omega_1 \subset \overline{\Omega}_1 \subset \Omega$, $\psi \in D_+(\Omega)$, $\psi(x) = 1$ in Ω_1. Let $h \in D(\Omega)$, supp $h \subset \Omega_1$. There is

$$(2.8) \qquad -\|h\|_{C(\overline{\Omega})}\, \psi \leq h \leq \|h\|_{C(\overline{\Omega})}\, \psi ,$$

and it follows

$$(2.9) \qquad |\langle G_n, h\rangle| \leq \langle G_n, \psi\rangle \|h\|_{C(\overline{\Omega})}'$$

hence G_n is a sequence of Radon measures. Let u_n be defined by

$$(2.10) \qquad \int_\Omega (\nabla u_n \nabla h + u_n h)dx = \langle G_n, h\rangle , \; \forall h \in W^{1,2}(\Omega) ,$$

$$\Omega_2 \subset \overline{\Omega}_2 \subset \Omega_1, \quad q > n, \; h \in W_0^{1,q}(\Omega_1) .$$

Then

$$(2.11) \qquad „-\Delta u_n + u_n = G_n" \in [W_0^{1,q}(\Omega_1)]'$$

and

$$(2.12) \qquad \|u_n\|_{W^{1,q'}(\Omega_2)} \leq C(\Omega_2) .$$

Because $W^{1,q}(\Omega_1) \hookrightarrow\hookrightarrow C(\overline{\Omega}_1) \Rightarrow \{u_n\}$ is convergent in $W^{1,q'}(\Omega_2)$ to u. An usual interpolation technic as well as the estimate

$$(2.13) \qquad \Big|\int_{\Omega\setminus\Omega_2} \nabla(u_n - u)\nabla h dx\Big| \leq \Big(\int_{\Omega\setminus\Omega_2} |\nabla(u_n - u)|^2 dx\Big)^{\frac{1}{2}} \cdot$$

$$\Big(\int_{\Omega\setminus\Omega_2} |\nabla h|^2 dx\Big)^{\frac{1}{2}} \leq \Big(\int_{\Omega\setminus\Omega_2} |\nabla(u_n - u)|^2 dx\Big)^{\frac{1}{2}} \cdot$$

$$\cdot \Big(\int_{\Omega\setminus\Omega_2} |\nabla h|^p dx\Big)^{\frac{1}{p}} \cdot |\Omega\setminus\Omega_2|^{\frac{1}{2} - \frac{1}{p}}$$

gives the result.

2.14 Theorem.

Let $E_h = \{u; \|u\|_{W^{1,2}(\Omega)} \leq C, |\nabla u|^2 \leq s_0$,

$\forall \varphi \in D_+(\Omega) : - \int_\Omega h(|\nabla u|^2)\nabla u \nabla \varphi dx \leq K \int_\Omega \varphi dx\}$.

Then E_h is compact in $W^{1,2}(\Omega)$.

Idea of the proof: Put $\langle G_n, h\rangle = K \int_\Omega h dx + \int_\Omega h(|\nabla u_n|^2) \cdot$

$\cdot \nabla u_n \nabla h dx$, $\langle G, h\rangle = K \int_\Omega h dx + \int_\Omega h(|\nabla u|^2)\nabla u \nabla h dx.$

It follows from the theorem 2.7: $\langle G_n - G, u_n - u\rangle \to 0$, provided $u_n \rightharpoonup u$. For the pairing $\langle G_n - G, u_n - u\rangle$ we use the Leray-Lions trick from the theory of monotone operators. □

3. Solution of the transonic problem by use of the alternating functional

The equation (1.5) is the Euler's equation to the functional

$$\Phi(u) = \frac{1}{2}\int_\Omega \Big(\int_0^{|\nabla u|^2} \rho(t)dt\Big)dx - \int_{\partial\Omega} guds. \tag{3.1}$$

Let $|\nabla u|^2 \le \frac{2a_0^2}{\varkappa - 1}$, and define $w \in W^{1,2}(\Omega)$ by

$$\int_\Omega \rho(|\nabla u|^2)\nabla w \nabla h dx = \int_{\partial\Omega} ghds \tag{3.2}$$

for $h \in W^{1,2}(\Omega)$, $h = 0$, on Γ_1. (In the case $\Gamma_1 = 0$, $\Gamma_t = 0$, we suppose $\int_\Omega wdx = 0$.)

Define the alternating functional by

$$\psi(u) = \Phi(u) - \Phi(w(u)) . \tag{3.3}$$

Because $\rho'(s) \le 0$, we have with $c > 0$ (for details see [3], [4], [5])

$$c\|u - w(u)\|^2_{W^{1,2}(\Omega)} \le \psi(u). \tag{3.4}$$

So u is a solution of the transonic problem iff $\psi(u) = 0$. But if V is the space od solutions in $W^{1,2}(\Omega)$ then we have

3.4. Theorem

The alternating functional attains on $E_h \cap V \cap \{u = u_0 \text{ on } \Gamma_t\}$ its minimum in some point u. If $\psi(u) = 0$, then u is a h-entropic solution of the transonic problem.

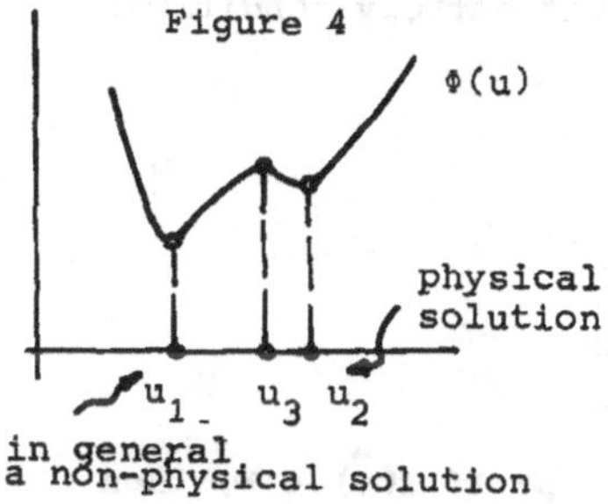

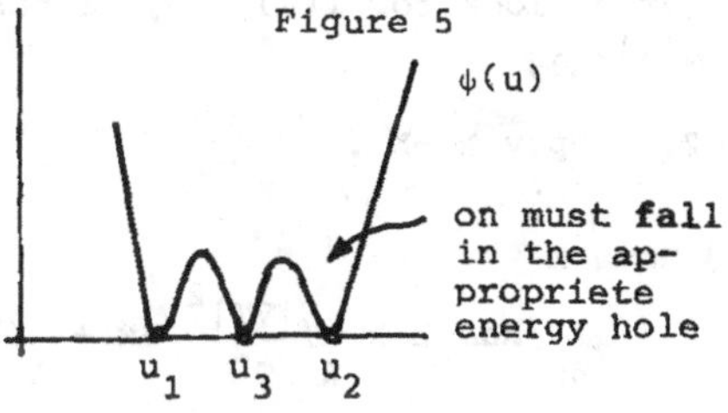

4. Viscosity method

Let us consider a complet system of gas: p - pressure, ρ-density, T -temperature, $\vec{v}$ - velocity vector, provided:

(4.1) $0 < T_1 \le T \le T_2 < \infty, \quad T \in W^{1,2}(\Omega),$

(4.2) $0 < \rho_1 \le \rho, \quad \rho \ln \rho \in L^2(\Omega),$

(4.3) $p = R\rho T,$ R is the gas-constant,

(4.4) $\vec{v} \in [W^{1,2}(\Omega)]^3, \quad |v|^2 \le \frac{2a_0^2}{\varkappa-1},$

(4.5) $\mu = \mu(T),\ \lambda = \lambda(T),\ \mu,\lambda \in C(R_+),\ \lambda = -\theta\frac{2}{3}\mu,\ \theta < 1,\ \Gamma_1 = 0,$
$\Gamma_t = 0,$

(4.6) $\operatorname{div}(\rho\vec{v}) = 0$ in Ω, $\rho v.\nu = g$ on $\partial\Omega$,

(4.7) $\rho v.\nabla v + \nabla p = \nabla(\lambda \operatorname{div} \vec{v}) + 2 \operatorname{div}(\mu \underset{\sim}{e})$, in Ω,

$$\underset{\sim}{e} = \{e_{ij}(v)\},\ 2e_{ij}(v) = \frac{\partial v_i}{\partial x_j} + \frac{\partial v_j}{\partial x_i},$$

(4.8) $\vec{v} = \vec{v}^0$ on $\partial\Omega$, $\|\vec{v}^0\|_{[W^{1,2}(\Omega)]^3} \le c,\ |\vec{v}^0| \le \frac{2a_0^2}{\varkappa-1},$

(4.9) $c_v \operatorname{div}(\rho.\ln \frac{T}{\rho^{\varkappa-1}}.\vec{v}) = k\frac{\Delta T}{T} + \frac{E(\vec{v})}{T}$ in Ω,

$$E(v) = \lambda(\operatorname{div} v)^2 + 2\mu \operatorname{tr} \underset{\sim}{e}\underset{\sim}{e},\ c_v,\ k \in R^1_+,$$

(4.10) $c_v g \ln \frac{T}{\rho^{\varkappa-1}} - k\frac{1}{T}\frac{\partial T}{\partial \nu} = h$ on $\partial\Omega$, $\|h\|_{L^1(\partial\Omega)} \le c$.

Let $\beta_n > 0$, and take $\mu_n = \mu.\beta_n$, $\lambda_n = \lambda\beta_n$, $k_n = k\beta_n$. We let go $\beta_n \to 0$ and look for an optimal control problem: let the cost functional, where $\vec{v} = \vec{\varepsilon} + \nabla u$, be

(4.11) $I(\vec{v}) = \int_\Omega |\vec{\varepsilon}|^2 dx + \int_\Omega [\rho - \rho_0 (1 - \frac{\varkappa-1}{2a_0^2}|\vec{v}|^2)^{\frac{1}{\varkappa-1}}]^2 dx$

and let us look for $I(v) \to 0$, "à posteriori" entropy condition: $p \in W^{1,1}(\Omega)$,

(4.12) $\nabla p.\vec{v} \ge -K$.

4.13 Remark

$$\int_{\partial\Omega} h dS \ge k \int_\Omega \frac{|\nabla T|^2}{T^2} dx + \int_\Omega \frac{E(\vec{v})}{T} dx.$$

Open problem: how to estimate the pressure p? If $I(\vec{v}) < \infty$, then

L^2 estimate of ρ follows.

4.14 Definiton

A sequence $\{u_n\}$, $\int_\Omega u_n dx = 0$, $|\nabla u_n|^2 \le s_0$ is h - entropic, if $\forall\varphi \in D_+(\Omega)$:

$$(4.15)\qquad -\int_\Omega h(|\nabla u_n|^2)\nabla u_n \nabla\varphi dx \le K\int_\Omega \varphi dx + \langle R_n,\varphi\rangle ,$$

where

$$R_n \to 0 \quad \text{in } [W^{1,2}(\Omega)]' .$$

4.16 Theorem

Let T_n, p_n, ρ_n, $\vec{v}^n$ be a sequence to solutions of (4.1) - (4.10) with $I(\vec{v}^n) \to 0$ and satisfying $\nabla p_n \vec{v}^n \ge -K$. Suppose, without the loss of generality, that $u_n \to u$. Then $\{u_n\}$ is $h(s) = \rho(s)s$ - entropic, $u_n \to u$, and u is a h-entropic solution to the transonic problem,

Provided $|\nabla u_n|^2 \le s_0 < \dfrac{6a_0^2}{3\varkappa-1}$.

Idea of the proof: Let $\varphi \in D_+(\Omega)$, multiply (4.7) by $\vec{v}^n\varphi$ and integrate by parts. We get

$$(4.17)\qquad \int_\Omega \rho_n v_j^n \frac{\partial v_i^n}{\partial x_j} v_i^n \varphi dx = -\frac{1}{2}\int_\Omega \rho_n |\vec{v}^n|^2 v_j^n \frac{\partial\varphi}{\partial x_j} dx =$$

$$= -\int_\Omega \frac{\partial p_n}{\partial x_i} v_i^n \varphi dx - \int_\Omega \lambda_n (\operatorname{div}\vec{v}^n)^2 \varphi dx - \int_\Omega \lambda_n \operatorname{div}\vec{v}^n . v_i^n \frac{\partial\varphi}{\partial x_i} dx -$$

$$- 2\int_\Omega \mu_n e_{ij}(\vec{v}^n) e_{ij}(\vec{v}^n)\varphi dx - 2\int_\Omega \mu_n e_{ij}(\vec{v}^n) v_i^n \frac{\partial\varphi}{\partial x_j} dx \le$$

$$\le K\int_\Omega \varphi dx - \int_\Omega \lambda_n \operatorname{div}\vec{v}^n v_i^n \frac{\partial\varphi}{\partial x_i} dx - 2\int_\Omega \mu_n e_{ij}(\vec{v}^n) v_i^n \frac{\partial\varphi}{\partial x_j} dx =$$

$$\stackrel{\text{def}}{=} K\int_\Omega \varphi dx + \langle S_n,\varphi\rangle .$$

But we have, because of 4.13

$$(4.18)\qquad \| S_n \|_{[W^{1,2}(\Omega)]'} \le c\beta_n^{\frac{1}{2}}$$

and replacing in $\int_\Omega \rho_n |\vec{v}^n|^2 v_j^n \frac{\partial\varphi}{\partial x_i}$, ρ_n by $\rho_0\left(1 - \frac{\varkappa-1}{2a_0^2}|\vec{v}^n|^2\right)^{\frac{1}{\varkappa-1}}$ and then $\vec{v}^n$ by ∇u_n, the result follows. □

Solving the system (4.1)-(4.10) with the cost functional (4.11), the inequality (4.12) can be expected, but must be supposed. For to cancel this condition, let us first formulate (4.6) in the weak sense:

$$(4.19)\qquad \int_\Omega \rho v_i \frac{\partial\varphi}{\partial x_i}\,dx = \int_{\partial\Omega} g\varphi dS .$$

Any approximate solution to (4.19) satisfies in fact

$$(4.20)\qquad \int_\Omega \rho v_i \frac{\partial\varphi}{\partial x_i}\,dx = \int_{\partial\Omega} g\varphi dS + \langle R,\varphi\rangle ,$$

where the term R represents small material sources in Ω and a small flux of the material through the $\partial\Omega$. If we choose R in an appropriate way, we get automatically (4.15) provided the cost functional

$$(4.21)\qquad J(\vec v) = \int_\Omega |\vec\varepsilon|^2 dx + \int_\Omega [\rho - \rho_0(1 - \frac{\varkappa-1}{2a_0^2}|\vec v|^2)^{\frac{1}{\varkappa-1}}]^2 dx + \int_\Omega [\ln\frac{T}{\rho^{\varkappa-1}} - \ln\frac{T_0}{\rho_0^{\varkappa-1}}]^2 dx$$

tends to zero; here $T_0 = \frac{1}{R}\frac{p_0}{\rho_0}$.

For to precise the conditions, let $\beta_n \to 0$, $\beta_n > 0$, λ_n, μ_n, k_n as before and put $\alpha_n \to 0$ in the way that $\alpha_n\beta_n^{-\frac{1}{2}} \to \infty$. Let us suppose $T_0 > \rho_0^{\varkappa-1}$. Put in (4.20)

$$(4.22)\qquad \langle R_n,\varphi\rangle = \int_\Omega \rho_n v_i^n \frac{\partial\varphi}{\partial x_i}dx - \int_\Omega \rho_n^{1-\alpha_n} v_i^n \frac{\partial\varphi}{\partial x_i}dx + \int_\Omega \rho_n\Big(1 - \frac{\ln\frac{T_n}{\rho_n^{\varkappa-1}}}{\ln\frac{T_0}{\rho_0^{\varkappa-1}}}\Big)v_i^n \frac{\partial\varphi}{\partial x_i}dx + \int_\Omega f_n\varphi dx + \langle G_n,\varphi\rangle + \langle H_n,\varphi\rangle ,$$

where

$$\|f_n\|_{L^\infty(\Omega)} \le K\alpha_n,\quad \frac{1}{\alpha_n}\|G_n\|_{[W^{1,2}(\Omega)]'} \to 0 ,$$

$$\|H_n\|_{[W^{1,2}(\Omega)]'} \to 0 \quad\text{and}\quad \langle H_n,\varphi\rangle \ge 0\ \forall\varphi \in D_+(\Omega).$$

4.23 Theorem.

Let $h(s) = -\rho(s)\ln(1 - \frac{\varkappa-1}{2a_0^2}s)$ (see examples (2.6′)) and T_n, p_n, ρ_n, $\vec v^n$ a sequence of solutions to (4.1)-(4.10), with $J(\vec v^n) \to 0$ and with (4.20), (4.22). Let us suppose without the loss of generality $u_n \to u$. Suppose $|\nabla u_n|^2 < s_0$ (see (2.6′), (3)). Then $\{u_n\}$ is h-entropic and u is a h-entropic solution to the transonic problem.

4.24 Remark

The term $\int_\Omega \rho^{1-\alpha} v_i \frac{\partial\varphi}{\partial x_i} dx$ gives "upwinding" in the continuity equation.

Idea of the proof to the theorem 4.23: We have from (4.9) as before

$$(4.25) \qquad \beta_n \int_\Omega \frac{1}{T_n^2} |\nabla T_n|^2 dx + \beta_n \int_\Omega |\nabla \vec{v}^n|^2 dx \le c < \infty .$$

Take $\varphi \in D_+(\Omega)$. It follows from (4.9)

$$(4.26) \qquad -c_v \int_\Omega \rho_n \ln \frac{T_n}{\rho_n^{\varkappa-1}} v_i^n \frac{\partial\varphi}{\partial x_i} dx = + k_n \int_\Omega \frac{|\nabla T_n|^2}{T_n^2} \varphi dx -$$

$$- k_n \int_\Omega \frac{\nabla T_n}{T_n} \nabla\varphi dx + \int_\Omega \frac{E_n(\vec{v}^n)}{T_n} \varphi dx .$$

Multiply (4.26) by $\frac{1}{c_v} \dfrac{1}{\ln \frac{T_0}{\rho_0^{\varkappa-1}}}$ and add to (4.20).

We get

$$(4.27) \qquad \frac{1}{\alpha_n} \int_\Omega (\rho_n - \rho_n^{1-\alpha_n}) v_i^n \frac{\partial\varphi}{\partial x_i} dx = - \frac{1}{\alpha_n} \int_\Omega f_n \varphi dx - \frac{1}{\alpha_n} \langle G_n, \varphi\rangle -$$

$$- \frac{1}{\alpha_n} \langle H_n, \varphi\rangle - \frac{k_n}{\alpha_n} \int_\Omega \frac{|\nabla T_n|^2}{T_n^2} \varphi dx - \frac{1}{\alpha_n} \int_\Omega \frac{E_n(\vec{v}^n)}{T_n} \varphi dx +$$

$$+ \frac{k_n}{\alpha_n} \int_\Omega \frac{\nabla T_n}{T_n} \nabla\varphi dx$$

and the result follows as in the theorem 4.16.

References

[1] R.Glowinski, *Numerical Methods for Nonlinear Variational Problems*, Springer-Verlag, 1984.

[2] F. Murat, *L'injection du cone positif de H^{-1} dans $W^{-1,q}$ est compacte pour tout $q < 2$*, J. Math. Pures. Appl. (9) 60 (1981),

309-322.

[3] M. Feistauer, J. Nečas, *On the Solvability of Transonic Potential Flow Problems*, Zeitschrift für Analysis und ihre Anwendungen, Bd. 4 (4) 1985, 305-329.

[4] M. Feistauer, J. Mandel, J. Nečas, *Entropy Regularization of the Transonic Potential Flow Problem*, Comm. Math. Univ. Carol. 25 (3) 1984, 431-443.

[5] J. Nečas, *Compacité par entropie d'écoulements des fluides*, Univ. Pierre et Marie Curie, Paris VI, 1985.

[6] Ph.G. Ciarlet, J. Mandel, J. Nečas, *On the convergence of Finite Element Approximations of thr Transonic Potential Flow Problem*, to appear.

THE GLOBAL EXISTENCE OF WEAK SOLUTIONS OF THE MOLLIFIED SYSTEM OF EQUATIONS OF MOTION OF VISCOUS COMPRESSIBLE FLUID

J. NEUSTUPA
Faculty of Mechanical Engineering, Czech Technical University
Suchbátarova 4, 166 07 Prague 6, Czechoslovakia

1. Introduction

It is known that weak solutions of the Navier-Stokes equations for incompressible liquid exist on a time interval of an arbitrary lenght (see e.g. [3] [10]). No analogous result has been derived in the case of equations of motion of viscous compressible fluid till now. Only the existence of solutions of such equations local in time was proved (see e.g. [1], [5], [8], [9]) and if some theorems about the global in time existence of solutions appeared, they contained assumptions of the type "the initial conditions are small enough" (see e.g. [4]), "the flow is one-dimensional" ([2]), etc. We study the existence of weak solutions of the equations of motion of viscous compressible fluid on a time interval of a given lenght in this paper, but the system of equations we deal with is rather modified in a comparison with a full general system of equations governing the motion of viscous compressible fluid. The modification consists in the following points:

a) We assume the dynamic viscosity coefficient μ to be a positive constant.

b) We do not take the energy equation into account and we use the relation between the pressure p and the density

$$p = c.\widetilde{\rho^{\varkappa}} \tag{1.1}$$

instead of it. c and $\varkappa$ are constants such that $c > 0$, $\varkappa \in (1,6)$. The tilda over $\rho^{\varkappa}$ represents a certain regularization (mollification). Its exact meaning is explained in the paragraph 2., but we can write in advance that $\widetilde{\rho^{\varkappa}}(x)$ is an average of $\rho^{\varkappa}$ considered with a proper smooth weight function on a neighbourhood $B_h(x)$ of x (where the radius h of this neighbourhood may be arbitrarily small).

c) We use the mollification denoted by ~ also in some terms in the Navier-Stokes equations for the system we deal with has the form

(1.2) $\rho_{,t} + (\rho\tilde{u}_j)_{,j} = 0$,

(1.3) $(\rho u_i)_{,t} + (\rho\tilde{u}_j u_i)_{,j} = - c.(\tilde{\rho^{\varkappa}})_{,i} + \frac{1}{3}\mu u_{j,ji} + \mu u_{i,jj}$

$(i = 1,2,3)$.

$U = (u_1,u_2,u_3)$ has a physical meaning of the velocity of the moving fluid. In [7], R. Rautmann used the similar mollification in the Navier-Stokes equations for the incompressible liquid in order to prove the global in time existence of strong solutions in three-dimensional domains. The notion of the velocitv of the fluid at the point x is usually introduced by means of an average of the velocities of all particles of the fluid contained in a small neighbourhood of x. So if h is small enough, $\tilde{u}_i$ is almost the same as u_i from the point of view of mechanics. The system (1.3) expresses the 2nd Newton law of mechanics applied to particles moving along the integral curves of the flow field $\tilde{U}$.

We shall use the Rothe method. We can give only a brief outline of the whole procedure here. Details may be found in [6].

2. Formulation of an initial-boundary value problem

Assume that Ω is a bounded region in R^3 with the boundary of the class $C^{2+(\alpha)}$ for some $\alpha \in (0,1)$. Let us choose $h > 0$ and put

$$\Omega_h = \{x \in R^3;\ \mathrm{dist}(x,\Omega) < h\} .$$

Assume that h can be chosen so small that $\partial\Omega_h$ is also of the class $C^{2+(\alpha)}$. Put

$$\omega_h(\xi) = K_h \exp\left(- \frac{|\xi|^2}{h^2-|\xi|^2}\right) \text{ for } \xi \in R^3,\ |\xi| < h ,$$

$$\omega_h(\xi) = 0 \quad \text{for } \xi \in R^3,\ |\xi| \geq h .$$

Let K_h be chosen so that the integral of ω_h over R^3 is equal to 1. If $f \in L^1(\Omega_h)$, put

(2.1) $$\tilde{f}(x) = \int_{\Omega_h} \omega_h(x - y)f(y)dy .$$

If f is defined in $\Omega_h \times R^1$ then we denote by $\tilde{f}$ the function regularized in the space variable only. If the regularization ~ is applied to any function def in the space variable on Ω only (like for example components of the velocity or their approximations), we deal with this function as if it is defined on Ω_h and is identically equal to zero on $\Omega_h - \Omega$.

We shall solve the equation (1.2) on $\Omega_h \times (0,T)$ and the system (1.3) on $\Omega \times (0,T)$ (where T is a given positive number). We consider the boundary condition

$$(2.2) \qquad u_i\big|_{\partial\Omega} \equiv 0 \quad (i = 1,2,3)$$

and the initial conditions

$$(2.3) \qquad \rho\big|_{t=0} = \rho_0 \,,$$

$$(2.4) \qquad (\rho u_i)\big|_{t=0} = \rho_0 u_{0i} \quad (i = 1,2,3) \,,$$

where ρ_0, $U_0 = (u_{01}, u_{02}, u_{03})$ are given functions such that $\rho_0 \in H^1(\Omega_h)$, $\rho_0 \geq 0$, $U_0 \in \mathring{H}^1(\Omega)^3$.

We shall call by the weak solution of (1.2), (1.3), (2.2), (2.3), (2.4) the couple of functions U, ρ such that

$$(2.5) \qquad \begin{aligned} &U \equiv (u_1, u_2, u_3) \in L^2(0,T;\mathring{H}^1(\Omega)^3), \\ &\rho \in L^\infty(0,T;H^1(\Omega_h)), \ \rho \geq 0 \,, \end{aligned}$$

$$(2.6) \qquad \int_0^T \int_\Omega \{\rho u_i \varphi_{i,t} + \rho\tilde{u}_j u_i \varphi_{i,j} + c(\tilde{\rho}^{\varkappa})\varphi_{i,i} - \tfrac{1}{3}\mu u_{j,j}\varphi_{i,i} - \mu u_{i,j}\varphi_{i,j}\}\,dxdt = - \int_\Omega \rho_0 u_{0i}(\varphi_i\big|_{t=0})\,dx$$

for all $\varphi \equiv (\varphi_1, \varphi_2, \varphi_3) \in C^\infty(\overline{\Omega} \times \langle 0,T\rangle)^3$ such that $\varphi_i\big|_{\partial\Omega} \equiv 0$, $\varphi_i\big|_{t=T} \equiv 0 \quad (i = 1,2,3)$,

$$(2.7) \qquad \int_0^T \int_{\Omega_h} \{\rho\psi_{,t} + \rho\tilde{u}_j\psi_{,j}\}\,dxdt = - \int_{\Omega_h} \rho_0(\psi\big|_{t=0})\,dx$$

for all $\psi \in C^\infty(\overline{\Omega}_h \times \langle 0,T\rangle)$ such that $\psi\big|_{t=T} \equiv 0$.

By means of a similar method as it is used in [1] in the case of the Navier-Stokes equations for the incompressible liquid, it can be proved that if U, ρ satisfy (2.5), (2.6), (2.7) then $\rho.U$ is a.e. in $\langle 0,T\rangle$ equal to a continuous function from $\langle 0,T\rangle$ into $H^{-1}(\Omega)^3$. Hence we can understand under $(\rho u_i)\big|_{t=0}$ (i=1,2,3) in (2.4) limits as $t \to 0+$ of the components of this function. Similarly, it may be shown that ρ is a.e. in $\langle 0,T\rangle$ equal to a continuous function from $\langle 0,T\rangle$ into $H^1(\Omega_h)^*$ (the dual of $H^1(\Omega_h)$). It gives a reasonable sence to the initial condition (2.3).

3. The time discretization

Let m be a natural number. Put $\tau = T/m$, $t_k = k.\tau$ $(k = -1,0,1,\ldots$ $\ldots,m)$. Denote $\rho^{(-1)} = \rho_0$, $u_i^{(0)} = u_{0i}$ $(i = 1,2,3)$ and let $\rho^{(k)}$, $U^{(k)} = (u_1^{(k)}, u_2^{(k)}, u_3^{(k)})$ denote an approximation of a solution on k-th time layer. A discrete version of (2.5), (2.6) and (2.7), which we use in the following, is: We look for $\rho^{(0)}, \rho^{(1)},\ldots,\rho^{(m)} \in H^1(\Omega_h)$, $\rho^{(k)} \geq 0$ $(k = 0,1,\ldots,m)$ and $U^{(1)},\ldots,U^{(m)} \in \mathring{H}^1(\Omega)^3$ so that

$$(3.1)_k \qquad \int_\Omega \{\rho^{(k-1)}u_i^{(k)}\Phi_i - \rho^{(k-2)}u_i^{(k-1)}\Phi_i - \tau\rho^{(k-1)}\tilde{u}_j^{(k-1)}u_i^{(k-1)}\Phi_{i,j}$$
$$- \tau c\tilde{\rho}^{(k)\varkappa}\Phi_{i,i} + \frac{1}{3}\tau\mu u_{j,j}^{(k)}\Phi_{i,i} + \tau\mu u_{i,j}^{(k)}\Phi_{i,j}\}dx = 0$$

for all $\Phi \equiv (\Phi_1,\Phi_2,\Phi_3) \in \mathcal{C}^\infty(\bar{\Omega})^3$ and $k = 1,\ldots,m$,

$$(3.2)_k \qquad \rho^{(k)} - \rho^{(k-1)} + \tau(\rho^{(k)}\tilde{u}_j^{(k)})_{,j} = 0$$

for $k = 0,1,\ldots,m$.

We can further proceed in such a way that we successively solve $(3.2)_0$ (for the unknown $\rho^{(0)}$), $(3.1)_1$ and $(3.2)_1$ (for the unknowns $U^{(1)}, \rho^{(1)}$),..., $(3.1)_m$ and $(3.2)_m$ (for the unknowns $U^{(m)}$, $\rho^{(m)}$). It can be done using standart methods of the functional analysis and the theory of the partial differential equations. The following inequalities may be also derived:

$$(3.3) \qquad \int_\Omega\{\tfrac{1}{2}\rho^{(k-1)}u_i^{(k)}u_i^{(k)} + \frac{1}{2}\sum_{s=1}^{k}\rho^{(s-2)}(u_i^{(s)}-u_i^{(s-1)})(u_i^{(s)}-u_i^{(s-1)}) +$$
$$+ \frac{1}{3}\tau\mu\sum_{s=1}^{k}(u_{j,j}^{(s)})^2 + \tau\mu\sum_{s=1}^{k}u_{i,j}^{(s)}\,u_{i,j}^{(s)}\}dx + \frac{c}{\varkappa-1}\int_{\Omega_h}\rho^{(k)\varkappa}dx \leq$$
$$\leq \int_\Omega \frac{1}{2}\rho_0 u_{0i}u_{0i}dx + \frac{c}{\varkappa-1}\int_{\Omega_h}\rho_0^\varkappa dx \qquad (k = 1,\ldots,m),$$

$$(3.4) \qquad \|\rho^{(k)}\|^2_{H^1(\Omega_h)} + \sum_{s=0}^{k}\|\rho^{(s)} - \rho^{(s-1)}\|^2_{H^1(\Omega_h)} \leq$$
$$\leq K_1\,\exp(4\tau\|U_0\|_{L^2(\Omega)}{}^3 + \int_\Omega \tfrac{1}{2}\rho_0 u_{0i}u_{0i}dx +$$
$$+ \frac{c}{\varkappa-1}\int_{\Omega_h}\rho_0^\varkappa dx).\|\rho_0\|^2_{H^1(\Omega_h)} \quad (k = 0,1,\ldots,m)$$

for an appropriate positive constant K_1, independent on k.

4. An approximate solution of (2.6), (2.7) and the limit process for $m \to +\infty$

Put

(4.1) $\quad {}^m\rho(t) = \rho^{(k)}$ for $t \in \langle t_k, t_{k+1}\rangle$ $(k = -1,0,1,\dots,m-1)$,

(4.2) $\quad {}^mU(t) = U^{(k+1)}$ for $t \in \langle t_k, t_{k+1}\rangle$ $(k = 0,1,\dots,m-1)$.

It follows from (3.3) and (3.4) that the sequence $\{{}^m\rho\}$ (resp. $\{{}^mU\}$) is uniformly bounded in $L^\infty(0,T;\ H^1(\Omega_h))$ (resp. in the space $L^2(0,T;\ \mathring{H}^1(\Omega)^3)$) and that $\{{}^m\rho|{}^mU|^2\}$ is uniformly bounded in $L^\infty(0,T;\ L^1(\Omega))$. Using the Hölder inequality, it can be also easily shown that $\{{}^m\rho\, {}^mU\}$ is uniformly bounded in $L^\infty(0,T;\ L^{12/7}(\Omega)^3)$ and in $L^2(0,T;\ W^1_{3/2}(\Omega)^3)$. There exist subsequences (denoted by $\{{}^m\rho\}$, $\{{}^mU\}$ again) and functions ρ, U so that ${}^m\rho \to \rho$ weakly - * in $L^\infty(0,T;\ H^1(\Omega_h))$, ${}^mU \to U$ weakly in $L^2(0,T;\ H^1(\Omega)^3)$, ${}^m\rho\ {}^mU \to \rho\, U$ weakly - * in $L^\infty(0,T;\ L^{12/7}(\Omega)^3)$ and weakly in the space $L^2(0,T;\ W^1_{3/2}(\Omega)^3)$. By means of other estimates of ${}^m\rho\ {}^mU$ and ${}^m\rho$ in $\mathcal{H}^\gamma(0,T;\ W^1_{3/2}(\Omega)^3,\ H^{-1}(\Omega)^3)$ and $\mathcal{H}^\gamma(0,T;\ H^1(\Omega_h),\ L^2(\Omega_h))$ (see e.g. [3] or [10] for the definition of these spaces), we can prove that even ${}^m\rho \to \rho$ strongly in $L^2(0,T;\ L^2(\Omega_h))$ and ${}^m\rho\ {}^mU \to \rho\, U$ strongly in $L^2(0,T;\ L^2(\Omega)^3)$.

The functions ${}^m\rho$, ${}^mU$ satisfy (2.6), resp. (2,7) with some errors E_1, resp. E_2. It is shown in [6] that $E_1 = O(\tau^{1/2})$ and $E_2 = O(\tau^{1/2})$ for $\tau \to 0+$ (i.e. $m \to +\infty$). These relations together with the types of convergences mentioned above are sufficient to prove that ρ, U satisfy (2.5), (2.6), (2.7).

If we use (3.3) and (3.4), we can also derive the estimate

$$(4.3)\qquad \|\rho\|^2_{L^\infty(0,T;\ H^1(\Omega_h))} \le K_1 \exp\Big(\int_\Omega \tfrac{1}{2}\rho_0 u_{0i} u_{0i}\,dx + \tfrac{c}{\varkappa-1}\int_{\Omega_h} \rho_0^\varkappa\,dx\Big).\|\rho_0\|^2_{H^1(\Omega_h)}$$

and the energy inequality

$$(4.4)\qquad \int_\Omega \tfrac{1}{2}\rho u_i u_i|_{t=t_1}\,dx + \tfrac{c}{\varkappa-1}\int_{\Omega_h} \rho^\varkappa|_{t=t_1}\,dx + \int_0^{t_1}\int_\Omega \{\tfrac{1}{3}\mu(u_{j,j})^2 + \mu u_{i,j}u_{i,j}\}\,dx\,dt \le$$

$$\le \int_\Omega \tfrac{1}{2}\rho_0 u_{0i}u_{0i}\,dx + \tfrac{c}{\varkappa-1}\int_{\Omega_h} \rho_0^\varkappa\,dx$$

(for every $t_1 \in \langle 0,T\rangle$).

While the estimate (4.3) depends on the parameter h (used in the reqularization in (1.2) and (1.3)) according to the dependance of K_1 on h, the energy inequality (4.4) is quite independent on h. But in spite of this fact, we are not able to prove that if $h \to 0+$, we can get a solution of (1.2), (1.3) without the mollification yet.

References

[1] ITAYA,N., *On the Cauchy problem for the system of fundamental equations describing the movement of compressible viscous fluids*, Kodai Math. Sem. Rep. 23, 1971, 60-120.

[2] KAZHIKOV,A.V., SHELUKIN,V.V., *Unique global solution in time of initial-boundary value problems for one-dimensional equations of a viscous gas*, Prikl. Math. Mech. 41, 1977, 282-291.

[3] LIONS,J.L., *Quelques méthodes de résolution des problèmes aux limites non linéaires*, Dunod. Paris. 1969.

[4] MATSUMURA.A.. NISHIDA.T.. *Initial-boundary value problems for the equations of motion of compressible viscous and heat-conductive fluids*, Comm on Math. Physics 89, 1983, 445-464.

[5] NASH.J..*Le problème de Cauchy pour les équations diff. d'un fluide général*, Bull. Soc. Math. France 90, 1962, 487-497.

[6] NEUSTUPA,J.. *The global weak solvability of an initial-boundary value problem of the Navier-Stokes type for the compressible fluid*, to appear.

[7] RAUTMANN,R., *The uniqueness and regularity of the solutions of Navier-Stokes problems*, Funct. Theor. Meth. for PDR, Proc. conf. Darmstadt 1976, Lecture N. in Math., 561, 1976, 378-393.

[8] SOLONNIKOV,V.A., *Solvability of the initial boundary value problem for the equations of motion of a viscous compressible fluid*, J. Soviet Math. 14, 1980, 1120-1133.

[9] TANI,A., *On the first initial-boundary value problem of compressible viscous fluid motion*, Publ. RIMS Kyoto Univ. 13, 1977, 193-253.

[10] TEMAM,R., *Navier-Stokes equations*, North-Holland Publ. Comp., Amsterdam - New York - Oxford, 1979.

BIFURCATIONS NEAR A DOUBLE EIGENVALUE OF THE RECTANGULAR PLATE PROBLEM WITH A DOMAIN PARAMETER

Z. SADOVSKÝ
Institute of Construction and Architecture of the Slovak Academy of Sciences
Dúbravská cesta, 842 20 Bratislava, Czechoslovakia

Let us consider the bifurcation problem of the Föppl-Kármán equations of a thin elastic rectangular plate having the length a and width b when the aspect ratio $\alpha=a/b$ varies near a value $\alpha=\alpha_c$ yielding a double buckling load of the plate. In a suitable non-dimensional formulation, the governing equations will refer to a common domain. On the other hand, there appears a small perturbation parameter say Θ in the equations which we introduce as (Matkowsky et al. 1980 [1])

$$\Theta = \frac{1}{\alpha_c^2} - \frac{1}{\alpha^2} . \tag{1}$$

Starting from a sample boundary conditions we define a variational solution to the boundary value problem using energy spaces H, V given as certain subspaces of the Sobolev space $W_2^2(\Omega)$. An introduction of suitable equivalent norms in H,V leads to the operator equations

$$w - \Theta M_1 w + \Theta^2 M_2 w - \lambda(\frac{1}{\alpha_c^2} - \Theta)\ Lw - (\frac{1}{\alpha_c^2} - \Theta)\ C(w,\emptyset) = 0 \tag{2a}$$

$$- \emptyset + \Theta A_1\emptyset - \Theta^2 A_2\emptyset - \frac{1}{2}\ (\frac{1}{\alpha_c^2} - \Theta)\ B(w,w) = 0\ , \tag{2b}$$

where λ is the load parameter and $w\in H$, $\emptyset\in V$ refer to the plate deflection and Airy stress function, respectively. The operators L,B,C are essentially those introduced by Berger 1967 [2] for a plate with a definite domain. In addition, we have obtained linear bounded and self-adjoint operators M_1, M_2 and A_1, A_2 acting from H into itself and from V into itself, respectively. Eq.(2b) can be uniquely solved for $\emptyset = \emptyset^*(\Theta,w)$. If Θ is sufficiently small, $\emptyset^*$ may be easily found in the form of power series in Θ. Substituting this solution into (2a) for $\emptyset$ and introducing the small load parameter χ,

$$\lambda = \frac{\lambda - \lambda_c}{\lambda_c}\ , \tag{3}$$

where λ_c corresponds to the double buckling load of the plate, we arrive at the resulting equation

$$F(w, \chi, \Theta) = w - \Theta M_1 - \lambda_c(1+\chi)\left(\frac{1}{\lambda_c^2} - \Theta\right) Lw + \frac{1}{2\alpha_c^4} C(w, B(w,w)) + \text{h.o.t.} = 0 \tag{4}$$

to be solved for $\{w, \chi, \Theta\}$ near the origin of the space $H \times R \times R$. Obviously, $w = w^*(\chi, \Theta) \equiv 0$ is always a solution to Eq. (4).

Let H_c be the eigensubspace of the double eigenvalue λ_c and P_c the operator of the orthogonal projection of H onto H_c. We assume that the following hypotheses hold:

(H1) For $u \in H_c$, $B\,u,u = 0$ only if $u = 0$.

(H2) The operator $P_c M_1$ restricted to $u \in H_c$ has only simple eigenvalues.

(H3) $F(w, \chi, \Theta)$ commutes with respect to the group

$_S = \{I, S, -I, -S\}$

of operators on H, where S possesses the action of one of the operators S_x, S_y, S_{xy}

S_x: $H \to H$, $u(x,y) \to (S_x u)(x,y) = u(1-x,y)$

S_y: $H \to H$, $u(x,y) \to (S_y u)(x,y) = u(x,1-y)$

$S_{xy} = S_x S_y = S_y S_x$

and H_c is spanned by $\varphi_1 \in H^+$, $\varphi_2 \in H^-$; $\|\varphi_i\| = 1$, $i = 1,2$ where

$H^+ \equiv \{u \in H: Su = u\}$, $\quad H^- \equiv \{u \in H: Su = -u\}$.

The hypothesis (H1) is commonly used. (H2) implies a transversal splitting of the double eigenvalue λ_c into simple ones appearing as a result of perturbation. Actually we can show that near $\lambda = \lambda_c$ the eigenvalues λ_i^Θ of the linearized equation are of the form $\lambda_i^\Theta = \lambda_c + \lambda_i'\Theta +$ + h.o.t., $i = 1,2$ with $\lambda_1' \neq \lambda_2'$. By the hypothesis (H3) restrictions upon the boundary and load conditions are imposed allowing for the occurence of the assumed symmetry of Eq. (4).

The study of Eq. (4) is constructive - via the Liapunov-Schmidt reduction and the implicit function theorem. Assuming

$$w = \zeta_1 \varphi_1 + \zeta_2 \varphi_2 + \omega \tag{5}$$

with $\zeta \in R^2$, $\omega \in H_c$ and following Vanderbauwhede 1982 [3] we obtain that the bifurcation equations admit the form

$$G_i(\zeta, \chi, \Theta) = \zeta_i H_i(\zeta, \chi, \Theta) = 0, \qquad i = 1,2 \tag{6}$$

with

$$H_i(\pm\zeta_1, \pm\zeta_2, \chi, \Theta) = H_i(\zeta_1, \zeta_2, \chi, \Theta), \qquad i=1,2. \tag{7}$$

Thus, a $Z_2 \oplus Z_2$ symmetry of Eqs. (6) is present. Moreover, we may distinguish one-mode and coupled-mode solutions to Eq. (4). The one-mode solutions $w \in H^+$ or $w \in H^-$ correspond to solutions of

$$\zeta_2 = 0, \qquad H_1(\zeta_1, 0, \chi, \Theta) = 0 \tag{8}$$

and

$$\zeta_1 = 0, \qquad H_2(0, \zeta_2, \chi, \Theta) = 0, \tag{9}$$

respectively, while coupled-mode solutions correspond to solutions of

$$H_1(\zeta, \chi, \Theta) = 0, \qquad H_2(\zeta, \chi, \Theta) = 0. \tag{10}$$

Departing from Chow-Hale 1982 [4], we estimate the small solutions to Eqs. 6 by an a priori bound based on (H1) and then scale Eqs. (6) ((8-10)) by

$$\zeta = \beta\mu, \qquad \chi = \mu^2 \sin\nu, \qquad \Theta = \mu^2 \cos\nu, \tag{11}$$

where $\beta = \beta(\nu, \mu)$, the angle determines a direction in the (χ, Θ) parameter plane and μ is a new small parameter. The scaled equations are

$$g_i(\beta, \nu, \mu) = \beta_i h_i(\beta, \nu, \mu) = 0, \qquad i=1,2 \tag{12}$$

where

$$h_1(\beta, \nu, \mu) = -\sin\nu + \frac{\lambda_1'}{\lambda_c}\cos\nu + a\beta_1^2 + b\beta_2^2 + \text{h.o.t.},$$

$$h_2(\beta, \nu, \mu) = -\sin\nu + \frac{\lambda_2'}{\lambda_c}\cos\nu + b\beta_1^2 + c\beta_2^2 + \text{h.o.t.}.$$

Due to (H1) it is $a > 0$, $c > 0$ and we assume that $b > 0$, too.

Definition 1. (Golubitsky-Schaeffer 1979 [5]) The bifurcation problem (6) is non-degenerate if $b/a \neq 1$, $b/c \neq 1$ and $b^2 \neq ac$.

According to Golubitsky-Schaeffer [5], the ratios b/a, b/c represent the modal parameters of bifurcation problem (6). The lines of deneracy divide the studied positive quadrant of b/a, b/c plane into six regions within each of which the local features of bifurcating solutions to Eqs. (6) are topologically equivalent. The nondegenerate cases of Eqs. (6) were analysed in [5] by means of the singularity theory. Our study comprises the degenerate cases while the employed tools are simpler.

Letting $\mu \to 0$, the reduced equations $g(\beta, \nu, 0) = 0$ the scaled equivalents of (8-10)) can be easily solved for $\beta = \beta^o$. If $b^2 \neq ac$, the non-trivial solutions appear at ν values forming an open subinterval within the considered interval $(-\pi/2, 3\pi/2)$ of ν values, with end points differing from $-\pi/2$, $3\pi/2$. If $b^2 = ac$, the reduced system is solvable only at a certain $\nu \in (-\pi/2, 3\pi/2)$, say $\nu = \nu^s$, having then a continuum of solutions given by the equation

$$a\,(\beta_1^o)^2 + b(\beta_2^o)^2 = \sin\nu^s - \frac{\lambda_1'}{\lambda_c}\cos\nu^s . \qquad (13)$$

Successive continuations of the solutions to $\mu \neq 0$ by the implicit function theorem succed only within the open set of regular points in the parameter plane. Thus, the description of the solution set needs to be completed in a beighbourhood of the singular points (potential bifurcation curves). The first step in the analysis which appears to be crucial is the choice of (ζ_1,θ) or (ζ_2,θ) or (ζ_1, ζ_2) as a new parameter plane if (8),(9) or (10), respectively, is solved. In this way we obtain that the small solutions to Eq. (4) form a connected set consisting of the trivial, one-mode and coupled-mode solution subsets.

The description of the solution set to Eq.(4) yields immediately primary and secondary bifurcation curves at cross-sections of trivial and one-mode or one-mode and coupled-mode solution subsets, respectively. Now the question on the possible additional bifurcation curves and the conditions of their appearing or non-existence is to be answered.

Theorem 2. Suppose (H1-3) hold and the coefficients a, b, c of bifurcation equations (6) of Eq.(4) satisfy:

(i) $b^2 \neq ac,\ b > 0$

(ii) $b^2 = ac,\ b > 0$

In the case (ii) let further be

$$\lim_{\beta_1^o \to 0} T(\beta^o),_{①①} \neq 0 \cap \lim_{\beta_2^o \to 0} T(\beta^o),_{②②} \neq 0,$$

where $T(\beta^o)$ is defined by

$$T(\beta^c) = bh_{1,\mu\mu}(\beta^o,\nu^s,0) - ah_{2,\mu\mu}(\beta^o,\nu^s,0)$$

over the ellipse (13)(the circles in the subscript positions denote a total differentiation of $T(\beta^o)$with respect to β_1 or β_2) and either

(a) it holds

$$T(\beta^o),_{①} = -\frac{a}{b}\frac{\beta_1^o}{\beta_2^o}\,T(\beta^o),_{②} = 0, \quad \forall\ \beta^o:\ \beta_1^o > 0,\ \beta_2^o > 0$$

or

(b) for certain $\beta^o = \beta^{o*}$: $\beta_1^o{}^* > 0 \cap \beta_2^o{}^* > 0$ it is

$$T(\beta^{o*}),_{①} = T(\beta^{o*}),_{②} = 0,$$

$$T(\beta^{o*}),_{①①} = \left(\frac{a}{b}\frac{\beta_1^*}{\beta_2^{o*}}\right)^2 T(\beta^{o*}),_{②②} \neq 0$$

and

$$\lim_{\beta_1^o \to 0} T(\beta^o) \neq \lim_{\beta_2^o \to 0} T(\beta^o).$$

Then near the origin of the (χ,Θ) parameter plane, the bifurcation diagram of Eq.(4) consists of four distinct bifurcation curves: two primary and two secondary bifurcation curves, and in addition in the case (ii)(b) of a unique curve of limit points. A crossing of the primary and secondary bifurcation curves changes the number of solutions to Eq.(4) by two and four, respectively. A crossing of the limit-point curve changes the number of solutions by eight.

Proof. In order to study the set of bifurcation curves, we solve the system consisting of the scaled bifurcation equations (12) together with the condition of vanishing of the corresponding Jacobian $J_g = J_g(\beta,\nu,\mu)$. At solutions to the reduced system ($\mu = 0$ the value of the corresponding Jacobian is always zero but one of Eqs.(12), say $g_1 = 0$, and the equation $J_g = 0$ may be uniquely solved for $\beta_1 = \tilde{\beta}_1(\beta_2,\mu)$, $\nu = \tilde{\nu}(\beta_2,\mu)$near such solution. Substituting $\tilde{\beta}_1, \tilde{\nu}$ for β_1, ν in the remaining equation $g_2 = 0$ we get an equation the small solutions $\beta_2 = \beta_2^*(\mu)$ of which can be studied by Newton´s polygon method. The primary and secondary bifurcation curves correspond to the tripple roots while the limit-point curve to the simple root of the remaining equation ($g_1 = 0$ or $g_2 = 0$). We note that $T(\beta^o)$is a polynomial of second degree in $(\beta_1^o)^2$ and $(\beta_2^o)^2$.

<u>Corollary 3.</u> There exist two pairs of one-mode (one from H^+ and the other from H^-) and in the cases (i),(ii)(a) no or two pairs while in the case ii b no, two or four pairs of coupled-mode solutions to Eq.(4), $\Theta = 0$ near $w = 0$ and any $\lambda > \lambda_c$ sufficiently close to λ_c.

Equation $F(w,\chi,0) = 0$ describes an important problem of plate having a double buckling load. A direct calculation of the buckled states of the plate bifurcating at the buckling load may be performed eliminating one of the unknowns and then applying the Newton polygon method to the remaining equation.

A necessary part of the analysis of the studied bifurcation problem is the investigation of stability of bifurcating solutions. Following the concept of linearized stability, a solution $w = w^*(\chi,\Theta)$ to Eq.(4) is stable if the eigenvalues ρ of the eigenvalue problem

$$F'(w^*,\chi,\Theta)\phi - \rho\phi = 0 \tag{14}$$

are positive. It is well known that the trivial solution is stable at any Θ and $\lambda > 0$ less than the first positive eigenvalue λ_+^Θ of the linearized equation. For $\lambda > \lambda_+^\Theta$ the trivial solution is always unstable.

McLeod and Sattinger 1973 [6] showed that the Liapunov-Schmidt reduction contains informations required for the stability analysis of bifurcating solutions. Later Sattinger 1979 [7] has shown that the stability of a one-parameter family of bifurcating solutions is determined, to the lowest order, by the eigenvalues of the Jacobian matrix of reduced bifurcation equations. Sattinger's theorem fails e.g. if $b^2 = ac$, since then the Jacobian matrix of the reduced equations has always one zero eigenvalue. The non-conforming degenerate cases can be treated by the following theorem:

Theorem 4. Suppose (H1-3) hold and the coefficient b of bifurcation equations (6) of Eq. (4) satisfies $b > 0$. Then, the linearized stability of any one-parameter family of isolated solutions to Eq. (4) bifurcating at the double eigenvalue λ_c is sufficiently close to the bifurcation point determined by the Jacobian of bifurcation equations (6) (positive value implies stability).

Proof. $F'(w^*, \chi, \theta)$ is an analytic and symmetric perturbation of the operator $F'(0,0,0)$. The spectrum of $F'(0,0,0)$ is discrete and non-negative with zero as a double eigenvalue having a positive isolation distance. Applying the Liapunov-Schmidt reduction to Eq. (14) we arrive at an eigenvalue problem in R^2 yielding the perturbation of zero eigenvalue. Now if we consider the eigenvalue problem for the Jacobian matrix of bifurcation equations (6) evaluated at $w = w^*$ we see that this equation differs from the former one only in the higher-order terms. Justifying the perturbation technique in both cases and comparing the perturbation equations we conclude the assertion.

Let us note that the stability analysis sometimes fails to indicate the energetically preferred equilibrium path of the plate and direct comparison of energy levels of buckled states is necessary. Such situation is encountered if $\theta = 0$, $b > a$, $b > c$, since then at $\lambda = \lambda_c$ there bifurcate two different pairs of stable solutions to Eq. (4).

References

1. B.J. Matkowsky, L.J. Putnick and E.L. Reiss, "Secondary states of rectangular plates," SIAM J.Appl.Math. 38(1980), 38-51.
2. M.S. Berger, "On von Kármán's equations and the buckling of a thin elastic plate, I, The clamped plate", Comm.Pure Appl. Math. 20(1967), 687-719.
3. A. Vanderbauwhede, "Local Bifurcation and Symmetry," Pitman, London, 1982.
4. S.N. Chow and J.K. Hale, "Methods of Bifurcation Theory", Springer-Verlag, Berlin, 1982.
5. D. Schaeffer and M. Golubitsky, "Boundary conditions and mode jumping in the buckling of a rectangular plate", Comm.Math.Phys. 69(1979), 209-230.
6. J.B. McLeod and D.H. Sattinger, "Loss of stability and bifurcation at a double eigenvalue", J. Funct. Anal. 14 1973 , 62-84.
7. D.H. Sattinger, "Group Theoretic Methods in Bifurcation Theory", Lecture Notes in Math., 762, Springer-Verlag, Berlin, 1979.

DELAY MAKES PROBLEMS IN POPULATION MODELLING

K. SMÍTALOVÁ
Department of Applied Mathematics, Comenius University
Mlynská dolina, 842 15 Bratislava, Czechoslovakia

The basic population model $x' = ax$, although it is very simple, plays an important role in the history of modelling. Contrary to mathematical models in physics, it is not designed to determine quantities. Here the mathemtics only exhibits a tendency. The purpose of such a model is not prediction, but insight [2].

The same is true for most population models. Important are the qualitative properties, such as stability of equilibrium states, existence of periodic solutions, etc.

Intuitively it is clear that the history of development of a modelled biological system is very important for his state in the future. Hence the use of delayed differential equations seems to be appropriate. Moreover, using the delay one can describe age and also the spatial structure of the population [3]. Generally speaking, a model involving the delay allows more complex types of behavior than models without delay.

By modelling usually the following assumptions are made:

1) Different histories determine different solutions.

2) Suitable choice of the history allows the solution to attain any prescribed value at a given time.

The first problem is a problem of injectivity of the shift operator (or solution map) [1]. Take e.g. the equation

$$x'(t) = -ax(t-1)[1-x(t)]$$

which arises in a natural way from a model

$$y'(t) = a(y-y^2)$$

of self-limited population. Then for every initial function φ satisfying the condition $\varphi(0) = 1$ we have $x(t) \equiv 1$ for $t \geq 0$.

Let $C(-1,0)$ be the space of continuous mappings from $[-1,0]$ to the set R of real numbers. For the equation

$$x'(t) = f(x(t-1)) \tag{1}$$

define an operator $T : C(-1,0) \to C(-1,0)$ by

$$T(\varphi)(t) = x_\varphi(t+1), \qquad t \in [-1,0]$$

where x_φ denotes the solution of (1) determined by the initial function $\varphi \in C(-1,0)$. Such an operator T is called shift operator or solution map for the equation (1). The above quoted example exhibits an equation for which the shift operator is not injective. The following result states that this is actually the typical case for equation (1).

Theorem 1. Let C be the metric space of continuous mappings from R to R, equipped with the metric $\rho(f,g) = \min\{1, \sup_x |f(x) - g(x)|\}$. Let $H \subset C$ be the set of those mappings $f \in C$, for which there are initial functions $\varphi \neq \psi$ with $\varphi(0) = \psi(0)$, generating the same solution $x_\varphi(t) = x_\psi(t)$ of (1) for $t \geq 0$. Then $C \setminus H$ is nowhere dense in C.

In other words, the shift operator for the solution (1) is generically not injective.

Proof. The set M of those $f \in C$, which attain strong local maxima at certain points, is clearly open and dense in C. Hence it suffices to show that $M \subset H$.

Choose, for any given $f \in M$, points $a < c < b$ with $f(a) = f(b) < f(c) = \max\{f(t);\ t \in [a,b]\}$. For any $d > 0$ choose a set $M(d)$ of points

$$a = a_0 < a_1 < \ldots < a_k = c = b_k < \ldots < b_1 < b_0 = b$$

(k is suitable integer) such that $f(a_i) = f(b_i)$ for any i, and

$$|f(t) - f(s)| < d \text{ whenever } t, s \in [a_j, a_{j+1}] \text{ or } t, s \in [b_{j+1}, b_j]$$

for some j. Let $\varphi(d), \psi(d) \in C(-1,0)$ be such that $\varphi(d)(i/k - 1) = a_i$, $\psi(d)(i/k - 1) = b_i$, and $\varphi(d)$, $\psi(d)$ be linear on every of the intervals $[i/k - 1, (i+1)/k - 1]$, $i = 0,\ldots,k$.

We follow the above construction for every $d = 1/n$, $n = 1,2,\ldots$. Without loss of generality we may assume that $M(1/n) \subset M(1/(n+1))$

$$\lim_{n\to\infty} \varphi(1/n) = \varphi \quad \text{and} \quad \lim_{n\to\infty} \psi(1/n) = \psi$$

uniformly. Moreover, $\varphi(0) = \psi(0)$ and $f(\varphi(t)) = f(\psi(t))$ for every $t \in [-1,0]$, i.e. φ and ψ are the desired initial functions, q.e.d.

Modifying the above argument for equation with continuously differentiable right side, we obtain the next

Theorem 2. Let C^1 be the set of continuously differentiable functions from R to R, with the usual C^1-netric. Let H^1 be the set of those f from C^1, for which there are initial functions $\varphi \neq \psi$ with $\varphi(0) = \psi(0)$, generating the same solution of (1). Then both

$$\text{Int } H^1 \neq \phi \quad \text{and} \quad \text{Int } (C^1 \setminus H^1) \neq \phi$$

Remark. Theorem 2 can be generalized to the equation

$$x'(t) = f(x(t), x(t-1)) \tag{2}$$

where f is continuously differenatiable. We conjecture that also Theorem 1 can be generalized to this case, although we are not able to give any proof.

Next consider the second problem. Given some $(\bar{t},\bar{x}) \in R^2$, where $\bar{t} > 0$, does there exist an initial function $\varphi \in C(-1,0)$ such that $x_\varphi(\bar{t}) = \bar{x}$, where x_φ is the solution of the equation (1)? THis is the problem of pointwise completeness of (1). In [4] we recently gave a sufficient condition for pointwise completeness, which for (1) can be formulated as follows:

Theorem 3. Let $\bar{t},\bar{x} \in R$ be given. If $f \in C$ satisfies the Lipschitz condition

$$|f(x) - f(y)| \le L|x - y| \quad \text{for any} \quad x,y \in R$$

and

$$L \cdot \bar{t} < 1 \tag{3}$$

then there is a function $\varphi \in C(-1,0)$ such that $x_\varphi(\bar{t}) = \bar{x}$.

Note that theorem is true also for $\bar{x} \in R^n$.

The following example shows that the condition (3) in general cannot be omitted.

Example. Let $f(x) \equiv 0$ for $x \leq 1$, and $f(x) = 1 - x$ for $x > 1$. Let $\varphi \in C(-1,0)$ be an initial function. Let x_φ be the corresponding solution of (1). Then $x_\varphi(t)$ is non-increasing in $[0,\infty)$, hence we have $x_\varphi(t) \geq x_\varphi(1)$, for $t \in [0,1]$. Since $x_\psi(t) \leq 1$ for $t > 1$, where $\psi(t) = x_\varphi(1) = \text{const}$ for $t \in [-1,0]$, we have $x_\varphi(2) \leq 1$. This is true for arbitrary φ.

The above quoted properties of delayed differential equations does not seem to be good for modelling. Theorem 1 e.g. indicates a structural non-stability for models involving the equation (1). But also the well-known Volterra's - Lotka's predator-prey differenatial model is structurally unstable though its role among the known models is of great importance.

References

[1] HALE,J., *Theory of Funtional Differential Equations*, Springer, Berlin - Heidelberg - New York 1977.

[2] HIRSCH,M.W., *The dynamical approach to differential equations*, Bull. Amer. Math. Soc. 11, No. 1 (1984), 1-64.

[3] MAC DONALD,N., *Time Lags in Biological Models*, Lecture Notes in Biomathematics 27, Springer, Berlin - Heidelberg - New York 1978.

[4] SMÍTALOVÁ,K., *On a problem concerning a functional differential equation*, Math. Slovaca 30 (1980), 239-242.

ol. 1034: J. Musielak, Orlicz Spaces and Modular Spaces. V, 22 pages. 1983.

'ol. 1035: The Mathematics and Physics of Disordered Media. 'roceedings, 1983. Edited by B.D. Hughes and B.W. Ninham. 'II, 432 pages. 1983.

'ol. 1036: Combinatorial Mathematics X. Proceedings, 1982. Edited y L.R.A. Casse. XI, 419 pages. 1983.

'ol. 1037: Non-linear Partial Differential Operators and Quantization 'rocedures. Proceedings, 1981. Edited by S.I. Andersson and I.-D. Doebner. VII, 334 pages. 1983.

ol. 1038: F. Borceux, G. Van den Bossche, Algebra in a Localic opos with Applications to Ring Theory. IX, 240 pages. 1983.

'ol. 1039: Analytic Functions, Błażejewko 1982. Proceedings. Edited y J. Ławrynowicz. X, 494 pages. 1983

'ol. 1040: A. Good, Local Analysis of Selberg's Trace Formula. III, 28 pages. 1983.

'ol. 1041: Lie Group Representations II. Proceedings 1982–1983. dited by R. Herb, S. Kudla, R. Lipsman and J. Rosenberg. IX, 40 pages. 1984.

ol. 1042: A. Gut, K. D. Schmidt, Amarts and Set Function Processes. , 258 pages. 1983.

ol. 1043: Linear and Complex Analysis Problem Book. Edited by P. Havin, S.V. Hruščëv and N.K. Nikol'skii. XVIII, 721 pages. 1984.

ol. 1044: E. Gekeler, Discretization Methods for Stable Initial Value roblems. VIII, 201 pages. 1984.

ol. 1045: Differential Geometry. Proceedings, 1982. Edited by .M. Naveira. VIII, 194 pages. 1984.

ol. 1046: Algebraic K-Theory, Number Theory, Geometry and Analy- s. Proceedings, 1982. Edited by A. Bak. IX, 464 pages. 1984.

ol. 1047: Fluid Dynamics. Seminar, 1982. Edited by H. Beirão da eiga. VII, 193 pages. 1984.

ol. 1048: Kinetic Theories and the Boltzmann Equation. Seminar, 981. Edited by C. Cercignani. VII, 248 pages. 1984.

ol. 1049: B. Iochum, Cônes autopolaires et algèbres de Jordan. , 247 pages. 1984.

ol. 1050: A. Prestel, P. Roquette, Formally p-adic Fields. V, 167 pages. 984.

ol. 1051: Algebraic Topology, Aarhus 1982. Proceedings. Edited by Madsen and B. Oliver. X, 665 pages. 1984.

'ol. 1052: Number Theory, New York 1982. Seminar. Edited by D.V. hudnovsky, G.V. Chudnovsky, H. Cohn and M.B. Nathanson. V, 309 ages. 1984.

ol. 1053: P. Hilton, Nilpotente Gruppen und nilpotente Räume. , 221 pages. 1984.

'ol. 1054: V. Thomée, Galerkin Finite Element Methods for Parabolic roblems. VII, 237 pages. 1984.

ol. 1055: Quantum Probability and Applications to the Quantum heory of Irreversible Processes. Proceedings, 1982. Edited by . Accardi, A. Frigerio and V. Gorini. VI, 411 pages. 1984.

'ol. 1056: Algebraic Geometry. Bucharest 1982. Proceedings, 1982. dited by L. Bădescu and D. Popescu. VII, 380 pages. 1984.

ol. 1057: Bifurcation Theory and Applications. Seminar, 1983. Edited y L. Salvadori. VII, 233 pages. 1984.

'ol. 1058: B. Aulbach, Continuous and Discrete Dynamics near lanifolds of Equilibria. IX, 142 pages. 1984.

ol. 1059: Séminaire de Probabilités XVIII, 1982/83. Proceedings. dité par J. Azéma et M. Yor. IV, 518 pages. 1984.

ol. 1060: Topology. Proceedings, 1982. Edited by L. D. Faddeev and . A. Mal'cev. VI, 389 pages. 1984.

ol. 1061: Séminaire de Théorie du Potentiel. Paris, No. 7. Pro- eedings. Directeurs: M. Brelot, G. Choquet et J. Deny. Rédacteurs: Hirsch et G. Mokobodzki. IV, 281 pages. 1984.

Vol. 1062: J. Jost, Harmonic Maps Between Surfaces. X, 133 pages. 1984.

Vol. 1063: Orienting Polymers. Proceedings, 1983. Edited by J.L. Ericksen. VII, 166 pages. 1984.

Vol. 1064: Probability Measures on Groups VII. Proceedings, 1983. Edited by H. Heyer. X, 588 pages. 1984.

Vol. 1065: A. Cuyt, Padé Approximants for Operators: Theory and Applications. IX, 138 pages. 1984.

Vol. 1066: Numerical Analysis. Proceedings, 1983. Edited by D.F. Griffiths. XI, 275 pages. 1984.

Vol. 1067: Yasuo Okuyama, Absolute Summability of Fourier Series and Orthogonal Series. VI, 118 pages. 1984.

Vol. 1068: Number Theory, Noordwijkerhout 1983. Proceedings. Edited by H. Jager. V, 296 pages. 1984.

Vol. 1069: M. Kreck, Bordism of Diffeomorphisms and Related Topics. III, 144 pages. 1984.

Vol. 1070: Interpolation Spaces and Allied Topics in Analysis. Proceedings, 1983. Edited by M. Cwikel and J. Peetre. III, 239 pages. 1984.

Vol. 1071: Padé Approximation and its Applications, Bad Honnef 1983. Prodeedings. Edited by H. Werner and H.J. Bünger. VI, 264 pages. 1984.

Vol. 1072: F. Rothe, Global Solutions of Reaction-Diffusion Systems. V, 216 pages. 1984.

Vol. 1073: Graph Theory, Singapore 1983. Proceedings. Edited by K. M. Koh and H. P. Yap. XIII, 335 pages. 1984.

Vol. 1074: E. W. Stredulinsky, Weighted Inequalities and Degenerate Elliptic Partial Differential Equations. III, 143 pages. 1984.

Vol. 1075: H. Majima, Asymptotic Analysis for Integrable Connections with Irregular Singular Points. IX, 159 pages. 1984.

Vol. 1076: Infinite-Dimensional Systems. Proceedings, 1983. Edited by F. Kappel and W. Schappacher. VII, 278 pages. 1984.

Vol. 1077: Lie Group Representations III. Proceedings, 1982–1983. Edited by R. Herb, R. Johnson, R. Lipsman, J. Rosenberg. XI, 454 pages. 1984.

Vol. 1078: A.J.E.M. Janssen, P. van der Steen, Integration Theory. V, 224 pages. 1984.

Vol. 1079: W. Ruppert. Compact Semitopological Semigroups: An Intrinsic Theory. V, 260 pages. 1984

Vol. 1080: Probability Theory on Vector Spaces III. Proceedings, 1983. Edited by D. Szynal and A. Weron. V, 373 pages. 1984.

Vol. 1081: D. Benson, Modular Representation Theory: New Trends and Methods. XI, 231 pages. 1984.

Vol. 1082: C.-G. Schmidt, Arithmetik Abelscher Varietäten mit komplexer Multiplikation. X, 96 Seiten. 1984.

Vol. 1083: D. Bump, Automorphic Forms on GL (3,IR). XI, 184 pages. 1984.

Vol. 1084: D. Kletzing, Structure and Representations of Q-Groups. VI, 290 pages. 1984.

Vol. 1085: G.K. Immink, Asymptotics of Analytic Difference Equations. V, 134 pages. 1984.

Vol. 1086: Sensitivity of Functionals with Applications to Engineering Sciences. Proceedings, 1983. Edited by V. Komkov. V, 130 pages. 1984

Vol. 1087: W. Narkiewicz, Uniform Distribution of Sequences of Integers in Residue Classes. VIII, 125 pages. 1984.

Vol. 1088: A.V. Kakosyan, L.B. Klebanov, J.A. Melamed, Characterization of Distributions by the Method of Intensively Monotone Operators. X, 175 pages. 1984.

Vol. 1089: Measure Theory, Oberwolfach 1983. Proceedings. Edited by D. Kölzow and D. Maharam-Stone. XIII, 327 pages. 1984.

Vol. 1090: Differential Geometry of Submanifolds. Proceedings, 1984. Edited by K. Kenmotsu. VI, 132 pages. 1984.

Vol. 1091: Multifunctions and Integrands. Proceedings, 1983. Edited by G. Salinetti. V, 234 pages. 1984.

Vol. 1092: Complete Intersections. Seminar, 1983. Edited by S. Greco and R. Strano. VII, 299 pages. 1984.

Vol. 1093: A. Prestel, Lectures on Formally Real Fields. XI, 125 pages. 1984.

Vol. 1094: Analyse Complexe. Proceedings, 1983. Edité par E. Amar, R. Gay et Nguyen Thanh Van. IX, 184 pages. 1984.

Vol. 1095: Stochastic Analysis and Applications. Proceedings, 1983. Edited by A. Truman and D. Williams. V, 199 pages. 1984.

Vol. 1096: Théorie du Potentiel. Proceedings, 1983. Edité par G. Mokobodzki et D. Pinchon. IX, 601 pages. 1984.

Vol. 1097: R.M. Dudley, H. Kunita, F. Ledrappier, École d'Éte de Probabilités de Saint-Flour XII – 1982. Edité par P.L. Hennequin. X, 396 pages. 1984.

Vol. 1098: Groups – Korea 1983. Proceedings. Edited by A.C. Kim and B.H. Neumann. VII, 183 pages. 1984.

Vol. 1099: C.M. Ringel, Tame Algebras and Integral Quadratic Forms. XIII, 376 pages. 1984.

Vol. 1100: V. Ivrii, Precise Spectral Asymptotics for Elliptic Operators Acting in Fiberings over Manifolds with Boundary. V, 237 pages. 1984.

Vol. 1101: V. Cossart, J. Giraud, U. Orbanz, Resolution of Surface Singularities. Seminar. VII, 132 pages. 1984.

Vol. 1102: A. Verona, Stratified Mappings – Structure and Triangulability. IX, 160 pages. 1984.

Vol. 1103: Models and Sets. Proceedings, Logic Colloquium, 1983, Part I. Edited by G.H. Müller and M.M. Richter. VIII, 484 pages. 1984.

Vol. 1104: Computation and Proof Theory. Proceedings, Logic Colloquium, 1983, Part II. Edited by M.M. Richter, E. Börger, W. Oberschelp, B. Schinzel and W. Thomas. VIII, 475 pages. 1984.

Vol. 1105: Rational Approximation and Interpolation. Proceedings, 1983. Edited by P.R. Graves-Morris, E.B. Saff and R.S. Varga. XII, 528 pages. 1984.

Vol. 1106: C.T. Chong, Techniques of Admissible Recursion Theory. IX, 214 pages. 1984.

Vol. 1107: Nonlinear Analysis and Optimization. Proceedings, 1982. Edited by C. Vinti. V, 224 pages. 1984.

Vol. 1108: Global Analysis – Studies and Applications I. Edited by Yu.G. Borisovich and Yu.E. Gliklikh. V, 301 pages. 1984.

Vol. 1109: Stochastic Aspects of Classical and Quantum Systems. Proceedings, 1983. Edited by S. Albeverio, P. Combe and M. Sirugue-Collin. IX, 227 pages. 1985.

Vol. 1110: R. Jajte, Strong Limit Theorems in Non-Commutative Probability. VI, 152 pages. 1985.

Vol. 1111: Arbeitstagung Bonn 1984. Proceedings. Edited by F. Hirzebruch, J. Schwermer and S. Suter. V, 481 pages. 1985.

Vol. 1112: Products of Conjugacy Classes in Groups. Edited by Z. Arad and M. Herzog. V, 244 pages. 1985.

Vol. 1113: P. Antosik, C. Swartz, Matrix Methods in Analysis. IV, 114 pages. 1985.

Vol. 1114: Zahlentheoretische Analysis. Seminar. Herausgegeben von E. Hlawka. V, 157 Seiten. 1985.

Vol. 1115: J. Moulin Ollagnier, Ergodic Theory and Statistical Mechanics. VI, 147 pages. 1985.

Vol. 1116: S. Stolz, Hochzusammenhängende Mannigfaltigkeiten und ihre Ränder. XXIII, 134 Seiten. 1985.

Vol. 1117: D.J. Aldous, J.A. Ibragimov, J. Jacod, Ecole d'Été d Probabilités de Saint-Flour XIII – 1983. Édité par P.L. Hennequi IX, 409 pages. 1985.

Vol. 1118: Grossissements de filtrations: exemples et application Seminaire, 1982/83. Edité par Th. Jeulin et M. Yor. V, 315 pages. 198

Vol. 1119: Recent Mathematical Methods in Dynamic Programmin Proceedings, 1984. Edited by I. Capuzzo Dolcetta, W.H. Flemin and T. Zolezzi. VI, 202 pages. 1985.

Vol. 1120: K. Jarosz, Perturbations of Banach Algebras. V, 118 page 1985.

Vol. 1121: Singularities and Constructive Methods for Their Treatme Proceedings, 1983. Edited by P. Grisvard, W. Wendland and J. Whiteman. IX, 346 pages. 1985.

Vol. 1122: Number Theory. Proceedings, 1984. Edited by K. Allac VII, 217 pages. 1985.

Vol. 1123: Séminaire de Probabilités XIX 1983/84. Proceedings. Ed par J. Azéma et M. Yor. IV, 504 pages. 1985.

Vol. 1124: Algebraic Geometry, Sitges (Barcelona) 1983. Procee ings. Edited by E. Casas-Alvero, G.E. Welters and S. Xamb Descamps. XI, 416 pages. 1985.

Vol. 1125: Dynamical Systems and Bifurcations. Proceedings, 198 Edited by B.L.J. Braaksma, H.W. Broer and F. Takens. V, 129 page 1985.

Vol. 1126: Algebraic and Geometric Topology. Proceedings, 198 Edited by A. Ranicki, N. Levitt and F. Quinn. V, 523 pages. 1985.

Vol. 1127: Numerical Methods in Fluid Dynamics. Seminar. Edited F. Brezzi, VII, 333 pages. 1985.

Vol. 1128: J. Elschner, Singular Ordinary Differential Operators a Pseudodifferential Equations. 200 pages. 1985.

Vol. 1129: Numerical Analysis, Lancaster 1984. Proceedings. Edit by P.R. Turner. XIV, 179 pages. 1985.

Vol. 1130: Methods in Mathematical Logic. Proceedings, 1983. Edit by C.A. Di Prisco. VII, 407 pages. 1985.

Vol. 1131: K. Sundaresan, S. Swaminathan, Geometry and Nonline Analysis in Banach Spaces. III, 116 pages. 1985.

Vol. 1132: Operator Algebras and their Connections with Topolo and Ergodic Theory. Proceedings, 1983. Edited by H. Araki, C. Moore, Ş. Strătilă and C. Voiculescu. VI, 594 pages. 1985.

Vol. 1133: K.C. Kiwiel, Methods of Descent for Nondifferentiable Op mization. VI, 362 pages. 1985.

Vol. 1134: G.P. Galdi, S. Rionero, Weighted Energy Methods in Flu Dynamics and Elasticity. VII, 126 pages. 1985.

Vol. 1135: Number Theory, New York 1983–84. Seminar. Edited D.V. Chudnovsky, G.V. Chudnovsky, H. Cohn and M.B. Nathans V, 283 pages. 1985.

Vol. 1136: Quantum Probability and Applications II. Proceeding 1984. Edited by L. Accardi and W. von Waldenfels. VI, 534 pag 1985.

Vol. 1137: Xiao G., Surfaces fibrées en courbes de genre deux. IX, pages. 1985.

Vol. 1138: A. Ocneanu, Actions of Discrete Amenable Groups on Neumann Algebras. V, 115 pages. 1985.

Vol. 1139: Differential Geometric Methods in Mathematical Physi Proceedings, 1983. Edited by H. D. Doebner and J. D. Hennig. VI, 3 pages. 1985.

Vol. 1140: S. Donkin, Rational Representations of Algebraic Grou VII, 254 pages. 1985.

Vol. 1141: Recursion Theory Week. Proceedings, 1984. Edited H.-D. Ebbinghaus, G.H. Müller and G.E. Sacks. IX, 418 pages. 198

Vol. 1142: Orders and their Applications. Proceedings, 1984. Edi by I. Reiner and K. W. Roggenkamp. X, 306 pages. 1985.

Vol. 1143: A. Krieg, Modular Forms on Half-Spaces of Quaternio XIII, 203 pages. 1985.

Vol. 1144: Knot Theory and Manifolds. Proceedings, 1983. Edited D. Rolfsen. V, 163 pages. 1985.